ISBN 978-0-266-59114-6
PIBN 10872060

THE

MATHEMATICAL QUESTIONS,

PROPOSED IN THE

LADIES' DIARY,

AND THEIR ORIGINAL ANSWERS,

TOGETHER WITH SOME NEW SOLUTIONS,

FROM ITS COMMENCEMENT IN THE YEAR

1704 TO 1816.

IN FOUR VOLUMES.

BY THOMAS LEYBOURN,

OF THE ROYAL MILITARY COLLEGE.

VOL. II.

London:

PRINTED BY W. GLENDINNING, HATTON GARDEN;

AND

PUBLISHED BY J. MAWMAN, LUDGATE STREET;

J. DEIGHTON AND SON, CAMBRIDGE; AND J. PARKER, OXFORD.

1817.

MATHEMATICAL QUESTIONS

PROPOSED IN THE

LADIES' DIARY,

AND THEIR ANSWERS.

Questions proposed in 1748, *and answered in* 1749.

I. QUESTION 295, *by Mr.* Heath.

With guineas and moidores the fewest, which way
Three hundred and fifty-one pounds can you pay?
If paid ev'ry way 'twill admit of, what sum
Do the pieces amount to?——My fortune to come.

Answered by Mr. John Turner, *of* Heath, *near* Wakefield.

Put $x =$ number of guineas, $y =$ number of moidores $= \frac{7020 - 21x}{27}$ by consequence, which must be a whole number. Whence $\frac{21x}{27}$, or $\frac{x}{9}$ is a whole number; consequently the least value of $x = 9$; whence $y = 253$, whose sum is 262 the least number of pieces; because there are taken the most moidores, expect paying the whole with them. Let $\frac{x}{9} = m$, then $x = 9m$, and $y = 260 - 7m$ by substitution; by which the greatest value of $m = 37$, when $x = 333$ guineas, and $y = 1$ moidore, the greatest number of pieces. The first term of an arithmetical progression being 9 (a) the last 333 (y), and the difference 9 (e) as it is evident, then the number of terms $= \frac{y + e - a}{e} = 37$, the ways to pay the sum of 351l. which being 37 times taken is $=$ 12987l. due to Mr. Heath.

N. B. The intermediate numbers are found by continually adding 9 to the guineas, and subtracting 7 from the moidores. There will be but 36 ways, if the payment with one moidore is not admitted.

II. QUESTION 296, *by* Amanda.

To find the least number of guineas, which being divided by 6, 5, 4, 3, and 2, respectively, shall leave 5, 4, 3, 2, and 1, respectively remaining?

Answered by Mr. Heath.

It is a maxim, *that whole numbers added to, subtracted from, or multiplied into whole numbers, shall produce whole numbers:* Upon which fundamentals these kind of questions are naturally resolved; though I have not seen the method clearly explained. Let $x =$ the number sought, then $\frac{x-5}{6}$, $\frac{x-4}{5}$, $\frac{x-3}{4}$, $\frac{x-2}{3}$, $\frac{x-1}{2}$ are all whole numbers by the question. Put $\frac{x-5}{6} = m$, then $x = 6m + 5$ a whole number, and by substitution in the second whole number for the value of x, $\frac{6m+1}{5} = m + \frac{m+1}{5}$ a whole number, $\therefore \frac{m+1}{5} = n$ a whole number, and $m = 5n - 1$; which substituted for m, in the first value of x, is $x = 30n - 1$ for the second value; $\therefore \frac{30n-4}{4}$ by substitution is the value of the 3d number, and $7n - 1 + \frac{2n}{4}$ is a whole number, and $n =$ a whole number, $\therefore$ substituting again the 2d value of x in the 4th whole number, $\frac{30n-3}{3} = 10n - 1 =$ a whole number, $\therefore n$ is a whole number (first supposed); and substituting again the 2d value of x in the 5th whole number, $\frac{30n-2}{2} = 15n - 1$ a whole number; so that $x = 30n - 1$ is the general value, after assuming $n = 2$, by which the least value of $x = 59$.

Mr. *Richard Gibbons* of Plymouth has shewn an easy method of finding the least number only, to answer these kind of questions. For he observes that $2 \times 3 \times 4 \times 5 \times 6 = 720$ is a number, which being divided respectively by 6, 5, 4, &c. will leave no remainder; and therefore $720 - 1 = 719$ will leave the required remainders, of 1 less than each factor. But 4 being a square of 2, and 6 a multiple of 2 and 3, $\therefore 2 \times 3 \times 2 \times 5 : -1 = 59$ the least number required.

III. QUESTION 297, *by Mr.* John Turner.

What is the day of the month, and hour of the day, at Rochester, in Kent, (at which time a couple are to be married this year) when the degrees of time from noon, to the degrees of time from sun rise, are in the proportion of 3 to 1, and the respective sines of those degrees in the proportion of 9 to 4?

Answered by Mr. Nich. Farrer.

Let $x =$ s. time from sun rise, then $3x - 4x^3 =$ s. time from noon. By quest. $9 : 4 :: 3x - 4x^3 : x$; hence $x = \frac{1}{4}\sqrt{3} = \cdot4330127$ $=$ s. $25^\circ\ 39'\ 32''$, and $3x - 4x^3 = \cdot9742786 =$ s. of $76^\circ\ 58'\ 36'' =$ $5^h\ 7'\ 54''$ time. Hence the ceremony begun at $6^h\ 52'\ 6''$ in the morning. ☉ rises $5^h\ 9'\ 27''$, ascensional diff. $= 12^\circ\ 38'\ 8''$, lat. Rochester $51^\circ\ 28'$, from which the sun's declination $= 9^\circ\ 53'$, answering to April 4th, or August 16th.

IV. QUESTION 298, *by Miss* Manlove.

All the different ways possible, in which a gentleman can place his servants, combining them by 1, 2, 3, &c. at a time are 960799; what number of servants does he keep?

Solved by Mr. Collingridge.

Let $x =$ the number of servants; then, by Stone's Mathematical Dictionary, $\dfrac{x^{x+1} - x}{x - 1} = 960799$, all the possible variations of the servants; which equation solved, $x = 7$ servants exactly.

V. QUESTION 299, *by Mr.* Bulman.

The top and bottom diameters of a tub in the form of the frustum of a cone are as 3 to 2, the diagonal 7 yards, and content a maximum. Required the top and bottom diameters of the tub, and how far from the earth's centre it must be placed to hold 3 gallons more?

Answered by Mr. James Terey, *of* Portsmouth.

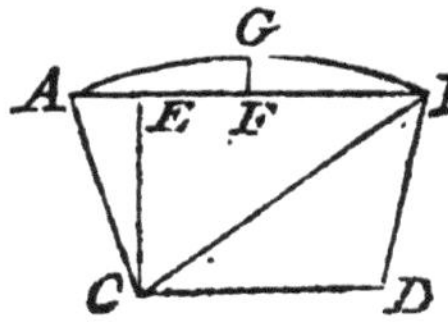

Let BC $= 7 = d$, AB $= x$, then CD $= \frac{2}{3}x$, per quest. also BE $= \frac{5}{6}x$. Then $\sqrt{(dd - \frac{25}{36}xx)} =$ CE. Now $\frac{19}{9}xx \times \sqrt{(dd - \frac{25}{36}xx)} \times 43\cdot313864$ $=$ content in ale gallons, a maximum; or $x^4\,(36dd - 25xx)$, a maximum; in fluxions and reduced, $x = \frac{1}{5}d\sqrt{24} = 6\cdot858571279$ yards $=$ AB; and CD $= 4\cdot572380852$, CE $=$

4·04145188 yards. The content in ale gallons = 17383·721448 &c. But to hold 3 gallons more AGB must be a spherical segment, containing 3 gallons. Let FG = z, AF = FB = b. By a theorem, ·523598z &c. × $(3bb + z^2)$ = cont. spher. seg.; in numbers $z^3 + 35{\cdot}28z =$ ·0346309367, where $z =$ ·00098160248 yards. The diameter of which sphere = 11980·410811 &c. yards. Consequently, if the brim of the tub is placed 5990·205405 yards from the earth's centre, it will hold 3 gallons more.

VI. QUESTION 300, *by* Rodomontado.

What is the least degree of velocity with which an iron ball of 12 pounds weight must be projected from the surface of our earth, at an angle of 40° elevation, whereby it shall not return?

Solved by the Excellent J. Landen, *near* Peterborough.

Let c be the earth's centre, ear the surface, ab the projectile's direction, and adf its trajectory. Suppose cd, cf indefinitely near each other, and call ca, (the earth's radius = 21000000 feet) a; cd, x; 32·2 feet, the velocity generated in a second at the earth's surface, b; v the velocity in d; v the required velocity. Then the centripetal force in d will be $\frac{a^2b}{x^2}$ (being reciprocally as the square of the distance from the earth's centre) and the force to retard the motion in the direction df, $\frac{a^2b\dot{x}}{x^2\dot{df}}$; this retarding force drawn into the fluxion of the time, being equal to the fluxion of the velocity, $\frac{a^2b\dot{x}}{vx^2}$ will be $= -\dot{v}$; therefore $v\dot{v} = -\frac{a^2b\dot{x}}{xx}$ and the fluent $\frac{vv}{2} = \frac{a^2b}{x}$: But in a, (v being $= \mathrm{v}$, and $x = a$) the correct fluent gives $v = \sqrt{(\mathrm{vv} + 2ab + 2a^2bx^{-1})}$. After an infinite time, x will be infinitely great, and $\sqrt{(\mathrm{vv} - 2ab + 2a^2bx^{-1})}$ infinitely small, and therefore may be put = 0, in which equation a is nothing in respect of the value of x: and therefore $\mathrm{v} = \sqrt{2ab}$. Hence, without regard to the angle of direction, if a body be projected from the earth's surface, in any direction whatever above the horizon, with such a velocity as will carry it above 7 miles per second, it will never return.

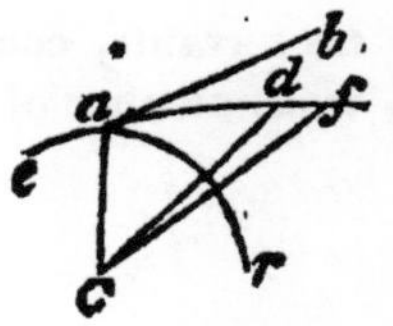

Mr. John Turner, *of* Heath, *sends us his Solution as follows:*

Let c represent the earth's centre, and let a body be supposed to

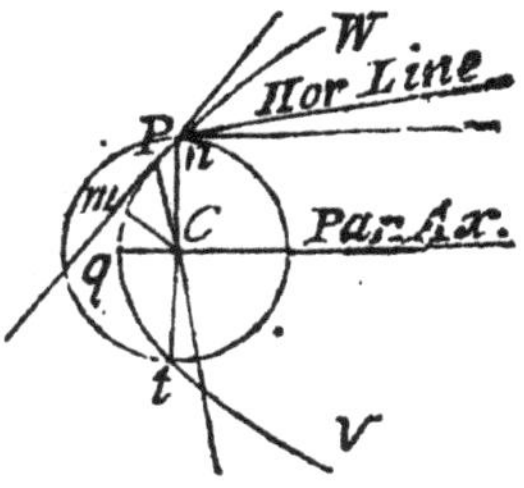

revolve in a circle at the superficies, the angle of the projectile's elevation being 40°, and a tangent to the parabolical curve QPW at P, the $\angle$ CPm will be 50°, (letting fall Cm perpendicular to Pm) whose complement to 180° = 130° = angle mPb. By prop. 17 and corol. of prop. 16, Newton's Principia, the axis of the parabolic trajectory will be parallel to Pb, passing through C; and likewise the latus rectum of the orbit = 2CP + 2Pn = 2·3472 semi-diameters of the earth (letting fall Cn perpendicular to Pb) and the focal distance from the vertex = ·5868 semi-diameters. Lastly, the velocity of a body moving in a circular orbit at the earth's superficies (by the prop. aforesaid) is such as would carry it through 4·92 miles uniformly, in each second; therefore, if x = velocity in the parabolic curve at P, we have (by prop. 15, 16, and 17 of the said Principia) as 24·2064 : ·5868xx : : 2 : 2·3472; whence x = 6·958 miles, or about 7 miles per second, the uniform velocity with which the body must be projected from P in the given angle of elevation, not to return. Or, since it is demonstrated (by the writers on physics) that the velocity of a body moving in a parabola is to the velocity of a body moving in a circle, at the same distance from the centre of force, as $\sqrt{2}$ to 1, ∴ 4·92 × 1·4142 = 6·958 miles, the projectile's velocity per second, as before.

The letter *b* is wanting in the figure.

VII. QUESTION 301, *by Mr.* Landen.

If an infinite number of perpendiculars be let fall from one end of the diameter of a semi-circle, upon an infinite number of tangents drawn about it, and a curve passes through all those angular points, what will be the length of that curve? The area of the space included betwixt it and the semi-circle? With the dimensions of the greatest ordinate, when the said diameter is = 20 inches?

Answered by Mr. Landen.

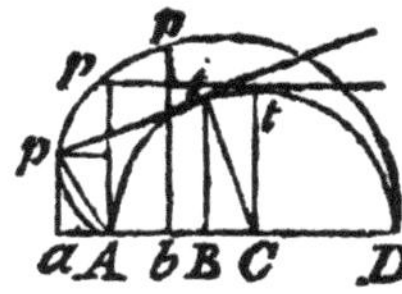

It is easily proved that the perpendicular Ap is always = its respective abscissa AB of the semi-circle, and the $\angle p$Ae = $\angle tct$. Then, if Aep be supposed infinitely near Ap, and Ap or AB be called x, and AD, $2a$; it follows, as $a : \frac{a\dot{x}}{\sqrt{(2ax - xx)}} :: x : \frac{x\dot{x}}{\sqrt{(2ax - xx)}}$ = the infinitely small arch pe, and the fluxion of the curve = $\sqrt{2a}\,\frac{\dot{x}}{\sqrt{(2a - x)}}$, the fluent of which corrected is

$4a - 2\sqrt{(4aa - 2ax)} =$ length of the curve; which, when $x = 2a$, in the present case, is equal to twice the diameter AD, or 40 inches.

But $\frac{1}{2}x \times (pe)$ $\frac{x\dot{x}}{\sqrt{(2ax - xx)}} = \frac{\frac{1}{2}x^{\frac{3}{2}}\dot{x}}{\sqrt{(2a - x)}}$, the fluxion of the area; whose fluent, by Mr. Cotes' 6th form, or Mr. Emerson's 10th and 11th forms, is $\frac{3}{2}aa \times$ arch of a circle whose radius $= 1$ and nat. sine $\sqrt{\frac{x}{2a}}$, $- \frac{3}{4}ax^{\frac{1}{2}}\sqrt{(2a - x)} - \frac{1}{4}x^{\frac{3}{2}}\sqrt{(2a - x)} =$ area sought; in the present case $= \frac{3}{2} \times$ area of the semi-circle, when $x = 2a$. Consequently the space ApDtA = quadrant AttCA = 78·54 inches.

For the greatest ordinate, the proportion is $a : \sqrt{(2ax - xx)} :: x : \frac{\sqrt{(2ax^3 - x^4)}}{a} =$ any ordinate. Its fluxion being made $= 0$, $x = \frac{3}{2}a = 15$ (though $= b$D), which being substituted for x in the expression of the ordinate above, then $\frac{3}{4}a\sqrt{3}$ will be the greatest ordinate in all cases; but in this case $= 12·9903$ inches $= bp$.

The letter e is wanting in the figure, which the reader may supply.

VIII. QUESTION 302, *by* Upnorensis.

Observing a horse tied to feed in a gentleman's park, with one end of a rope to his fore foot, and the other end to one of the circular iron rails, inclosing a pond, the circumference of which rails being 160 yards, equal to the length of the rope, what quantity of ground, at most, could the horse feed?

Answered by Mr. Heath.

Let all the rope be wrapt round the rails of the pond $vopqtv$, and the horse begin to move, or unwrap his rope from v (where it is fixed) in the track of vnmFGHIFv; then the space FGHIF, deducting the pond's area, is the greatest he can feed. When his rope is completely unwrapt at G, he describes the semi-circle GHI, with rad. vG $= 160$ yards, the area of which $= 40212·48$ yards square. Then he begins to wind his rope round the rails the contrary way to the former winding; describing the track IFv, similar and equal to the track vnmFG, at the unwinding of his rope. To

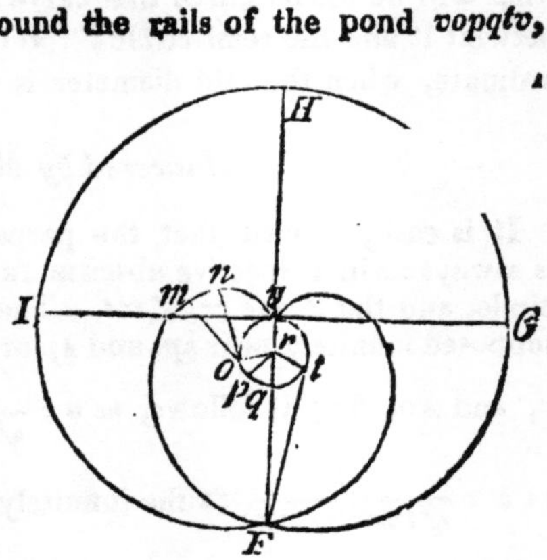

find the area he describes, at unwinding, or winding? Put $r =$ pond's rad. $= 35{\cdot}464731$ yards; $z =$ any arch of the rails unwound, suppose $vo = on$, and $\dot{z}$ its fluxion $= op$. Now, little sector rop is the fluxion or next increment of circular space rvo, as little sector pnm (with radii of curvature pn, pm) is the next increment of the space von; both sectors, when infinitely little, being similar, say, $r\ (ro) : \dot{z}\ (op) :: z\ (on) : z\dot{z} \div r = nm$, the fluxion of the involute arch vn, whose fluent is $zz \div 2r$; but sector $pnm = zz\dot{z} \div 2r$, the fluxion of the space von, whose fluent is $z^3 \div 6r =$ (when $z = 160$) to 26808·317 yards square, being area space vnmFG$vtqpov$, when the rope is quite unwound. Also Ft $= tqpov = 114{\cdot}42$ yards, best found by trial and a table of logarithms (series and the reversion of series being less certain and expeditious); whence area vnmF$tqpov$ $= 9804{\cdot}2$ square yards; whence area GvtFG $= 17004{\cdot}117$ by substitution, to which adding area tqF $= 1018{\cdot}58$ gives $18022{\cdot}69 =$ area GvtqFG, which doubled and added to semi-circle GHI make 76257 86 square yards = 15 a. 2 r. 12 p. And length of the whole track $= 1507{\cdot}96$ yards.

N. B. vnmFG $=$ GHI $= 502{\cdot}656$.

IX. QUESTION 303, *by Mr.* John Turner.

If the axis of the penumbral cone, falling upon the disk of the earth, makes an angle with the earth's diamoter at the surface of 24°, (the angle at the cone's vertex being 32′ 46″) and from a point in that axis, at the distance of 58·5 semi-diameters of the earth from the vertex, it is 64 semi-diameters to the earth's centre, how much of the earth's surface is included in the penumbral shadow?

Answered by the Rev. Mr. Baker.

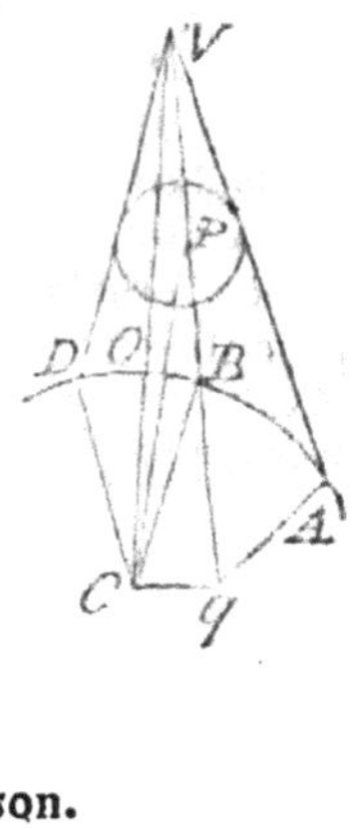

1. In △CBq there is given ∠CBq = 24°, and CB = 1 semi-diameter; whence (by trigon.) Bq = ·913545, and Cq = ·406736.

2. In △ PCq, PC = 64, and Cq = as above; whence Pq = 63·9987 and Vq = 122·4987. Now ∠qVA = 16′ 23″, ∓(∠CVq =) 11′ 27″ give the ∠OVD = 4′ 56″, ∠CVA = 27′ 50″.

3. In △s CVD and CVA, CV = 122·4993, CD = CB, and ∠s CVD and CVA as above; thence ∠DCV = 9° 59′ fere, and ∠VCA = 81° 12′, which added = 91° 11′ = arch DO + arch OA, the quantity of the earth's surface included in the penumbral shadow.

X. QUESTION 304, *by Mr.* John Hampson.

Required to find three such fractional numbers, that when each is lessened by the cube of their sum, three cube numbers shall remain?

Answered by Mr. Landen.

This gentleman puts zy for the 1st number, zx for the 2d, and zv for the 3d, p^2z for the sum of the three; zs for the side of the 1st cube, zr for the side of the 2d, and zq the side of the 3d, and so proceeds to equations; from whence, in a curious manner, he determines the 3 cube numbers $\frac{341}{4913}$, $\frac{854}{4913}$, and $\frac{250}{4913}$. But we must reserve his method for more room than we have at present.

Mr. *Farrer* informs us, that this question was taken from Branker's Algeb. where the solution may be seen at large; and says that one set of numbers is $\frac{494424}{2352637}$, $\frac{248696}{2352637}$, $\frac{448000}{2352637}$.

Mr. *Hampson's* numbers are $\frac{854}{4913}$, $\frac{341}{4913}$, $\frac{250}{4913}$, besides other of his numbers sent.

Solution from the 5th book of Diophantus, *quest.* 20.

To find rational values of x, y, z, u, v, w such that
$x^3-(x+y+z)^3=u^3$, $y^3-(x+y+z)^3=v^3$ $z^3-(x+y+z)^3=w^3$,
put $x+y+z=n$, $x=2n^3$, $y=9n^3$, $z=28n^3$.

It remains that the sum of these three should be n, but they are together equal to $39n^3$. Therefore $39n^3=n$, or $39n^2=1$.

And if the number of squares were a square number the problem would be solved. But 39 is the sum of three cubes increased by the number three (i. e. $39=2+9+28=1+1^3+1+2^3+1+3^3=3+1^3+2^3+3^3$. We must therefore find three cubes, the sum of which increased by the number 3, may make a square.

Put the side of the first $=n'$, second $=3-n'$, third $=1$.

Then the sum of the three cubes is $9n'^2+28-27n'$, to which adding 3 we have $9n'^2+31-27n'=(3n'-7)^2$ and $n'=\frac{6}{5}$, that is the side of the first cube, the side of the second is $\frac{9}{5}$, and of the last 1.

Farther, to the cube of every one of these I add unity, and come to the original question. I put each (number) of so many cubes (i. e. I put $x=((\frac{9}{5})^3+1)n^3$, &c.). It remains that the sum of them should be equal to n. But the sum of them is $11\frac{14}{25}n^3$. Therefore $n=11\frac{14}{25}n^3$, therefore $n=\frac{5}{17}$. Ad positiones. L.

Another Additional Solution.

Put $(a^3+u^6)v^3$, $(b^3+u^6)v^3$, and $(c^3+u^6)v^3$ to denote the numbers sought, and u^2v their sum; then if each number be lessened by the cube of their sum, there remains the three cubes a^3v^3, b^3v^3, and c^3v^3; and we have only to satisfy the condition that the sum of the numbers be $=u^2v$, or that $(a^3+b^3+c^3+3u^6)v^3$ be $=u^2v$, therefore, dividing by v, $(a^3+b^3+c^3+3u^6)v^2=u^2$, and consequently that v may be rational, $a^3+b^3+c^3+3u^6$ must be a square. Now by trying small whole numbers it will be found that the values $a = 5$, $b = 8$, $c = 9$, and $u = 1$ will succeed, for $a^3 + b^3 + c^3 + 3u^6$ is then $= 125 + 512 + 729 + 3 = 1369 = 37^2$, therefore $v = \frac{1}{37}$ and the numbers are

$$(a^3 + u^6)v^3 = (125 + 1) \times \frac{1}{37^3} = \frac{126}{50653}$$
$$(b^3 + u^6)v^3 = (512 + 1) \times \frac{1}{37^3} = \frac{513}{50653},$$
$$(c^3 + u^6)v^3 = (729 + 1) \times \frac{1}{37^3} = \frac{730}{50653}.$$

But to find other answers in a direct manner, without trial, put $a = pv$, $b = (3q^2 - p)v$, $c = sv$, then $a^3 + b^3 + c^3 + 3u^6 = 9q^4p^2 - 27q^4p + 27q^6 + s^3 + 3u^6 =$ a square $= (m - 3qp)^2 = m^2 - 6qmp + 9q^2p^2$, therefore $p = \frac{m^2 - 27q^6 - s^3 - 3u^6}{6qm - 27q^4}$, where m, q and s may be taken at pleasure, within certain limits which are easily determined for any assumed value of u.

Ex. Suppose $q = 1$, $m = 7$, $s = 1$, and $u = 1$, then $p = \frac{49 - 27 - 1 - 3}{42 - 27}$ $= \frac{18}{15} = \frac{6}{5}$, $b = 3q^2 - p = \frac{9}{5}$, $v = \frac{u}{m - 3qp} = \frac{5}{17}$, and the three numbers are

$$(a^3 + u^6)v^3 = \frac{6^3 + 5^3}{5^3} \times \frac{5^3}{17^3} = \frac{341}{4913}$$
$$(b^3 + u^6)v^3 = \frac{9^3 + 5^3}{5^3} \times \frac{5^3}{17^3} = \frac{854}{4913}$$
$$(c^3 + u^6)v^3 = \frac{5^3 + 5^3}{5^3} \times \frac{5^3}{17^3} = \frac{250}{4913}.$$ L.

XI. QUESTION 305, *by* Rosamond.

To find two (or more such pair of amiable, but unequal) numbers, that each shall be mutually equal to the sum of the aliquot parts of the other? And also to find the least number, whose aliquot parts summed up, shall exceed it by 7?

Answered by Mr. Landen.

Put $4x$ for one of the amiable numbers, and $4yz$ for the other, x, y, and z being primes; we shall have $7 + 3x = 4yz$, or $x = \frac{4yz - 7}{3}$, and $4x = 7 + 7y + 7z + 3yz = \frac{16yz - 28}{3}$; hence $z = 3 + \frac{16}{y - 3}$. Now taking $y = 5$, z will be $= 11$, and $x = 71$, $4x = 284$, and $4yz = 220$, the first and least amiable pair.

This gentleman proceeds in a new method of substitution (shewing the defect of the present method) and finds 18416 and 17296 the next amiable pair, and 9437056 and 9363584 for a third amiable pair. And then gives this general rule: If 2^n (n being an affirmative integer) be taken such, that $3 \times 2^n - 1$, $6 \times 2^n - 1$, and $18 \times 2^{2n} - 1$ be primes, then $2^{n+1} \times 18 \times 2^{2n} - 1$ shall be an amiable number, and $2^{n+1} \times 3 \times 2^n - 1 \times 6 \times 2^n - 1$ its amiable correspondent or partner.

N. B. Mr. Stone's theorem, in his Mathematical Dictionary, for finding amiable numbers, is erroneous. The above true rule is said by Mr. Landen to be first shewn by the famous Descartes.

The same ingenious gentleman proceeds to the solution of the 2d part of the question, putting $4x^n =$ the number sought, x being a prime, and n an affirmative integer.

$7 + 7x + 7x^2 + \ldots 7x^{n-1} + 3x^n$ is the sum of its aliquot parts, which must be $= 4x^n + 7$. Therefore $x^n - 7x - 7x^2 - \ldots 7x^{n-1}$ $= \frac{x^{n+1} - 8x^n + 7x}{x - 1} = 0$. Whence $x^n - 8x^{n-1} + 7 = 0$. Herein, if $n = 2$, x will be $= 7$. Whence $4 \times 7^2 = 196$ the least whole number, whose aliquot parts summed up shall exceed it by 7.

XII. QUESTION 306, *by Mr.* Heath.

If the long-disputed prize money, between the officers of the ship Centurion and Gloucester, had been divided in the proportion of two numbers [each of which being raised to a power expressed by the natural logarithm of the other, shall be equal to the sum and difference of those numbers]. What would be the odds of advantage allotted to Lord Anson's officers belonging to the said ship the Centurion?

Answered by Mr. Heath *only.*

Instead of each proportional number raised to the power expressed by the natural logarithm of the other, it should have been expressed *root extracted*, or each proportional number raised to a power expressed also by the logarithm root of the other. Then putting x and z for the numbers, $x^{\frac{1}{\text{l.}z}} = x + z$, and $z^{\frac{1}{\text{l.}x}} = x - z$; which reduce to $x = (x + z)^{\text{l.}z}$ and $z = (x - z)^{\text{l.}x}$; and if $b =$ 2·3025 &c. then $x = (x + z)^{\text{l.}z \times b}$, and $z = (x - z)^{\text{l.}x \times b}$; whence, by the table of common logarithms (performing beyond the art of series) $x = 3{\cdot}8761$, and $z = 2{\cdot}1292$ fere. And hence the odds of proportion of payment are as 1·8204 to 1. As the question was expressed $x^{1.z} = x + z$, and $z^{1.x} = x - z$, where, if $z = 0$, the Gloucester's share of prize money, in the first equation, then x, in the same equation has an impossible value for the Centurion's lot.

XIII. QUESTION 307, *by Mr.* Landen.

Let a ball of heavy metal be laid upon one end of an horizontal plane, of an indefinite length, round which end let the plane be made to revolve downwards, with such an uniform motion, that the angle of inclination may increase at any given rate: It is required to find what length the ball will descend along the plane, before it acquires such a velocity as will cause it to fly off, and ceases touching the plane?

Solution by Mr. Landen, *the proposer, taken from the Appendix to Dr.* Hutton's *Edition of the Diaries; his original solution being false.*

Let b be the angular velocity of the plane per second, measured on an arc of a circle whose radius is 1; y the sine of the inclination of the plane to the horizon, to the same radius; z the arc whose sine is y; and x the distance of the ball from the axis about which the plane revolves.

Then bx will be the circulatory, or paracentric, velocity of the ball; and bbx will be its centrifugal force.

Moreover, denoting $16\frac{1}{12}$ feet by g, $2gy$ will be the force of gravity urging the ball down the plane. Therefore $bbx + 2gy$ will be the whole force urging the ball from the axis of motion. Suppose its velocity from that axis to be v: then will $(bbx + 2gy)\frac{\dot{z}}{b}$ be $= \dot{v}$; and $\frac{\dot{z}}{b} = \frac{\dot{x}}{v}$, or $v = \frac{b\dot{x}}{\dot{z}}$. Hence, $\dot{v}$ being $= \frac{b\ddot{x}}{\dot{z}}$ when $\dot{z}$ is supposed invariable, we get by substitution $(bbx + 2gy)\frac{\dot{z}}{b} = \frac{b\ddot{x}}{\dot{z}}$. It appears therefore that $\frac{2g}{bb}y\dot{z}^2$ is $= \ddot{x} - x\dot{z}^2$; whereof the equation of the fluents is $n^z - n^{-z} = \frac{2bbx}{g} + 2y$, n being the number whose hyp. log. is 1; as appears by what follows.

Multiplying each side of the equation $\frac{2g}{bb}y\dot{z}^2 = \ddot{x} - x\dot{z}^2$ by n^z, we have $\frac{2g}{bb}yn^z\dot{z}^2 = n^z\ddot{x} - xn^z\dot{z}^2$: whence, by taking the fluents, we find the fluent of $\frac{2g}{bb}yn^z\dot{z}^2 = n^z\dot{x} - xn^z\dot{z}$.

Now, denoting the cosine $\sqrt{(1-yy)}$ by w, $\dot{w}$ is $= -y\dot{z}$, and $\dot{y} = w\dot{z}$; therefore $y\dot{z}$ is $= -\dot{w} - w\dot{z} + \dot{y}$, and it is obvious that, by adding $y\dot{z}$ on each side and multiplying by $\frac{gn^z\dot{z}}{bb}$, we shall get $\frac{2g}{bb}yn^z\dot{z}^2 = \frac{g}{bb}(yn^z\dot{z}^2 + \dot{y}\dot{z}n^z - wn^z\dot{z}^2 - \dot{w}\dot{z}n^z)$. Whence it appears that the fluent of $\frac{2g}{bb}yn^z\dot{z}^2$ is $= \frac{g}{bb}(yn^z\dot{z} - wn^z\dot{z} + \dot{z}) = n^z\dot{x} - xn^z\dot{z}$ by what is said above. Hence, by division, we have $n^{-z}\dot{x} - xn^{-z}\dot{z} = \frac{g}{bb}(n^{-2z}\dot{z} + yn^{-z}\dot{z} - wn^{-z}\dot{z}) = \frac{g}{bb}(n^{-2z}\dot{z} + yn^{-z}\dot{z} - n^{-z}\dot{y})$: from which equation, by again taking the fluents, we get $xn^{-z} = \frac{g}{bb}(\frac{1}{2} - \frac{1}{2}n^{-2z} - yn^{-z})$; and consequently $n^z - n^{-z} = \frac{2bbx}{g} + 2y$, as expressed above.

By means of this equation of the curve described by the ball, the velocity $v\ (= \frac{b\dot{x}}{\dot{z}})$ is found $= \frac{g}{2b}(n + n^{-z} - 2w)$.

To find the pressure against the plane: let p be the perpendicular from the centre of motion to the tangent to the curve described by the ball, the ray from that centre to the point of contact being x; and let q be the absolute velocity of the ball in the curve at that point. If then the gravity ceased to act on the ball, it would proceed in the direction of the said tangent, and (p and q being in that case invariable) the fluxion of $\frac{pq}{x}$ its circulatory velocity would be $= \frac{-pq\dot{x}}{xx}$; which expression is $= -b\dot{x}$, $\frac{pq}{x}$ being $= bx$. But the gravity acting, and the ball pressing against the plane, the fluxion of (bx) its circulatory velocity is $= b\dot{x}$. Therefore $2b\dot{x} = 2v\dot{z}$ is the fluxion of the circulatory velocity arising from $(2gw - P)$ the excess of the force of gravity at right angles to the ray x above P the pressure against the plane. Consequently $2v\dot{z}$ is $= (2gw - P)\frac{\dot{z}}{b}$: whence $P = 2gw - 2bv = g(4w - n^z - n^{-z})$.

When the ball quits the plane P is $= 0$, and $4w = n^z + n^{-z}$; from which equation z is found to be, at that instant, equal to (·823766) an arc of 47° 11′ 54″. Mr. Simpson, finding the fluents by infinite series, makes the angle 47° 9′.

Remark. If, instead of the plane, a *tube* be substituted, continued both ways from the axis of motion, just capable of receiving the ball so that it may move freely therein; and, the instant the tube begins to revolve from a horizontal position, the ball be made to move therein from the said axis, along the ascending part of the tube, with a velocity $= \frac{g}{b}$; the fluxionary equation adapted to this case will be $\frac{2g}{bb}y\dot{z}^2 = x\dot{z}^2 - \ddot{x}$: and, by correcting the fluents accordingly, x will from thence be found equal to the simple expression, $\frac{gy}{bb}$! Which equation suggests this *very remarkable inference:* The ball being at first put in motion with the velocity $\frac{g}{b}$, it will revolve uniformly in a circle (whose diameter is $\frac{g}{bb}$) touching the horizontal line with which the tube at first coincides! and it will continue so to revolve (moving up and down alternately in the different branches of the tube) so long as the motion of the tube is continued, making two complete revolutions whilst the tube makes one revolution! and the uniform velocity $\left(\frac{g}{b}\right)$ wherewith the ball so revolves in the circle will be to its velocity along the tube every where as 1 to w.

XIV. QUESTION 308, *by* Upnorensis.

A lady paid twice as much a-piece for geese, as she paid for ducks; and twice as much a-piece for ducks, as she paid for chickens, which cost together 1l. 13s. 4d. the sum of the squares of the number she bought of each sort was 326. What number of geese, ducks, and chickens did she buy? And what was the price of each?

Answered by Mr. James Terey.

All the numbers whose squares equal 326, are easily determined.

18	1	1
17	6	1
15	10	1
14	11	3
14	9	7
13	11	6

Putting x, y, and z for the number of geese, ducks, and chickens, and v the pence paid for the chicken; then, among some of those 3 numbers (varied for geese, ducks, and chickens) $4vx + 2vy + vz = 400$; whence $v = \frac{400}{4x + 2y + z}$ a whole number; which only admits of

		s.	d.				s.	d.
13 geese	price of each	1s.	8d.	and	9 geese	price of each	2s.	1d.
11 ducks		0	10		7 ducks		1	0½
6 chickens		0	5		14 chickens		0	6¼

Amounting each way to 1l. 13s. 4d.

XV. QUESTION 309, *by* Hurlothundro.

What are the odds of battle, or the different probabilities of success of two armies going to engage, the chances of each army for victory being respectively equal to the sum raised to a power expressed by the difference, and the difference raised to a power expressed by the sum of those chances?

Answered by the Proposer.

Raised to a power, &c. should have been expressed *root extracted, &c.* Then, if v and y represent the respective chances of each army for victory, $(v+y)^{\frac{1}{v-y}} = v$, and $(v-y)^{\frac{1}{v+y}} = y$; whence $v + y = v^{v-y}$, and $v - y = y^{v+y}$; where $v = \cdot 5806$ and $y = \cdot 2590$ very correctly, as may be proved by a table of logarithms; which numbers were not a little curious to determine. Hence the odds of battle are as 2·2417 to 1.

As most persons of science are but little conversant with these sort of equations, it may not be improper here to unfold the mystery of raising powers of all sorts. And 1. A decimal raised to a decimal power produces a greater value than the root. 2. A decimal raised to an integral power produces a less value than the root. 3. An integer raised to a decimal power produces a less value than the root.

$x^0 = 0^0 = 1$; all very small powers of quantities approaching the value of unity. x^x is least $= \cdot 6922$ correctly, when $x = \cdot 3678798$ &c. $= \dfrac{1}{2\cdot 30258509 \text{ \&c.}}$. The logarithm of a decimal is negative, or so much less than nothing.

To determine the value of the unknown quantities in all exponential equations, it is convenient to suppose the least quantity unknown $= 0$, and thence to find the value or values of the next greater, noting the error in the next equations. Again, suppose the value of the least unknown quantity $= 1$, and thence find the value or values of the next greater: noting the quality of the error, in the next equations, as before; and so on to 10, 20, &c. for the least value, if need requires. Or supposing the value of the greatest unknown quantity to be 1, 10, 100, &c. determining the value of the next less unknown quantity to each supposition; at the same time always denoting the quality of errors, by which the true values of the unknown quantities are determined by a table of logarithms very exactly.

N. B. It is often very easy and convenient to suppose the value of one unknown quantity in two equations, in such a manner as that by it the other may be determined in a whole number.

XVI. QUESTION 310, *by Mr.* Ash.

A spider, at one corner of a semi-circular pane of glass, gave uniform and direct chase to a fly, moving uniformly along the curve before him: the fly was 30° from the spider at the first setting out, and was taken by him at the opposite corner. What is the ratio of both their uniform motions?

Answered by Nobody.

Mr. Farrer sent us a solution which was not true. Mr. Lauden sent us a true method; but the calculus being so operose, it was not wrought out. And no method appearing to us yet elegant enough for a place, it will be next year before we shall have time to catch the solution to this famous spider and fly question. The ratio of the motion of the insects are little different from an equality; though a certain gentleman makes the motion of the chasing insect the slowest, to overtake the fly.

Solution by ΦΙΛΟΠΟΝΟΣ, *taken from Turner's Exercises, where this question was afterwards proposed and answered as follows:*

Question. To find the nature of the curve, described by a body giving uniform and direct chase to another, supposing the body pursued to move uniformly in the periphery of a given circle.

Solution. Let CAGM be the proposed circle, and AEM the required curve; also let E and G be any cotemporary positions of the two bodies, and from the centre C of the given circle suppose CG, CE and CF to be drawn, the last of them perpendicular to EG. Put CG $= a$, GE $= x$, GF $= y$, and let the velocity of the pursuing body be denoted by m, and that of the pursued by n. Then the relative celerity of the latter in the direction EGK, with which it moves from the former body at E, will be expressed by n (CF $\div$ CG), or its equal $(n \div a)\sqrt{(aa - yy)}$; which, taken from the velocity m of the pursuing body, leaves $m - (n \div a)\sqrt{(aa - yy)}$, for the true celerity with which this last body gains upon the other, or that wherewith the distance EG decreases. But CE : EF :: $m : m$ (EF$\div$CE) $= m(x - y) \div \sqrt{(aa + xx - 2xy)} =$ the velocity with which CE decreases.

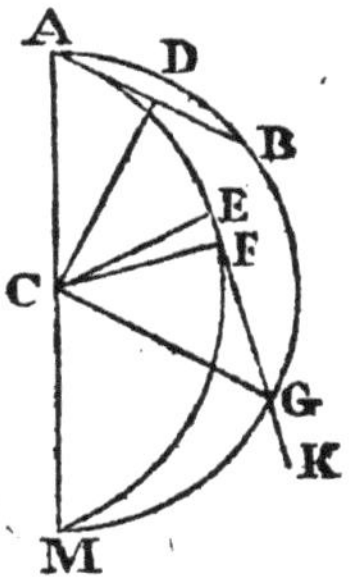

Consequently the fluxion of EG is to that of CE, as $m - (n \div a)\sqrt{(aa - yy)} : (mx - my) \div \sqrt{(aa + xx - 2xy)}$; that is, $\dot{x} : (x\dot{x} - x\dot{y} - y\dot{x}) \div \sqrt{(aa + xx - 2xy)} :: m - (n \div a)\sqrt{(aa - yy)}$

: $(\dot{m}x - my) \div \sqrt{(aa + xx - 2xy)}$. Whence, we get the following equation $y\dot{x} + x\dot{y} - x\dot{x} = ma\dot{x}\dot{y} \div n\sqrt{(aa - yy)}$; which, by putting $m \div n = r$, and converting the radical quantity into an infinite series, becomes $y\dot{x} + x\dot{y} - x\dot{x} = rx\dot{y}$ into $1 + \frac{y^2}{2a^2} + \frac{3y^4}{8a^4}$, &c. From which, by the method of resolving fluxional equations (see sect. 2, part 2, of Simpson's Fluxions), the value of y will be found in terms of x. Whence EF $(x - y)$ and CE $(\sqrt{(aa + xx - 2xy)})$ will also be had in terms of x. But half the fluxion of CE², divided by EF (or its equal $\dot{x} - x\dot{y} \div (x - y)$) is the fluxion of ME (see art. 136 of the said book) and consequently $\dot{x} \div r - x\dot{y} \div r\,(x - y) =$ that of the arch MG, whose fluent will consequently be equal to the arch itself; by means whereof ME will likewise be given.

But in cases where the position of both the bodies is given when the chase begins it may sometimes be necessary (when the series does not converge sufficiently fast) to find the values of FG and the described arch GB, corresponding to any given value of EG, at two, or more operations.

Thus, supposing b, c, and d to denote any three corresponding values of EC, FG, and the arch BG, let $b - z$, $c + v$, and $d + w$ be three other cotemporary values of the said quantities respectively.

Then by substituting $b - z$, and $c + v$, for x and y respectively, in the fluxional equation given above, we have $(b - c)\dot{z} + b\dot{v} - z\dot{z} - v\dot{z} - z\dot{v} = ra\dot{v}\,(b - z) \div \sqrt{(aa - cc - 2cv - vv)} = rab\dot{v} \div f - raz\dot{v} \div f + rabcv\dot{v} \div f^3$, &c. (by putting $f = \sqrt{(aa - cc)}$, and converting the radical quantity into a series. Whence by reduction, and putting $b - c = g$, $ra \div f - 1 = h$, and $rabc \div f^3 = k$, there arises $g\dot{z} - bh\dot{v} - z\dot{z} - v\dot{z} + hz\dot{v} - kv\dot{v}$, &c. $= 0$.

From this equation v will be found in terms of z: which value being supposed small, two terms of the resulting series will be sufficient. And then, by substituting these new values of $b - z$, and $c + v$, instead of x and y in $\dot{x} - x\dot{y} \div (x - y) = \dot{w}$, and taking the fluent, the corresponding value of w will also become known.

And so we shall have three new values of EG, FG, and the arch BG; with which the operation may be repeated at pleasure, and other corresponding values of those quantities may be found, by the very same equation above derived.

As to the particular case in the Diary question, the first value of EG being given equal to AB the chord of 30°, and that of FG equal to the half of it BD; the velocity of the pursuing body to that of the body pursued will be found to be in the ratio of CD ÷ AC + AB ÷ BM to unity, very near; that is in numbers, as 1·16 to 1.

Note. When the velocity of the body pursued is the greatest of the two, the required curve will be a spiral converging continually, nearer and nearer to the circumference of a circle concentric with the given one; whose radius is to that of the given one in the ratio of the lesser velocity to the greater.

PRIZE QUESTION, *by Mr.* R. Heath.

On what days of the year does the city of London travel the greatest and least number of miles, by the diurnal and annual motion of the earth? And how many miles per day, and also per hour about noon does it travel, when the days are longest and shortest in that place?

Answered by Mr. Heath, *the Proposer.*

Anno 1748, June $18^d\ 10^h\ 17^m$, and Dec. $18^d\ 1^h\ 22^m$ by authentic tables the earth was in the aphelion and perihelion points of her orbit, or at the greatest and least dist. à ☉; her motion (by describing equal areas in equal times round him) being at those times of aphelion and perihelion, (later by 9 hours than according to Street's Tables) the slowest and quickest respectively.

By the equation table of the earth's orbit, her true diurnal motion round ☉ at coming to aphelion is $57'\ 12''$ (earth accelerating variously from aph. to perih.) and true diurnal motion coming to perih. $61'\ 10''$ (earth retarding variously from perih. to aph.). Now allowing $10''$ one-half ☉'s mean parallax, or angle at ☉, subtended by earth's semi-diameter, then by trigon. 19644 earth's semi-diameters is her mean dist. à ☉; being then at the conjugate end of her orbit; which dist. à ☉ = length of the semi-transverse.

Mr. Flamsteed's eccentricity of the earth's orbit, or ☉'s focal dist. à centre of her orbit is 1692 such parts as semi-transverse or mean dist. is 100000, consequently 100000 : 1692 :: 19644 : 332·37 earth's semi-diameters, the ☉'s true dist. à centre earth's orbit: whence 19976·37 and 19311·63 semi-diameters are the earth's greatest and least dist. à ☉. And by her describing equal elliptical sectors round him each day, with the respective angles $57'\ 12''$ and $61'\ 10''$, as before observed, the correspondent elliptical arches, which may be considered as circular for a day, will be 332·28 and 343·6 &c. semi-diameters which the earth's centre goes over in 24 hours, when her motion is slowest and fastest. Whence, by allowing the earth's semi-diam. = 3967 miles, her centre goes over at fastest 56794 miles per hour, 946 per minute, and almost 16 miles per second! an amazing swiftness! Also at her slowest rate, in aphelion, 54923 miles per hour, 915 per minute, and 15 per second.

Anno 1748, June $9^d\ 16^h\ 34^m\ 46^s$, and Dec. $9^d\ 21^h\ 8^m$. equal time, ☉ enters Cancer and Capricorn, the respective mean anomalies being then 11 S. $21°\ 23'\ 15''$, and 5 S. $21°\ 56'\ 28''$ (by Halley's and Flamsteed's numbers); whence the proportional distances of earth à ☉ are 101675 and 98322 respectively, and thence the true distances 19973 and 19314 earth's semi-diameters (by multiplying 19644 a constant multiplier, and cutting off 5 fig.); and the earth's diurnal angular motion round ⊙ being 57′ 11″ and 61′ 9″ respectively (by tables of the earth's orbit) to the said radii; consequently the arches respectively moved over by the earth's centre, when the days are longest and shortest, are 332·230 and 343·554 &c. semi-diameters of the earth; nearly equal to her slowest and fastest motions as above: she being at those times but a few days from the aphelion and perihelion points.

N. B. Anno 1748, March $19^d\ 2^h\ 41^m$, and Sept. $17^d\ 17^h\ 51^m$ the mean anomalies of the earth are respectively 9 and 3 degrees when her mean and true places differ the most, viz. 1° 56′ 20″; about which times she neither accelerates nor retards for some days.

To find the distance travelled over by a spot on the earth's surface for a day.

The earth's radius bearing so small a proportion to her dist. à ⊙, she may be considered as revolving forward in the direction of a straight line, for a small interval of time, with a progressive motion as p, to 1 rotatory, or over 332·23 and 343·554 semi-diameters respectively, on the days about the solstices of Cancer and Capricorn; to each of which distances gone forward she makes but one revolution.

Put $a =$ ÆQ, the diameter of a circle generating a cycloidal curve, with progressive motion p, and rotatory 1; $x =$ Æm; then $om = \sqrt{(ax - xx)}$ per circle. And fluxion arch Æo =

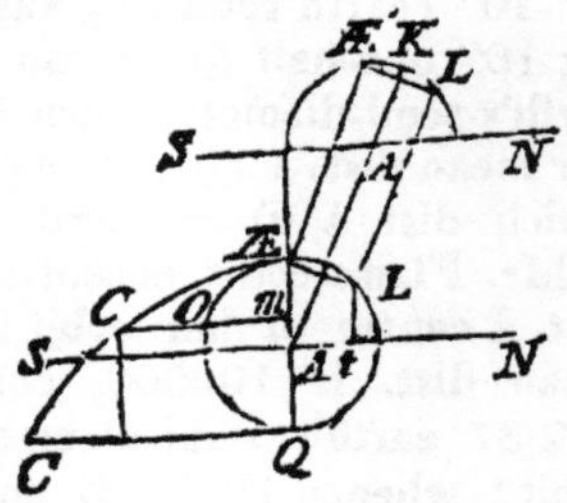

$\frac{a\dot{x}}{2\sqrt{(ax - xx)}}$; whence $1 : p ::$

$\frac{a\dot{x}}{2\sqrt{(ax - xx)}} : \frac{pa\dot{x}}{2\sqrt{(ax - xx)}} =$ flux. co; and flux. $om = \frac{a - 2x}{2\sqrt{(ax - xx)}} \times \dot{x}$ whose sum $\frac{ap + a - 2x}{2\sqrt{(ax - xx)}} \times \dot{x} =$ fluxion cm. But $\sqrt{((\text{flux. c}m)^2 +}$ $(\text{flux. Æ}m)^2) =$ fluxion arch cÆ $= \frac{\dot{x}}{4x} \times \frac{\sqrt{((p+1)^2 \times a^2 - 4pax)}}{a - x}$; whose fluent is $\sqrt{ax} \times : p + 1 + \frac{(p+1)^2 - 4p}{2.3.(p+1)} \times \frac{x}{a} + \frac{(p+1)^2 - 4p}{2.5.(p+1)}$ $- \frac{((p+1)^2 - 4p)^2}{2.4.5.(p+1)^3} \times \frac{x^2}{a^2} + \frac{(p+1)^2 - 4p}{2.7.(p+1)} - \frac{((p+1)^2 - 4p)^2}{2.2.7.(p+1)^3} +$ $\frac{(p+1)^2 - 4p}{2.2.4.7.(p+1)^3} \times \frac{x^3}{a^3}$ &c.

The poles of the earth S, N, moving nearly parallel, for a short time, the earth's centre A is carried with an oblique direction AK, in an angle of 23° 30′ with the poles of the equator ÆQ; and the city of London, on the surface, at L, in the parallel rotatory direction to ÆQ on the surface; the distance betwixt the finishing of each revolution being LL = AA = ÆÆ = kK; each town on the earth's surface describing cycloidal curves whose bases are equal. Lt the dist. of the city of London from the earth's axis = ·622059 semi-diam. being nat. s. co-lat. London 88° 28′ (vid. curve of the tack, D. 1747). Now, when $x = 2$, Lt = 1·244118 earth's semi-diam. $= a$, $p = 85$, and $p = 87·9$ fere (dividing 332·230 and 343·554 by 3·90852 = lesser circle's circumference whose diam. $= a = 2$Lt) the semi-cycloids = 166·12 and 171·78 fere; whence 332·24 and 343·56 earth's semi-diam. = distances travelled by the city of London in 24 hours, when the days are longest and shortest; consequently it travels about 54916 and 56787 miles per hour, respectively at those times.

N. B. This being but in consequence of theory, those who are more curious may rectify the cycloidal curve described by the progressive motion of a point on a revolving globe, proceeding uniformly, in a circular or elliptical direction, forward as p, with a rotatory motion as 1, at the same time; though a real acceleration or retardation in either, would perplex the motion so, as to render the solution of the track next to impossible.

Questions proposed in 1749, *and answered in* 1750.

I. QUESTION 311, *by Mr.* Landen, *near Peterborough.*

To find three such numbers, that the sum or difference of any two of them shall be a square number?

Answered by Mr. C. Bumpkin.

By a method of substitution, too tedious to insert, he finds 1873432, 2399067, 2288168, the three numbers, answering the conditions of the question.

A solution to this question may be seen in Ozanam's Dictionary, and in Euler's Algebra. We shall also here insert the three following solutions to it, from the Appendix to Dr. Hutton's edition of the Diaries.

I. *By Mr.* J. Landen, F. R. S.

To find three such numbers, that the sum and difference of any two of them shall be square numbers. Supposing x, y, and z to be the required numbers; we may assume

$$1\text{st}, \; x = \tfrac{1}{2} \times (f^4g^4 + g^4 + f^4 + 1),$$
$$y = \tfrac{1}{2} \times (f^4g^4 - g^4 - f^4 + 1) + 2ffgg,$$
$$z = \tfrac{1}{2} \times (f^4g^4 - g^4 - f^4 + 1) - 2ffgg;$$

whence the values of $x + y$, $x - y$, $x + z$, $x - z$, and $y - z$, are manifestly squares. Therefore, to solve the question, it only remains to find $f^4g^4 - g^4 - f^4 + 1$ (the value of $y + z$) a square.

Or 2dly we may assume $x = ffgg + 1$, $y = ff + gg$, $z = 2fg$; where the values of $x + z$, $x - z$, $y + z$, $y - z$, are manifestly squares. Therefore, after this assumption, we have to find $ffgg + gg + ff + 1$ (the value of $x + y$) a square, and $ffgg - gg - ff + 1$ (the value of $x - y$) a square.

It is obvious, that, according to this assumption, the value of the expression $f^4g^4 - g^4 - f^4 + 1$ ($= (x + y)(x - y)$) must be a square as in the 1st assumption where it is $= y + z$.——To find that expression a square, substitute $f + r$ instead of g; it will then become $(f^4 - 1)^2 \times (1 + \dfrac{4f^3r + 6ffrr + 4fr^3 + r^4}{f^4 - 1})$. It appears therefore that $1 + \dfrac{4f^3r + 6ffrr + 4fr^3 + r^4}{f^4 - 1}$ must be a square: supposing it $= (1 + \dfrac{2f^3r}{f^4 - 1} + \dfrac{(f^6 - 3ff)\,rr}{(f^4 - 1)^2})^2$, from that equation r will be found $= \dfrac{4f \times (f^8 - 1)}{1 + 6f^4 - 3f^8}$, and $g\ (= f + r) = \dfrac{f(f^8 + 6f^4 - 3)}{1 + 6f^4 - 3f^8}$. Therefore f being any number whatever in the first assumption, and g as here found, the question will be answered.

It is plain that, $f^4g^4 - g^4 - f^4 + 1$ being a square, its factors $ffgg + gg + ff + 1$ and $ffgg - gg - ff + 1$ will be squares if $\dfrac{gg \pm 1}{ff \pm 1}$ be a square. Now, taking g equal to its value found above, it will appear that the value of $\dfrac{gg \pm 1}{ff \pm 1}$ is $= \left(\dfrac{f^8 \pm 4f^6 - 6f^4 \pm 4f^2 + 1}{1 + 6f^4 - 3f^8}\right)^2$. Consequently taking f any number whatever, and g as just now mentioned, the question will be answered by the 2d assumption in much lower terms than by the 1st assumption.——If f be taken $= 2$, the numbers given in the Diary 1750 will be obtained.

There is another way to find $f^4g^4 - g^4 - f^4 + 1$ ($= (f^4 - 1)(g^4 - 1)$) a square. Assume it $= (f^4 - 1)^2 (g^2 + 1)^2$, or $= (f^4 - 1)^2 (g^2 - 1)^2$: whence $g = \dfrac{ff}{\sqrt{(2 - f^4)}}$ or $= \dfrac{ff}{\sqrt{(f^4 - 2)}}$. Consequently, if, in the 1st assumption, f be so taken that $2 - f^4$, or $f^4 - 2$, be a square, and $g = \dfrac{ff}{\sqrt{(2 - f^4)}}$, or $\dfrac{ff}{\sqrt{(f^4 - 2)}}$ respectively, the quest. will be answered.

To find $2 - f^4$ a square, I write $1 - d$ for f: by which means the expression becomes $1 + 4d - 6d^2 + 4d^3 - d^4$, which being supposed $= (1 + 2d - 5dd)^2$, we find $d = \frac{12}{13}$, $f (= 1 - d) = \frac{1}{13}$, and $g = \frac{1}{239}$. Therefore, if, in the 1st assumption, f be $= \frac{1}{13}$ and $g = \frac{1}{239}$, or $f = 13$ and $g = 239$, the quest. will be answered.

To find $f^4 - 2$ a square, I suppose it $= (ff - hh)^2$; whence $ff = \frac{4 + 2h^4}{4hh}$. Therefore $4 + 2h^4$ must be a square.

To find it so, I suppose it $= (2 + bbhh)^2$; and, by this supposition, I find $hh = \frac{4bb}{2 - b^4}$; where, if b be $= 1$, hh will be $= 4$; and consequently $f = \frac{3}{2}$, $g = \frac{9}{7}$.

Having found $2 - f^4$ a square when f is $= \frac{1}{13}$, it is obvious that hh (and consequently $f^4 - 2$) will be found a square by taking $b = \frac{1}{13}$. And other values of f and g may be found in like manner.

II. *By the Rev. Mr.* Wildbore.

Suppose $2abxy$, $aaxx + bbyy$, and $aayy + bbxx$ to be the three numbers required. Then the sums and differences $aaxx \pm 2abxy + bbyy$ and $aayy \pm 2abxy + bbxx$ being necessarily squares, it only remains to make $aaxx + bbyy + aayy + bbxx$ and $aaxx + bbyy - aayy - bbxx$ squares; their product therefore $= a^4x^4 - b^4x^4 - a^4y^4 + b^4y^4$ must be a square: make $x = z - ay \div b$; then substituting this value for x in the last expression, it will thence appear that $a^8y^4 - 2a^4b^4y^4 + b^8y^4 - 4a^7by^3z + 4a^3b^5y^3z + 6a^6bbyyzz - 6aab^6yyzz - 4a^5b^3yz^3 + 4ab^7yz^3 + a^4b^4z^4 - b^8z^4$ must be a square. Suppose its side $= (a^4 - b^4)\ yy - 2a^3byz + \frac{a^4 - 3b^4}{a^4 - b^4} aabbzz$; this squared and made equal to the other, will give, by reduction, $-4a^9y + 4ab^8y = -3a^8bz + 6a^4b^5z + b^9z$; whence $z = 4a^9 - 4ab^8$, and $y = 3a^8b - 6a^4b^5 - b^9$, and consequently $x = a^9 + 6a^5b^4 - 3ab^8$. These values being substituted in the assumed expressions, they will become respectively

$$6a^{18}b^2 + 24a^{14}b^6 - 92a^{10}b^{10} + 24a^6b^{14} + 6a^2b^{18},$$
$$a^{20} + 21a^{16}b^4 - 6a^{12}b^8 - 6a^8b^{12} + 21a^4b^{16} + b^{20},$$
$$10a^{18}b^2 - 24a^{14}b^6 + 60a^{10}b^{10} - 24a^6b^{14} + 10a^2b^{18}.$$

Which are general theorems for infinite answers, where a and b may be taken any numbers at pleasure; the only limitation being that $6a^4b^4 + b^8$ must be less than $3a^8$. And when $a = 2$, $b = 1$, the three numbers answering the question are 2288168, 2399057, and 1873432, as put down at page 19.

III. *By the* Editor, (*Dr.* Hutton.)

Three numbers answering the conditions may be found by the following easy process.

Assume $4x$, $4+xx$, and $1+4xx$ for the three numbers; where the sum and diff. of the first and each of the other two being all four squares, we have only to make the sum and diff. of the two latter to be squares. viz. $5xx+5$, and $3xx-3$ each a square. Their product $15x^4-15$ will therefore be a square. Which it will evidently be when $x=2$; for then it is $15^2=225$. And then the three assumed numbers become 8, 8, and 17, which answer the conditions. But as two of the numbers are equal to each other, make $x=z-2$; then $15x^4-15$ is $=225-480z+360z^2-120z^3+15z^4=$ a square $=$ suppose $(15-az+bzz)^2=225-30az+(30b+aa)z^2-2abz^3+b^2z^4$. Here equate the 2d term of the one of these to the second of the other, and a is found $=16$; then equate the 3d to the 3d, and there results $b=\frac{59}{15}$; and finally the last two terms of the one equated to the last terms of the other, we have $z=\frac{2040}{671}$. Hence $x=z-2=\frac{698}{671}$; which substituted in the assumed expressions, they become $4\times\frac{698}{671}$, $4+\frac{698^2}{671^2}$, and $1+\frac{4\cdot698^2}{671^2}$; or, by multiplying each by 671^2, they become 1873432, 2288168, and 2399057, the very same numbers as before.

Another Additional Solution.

This problem easily reduces to another which is better known, namely to find three square numbers such that the difference between each two may be a square; for if the numbers sought be denoted by x, y, and z, and $x+y$ be put $=u^2$, $x+z=v^2$, and $y+z=w^2$, then by subtraction $y-z=u^2-v^2$, $x-y=v^2-w^2$, and $x-z=u^2-w^2$; therefore all the conditions will be satisfied if u^2-v^2, v^2-w^2, and u^2-w^2 are squares. Now these differences will be squares when $u=(m^2+n^2)(r^2+s^2)$, $v=2mn(r^2-s^2)+2rs(m^2-n^2)$ and $w=2mn(r^2+s^2)$, m being $=r^4+6r^2s^2+s^4$, $n=4rs(r^2-s^2)$, r and s being any numbers whatever, (see Mathematical Repository, Quest. 310); and these values of u, v and w being found we shall have

$$x=\tfrac{1}{2}(u^2+v^2-w^2),\ y=\tfrac{1}{2}(u^2+w^2-v^2),\ z=\tfrac{1}{2}(v^2+w^2-u^2).$$

Ex. Take $r=3$ and $s=1$, then $m=136$ and $n=96$, or, dividing by 8, $m=17$, and $n=12$: whence $u=2165$, $v=2067$, $w=2040$, and therefore

$$x=\tfrac{1}{2}(2165^2+2067^2-2040^2)=4798114$$
$$y=\tfrac{1}{2}(2165^2+2040^2-2067^2)=4576336$$
$$z=\tfrac{1}{2}(2067^2+2040^2-2165^2)=3746864,$$ or dividing by 2, $x=2399057$, $y=2288168$, and $z=1873432$. L.

II. QUESTION 312, *by the Rev. Mr.* Baker, *at Stickney, Lincolns.*

A bowl, by its bias, describes a spiral expressed by z, whose equation is $\frac{2}{3}y^{\frac{5}{3}} - \frac{3}{2}z^{\frac{3}{2}} + y^{\frac{2}{3}}z^{\frac{3}{5}} = 0$. In what direction must the bowl set out to fall upon the jack, when the length of the cast is 47 yards.

Answered by Mr. W. Jepson, *of Lincoln.*

In the equation $\frac{2}{3}y^{\frac{5}{3}} - \frac{3}{2}z^{\frac{3}{2}} + y^{\frac{2}{3}}z^{\frac{3}{5}} = 0$, if we write v^5 and v^2 for $z^{\frac{3}{2}}$ and $z^{\frac{3}{5}}$ it will become $v^5 - 8{\cdot}68245v^2 = 272{\cdot}05$ when $y = 47$; whence by converging series, $v = 3{\cdot}252325$, and thence $z = 50{\cdot}9695$.

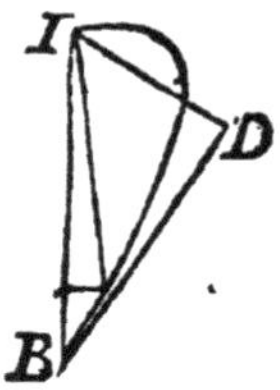

To the tangent BD, which is the line of direction, draw the perpendicular ID. Then, $\dot z : \dot y :: y : y\dot y \div \dot z =$ BD. And the above equation in fluxions is $\frac{10}{9}y^{\frac{2}{3}}\dot y - \frac{9}{4}z^{\frac{1}{2}}\dot z + \frac{2}{3}z^{\frac{3}{5}}y^{-\frac{1}{3}}\dot y + \frac{3}{5}y^{\frac{2}{3}}z^{-\frac{2}{5}}\dot z = 0$, which multiplied by $y^{\frac{1}{3}}z^{\frac{2}{5}}$ gives $\dot y = \frac{\frac{9}{4}y^{\frac{1}{3}}z^{\frac{9}{10}} - \frac{3}{5}y}{\frac{10}{9}yz^{\frac{2}{5}} + \frac{2}{3}z} \times \dot z$, whence BD $= \frac{\frac{9}{4}y^{\frac{1}{3}}z^{\frac{9}{10}} - \frac{3}{5}y}{\frac{10}{9}yz^{\frac{2}{5}} + \frac{2}{3}z} \times y$. Now BI $(= y) : 1 ::$ BD $: \frac{\frac{9}{4}y^{\frac{1}{3}}z^{\frac{9}{10}} - \frac{3}{5}y}{\frac{10}{9}yz^{\frac{2}{5}} + \frac{2}{3}z} = {\cdot}879268$ &c. the cosine of 28° 27′ (nearly) the angle of direction required.

The same answered by Mr. J. Powle.

It is evident that the line of direction will be a tangent to the curve where the bowl is delivered, the position whereof, with a line drawn from thence to the jack is what is required? The equation of the curve in fluxions is,

$\frac{10}{9}\dot y y^{\frac{2}{3}} - \frac{9}{4}\dot z z^{\frac{1}{2}} + \frac{2}{3}z^{\frac{3}{5}}y^{-\frac{1}{3}}\dot y + \frac{3}{5}y^{\frac{2}{3}}z^{-\frac{2}{5}}\dot z = 0$, therefore $\frac{\dot y}{\dot z} = \frac{\frac{9}{4}z^{\frac{1}{2}} - \frac{3}{5}z^{-\frac{2}{5}}y^{\frac{2}{3}}}{\frac{10}{9}y^{\frac{2}{3}} + \frac{2}{3}z^{\frac{3}{5}}y^{-\frac{1}{3}}}$, consequently $\frac{\dot y y}{\dot z}$ the subtangent BD $= \frac{\frac{9}{4}z^{\frac{1}{2}} - \frac{3}{5}z^{-\frac{2}{5}}y^{\frac{2}{3}}}{\frac{10}{9}y^{-\frac{1}{3}} + \frac{2}{3}z^{\frac{3}{5}}y^{-\frac{4}{3}}}$. But y being given, z is known from the equation of the curve. Therefore BD $= 41{\cdot}47$ nearly. Then, by trigonometry, as BI : BD :: rad. : cosine angle of direction, IBD $=$ 28° 8′ nearly.

Mr. Landen (in the Diary for the year following) says, that Mr. Jepson and Mr. Powle have both absurdly considered the bowl running backwards to the point where the spiral begins, and call their solutions erroneous; though the proposer, Mr. Baker, meant that it should do so, and solved the question himself the same way. Mr. Landen thinks it too easy a question, in the case of drawing a tangent to a curve, whose equation is given; and therefore correcting the fault, proposes it should be solved by drawing a tangent to] the spiral, at the point where it begins, which will make it the more hard: and says it is a case not taken notice of by authors. He has gone through part of the solution this way himself. H.

III. QUESTION 313, *by Mr.* Landen.

There are four remarkable high trees growing in a straight hedgerow, the distance of the 1st and 2d is 60 yards, of the 2d and 3d 40 yards, and of the 3d and 4th 20 yards. Where must I stand to observe them, so that the three intervals may appear equal?

Answered by Mr. Terey, *of Portsmouth.*

Let A, B, C, D, represent the four trees, and E the station sought of the observer.

AB $= 60$ yards $= a$, BC $= 20 = b$, CD $= 40 = c$.

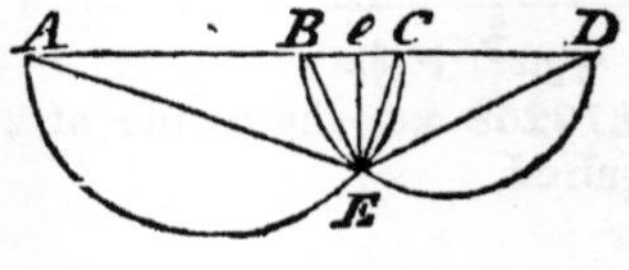

Geometrically. In this particular case, draw two semi-circles AEC and BED, on the diameters AC and BD, and the point of intersection E will be the observer's station: For then AB : BC :: AD : CD; agreeing with the data; and AB : BC :: AE : EC; also CD : BC :: ED : EB, 3 Euc. VI. (vide Univer. Arith.) Letting fall the perpendicular Ee $= y$, and putting $x = $ eC, per property of the circle $\frac{2cb}{c-b}x - xx = \frac{2ab}{a-b}(b-x) - (b-x)^2$, whence $x = \frac{1}{2}b \pm \frac{c-a}{2ac-bb}bb =$ 12 or 8 yards; whence $y = 24$ yards, and each of the angles of interval 45°.

General Solution, by Mr. Ch. Smith.

Let E be the place of the observer; A, B, C, D, the places of the trees; put $y =$ the perpendicular Ee, $a = 20 =$ BC, $x =$ eC, then $4a - x =$ Ae, $a - x =$ Be, and $2a + x =$ eD; whence the tangents of the angles AEB, BEC, and CED are $\frac{3ay}{yy + 4aa - 5ax + xx}$,

$\frac{ay}{yy - ax + xx}$, and $\frac{2ay}{yy + 2ax + xx}$, which must be all equal by the question. Whence $yy = 4ax - xx = 2aa - ax - xx$, $x = \frac{2}{5}a = ce$ $= 8$ yards, $Ee = 24$, and each angle $= 45°$.

IV. QUESTION 314, *by the Rev. Mr.* Baker.

A hare sets out 50 yards before a grey-hound, at the rate of 31 yards per second, and continues a straight course in the subquadruplicate inverse ratio of the time taken up in running: The dog sets forward only at the rate of 26 yards per second, and maintained his pace in the subquintuplicate inverse ratio of his time spent in running: How far had the dog run when his speed was equal to that of the hare's? Also when he was again as near to the hare as at first? And lastly, when he killed her?

Answered by the Rev. Mr. Baker, *the Proposer.*

Let Bb, bb, &c. represent equal portions of time, indefinitely small; BC, bc, &c. the celerities of the dog; BD, bd, &c. the celerities of the hare, at each of those times. Then, by mechanics, the areas $BCcb$ and $BDdb$, will denote the respective distances run by the dog and hare in any given time. Putting $d = 50$, $AB = 1$, $BC = 26$ yards $= a$, $BD = 31 = b$, $Bb = x$, $bc = y$, and $bd = v$. we have, by the question, $(1 + x)^{\frac{1}{5}} : 1^{\frac{1}{5}} :: a : bc = y = a(1 + x)^{-\frac{1}{5}}$, $\therefore \dot{x}y = a\dot{x}(1 + x)^{-\frac{1}{5}} =$ fluxion

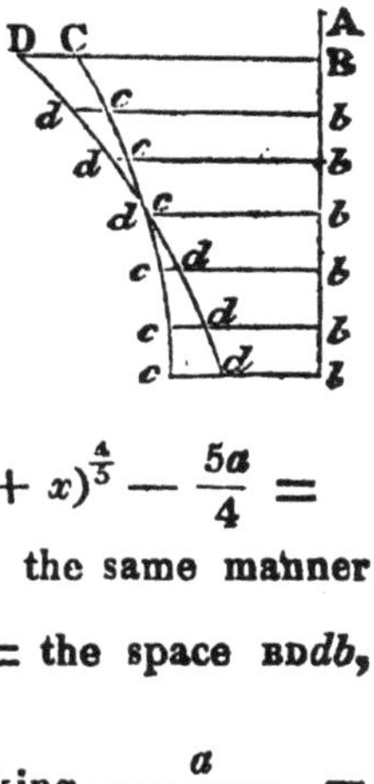

of the space $BCcb$, whose corrected fluent is $\frac{5a}{4}(1 + x)^{\frac{4}{5}} - \frac{5a}{4} =$ dist. run by the greyhound in the time x. After the same manner we get $v = \frac{b}{(1 + x)^{\frac{1}{4}}}$, and $\frac{4b}{3}(1 + x)^{\frac{3}{4}} - \frac{4b}{3} =$ the space $BDdb$, run by the hare in the same time. Hence, by making $\frac{a}{(1 + x)^{\frac{1}{5}}} = \frac{b}{(1 + x)^{\frac{1}{4}}}$, we have $x = \left(\frac{b}{a}\right)^{20} - 1 = 32{\cdot}7107$ seconds, and $509{\cdot}623$ yards, the distance run by the dog, when his pace equalled the hare's. Again, making $\frac{5a}{4}(1 + x)^{\frac{4}{5}} - \frac{5a}{4} = \frac{4b}{3}(1 + x)^{\frac{3}{4}} - \frac{4b}{3}$, we have $x = 106{\cdot}8167$ seconds, and $1341{\cdot}577$ yards run by the greyhound, when he had regained his lost ground. Lastly, making $\frac{5a}{4}(1+x)^{\frac{4}{5}}$ $-\frac{5a}{4} = \frac{4b}{3}(1+x)^{\frac{3}{4}} - \frac{4b}{3} + d$, we have $x = 181{\cdot}3635$ seconds, and consequently 2059·846 yards = distance run by the dog when the hare died.

We have received no other answer to this question, nor was the application of the inverse ratio, as expressed, clearly understood by any. One of the greatest mathematicians of the age, and a fluxionist, has asserted it unintelligible, as he does of the exhalation question, this year inserted.

Mr. *Landen* corrects this question (in the year following) by adding *speed*, for rate of going forward: which speed, at the end of the first second, according to his Commentary, was 31 yards, &c. solving this question throughout, except giving the numbers; and says, the proposer's solution would have agreed exactly the same with his, had he brought the spaces passed over in the first second into consideration. He says, it will be as $1 : 31 :: 1 \div x^{\frac{1}{4}} : 31 \div x^{\frac{1}{4}}$ the hare's speed; and $1 : 26 :: 1 \div x^{\frac{1}{5}} : 26 \div x^{\frac{1}{5}}$ the dog's speed. And the speed into $\dot{x}$, the fluxion of the time, will be equal to the fluxion of the distance run, &c. H.

V. QUESTION 315, *by Mr.* Philip Stevens, *of Bristol.*

If the diurnal rotation of the earth was stopped from the 10th of December, 1748, at midnight, to the 10th of December, 1749, what time of the year would it be day-break, sun-rise, and mid-day, at London?

Answered by Mr. W. Sutton.

Let e, e, ♈, E represent the pole of the earth's ecliptic, in the respective positions of the earth, when the required appearances happen, as the earth moves along in the plane of the ecliptic from A to B, ♈, and E about ☉, the sun's centre. Let A be the place of the earth on Dec. 10, 1748, at midnight, p the pole of the equator, and v the vertex of London, in the obscure hemisphere of the disk, at which time the vertex of London is distant 61° 57′ from the horizon of the disk (noted b, e, d) = sum of the co-latitude and sun's greatest declination.

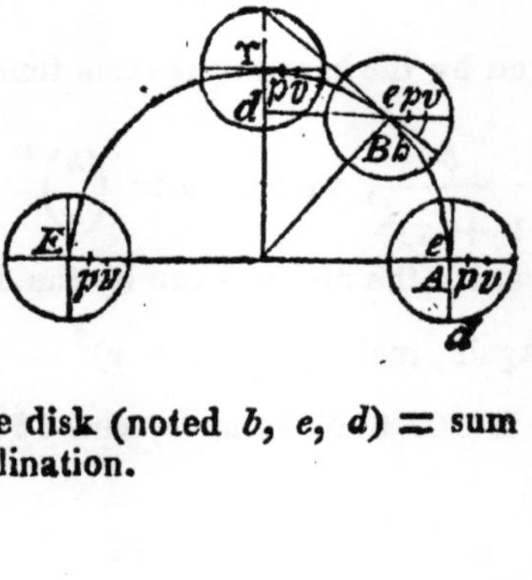

2. Conceive the earth with its axis, still keeping its parallelism, carried to B, where the distance of the vertex and horizon is 18° = *vh*, which sides *vh*, *ev*, and *eh* constitute a right-angled spherical triangle, right-angled at *h*, in which is given *vh* = 18° 0′, and *ev* = 61° 57′, to find the ∠*veh*, = ♈*ed* = *d* ☉ *e*, the cos. of the diff. of long. from Dec. 10, = 69° 31′, which added to the sun's longitude on Dec. 10, gives the longitude for the day 11s. 10° 8′ when it is day-break first at London, according to the question, answering to Feb. 17.

3. 'Tis evident, that when the earth comes to the right angle in the ecliptic, A ☉ ♈ or 90°, from its place Dec. 10, at ♈, the vertex of London at *v* will first arrive in the horizon of the disk, where the sun will first appear to rise: Therefore the longitude answering that appearance 0s 0° 37′ answering with March 10.

4. When the sun comes to the opposite point of the ecliptic, at E, or 180° from its long. Dec. 10, the vertex of London will transit the meridian at *v*, whence the sun's longitude then is 3s. 0° 37′, to which agrees June 11, 1749.

Mr. J. Powle, drawing a scheme, says, that since on Dec. 10 the sun enters Capricorn, and that sign being on the meridian at midnight, it is evident, on the earth's rotation being stopped, that when the sun is depressed below the horizon 18° in his progress through the ecliptic, day will break.

In Aries he will rise; in Cancer it will be mid-day, *i. e.* sun-rising and mid-day are on the 10th of March and 10th of June respectively.

To find day-break. Say, sine sum complement lat. and declin. 61° 58′ : sine sun's depression 18° : : radius : sine sun's distance from vern. equinox 20° 29′, answering to the 20th of April, the time of day-break required.

These being the only answers received, we thought fit to insert both, that each gentleman may be convinced of the truth.

Mr. Landen (in the year following) says, Mr. Powle, in his solution to this question, is right, by reckoning 20° 29′ from Aries into Pisces; but, by mistake, makes day-break to follow sun-rise; otherwise his solution had been like Mr. Sutton's. H.

⁂ ☉ is wanting at the centre of the last figure.

VI. QUESTION 316, *by Mr* John Hampson, *Leigh, Lancashire.*

A line of 11 chains in length from one of the angles of a triangular piece of land, to the opposite side divides the angle into the parts $60°\frac{10}{13}$, $18°\frac{3}{13}$ and the area is the least possible under these circumstances: Required the sides, and area?

Answered by Mr. W. Jepson.

Let s. ∠APC $= s$, s. ∠APB $= n$, [See fig. page 261, vol I.] s. ∠BPC $= m$, BP $= a$, AP $= x$, DC $= y$; then, by a well-known theorem, $sxy = nax + may$, ∴ $sxy - may = nax$, and $y = \frac{nax}{sx-ma}$; whence $\frac{snaxx}{sx - ma}$ or $\frac{xx}{sx - ma}$ is a minimum by the question; in fluxions $2x\dot{x}\,(sx - ma) - sxx\dot{x} = 0$; ∴ $2sx - 2ma - sx = 0$, or $sx = 2ma$, and $x = 2ma \div s = 19{\cdot}558$ &c. Hence $y = 2na \div s = 7{\cdot}012$ &c. and the area $= 2mnaa \div s = 6{\cdot}731$ acres $= 6$ a. 2 r. $36{\cdot}96$ p.

Corol. The area APB = area BPC $= nmaa \div s$, and ∴ AB = BC.

Remark. When the given dividing line is considered as making given angles with the two sides of the triangle about a given vertical angle, it is constructed by theor. 8, p. 196, Simpson's Geom. And when the area is a given quantity instead of a min. it is constructed in prob. 5, pa. 214. H.

VII. QUESTION 317, *by Mr.* James Collingridge.

A marble table of six equal hexagon sides, each $1\frac{1}{2}$ feet length, and the table $1\frac{1}{2}$ inch thickness, is levelly suspended by a point on the under surface, and there hangs 7, 11, 15, 19, 23, and 27 pound weights, in a successive order, from each corner of the table. Quere the point of suspension?

Answered by Mr. Turner, *of Brompton, Kent.*

The solidity of the hexagon $= 631{\cdot}33$ inches $= 26{\cdot}5$ pounds $= w$; and let the weights be represented by 27, c; 7, d; 11, e; 15, f; 19, g; 23, h; whose sum with $w = 128\frac{1}{2}$ pounds $= s$: Let R be the centre of gravity, draw QR ∥ ch, and Rm ⊥ QR: put $2p =$ side hexagon $= ch$, P$k = a = 15{\cdot}5884$, PR $= x$, $gk = y$. Then, by mechanics $sx = (d + g + w) \times a + (e + f) \times 2a$, and $sy = (h + f) \times p + w \times 2p + (c + e) \times 3p + d \times 4p$. These equations, reduced, are

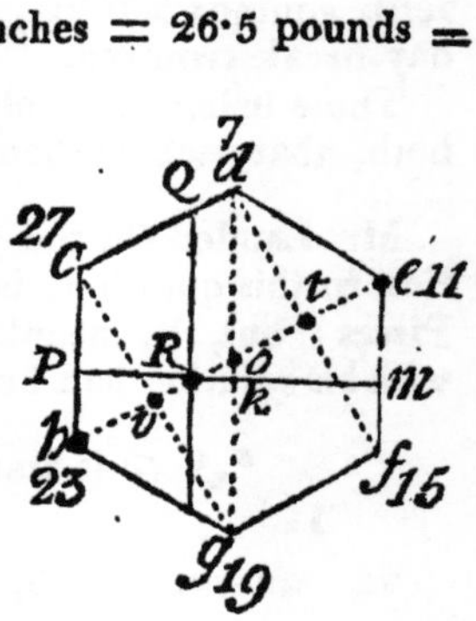

$$x = \frac{d+g+w+2e+2f}{s}a = 12{\cdot}677 \text{ inches},$$

$$y = \frac{h + f + 2w + 3c + 3e + 4d}{s}p = 16{\cdot}319 \text{ inches}.$$

Whence the point R is determined by a general method, let the weights be what they will.

Answered by Mr. Heath.

With the given weights, the centre of gravity R will fall on the axis *eh* at RO dist. from the table's centre $=$ 3·31618 inches (which would be 4·23529, as answered by Mr. Baker, the weight of the table $26\frac{1}{2}$lb. not being considered) for as the sum of the weights suspended on each side of the axis *eh* are equal, the centre of gravity must needs fall on that line. The opposite weights 27 and 19 may be considered in one sum $=$ 46 pounds, suspended from the point *v*; and the weights 7 and 15 $=$ 22 suspended from *t*, and placing the weight of the table $= 26\frac{1}{2}$, in the middle at the point *o*, the question will be reduced to find the centre of gravity to five weights 23, 46, $26\frac{1}{2}$, 22, and 11, placed at the equal dist. of 9 inches on the line *he* $=$ 36. Then substituting for the distance of the centre of gravity from *o*, and multiplying the repective distances therefrom into the respective weights suspended, making the products on each side the centre of gravity, R, equal, the equation will give the value of *o*R as before.

VIII. QUESTION 318, *by Mr.* James Terey.

Two roads, perpendicular to each other, issue from the extremity of a semi-elliptical enclosure, next to a gentleman's seat; one road runs along the transverse close by the side of the straight paling, some distance beyond the other end of the enclosure, and the other road proceeds forward next the crooked paling: A straight diagonal visto of 20 yards breadth is to be made, so as to communicate between the two roads, with one of its fences touching the curved fence of the elliptic enclosure; how must the visto fences be drawn, and of what dimensions, so that the visto may take up the least quantity of ground possible; the transverse axis of the elliptic enclosure being 100, and its semi-conjugate 40 poles?

Answered by the Rev. Mr. Baker.

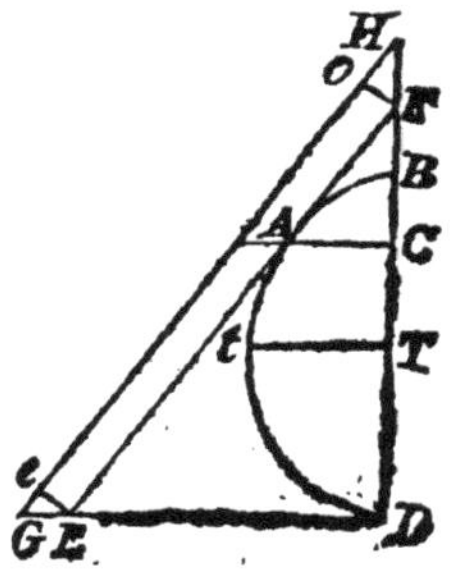

Seeing that on every variation of the visto fence EF, the ▭ *eOFE*, increases or decreases, in magnitude, much faster than the sim. Δs *Ge*E and *O*HF, therefore when the visto GHFE is a minimum, the fence EF will be also a minimum. Put $b =$ TB $=$ 50, $c =$ T*t* $=$ 40, $x =$ BC; then CA $= \frac{c}{b} \times \sqrt{(2ax - xx)}$, FC $= \frac{2bx - xx}{b - x}$, FD $= \frac{2bb - bx}{b - x}$: Now by sim. Δs, FC : CA :: FD : DE $= \frac{c}{x} \times \sqrt{(bx - xx)}$, and by 47 Euc. I. $\frac{cc}{x}(2b - x) +$

$bb\,\frac{(2b-x)^2}{(b-x)^2} = $ EF^2, a minim.; or by reduction, $\frac{2cc}{bx} + \frac{(2b-x)^2}{(b-x)^2}$ is a minimum; whose fluxion made $= 0$, and reduced, gives $x^3 + \frac{3cc-2aa}{aa-cc}axx - \frac{3aacc}{aa-cc}x = \frac{-a^3cc}{aa-cc}$, where $x = 16{\cdot}785$ fere, whence EF $= 153{\cdot}702$ required.

Solution by the Rev. Mr. Wildbore, *taken from the Appendix to Dr.* Hutton's *Edition of the Diaries.*

As no notice whatever is taken of the given breadth of the visto, in the original solution, it cannot be right. For though the $\triangle$s GeE, OHF in this particular example be small in respect of the $\square$eoFE, the breadth of the visto may be supposed increased till they even infinitely exceed it. These three things (the breadth and two $\triangle$s) must not therefore be omitted in the solution, which may be thus:

The area of the visto is evidently equal to the sum of these two $\triangle$s and $\square$. And Ge being, by sim. triangles, $= \frac{\text{ED}\,.\,\text{E}e}{\text{FD}}$, and HO $= \frac{\text{FD}\,.\,\text{E}e}{\text{ED}}$; the visto $= \frac{\text{ED}\,.\,\text{E}e^2}{2\text{FD}} + \frac{\text{FD}\,.\,\text{E}e^2}{2\text{ED}} + \text{FE}\,.\,\text{E}e$ a minimum, or $\frac{\text{E}e}{2} \times \left(\frac{\text{ED}}{\text{FD}} + \frac{\text{FD}}{\text{ED}}\right) + \sqrt{(\text{FD}^2 + \text{ED}^2)}$ a min. At B erect BS $\perp$ FD meeting FE in S. Then, by sim. $\triangle$s, SB $=$ ED $\times \frac{\text{FD}-\text{BD}}{\text{FD}}$ and, by the conics, SB $= \frac{\text{T}t^2}{\text{ED}}$; hence ED$^2 = \frac{\text{FD}\,.\,\text{T}t^2}{\text{FB}} = \text{T}t^2 + \frac{\text{BD}\,.\,\text{T}t^2}{\text{FB}}$.

Make now FB $= u$, BD $= 2b$, T$t = c$, and E$e = d$. Then FD $= 2b + u$, ED $= c\sqrt{\frac{2b+u}{u}}$, $\frac{\text{ED}}{\text{FD}} = \frac{c}{\sqrt{(2bu+uu)}}$, and $\frac{dc}{2\sqrt{(2bu+uu)}} + \frac{d\sqrt{(2bu+uu)}}{2c} + \sqrt{\left((2b+u)^2 + cc\,.\,\frac{2b+u}{u}\right)}$ is a min. The flux. made $= 0$, &c. there results $d \times \frac{b+u}{2c} \times \frac{2bu+uu-cc}{2b+u} = \frac{bcc - uu\,(2b+u)}{\sqrt{(2bu+uu+cc)}}$. Whence u may be found by an equation of the 8th power, or by the well-known rule of position; and thence the rest. Thus in the present case $u = 24{\cdot}863187$. Whereas when EF only is a min. u is $= 25{\cdot}26757$, EF $= 153{\cdot}702$, and the mean length $= 157{\cdot}552$. But the mean length determined as above is $= 157{\cdot}545$, and the quantity of land in the visto $\cdot 02545$ parts of a square pole less than when EF is a minimum.

IX. QUESTION 319, *by* Harmonicus.

There are 13 musical cords of equal thickness and tension, each an inch longer than the other, from the shortest of 12, to the longest of 24 inches; the tone of the longest cord to the shortest is as 2 to 1 Quere the proportion of the tones of all the intermediate cords.

Answered by Mr. Turner, *of Brompton, near Rochester.*

The longest cord to the shortest being as 2 to 1, which is as 24 to 12, consequently the second cord will be as 24 to 13, the third as 24 to 14 or 12 to 7, the fourth as 24 to 15 or 8 to 5, the fifth as 24 to 16 or 3 to 2, the sixth as 24 to 17, the seventh as 24 to 18 or 4 to 3, the eighth as 24 to 19, the ninth as 24 to 20 or 6 to 5, the tenth as 24 to 21 or 8 to 7, the eleventh as 24 to 22 or 12 to 11, the twelfth as 24 to 23, the thirteenth as 24 to 24 or 1 to 1.

X. QUESTION 320, *by Mr.* Heath.

At what time (next ensuing) will Mars and Venus, Sol and Terra, be conjunctly in a right line?

Answered by Mr. T. Cowper.

The last mean opposition of the Sun, Mars, and Venus (in the superior part of her orbit) was January 20, 1695, and I cannot find (by the proportion of the one conjunction of the sun and another with those planets) that they will be so conjoined again till the 11th of January 2942, though they happen very nearly in a right line about the 20th of October, anno 2006.

XI. QUESTION 321, *by Mr.* Farrer.

A musket ball being shot $3\frac{1}{2}$ furlongs perpendicularly upwards, at what distance in its return will the force of the ball be equal to the weight of 9 pounds? And what will be its force at coming to the earth's surface, supposing the ball to weigh an ounce when at rest?

Answered by Mr. John Turner, *of Heath, Yorkshire.*

1. Putting $b = 2310$ feet in $3\frac{1}{2}$ furlongs, $c = 16\frac{1}{12}$ feet, then $2\sqrt{bc}$ = the celerity of the musquet ball when it falls to the ground, which being multiplied by its weight (viz. 1 ounce) produces 358 ounces, or 24 pounds, equal to its absolute force.

2. Let $x =$ the time of its fall in seconds, when its absolute force $= 9$ pounds, or 144 ounces. Say, $1'' \times 1'' : c :: x \times x : cxx$ the space descended; whence $2cx =$ the celerity at that time, which multiplied by 1 ounce, is $144 = 2cx$; whence $x = 144 \div 2c = 4\frac{1}{2}$ seconds nearly, and the space descended, when the ball's force $= 9$ pounds, is 322·321 feet.

The Rev. Mr. Baker's *Solution.*

Putting $r = 21000000$ feet $=$ the earth's rad. $d = 2310$ feet $=$ $3\frac{1}{2}$ furlongs, $b = 9$ pounds, $c = \frac{1}{16}$ of a pound, the ball's weight at rest, $s = 16\frac{1}{12}$ feet, $x =$ space descended. The velocities of falling bodies being in the subduplicate ratio of the spaces descended through, we have $\sqrt{s} : 2s :: \sqrt{x} : 2\sqrt{sx} =$ the ball's velocity at x distance descended. By mechanics, and the condition of the question, $2c\sqrt{sx} = b$, $\therefore x = bb \div 4ccs = 322{\cdot}321$ feet, exactly agreeing with the foregoing number by Mr. Turner. And when $x = d$, then $2c\sqrt{sd} = 24{\cdot}0937$ pounds, the ball's force coming to the earth's surface. But accurately thus. From p. 369, of Mr. Emerson's Flux. we get $2r\sqrt{\frac{sx}{aa - ax}} =$ the velocity, whence $2cr\sqrt{\frac{sx}{aa - ax}} = b$, and $x = \frac{aabb}{4sccrr + abb} = 322{\cdot}3873$ feet (where $a = d + r$). And when $x = d$, then $2c\sqrt{(sdr \div a)} = 24{\cdot}0924$ pounds, &c.

Mr. Landen (in the year following) says, Mr. Turner and Mr. Baker are both wrong in their solutions to this question. This gentleman solves it, by assuming gratis, that 1100 grains, descending one-fourth of a foot, acquires a force $= 4660$ grains weight; and the rest upon the same principles with others, supposing the force to be as the velocity and quantity of matter; and so it may be solved by as many different suppositions as any one pleases. But *Newtoniensis* says, there can be no proper solution to this question, for want of proper data, like the exhalation question. H.

XII. QUESTION 322, *by Mr.* Landen.

A cylindric pillar of stone, of 2 yards in circumference being drawn up 20 yards above ground at a building, a rope being at the same time wrapt ten times round its convexity, with its lower end fixed to a hole upon the middle part where it begun to wrap; but before the pillar could be lodged upon the scaffolding, as it was drawn up with the rope's upper part passing through pulleys above, the rope unfixed its security at the tenth round, and the suspended pillar by its weight then unwrapped itself of the rope, and descended to the ground. How long was the time of its descent?

Answered by Mr. C. Bumpkin *only.*

If the stone were suspended by two ropes hanging perpendicular, with one fixed at a point in the middle of its convex surface, and the other, at the centre of percussion of that circular section in which the aforesaid point is situated (which point, in computing the place of the centre of percussion, is to be considered as the point of suspension) and likewise the diam. in which the centre of percussion is found, were parallel to the horizon, the tension of the rope fixed at the centre of percussion would then be equal to the motive force bringing the stone down, if that rope was unfixed, and the stone left to descend in the manner described in the question. Consequently, putting w for the weight of the stone, a for the rad. thereof, then $1\frac{1}{2}a$ will be the dist. of the centre of percussion from the centre of suspension; and $\frac{2}{3}$w that motive force which would act continually upon the stone. Moreover, w : 32·2 ($= s$) : : $\frac{2}{3}$w : $\frac{2}{3}s$, the accelerating force, or velocity generated per second, by the descending stone. Putting v for the velocity of the stone's centre of gravity, and z the space descended, then $\frac{\dot{z}}{v}$ being the fluxion of the time $\frac{2s}{3} \times \frac{\dot{z}}{v}$ will be $= \dot{v}$, or $\frac{2s}{3} \times \dot{z} = \dot{v}v$, whence $v = 2\sqrt{\frac{sz}{3}}$, and applying $\dot{z}$, we have $\frac{\dot{z}}{2}\sqrt{\frac{3}{sz}}$ = fluxion of the time, whose fluent $\sqrt{\frac{3z}{s}}$, when $z = 60$, is $= 2'' 21''' 51''''$.

XIII. QUESTION 323, *by Mr.* John Corbet, *Surveyor.*

How many acres of the moon's surface are seen enlightened 10 days after her conjunction with the sun? And how many acres are contained on the convex superficies of a lunar mountain (part of a gentleman's estate) its height being 3 furlongs, and its superficies equal to that by the rotation of the semi-cycloid of that height about its axis?

Answered by Mr. Heath.

Say $14^d\ 18^h\ 22^m\ 1\frac{1}{2}^s$ (the time of half a mean lunation) : $\frac{1}{2}$ (the moon's surface at most seen enlightened) : : 10 days (the time after conjunction with the sun) : $10 \div 14{\cdot}765299 = {\cdot}3386319$ &c. parts of the moon's surface then seen enlightened $= 3245609437$ acres; the moon's whole surface being 9584476353 acres.

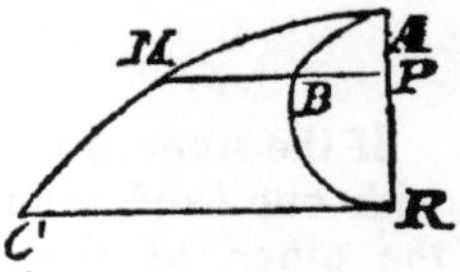

Put a = lunar mountain's height = AR = 3 furlongs, x = AP, any indefinite part of the cycloid's axis, then BP $= \sqrt{(ax - xx)}$, whose fluxion is $\frac{a - 2x}{2\sqrt{(ax - xx)}} \times \dot{x}$ and fluxion arch

$\dot{A}B$ = flux. MB $= \frac{a\dot{x}}{2\sqrt{(ax - xx)}}$, whence flux. (MB + BP) = flux. MP $= \dot{y} = \dot{x}\sqrt{(a - x)} \div x$, whose fluent $y = \sqrt{ax} \times : 2 - \frac{x}{3a} - \frac{x^2}{20a^2} - \frac{x^3}{56a^3} - \frac{5x^4}{576a^4}$ — &c. by series; put $c = 2 \times 3{\cdot}1416$, then it being known that $\dot{x}\sqrt{(a \div x)}$ is the flux. of the cycloid's arch, $cy\dot{x}\sqrt{(a \div x)}$ will be the fluxion of a curved surface, by the rotation of the space AMP about AP, whose fluent is found is $2ax - \frac{x^2}{6} - \frac{x^3}{60a} - \frac{x^4}{224a^2} - \frac{5x^5}{2880a^3}$ — &c. $\times c$ = the curved superficies; which when $x = a$, becomes $2aa - \frac{aa}{6} - \frac{aa}{60} - \frac{aa}{224} - \frac{5aa}{2880}$ — &c. $\times 6{\cdot}2832 = aa \times : 2 - \frac{1}{6} - \frac{1}{60} - \frac{1}{224} - \frac{5}{2880}$ — &c. $\times 6{\cdot}2832 = aa \times 11{\cdot}3617 = 1022{\cdot}553$ acres, or 1022 a. 2 r. 8·48 p. the quantity of the lunar mountain's surface required.

Solution by Curiosus.

Let s = surface, AP = x, PM = y, AM = z, AB = v, $c = 2 \times 3{\cdot}1416$; then will $\dot{z} = \dot{x}\sqrt{(a \div x)}$, and $\dot{s} = c \times$ PM $\times \dot{x}\sqrt{(a \div x)}$; but PM $= \sqrt{(ax - xx)} + v$, whence $\dot{s} = c\dot{x}\sqrt{(aa - ax)} + cv\dot{x}\sqrt{(a \div x)}$. And $s = \frac{2}{3}c\sqrt{a}\,(a - x)^{\frac{3}{2}}$ + flu. $cv\dot{x}\sqrt{(a \div x)}$. But (by rule 8, p. 50, Emerson's Fluxions) this fluent is $= 2c\sqrt{ax} + 2ca\sqrt{(aa - ax)}$, because $\dot{v} = a\dot{x} \div (aa - ax)$. Whence $s = -\frac{2}{3}c\sqrt{a}\,(a - x)^{\frac{3}{2}} + 2cv\sqrt{ax} + 2ca\sqrt{(aa - ax)}$, and corrected, $s = (\frac{4}{3}ca + \frac{2}{3}cx)\sqrt{(aa - ax)} + 2cv\sqrt{ax} - \frac{4}{3}caa$; and when $x = a$, the whole surface $= (3c - 8)caa \div 6 = 11{\cdot}36169aa$.

THE PRIZE QUESTION, *by the Excellent Mr.* J. Landen.

An eagle (200 yards above) stooped to a kite, then taking flight from the ground with a chicken, at an angle of 60°, the eagle soaring directly towards the kite, then flying from her, at an uniform rate of swiftness of 3 to 1, the ratio of both their uniform motions; when, after some time flying, the kite finding herself closely pursued, quits her little captive, which fell to the ground at the instant the bird of Jove seized her prisoner, who was then just as high from the ground as at her first stooping. The eagle's distance from the kite, at her first setting out? Her nearest approach to the ground during the pursuit? Her height above it, and distance from the kite, when the chicken begun to fall? And also the time of flight, are from hence required?

Answered by his Excellency Sir Stately Stiff.

The equation of the curve described by the eagle, is $2x = \frac{a^n y^{1-n}}{1-n} - \frac{a^{-n} y^{n+1}}{1+n}$, computing x from the point where the pursuit ended, y being an ordinate at right angles, $n = \frac{1}{3}$, and a an invariable quantity to be determined. Moreover, if the variable distance of the eagle and kite be $= d$, and the distance of the kite from the point where x begins $= z$, and s and c be the sine and cosine of 60°, we have, by the nature of the curve, $(z \div n) - nx = d$, $xx - 2zx + zz + yy = dd$; and at the beginning of flight, when $z = 230{\cdot}9$ yards, $cy = sx$; at which time, as it is proved from these three last equations, $d = 582{\cdot}3$ yards, the eagle's distance from the kite at first setting out, $x = 331{\cdot}2$, $y = 573{\cdot}6$. Now by putting these values of x and y in the first equation, a is determined $= 967{\cdot}1$.

At the lowest point of the curve, $s : y :: 1 : y \div s = d$, and $s : y :: c : cy \div s = x - z$; therefore $z = (sx - cy) \div s$, by means of which, and the former equations, x and y are found corresponding to that point; and then, by a short computation, the eagle's nearest approach to the ground $= 117{\cdot}8$. Her height above it when the chicken began to fall, is found $= 119{\cdot}7$ (by means of the former equations, trigonometry, and solving a cubic equation) and the eagle's distance from the kite, at the same instant is found $= 251{\cdot}7$ yards; and lastly, the whole time of flight $= 9{\cdot}44$ seconds.

N. B. The data should be corrected by writing, *when after* 5 *seconds flying*, instead of, *when after some time flying*. The whole operation requires too much room to be inserted at length.*

---

* The above solution will be evident by reading prob. 15, Simpson's Fluxions.

Questions proposed in 1750, *and answered in* 1751.

I. QUESTION 325, *by Mr.* T. Cowper, *of Wellingborough.*

There is a cottage at Wellingborough in Northamptonshire, lat. 52° 20′, from whence can be seen the steeples of Irchester, Rushden, and Higham Ferrars. At this cottage on the 6th of January, 1749, the sun appeared to rise exactly behind the steeple of Irchester, the nearest; and on the 14th of February the sun appeared to rise behind the steeple of Rushden, the most remote; and on the 13th of March behind the steeple of Higham Ferrars. A person measured 24 chains from the cottage, in the direction of the steeple of Higham Ferrars; and from this point the sun would appear to rise exactly behind the steeple of Rushden on the 11th of February. Moreover it is known that a right line joining the steeples of Higham Ferrars and Rushden is perpendicular to the right line joining the cottage and Higham Ferrars. Required the distance from the cottage to each of the steeples?

Answered by the Proposer.

Let I, R, and H represent the situation of the spires, Irchester, Rushden, and Higham; and W, B, the places of observation at Wellingborough, and next it and Higham. The declination of ⊙ on 6th of Jan. last at rising was 20° 42′ S. with which, the complement of lat. of Wellingborough 37° 40′, and ⊙'s zenith distance 90° 34′, I find (per spherics) the opposite angle, or ⊙'s azimuth from N. when his centre appears in the horizon = 124° 27′. In like manner, the ⊙'s apparent azimuth from N. at rising Feb. 14th = 104° 29′, the difference of these 19° 58′ = ∠IWR. The ⊙'s apparent amplitude March 13th was 3° 7′ N. Therefore the ∠RWH = 17° 36′. Sun's apparent amplitude on the winter solstice is 39° 44′, and on the 11th of Feb. 15° 53′ S. Consequently the ∠IBH = 42° 51′ and ∠RBH = 19° 0′; from whence with the measured distance BW (by plain trigonometry) WI is found = 2 miles, 1 furl. 29 pol. WR = 3 m. 7 f. 39 p. and WH = 3 m. 6 f. 19 p. required.

Mr. William Sutton's *Answer to the same.*

The visible amplitudes of the sun at rising, viz.

Dec. 10th	39°	45′	south.
Jan. 6th	34	25	
Feb. 11th	15	48	
Feb. 14th	13	57	
Mar. 13th	3	18	north.

From whence the diff. of amplitudes from Jan. 6th to Feb. 14th = 20° 28′ = ∠RWI. And the sum of amplitudes from Feb. 14th to March 13th = ∠RWH = 17° 15′.

Also diff. amplitudes in the right line from Wellingborough to Higham, between Dec. 10th and Feb. 11th $= 23° 57' = \angle$ IBR. The diff. between Feb. 11th and 14th is $\angle$ BRW $= 1° 51'$, from whence the $\angle$ HBR $= 19° 6'$, $\angle$ BIW $= 5° 20'$, and consequently WI $= 2$ m. 1 f. 25 p. WR $= 3$ m. 0 f. 13 p. WH $= 2$ m. 7 f. 9 p.

II. QUESTION 326, *by Mr.* Christopher Mason, *Surveyor to the Right Hon. the Earl of Northampton.*

A heavy body being let fall from the top of a castle, it was observed that the time of its descent was exactly equal to the time of the ascent of sound from the bottom to the top. Required the height of the castle; the velocity of the body when it reached the ground; the respective gravities of the body at the top and bottom of the castle; and the distance upon the earth's surface, which may be seen from the top of the castle, the radius of the earth being 4000 miles?

Answered by Mr. Chr. Mason, *the Proposer.*

To find the height of the castle. Let $x =$ seconds required, either in the acceleration of a body or velocity of sound propagated, $b = 16\frac{1}{12}$ feet $=$ acceleration of a body the first second, $c = 1142$ feet the distance moved over by sound in a second. Then, $bxx = cx$ $=$ castle's height, per question. Hence $x = c \div b = 71$ seconds, and the degree of velocity 141.

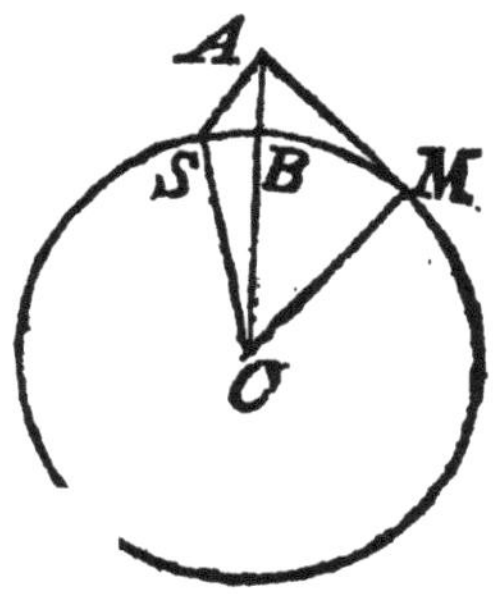

To find the proportion of gravity. Let CB $= r = 4000$ miles $=$ radius given, CA $= s$, then $rr : 1 :: ss : rr \div ss$, being in a reciprocal proportion. If the weight upon the earth be unity, the same weight at the castle will be ·99 or as 100 to 99 nearly.

Lastly, to find the visual, or tangent line AM. By 36 Euc. III. and 47 Euc. I. $AC^2 - CM^2 = AM^2$. Hence AM $= 350$ miles and 1745 yards.

Answered by Mr. Roger Widger, *of Plymouth.*

Put $a = 1142$ feet sound moves per second, $b = 16\frac{1}{12}$ feet descended by a heavy body in a second, $x =$ dist. of the castle from the earth's surface. Then, $b : 1 :: x : x \div b =$ square time of the falling body. As $a : 1 :: x : x \div a =$ time of sound returning, whence $\sqrt{(x \div b)} = x \div a$ per quest. Whence $x = aa \div b = 81087{\cdot}917$ &c. feet; the rest following as in the above answer.

III. QUESTION 327, *by Mr.* W. Jepson, *of Lincoln.*

Required two general theorems, with their investigation, to determine the least triangle, and least cone, that will circumscribe any segment of an ellipsis, and frustum of a spheroid, when the dividing ordinate is parallel, and in any given ratio to the conjugate axis?

Answered by the Proposer, Mr. Jepson.

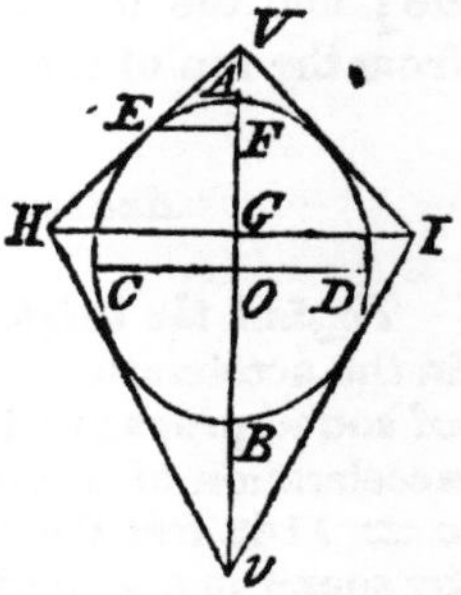

Let AB $= 2a$, CD $= 2b$, GO (which by the data will always be a known quantity) $= c$, HI $= 2y$, GV or Gv $= x$, OV or Ov $= x \pm c$. Then, by the properties of the figure, VO : AO :: AO : FO; viz. $x \pm c : a :: a : \frac{aa}{x \pm c} =$ FO; $\therefore$ BF $= a + \frac{aa}{x \pm c}$, AF $= a - \frac{aa}{x \pm c}$; and, by the properties of an ellipsis, $aa : bb :: \left(a + \frac{aa}{x \pm c}\right)\left(a - \frac{aa}{x \pm c}\right) : bb - \frac{aabb}{(x \pm c)^2} =$ EF^2, therefore EF $= b\sqrt{\left(1 - \frac{aa}{(x \pm c)^2}\right)} = b\sqrt{\frac{(x \pm c)^2 - aa}{(x \pm c)^2}}$, and VF $= x \pm c - \frac{aa}{x \pm c} = \frac{(x \pm c)^2 - aa}{x \pm c}$: Now, by sim. Δs, $\frac{(x \pm c)^2 - aa}{x \pm c} : x :: b\sqrt{\frac{(x \pm c)^2 - aa}{(x \pm c)^2}} : bx\sqrt{\frac{1}{(x \pm c)^2 - aa}} = y$; but xy is a minimum by the question, $\therefore bxx\sqrt{\frac{1}{(x \pm c)^2 - aa}}$ or $\frac{x^4}{xx \pm 2cx + cc - aa}$ is also a minimum, which in fluxions is $2x^5\dot{x} \pm 6cx^4\dot{x} + 4c^2x^3\dot{x} - 4a^2x^3\dot{x} = 0$, $\therefore x^2 \pm 3cx + 2cc - 2aa = 0$, or $xx \pm 3cx = 2aa - 2cc$, whence $x = \sqrt{\left(2aa + \frac{cc}{4}\right)} \pm \frac{3c}{2}$, the theorem for finding the least triangle.

Again, let $d = 3{\cdot}1416$, and $\frac{b^2dx^3}{3} \times \frac{1}{(x \pm c)^2 - a^2}$, or $\frac{x^3}{(x \pm c)^2 - a^2}$ is a minimum. In fluxions $x^4\dot{x} \pm 4cx^3\dot{x} + 3c^2x^2\dot{x}$ —

$3a^2x^2\dot{x} = 0$, $\therefore x^2 \pm 4cx + 3c^2 - 3a^2 = 0$, or $x^2 \pm 4cx = 3a^2 - 3c^2$, whence $x = \sqrt{(3a^2 + c^2)} \pm 2c$, the theorem for finding the least cone.

The same answered by Mr. William Bevil.

The ratio of the dividing ordinate to the conjugate being given, their distance from each other is readily found, which distance call d, and let a = semi-transverse, b = semi-conjugate of that ellipsis, and x = OV. Then, per conics, $x : a :: a : \frac{aa}{x} =$ OF, $a - \frac{aa}{x} =$ GF, $x - \frac{aa}{x} =$ FV, $aa : bb :: aa - \frac{aa}{xx} : \frac{bb}{aa}\left(aa - \frac{aa}{xx}\right) =$ EF^2, $\frac{xx - aa}{x} : \frac{b}{x}\sqrt{(xx - aa)} :: x \pm d : \frac{x \pm d}{xx - aa} b\sqrt{(xx - aa)} =$ HG, then $\frac{(x \pm d)^2}{xx - aa} b\sqrt{(xx - aa)} =$ area Δ HVI or HIv, which is a minimum; squared and put into fluxions $4x^3\dot{x} + 12dx^2\dot{x} + 12d^2x\dot{x} + 4d^3\dot{x}(xx - aa) - 2x\dot{x}(x \pm d)^4 = 0$. By reduction, $xx \pm dx = 2aa$, whence $x = \frac{1}{2}\sqrt{(8aa + dd)} \pm \frac{1}{2}d$, a theorem for the least triangle.

Now put ·2618 = c, then $\frac{(x \pm d)^3}{(xx - aa)^2}(xx - aa)4bbc =$ the cone's solidity, which, or $\frac{x \pm d}{(xx - aa)^{\frac{1}{3}}}$, is a minimum. In fluxions, and reduced, $xx \pm 2dx = 3aa$, whence $x = \sqrt{(3aa + dd)} \pm d$, a theorem for the least cone.

[IV. QUESTION 328, *by the Rev. Mr.* Baker, *of Stickney, Lincolnshire.*

To what height will an exhalation ascend, whose specific gravity is, at the earth's * surface, equal to half that of common air, but decreases in the subtriplicate ratio of the spaces ascended?

* Fluxioniensis says, it must be a mile, or some distance from the surface, to make it consistent.

Answered by the Proposer, the Rev. Mr. Baker, *only.*

Put r = 4000 miles = earth's radius, $68444 \div (3 \times 1760) = c$, space ascended = x, air's density at the earth's surface = d. Then, per question, that of the vapour there will be = $\frac{1}{2}d$. And $x^{\frac{1}{3}} : 1^{\frac{1}{3}}$

:: $\frac{1}{2}d : d \div 2x^{\frac{1}{3}}$ = its density at x height. But from page 96, of Mr. Emerson's Fluxions, we have $d \times$ number belonging to this log. $\frac{-rx}{cr+cx} = \frac{d}{2x^{\frac{1}{3}}}$, which reduced, according to the nature of logarithms, gives $3rx \div (cr + cx) - \text{l. } 8 = \text{l. } x$, whence by a table of logarithms, $x = 7{\cdot}763$ miles, the height required.

V. QUESTION 329, *by Mr.* J. Powle, *of Salop.*

Three spheres of brass in contact, whose diameters are 8, 9, and 10 inches respectively, support a fourth sphere, weighing 12 pounds; what quantity of weight does each supporting sphere sustain?

Answered by Mr. Widger.

For want of room for the process, we only insert the numbers, viz.

6·0614	pounds sup-	8	
3·2696	ported by the	9	inches diameter.
2·669	globe of	10	

Solution.

Let A, B, and C be the centres of the supporting spheres, and D the centre of that which is supported; r, r', r'' and R their respective radii: then the pressures will evidently be in the directions of the lines AD, BD, and CD. But each of these pressures may be resolved into two other, viz. a vertical pressure and a horizontal one. The sum of the vertical pressures must evidently be equal to the weight (W) of the supported sphere, and the horizontal pressures must be such as to balance each other when applied to the same point, or else the supported sphere could not be at rest but would have a lateral motion in consequence of the unequal pressure. Put a, b, and c for the angles that the lines AD, BD, and CD make with the horizon, and P, P′, P″ for the respective pressures in those directions; then P sin a, P′ sin b, and P″ sin c will be the vertical pressures, and P cos a, P′ cos b, and P″ cos c the horizontal ones; therefore

$$P \sin a + P' \sin b + P'' \sin c = W;$$

and to establish an equilibrium between the horizontal pressures, conceive vertical planes to pass through the lines AD, BD, and CD, and meet the horizontal plane passing through D in the lines ED, FD, and GD; then forces equivalent to the horizontal pressures, and applied in the directions of those lines, must be in equilibrio at the point D; therefore if α, β, and γ be put for the angles EDF, FDG, and GDE respectively, we have

$$\text{P} \cos a : \text{P}' \cos b :: \sin \beta : \sin \gamma$$
$$\text{and } \text{P} \cos a : \text{P}'' \cos c :: \sin \beta \quad \sin \alpha;$$

therefore $\text{P}' = \text{P} \times \frac{\cos \alpha \sin \gamma}{\cos b \sin \beta}$, and $\text{P}'' = \text{P} \times \frac{\cos a \sin}{\cos c \sin \beta}$;
and these values being substituted in the expression for the sum of the vertical pressures, we have

$$\text{P} \times \left(\sin a + \frac{\cos a \sin \gamma \sin b}{\cos b \sin \beta} + \frac{\cos a \sin \alpha \sin c}{\cos c \sin \beta}\right) = \text{W},$$

or, if R be put $= \sin a \cos b \cos c \sin \beta + \cos a \cos c \sin \gamma \sin b + \cos a \cos b \sin \alpha \sin c$, then

$$\text{P} \times \frac{\text{R}}{\cos b \cos c \sin \beta} = \text{W};$$

therefore

$$\text{P} = \frac{\cos b \cos c \sin \beta}{\text{R}} \times \text{W}.$$

$$\text{P}' = \frac{\cos a \cos c \sin \gamma}{\text{R}} \times \text{W}.$$

$$\text{P}'' = \frac{\cos a \cos b \sin \alpha}{\text{R}} \times \text{W}.$$

We have now only to calculate the values of the several angles a, b, and c; α, β, and γ: but the analytical expressions for those angles in terms of the radii of the spheres being rather complex, we shall not go through the whole calculation, but be content with briefly explaining the method of obtaining them. Now the lines AD, BD, and CD are the sides of a pyramid the base of which is the triangle ABC, and these sides being given, equal $r + \text{R}$, $r' + \text{R}$, and $r'' + \text{R}$ respectively, the angles that they make with the plane of the base will be easily found by trigonometry, and then the angles which they make with the horizontal plane will also be known; for this plane makes a given angle with the base of the pyramid. The angles a, b, and c being thus found, we can then easily find the angles α, β, and γ. For if perpendiculars be drawn from the centres A, B, and C to the horizontal plane which passes through D to meet it in the points H, K, and L, the distances HK, KL, and LH will obviously be $= 2\sqrt{rr'}$, $2\sqrt{rr''}$, and $2\sqrt{r'r''}$, and the distances DH, DK, and DL $= (r + \text{R}) \cos a$, $(r' + \text{R}) \cos b$, and $(r'' + \text{R}) \cos c$; therefore by trigonometry

$$\cos \alpha = \frac{(r + \text{R})^2 \cos^2 a + (r' + \text{R})^2 \cos^2 b - 2\sqrt{rr'}}{(r + \text{R})(r' + \text{R}) \cos a \cos b},$$

$$\cos \beta = \frac{(r' + \text{R})^2 \cos^2 b + (r'' + \text{R})^2 \cos^2 c - 2\sqrt{r'r''}}{(r' + \text{R})(r'' + \text{R}) \cos b \cos c},$$

$$\cos \gamma = \frac{(r'' + \text{R})^2 \cos^2 c + (r + \text{R})^2 \cos^2 a - 2\sqrt{rr''}}{(r'' + \text{R})(r + \text{R}) \cos c \cos a}.$$

Or when two of these angles have been found, the other may be obtained by subtracting their sum from four right angles. J. L.

VI. QUESTION 330, *by* Upnorensis.

To determine the path which a philosopher must describe, passing between two fires, at d distance from each other, and one fire n times as big as the other, so as to feel the least heat possible?

Answered by Waltoniensis.

Let F be the greater fire, n times bigger than f the lesser; $d = \text{F}f$, their distance.

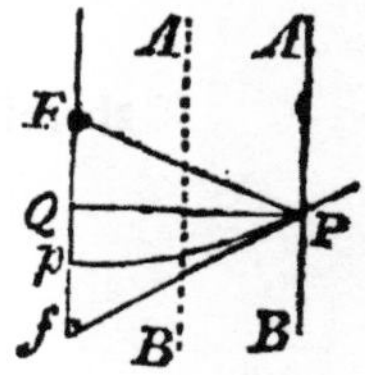

At any distance, draw AB parallel to Ff, then, since the philosopher must pass every line AB, we have only to find that point therein, at which if he stood still he should feel the least heat possible from both the fires. Suppose P to be that point, and drawing PQ $\perp$ Ff; call FQ, x; PQ, y; then $\frac{n}{xx + yy} + \frac{1}{(d - x)^2 + yy}$, expressing the heat of both fires, must be a minimum, y being given, or some constant quantity, while x is variable, the fluxion made $= 0$ will always show the relation betwixt x and y. The fluxion of this expression, when $y = 0$, is $\frac{-2nx\dot{x}}{x^4} + \frac{2d\dot{x} - 2x\dot{x}}{(d - x)^4} = 0$, where $x = \frac{dn^{\frac{1}{3}}}{1 + n^{\frac{1}{3}}}$; whence $\frac{d}{1 + n^{\frac{1}{3}}}$ is the philosopher's distance from the least fire, directly betwixt both fires, moving along the curve Fp, to be roasted on both sides alike.

N. B. The point p being found for the vertex of the curve and Fp and fp being in a given ratio to each other, if any other distances from the curve to the greater or lesser fires, PF and Pf, be supposed in the same ratio, the path of the curve will be a circle, as observed by Fluxioniensis.

VII. QUESTION 331, *by Mr.* Christ. Mason.

There are two bridges over two different channels, having flood-gates underneath them; one has four gates, each 4 feet 2 inches wide; the other has two, each 3 feet 9 inches wide; there is 100l. a year paid as water-scot by lands which these channels help to drain: A mean depth of 45 inches was taken at the greater bridge, and 24 inches at the lesser; the beds of both channels are supposed to incline alike in their level, or declivity; what part of the 100l.

must be allotted to each channel, according to the proportion of water which they respectively discharge at the aforesaid depths?

Answered by Britannicus.

Two hundred inches, the breadth of the greater channel by 45 its depth $= 9000$ square inches the area of the section; and 90 inches the breadth of the lesser channel, by 24 its breadth $= 2160$ square inches the area of its section; the velocity of water moving along each channel is as the square roots of its depth, repectively, viz. as $\sqrt{45} = 6{\cdot}708$ and $\sqrt{24} = 4{\cdot}899$ fere; therefore $9000 \times 6{\cdot}708$ and $2160 \times 4{\cdot}899$, or 60372 and 10582 the momenta, are nearly as the water respectively discharged by each channel, in the same time; therefore the greater channel pays 85l. 1s. 8d. $\frac{47380}{70954}$, and the lesser 14l. 18s. *3d.* $\frac{23634}{70954}$, required.

VIII. QUESTION 332, *by Mr.* Powle.

An equation of a curve is expressed by $y = \frac{\mathrm{x}x}{\sqrt{(rr + xx)}}$ (where x is the hyp. log. of x): Required an expression of its area in finite terms?

Answered by Newtoniensis.

Since $y = \frac{\mathrm{x}x}{\sqrt{(rr + xx)}}$, $\therefore y\dot{x} = \frac{\mathrm{x}x\dot{x}}{\sqrt{(rr + xx)}}$, and fluent of $y\dot{x} = \mathrm{x}\sqrt{(rr + xx)} - s$, whence (p. 58, Emerson's Flux.) $\dot{s} = \dot{\mathrm{x}}\sqrt{(rr + xx)} = \frac{\dot{x}}{x}\sqrt{(rr + xx)} = \frac{rr\dot{x}}{x\sqrt{(rr + xx)}} + \frac{x\dot{x}}{\sqrt{(rr + xx)}}$; and $s = \sqrt{(rr + xx)} +$ fluent $\frac{rr\dot{x}}{x\sqrt{(rr + xx)}}$; and fluent $\frac{rr\dot{x}}{x\sqrt{(rr + xx)}}$ or $\frac{rrx^{-2}\dot{x}}{\sqrt{(1 + rrx^{-2})}}$ by the table is $= \frac{-\mathrm{L}}{r} \times \log.\left(\frac{r}{x} + \sqrt{\frac{rr + xx}{xx}}\right)$.

Therefore the fluent of $\frac{\mathrm{x}x\dot{x}}{\sqrt{(rr + xx)}} = (\mathrm{x} - 1)\sqrt{(rr + xx)} + \frac{2{\cdot}3024}{r} \times \log.\left(\frac{r}{x} + \sqrt{\frac{rr + xx}{xx}}\right)$.

N. B. There is no more difficulty where surds are concerned, the process being the same as in simple quantities.

Mr. *Powle* sends $(\mathrm{x} - 1).\sqrt{(rr + xx)} + 2{\cdot}3025r \times \log. \frac{r + \sqrt{(rr + xx)}}{x} =$ area.

Remark. Mr. Powle's fluent is right, and Newtoniensis's would be the same if the latter part of it were drawn into r^2 as it ought. H.

IX. QUESTION 333, *by the Rev. Mr.* Baker.

What is the content of a cask, whose head and bung diameters are 30 and 40 inches respectively, supposed to be formed by the cassinian ellipse revolving on its principal axe, which is just four-thirds of the cask's length?

Answered by the Proposer only.

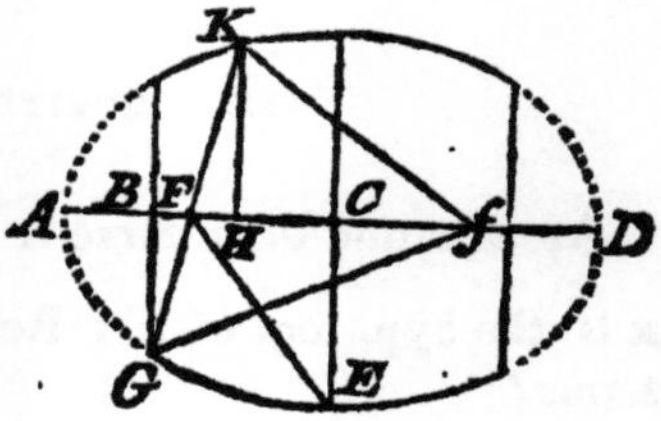

Let F, f, be the foci of the generating ellipse, BG $= h$, CE $= b$, $\frac{3}{4} = n$, $3{\cdot}1416 = p$, AF $\times$ DF $= r$, AC $= v$, FC $= z$, BF $= x$, BC $= y$. Then, by the figure, $v^2 - z^2 =$ EF^2 $= b^2 + z^2$, $\therefore z = \sqrt{(\frac{1}{2}v^2 - \frac{1}{2}b^2)}$, and $r = \frac{1}{2}v^2 + \frac{1}{2}b^2$; and because FG $\times$ Gf = AF $\times$ DF, and $x^2 + y^2 =$ FG^2, also $(2z + x)^2 + y^2 =$ GF^2, by elliptic property; therefore $(x^2 + y^2)((2z + x)^2 + y^2) = r^2$. Hence $y^2 = \sqrt{(r^2 + 4z^4 + 8z^3x + 4z^2x^2)} - 2z^2 - 2zx - x^2$, is the equation of the curve. But when $y = h$, then, per quest. $x + z = nv$, $\therefore x = nv - z$. And substituting the above values of r, x, y, and z, in the said equation, we get, by reduction,

$$v^4 + \frac{b^2 - n^2b^2 - 2n^2h^2 - h^2}{n^2 - n^4}v^2 = \frac{h^4 - b^2h^2}{n^2 - n^4},$$

whence, in the present case, $v = 12{\cdot}148$, or $32{\cdot}9267$, which last value is the true one, because v cannot be less than b. Hence we have $r = 742{\cdot}08365$, $z = 18{\cdot}4955$; and by substituting for the known quantities $y^2 = \sqrt{(a^2 + bx + c^2x^2)} - d - cx - x^2$. Wherefore $py^2\dot{x} = p\dot{x}\sqrt{(a^2 + bx + c^2x^2)} - dp\dot{x} - cpx\dot{x} - px^2\dot{x} =$ the fluxion of the solidity between A and F; whose fluent, when properly corrected, is $\frac{2c^2x + b}{4c^2}$

$$\times p\sqrt{(a^2 + bx + c^2x^2)} - \frac{abp}{4c^2} + \frac{4a^2c^2 - b^2}{8c^3}p \times \text{hyp. log.}$$

$$\frac{2c^2x + b + 2c\sqrt{(a^2 + bx + c^2x^2)}}{2ac + b} - dpx - \tfrac{1}{2}cpx^2 - \tfrac{1}{3}px^3,$$ which,

when $x =$ BF, will be $5445{\cdot}9815$ inches. Again, putting $x =$ FH, $y =$ HK, and proceeding as before, we get $py^2\dot{x} = p\dot{x}\sqrt{(a^2 - bx + c^2x^2)} - dp\dot{x} + cpx\dot{x} - px^2\dot{x} =$ the fluxion of the solidity between F and C, which differs from the other fluxion in nothing but the signs, consequently its correct fluent is discovered to be $\frac{2c^2x - b}{4c^2} \times$

$$p\sqrt{(a^2 - bx + c^2x^2)} + \frac{abp}{4c^2} + \frac{4a^2c^2p - b^2p}{8c^3} \times \text{hyp. log.}$$
$$\frac{2c^2x - b + 2c\sqrt{(a^2 - bx + c^2x^2)}}{2ac - b} - dpx + \tfrac{1}{2}cpx^2 - \tfrac{1}{3}px^3 = 22116{\cdot}2054$$
when $x = $ CF. Hence, the content of the whole cask is 195·4758 ale gallons, or 238·63278 wine gallons.

X. QUESTION 334, *by* Dictator Roffensis.

Three Irish evidences, namely, a pedant, a priest, and an alderman, offer their attendance to the plaintiff's attorney, on a trial at Westminster-hall, for the reward of half a hogshead of wine; the pedant can drink it out by himself in 12 days, the priest in 10, and the alderman in 15, when the days are 12 hours long: Quere, in what time can the pedant, priest, and alderman drink out the whole, drinking together, when the days are 10 hours long? And what will be each evidence's share?

Answered by Mr. Steph. Hodges, *of Wellingborough.*

The proportion of the quantities of wine drank by the pedant, priest, and alderman, in the same time, are as 1, 1·2, ·8, whose sum = 3. Then say, inversely, as 1 : 12 :: 3 : 4 days = 48 hours when the days are 12 hours long = 4 days 8 hours when the days are but 10 hours.

By direct proportion, say,

$$\text{As } 3 : \underset{\text{gall.}}{31{\cdot}5} :: \left\{\begin{matrix} 1{\cdot} \\ 1{\cdot}2 \\ {\cdot}8 \end{matrix}\right\} : \left\{\begin{matrix} \text{gall.}\ 10{\cdot}5 \\ 12{\cdot}6 \\ 8{\cdot}4 \\ \hline 31{\cdot}5 \end{matrix}\right\} \left\{\begin{matrix} \text{pedant's} \\ \text{priest's} \\ \text{alderman's} \end{matrix}\right\} \text{allowance.}$$

The same answered by Mr. Will. Smith, *at Churchdown, near Gloucester.*

Put x = time they will all drink it in? $12 \times 12 = 144 = c$, $12 \times 10 = 120 = d$, $12 \times 15 = 180 = h$, w = half the hogshead.

Say, $c : w :: x : \frac{wx}{c}$; $d : w :: x : \frac{wx}{d}$; $b : w :: x : \frac{wx}{b}$, the several shares collected $= \frac{wx}{c} + \frac{wx}{d} + \frac{wx}{b} = w$. Whence $x =$

$\frac{ebd}{bd + cd + cb} = 48$ hours when the day is 12 hours = 4 days 8 hours when the day is 10 hours.

Hence $\left\{\begin{array}{c}\text{gall.}\\ \frac{48}{144}\\ \frac{48}{120}\\ \frac{48}{180}\end{array}\right\}$ pedant's, priest's, alderman's $\}$ Share $\left\{\begin{array}{r}\text{gall.}\\ 10{\cdot}5\\ 12{\cdot}6\\ 8{\cdot}4\end{array}\right\}$ as above.

XI. QUESTION 335, *by Master* Dickey.

If 10 packs of cards and 3 packs of knaves are of equal value with 9 packs of knaves and 4 packs of cards, what will be the value of one pack of knaves?

Answered by Master Billy Branch, *of Rochester.*

If 10 p. cards + 3 p. knaves = 9 p. knaves + 4 p. cards. Then, by reduction, 6 packs of knaves is of equal value with 6 packs of cards; whence the value of a pack of knaves is only a pack of cards.

THE PRIZE QUESTION, *by Mr.* Turner, *of Brompton.*

Three towns, A, B, C, at which make no wonder,
Seven, eight, and ten miles are exactly asunder;
Four thousand good people in A live alert,
In B and C two and three thousand expert;
Religiously bent, must on Whitfield attend,
And wou'd chuse him a place, a-la-mode, for that end;
Where must he hold forth, that, in preaching to those,
All walking to hear him shall wear out least shoes?

Answered by Newtoniensis.

Let A, B, C, be the three towns, D the place of meeting sought; and suppose any of the distances, as AD to be given, and with the radius AD, describe the circular arch GD*n*H, and let EDF be a tangent at D; draw AD*m*, and take D*n* infinitely small, and draw the lines BD, B*n*, CD, C*n*, and with the radii B*n*, C*n*, describe the small arcs *ne*, *nf*; then D*e* is the increment of BD, and D*f* the decrement of CD.

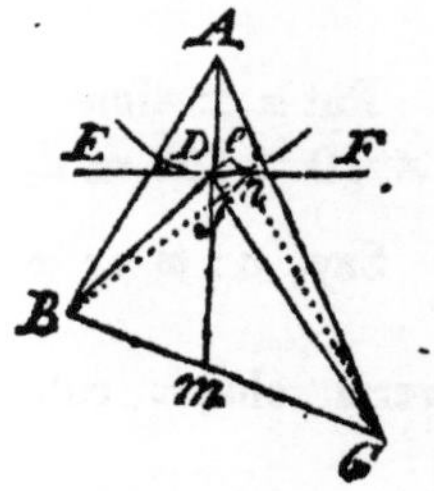

Let BD = x, CD = y, AD = z, and a, b, c, three given numbers 2, 3, 4, in proportion as per question, so that $ax + by + cz$ may be a minimum. Then, since z is supposed constant, we have $a\dot{x} + b\dot{y} = 0$, and $a\dot{x} = -b\dot{y}$, or $\dot{x} : -\dot{y} :: b : a$. In the two right-angled triangles Dne, Dnf, whose common hypothenuse is Dn, it is as De ($\dot{x}$) : Df ($-\dot{y}$) :: s. Dne : s. Dnf :: b : a; but $\angle$Dne = eDA = BDm, and $\angle$Dnf = mDf or mDC. Whence s. BDm : s. mDC :: b : a, and s. CD$m \div a$ = s. BDm $\div b$.

After the same manner it may be proved, that if y be supposed constant, s. CD$m \div a$ = s. CDB $\div c$; $\therefore$ when $ax + by + cz$ is a min. s. CD$m \div a$ = s. BD$m \div b$ = s. CDB $\div c$. Therefore, if s. CD$m = v$, then $(b \div a)v$ = s. BDm, and $(c \div a)v$ = s. CDm + BDm their sum. Therefore the problem comes to this, To find the $\angle$CDm whose sine is v, and $\angle$BDm whose sine is $(b \div a)v$, so that $(c \div a)v$ may be the sine of the sum CDm + BDm. And $\angle$CDm is easily found by Mr. Heath's method in the Diary, 1738.

N. B. $\angle$CDm is nearly 29°.

Now, all the parts of $\triangle$ABC being given, together with the angles at D, all the distances AD, BD, CD are easily found, viz. AD = 2·6, BD = 5·02, CD = 7·64. (See quest. 100.)

The same answered by Waltoniensis.

Let ABC represent the three towns, and let the number of people in A be $a = 4000$; in B, $b = 2000$; in C, $c = 3000$.

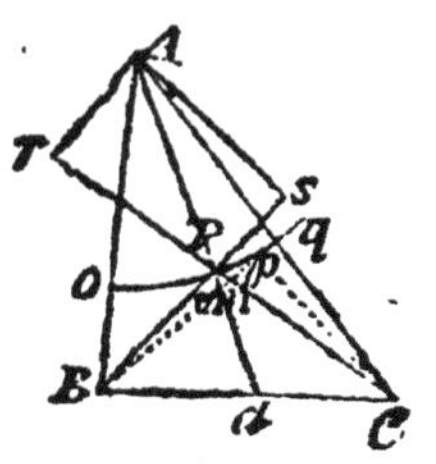

P the place of meeting sought. Draw AP, BP, CP, and with the rad. AP, on the centre A, describe the arc Opq. A point being now supposed to move along the said arc, lines drawn to it from B and C will be continually variable. Let the point have moved from P to p, over the indefinitely small arc Pp, draw Bp, Cp, and with the radii Bp, Cp, describe the small arcs Pm, Pn.

Then b times BP plus c times CP being a minimum, b times mp, the increment of BP, will be equal to c times Pn, the decrement of CP; therefore, the right-angled triangles APr, APs, made by letting fall the perpendiculars Ar, As, being respectively similar to the small right-angled triangles Ppn, pPm, the sine of the angle APr, or CPd, will be to that of $\angle$APs, or BPd, as b to c. In like manner it is proved, that the sine of $\angle$CPs is to that of $\angle$CPd, as a to b. Whence it follows, that if a $\triangle$EFG be constructed, whose sides FG, EF, and EG, are as a, b, and c repectively, and two circular arcs described without $\triangle$ABC, one upon the side BC, capable of the $\angle$F, and the other

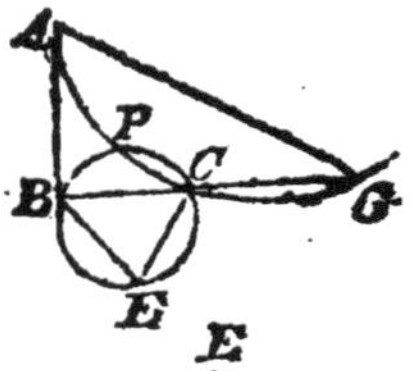

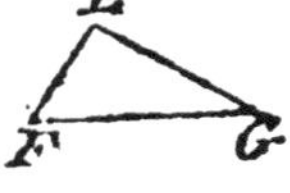

upon AC, capable of the $\angle$G, the supplement of those arcs when completed into circles, will intersect each other at the point P sought: And AP will be found = 2·596 miles, BP = 5·009, CP = 7·623.

Corollary. If a, b, and c be equal, then two segments of circles described within the $\triangle$ABC, on any two sides, each capable of an angle of 120°, will intersect at the point required, according to Mr. Simpson's new doctrine and application of Fluxions, p. 26 and 27: He has inserted and solved this our question, at p. 505, of his Doctrine aforesaid.

Questions proposed in 1751, *and answered in* 1752.

I. QUESTION 337, *by Mr.* T. Cowper, *of Wellingborough.*

The latitude explore, and time last winter, when
Day broke exact at four, and the sun rose at ten.

Answered by Mr. T. Cowper, *the Proposer.*

Put a = sine ⊙'s ascensional difference, 60°; b = sine hour from 6, at day-break, or 30°; d = sine ⊙'s depression at 18°; and x and y the sine and cosine of the latitude: also e and v the sine and cosine of ⊙'s declination: By spherics $bvy + ex = d$ and $ex = avy$, and substituting avy for ex in the first equation $vy = d \div (a + b)$: And also in the other equation putting $d \div (a + b)$ for vy, and we have $ex = ad \div (a + b)$; therefore $d(1 - a) \div (a + b) = \cdot 030074$ = cos. sum of lat. and sun's declin. 88° 15′ 48″. And $d(1 + a) \div (a + b) = \cdot 4221252$ the cos. of their diff. or 65° 1′ 52″. Hence the lat. 76° 38′ 50″, and declin. 11° 36′ 58″.

Theorem. As the sum of the sines of the sun's ascensional difference and arch of time from day-break to 6 o'clock, is to the sine of the sun's depression at day-break, so is the versed sine of the arch of time from sun-rise to noon, to the sine of the meridian altitude; and so is the versed sine of the time from midnight to sun-rise, to the sine of the sun's depression at midnight.

The same solved by Mr. Charles Smith, *of Rugby.*

Put r and n for the cosines of the hour angles from midnight till day-break, and from sun-rise till noon, respectively; d = cos. of 108° = Z⊙, x and y = sine and cos. of the required latitude; u and

z = those of declination. Then, in the spheric triangles ⊙PZ and OZP, by common theorems, $rzy + ux = d$, and $nzy - ux = 0$, from whence $zy - ux = d(1-n) \div (r+n) = \cdot 0303072$ = cos. 88° 15′ 48″ the sum of lat. and declin. and $zy + ux = d(1+n) \div (r+n) = \cdot 422125$ = cos. 65° 1′ 52″ the diff. Hence the lat. 76° 38′ 50″ N. and the declin. 11° 36′ 58″ S. (answering to 7th Feb.) required; proving the truth of Mr. Cowper's answer.

II. QUESTION 338, *by Mr.* William Leighton.

Two persons, A and B, playing at hazard, A wins from B a certain number of guineas, consisting of three places, whose digits are in arithmetical progression, in such manner, that, if the number of guineas be divided by the sum of their digits, the quotient will be 48; and, if from the said number of guineas you take 198, the digits will be inverted: Quere the number of guineas won.

Answered by Mr. Rich. Gibbons.

Let x, y, and z represent the three digits; then, by the question, we have $x + z = 2y$, $\frac{100x + 10y + z}{x + y + z} = 48$, and $99x - 99z = 198$; whence $x = 4$, $y = 3$, and $z = 2$; also number of guineas 432, required.

Philotheoros, putting $a = 198$, $c = 48$, makes $x = \frac{a}{198} \times \frac{c-4}{c-37} = \frac{44}{11} = 4$; whence the number $= 432$.

III. QUESTION 339, *by Mr.* William Bevil.

From what height must a ball of 4 ounces weight fall, to have $49\frac{67}{100}$ pounds force, on an inclining plane, whose angle of incidence is 40°?

Answered by Mr. John Ash.

Ecce homo!

As sine 40° : rad. :: 49·67 pounds force : 77·2728 pounds, the momentum or force of the falling body $= m$. Put n for the given weight $= \cdot 25$ pounds, and x for the required height; then, by the laws of motion $m \div n$ will be the velocity of the ball arrived at the plane of the horizon; and (if Desaguliers's experiment, Philos. Trans.

actions, No. 375, p. 269, can be depended on) we have $\sqrt{x} = m \div n$; whence $x = m^2 \div n^2 = 9553{\cdot}7$ feet, required.

Mr. *Richard Gibbons* solves this question in the same manner: Thus,

As the sine of the angle of incidence 40° : 49·67 pounds force :: rad. : 77·273 pounds force on the plane of the horizon, being let fall from the same height. By Dr. Desaguliers's experiments, an heavy body descending four feet will have twice the quantity of motion it had when it began to fall (i. e. *we observe at the end of one foot fallen)* the time of its falling half a second. Now, the force is always as the velocity and quantity of matter, *i. e.* $\sqrt{\text{(space)}} \times$ matter, perpendicularly descended; putting $m = 77{\cdot}273$ the momentum, perpendicularly descended; $q = 0{\cdot}25$ pounds the quantity of the ball; and $s =$ space required to run through: Then $q\sqrt{s} = m$; whence $\sqrt{s} = m \div q$, and $s = m^2 \div q^2 = 9553{\cdot}7$ feet, as before.

IV. QUESTION 340, *by Mr.* Davis, *Teacher of the Mathematics, at Painswick, Gloucestershire.*

A monument was finished on the 1st of June, at 5 in the afternoon, in lat. 48° north; at which time the length of its shade, upon a plane which inclined 9° to the horizon, was to its height, in the ratio of 23 to 7: How many years before the Christian æra was the monument erected.

Answered by the Proposer, Mr. Davis.

Let BC represent the monument, AB the shade's length, HO the horizon; then ∠OH☉ = sun's alt. The sides AB to BC as 92 to 28, and ∠ABC = 99°. By trigon. ∠OH☉ = 25° 0′ 37″ sun's alt. and allowing 17′ 37″ for sun's semi-diam. and refraction, 25° 0′ 37″ — 17′ 37″ = 24° 43′ = sun's true alt.

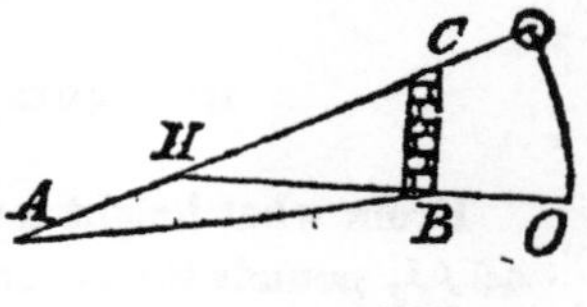

Now, from comp. lat. 42° 0′, comp. sun's alt. 65° 17′, and hour angle from noon of 5ʰ = 75°, the complement of sun's declin. will be found 69° 53′ 26″, and declin. 20° 6′ 34″, which answers to ♉ 29° 38′ the sun's place in the ecliptic, or longitude from ♈ 59° 38′; and, by making proportion, I find, June 1st, 5 hours P. M. *anno 965 ante christum*, the sun's place is ♉ 29° 38′ 0″, as may be proved from Leadbetter's tables; being the time when the monument was erected.

V. QUESTION 341, *by Mr.* Bevil.

Two men bought an equal number of sheep; and it being demanded of them what they gave a-piece for their sheep in each parcel, it was answered, that if the number of sheep each of us bought be severally multiplied by $\frac{24}{27}$ and $\frac{54}{27}$, 49 being respectively added to or subtracted from each product, both the sum and remainder will be equal to the square of the shillings given a-piece for sheep in each respective parcel. How many sheep did each person buy? And what did each parcel cost, at the cheapest price? for so every man would buy.

Answered by the Proposer, Mr. Bevil, *of Harpswell.*

Put x = number of sheep; then $\frac{24}{27}x + 49$ and $\frac{54}{27}x - 49$ are square numbers, whose roots are the shillings a-piece the sheep in each respective parcel (of different value, though equal number) cost.

But a square number, multiplied by a square number, will produce a square number. The expressions being multiplied respectively by 9 and 4, two square numbers, will be $\frac{216}{27}x + 441$, and $\frac{216}{27}x - 196$, whose difference is 637.

To find two square numbers having that difference.

Rule. Resolve the given difference into any two factors; then the half sum and half difference of those factors will be the sides of the squares having the difference given: $637 = 13 \times 49 = 7 \times 91$. Therefore $\frac{49 + 13}{2} = 31$, and $\frac{49 - 13}{2} = 18$ will be the sides of the squares: Consequently $\left(\frac{24}{27}x + 49\right) 9 = 31^2$, and $\left(\frac{54}{27}x - 49\right) 4 = 18^2$; from either of which equations $x = 65$ the number of sheep: And, consequently, $\frac{24}{27} \times 65 + 49 = \frac{31^2}{9}$, and $\frac{54}{27} \times 65 - 49 = \frac{18^2}{4}$, whose square roots are $\frac{31}{3}$ and $\frac{18}{2}$, or 10s. 4d. and 9s. the sheep cost a-piece, in each parcel of 65; whence $65 \times$ 10s. 4d. = 33l. 11s. 8d. one parcel cost; and $65 \times$ 9s. = 29l. 5s. the other parcel cost; the true answer.

The same method may be pursued with the factors $7 \times 91 = 637$, when the sides of the squares will be 49 and 42, and the number of sheep in each parcel 245; consequently $\frac{49}{3}$ and $\frac{42}{2}$, or 16s. 8d. and

21s. the sheep in each parcel cost a-piece, and 257l. 5s. and 204l. 3s. 4d. the price of each parcel; being dearer, and therefore not the true answer.

VI. QUESTION 342, *by Mr.* Stephen Hodges, *the younger.*

There is a cistern, the sum of one side and one end of which is 84 inches, its diagonal 60 inches, and the ratio of the breadth to the depth as 25 to 7: what are its dimensions?

Answered by Mr. Joseph Orchard, *of Gosport.*

Given $AB + AD = 84 = a$; $AE = 60 = d$; and $BE \div AD = 7 \div 25 = \cdot 28 = r$. Let $AD = x$, then $AB = a - x$, and $BE = rx$: But $AB^2 + BE^2 = AE^2$, i. e. $aa - 2ax + xx + r^2x^2 = dd$, $\therefore xx - 2ax \div (rr + 1) = (dd - aa) \div (rr + 1)$, and $x = \dfrac{a - \sqrt{(aa + (dd - aa)\,rr)}}{rr + 1} = 24\cdot 39$ nearly,

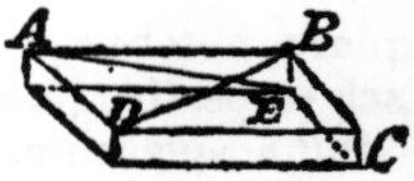

the breadth; and the length is $59\cdot 61$; also depth $6\cdot 8292$: whence the content is $35\cdot 208$ ale gallons, or $42\cdot 982$ wine gallons.

But, if by "*diagonal*" is meant DB at the bottom or top of the cistern, then this is the solution:

Given $AB + AD = 84 = a$, $DB = 60 = d$; let $AD = x$. Then $AB = a - x$: But $AB^2 + DA^2 = DB^2$, i. e. $aa - 2ax + 2xx = dd$; solved $x = \frac{1}{2}(a - \sqrt{(2dd - aa)}) = 36$ the breadth; the length is 48; and depth $10\cdot 08$; whence the content $65\cdot 312$ ale, or $75\cdot 403$ wine gallons.

Mr. *T. Cowper* solves it trigonometrically: As 60 : 84 :: cos. 45° : cos. 8° 8′ half the diff. of $\angle$s; consequently the $\angle$s are 53° 8′ and 36° 52′: Whence the length 48; breadth 36; depth $10\cdot 08$; and content $8\cdot 1$ bushels.

VII. QUESTION 343, *by Mr.* John Randle, *of Wem.*

A gentleman has a piece of ground in form of a geometrical square, the difference between whose sides and diagonal is 10 poles; he would convert two-thirds of the area into a garden of an octagonal form, but would have a fish-pond at the centre of the garden, in the form of an equilateral triangle, whose area must be equal five poles. Required the length of each side of the garden, and of each side of the pond?

Answered by Mr. Orchard.

Let $d = 10$ the difference between the side and diagonal of the square; then $(1 + \sqrt{2})\, d =$ the side of the square; two-thirds of the square of which is 388·5618 &c. = the area of the octagonal garden: And, if x be the side thereof, then $xx \times 4{\cdot}8284$ &c. = 388·5618 &c. the said area (*see Palladium for* 51, *p*. 24), $\therefore x = \sqrt{(388{\cdot}5618 \text{ \&c.} \div 4{\cdot}8284 \text{ \&c.})} = 8{\cdot}9707$ poles, each side of the garden: And each side of the pond is $2\sqrt{(5 \div \sqrt{3})} = 3{\cdot}398$ poles required.

Mr. *T. Cowper* answers it thus, very concisely and elegantly: As $3 - 2\sqrt{2} : 10 :: 10 : 100 \div (3 - 2\sqrt{2}) = 582{\cdot}842696$; two-thirds of which $= 388{\cdot}561797 =$ area of the octagonal garden; then $\sqrt{(388{\cdot}561797 \div 4{\cdot}8284272)} = 8{\cdot}9707$ poles, the side thereof; and $3{\cdot}398 =$ side of the triangular pond.

VIII. QUESTION 344, *by* Upnorensis.

To determine the sides of the least right-angled triangle in whole numbers, whose legs are in proportion as 7 to 11?

Answered by Upnorensis, *the Proposer, only.*

1. To find two such square numbers, whose roots may represent one leg and the hypothenuse of a right-angled triangle; and the difference of those squares to be a square number, whose root may represent the other leg.

Put x for one leg, or side of the square, $x + d$ for the hypothenuse, or side of the other square; then the squares will be denoted by xx and $xx + 2xd + dd$, whose difference will be $2xd + dd = (2x + d) \times d = yy$ for the square of the other leg, by question. It is evident, that $2x + d$ and d must be square numbers. Let $2x + d = n$, then the leg $x = \frac{1}{2}(n - d)$, and hypothenuse $x + d = \frac{1}{2}(n + d)$. Now, if $rr = n$, $ss = d$, then $\frac{1}{4}(rr + ss)^2 - \frac{1}{4}(rr - ss)^2 = rrss = yy$, per 47 Euc. I. and, by transposition and reduction, $(rr + ss)^2 = 4rrss + (rr - ss)^2$. Whence we have this THEOREM: $2rs$ and $rr - ss$ will express the legs of a right-angled triangle, and $rr + ss$ the hypothenuse, r and s being assumed any rational or whole numbers at pleasure.

The *ratio* of the legs, as 7 to 11, being given so far as in whole numbers (*for, exactly given,* it would be *no question, and an impossible one, if the sum of their squares were not a square*) by a *trial* or two, r will be found $= 3\frac{1}{2}$, and $s = 1$, by the theorem; when the complete *ratio* of the legs will be 7 to $11\frac{1}{4}$, the nearest to the given numbers, and the corresponding hypothenuse as $13\frac{1}{4}$, four times which values will be 28, 45, and 53, the sides of the *least* right-angled triangle, in whole numbers, required. See p. 186 of Dodson's Mathematical Repository, requiring *two numbers in the complete ratio of* 8 *to* 15, *the sum of whose squares shall be a square number ;* where the required is given, and a *superfluous* theorem that finds the numbers 576 to 1080, being 72 times 8 to 15 ; whereas 2, 3, 4, 5, &c. 8 to 15, had been a direct answer. And there was no way to propose this question, but as it was proposed, without giving what was required (or to the same effect) or else proposing an impossibility.

IX. QUESTION 345, *by* χρονουμονονυβικινος.

If a bookseller buys a copy for 21l. pays 21l. for paper, 21l. for printing 500 impressions, and 10l. for advertisements and other contingent expences, amounting in all to 73l. and sells 100 books yearly of the history, at 5s. each : What is his gain per cent. allowing compound interest, for the time he lies out of his money ?

Answered by Upnorensis.

The bookseller buys an annuity of 25l. a year, to continue five years, for 73l. ready money.—To find his gain per cent. according to the allowance of compound interest.

Let a signify the annuity.
r 1l. and its interest for one time or year.
t the number of times the annuity is to be paid.
z the whole amount of the annuity's present worth.

Say, $r : 1 :: a : \frac{a}{r}$ present worth of a payable at the end of 1st time.

$r : 1 :: \frac{a}{r} : \frac{a}{r^2}$, present worth of a payable at the end of 2d time.

$r : 1 :: \frac{a}{r^2} : \frac{a}{r^3}$ present worth of a payable at the end of 3d time.

Consequently $\frac{a}{r^t}$ present worth of a payable at the end of tth time.

The sum of all which progressionals $\frac{1}{r}, \frac{1}{r^2}, \frac{1}{r^3}$, &c. $\frac{a}{r^t}$, multiplied into a, $\left\{ \frac{(r^t - 1)\,a}{(r-1)r^t} = z \right\}$ the whole present worth of all the payments of a, from 1 to t times.

For, if z be the greatest term, a the least, and r the ratio, or any term decreased by the next lesser, $(rz - a) \div (r - 1) = s$ the sum universally.

The aforesaid equation reduces to $r^{t+1} - \frac{a+z}{z}r^t + \frac{a}{z} = 0$; in which, according to Dr. Halley, if t the number of years be great, 40 or upwards, and the rate of interest be high, $1 + \frac{a}{z}$ will be nearly, or more accurately $\frac{z+a}{z} - \left(\frac{z}{z+a}\right)^t \times \frac{a}{z}$, the value of r, when $\frac{a}{r^t(r-1)}$ will be exceedingly near the value of the reversion which, if it be called x, then $1 + \frac{a}{z+x}$ will approach the value of r sufficiently. *See Dr. Halley's method, at p.* 33 *and* 34 *appendix to Sherwin's Logarithms.* But t the years being small, this rule fails; in which case, if $\left(\frac{at}{z}\right)^{\frac{2}{t+1}} - 1 = y$, and $\frac{6}{t-1} = b$, then $1 + b - \sqrt{(bb - 2by)}$ is sufficiently near to r, and will still be nearer the truth, as t the years be of the smaller value, the small error being always in excess; viz. $r = 1 + b - \sqrt{\left\{b\left(b - 2 \times \left(\frac{at}{z}\right)^{\frac{2}{t+1}} - 1\right)\right\}}$. And, putting capitals for the logarithms of quantities denoted by small letters, then $2 \times \frac{A + T - Z}{t+1} = D$, and $\frac{B + L(b - 2d - 1)}{2} = E$, $\therefore r = 1 + b - e$; and consequently $r - 1 = b - e = \cdot 21094$ &c. the rate of interest of 1l. per year, and 21l. 1s. 10½d. for 100l. a year, bookseller's profit.

Mr. *Flitcon* elegantly considers z paid down as a principal put to interest, whose account at first year's end $= zr$; when a becoming due, $zr - a$ is the principal running on; which, drawn into r, is $zr^2 - ar$ the amount at second year's end; and, a being again paid, $zr^2 - ar - a$ will be the principal running on: which at the end of five years will be $zr^5 - ar^4 - ar^3 - ar^2 - ar - a = 0$ principal running on: the bookseller then being repaid all his money at first laid out, with the interest thereof running on as a principal; consequently the value of r in this equation shews the rate of interest as before.

N. B. The sum of all the terms, except the first $= a\frac{r^5 - 1}{r - 1}$, by the universal rule aforesaid for summing geometrical progressions; and $\therefore zr^5 - a\frac{r^5 - 1}{r - 1} = 0$, which reduces to $zr^6 - (z + a)r^5 + a = 0$ the same equation with the first.

THE PRIZE QUESTION, *by Mr.* T. Cowper, *Surveyor.*

Admit the moon, on the 17th of February, 1750, rose four minutes sooner in the latitude 51° 32′ north, than in the latitude 52° 20′ north, and was observed to come upon the meridian in the former latitude on the same morning 42 minutes after four, and the preceding morning 54 minutes after three, from whence her longitude and latitude at rising, in the latitude of 51° 32′, are required?

Answered by the Proposer.

Put $t =$ co-tang. 51° 32′, $T =$ co-tang. 52° 20′, $x =$ sine, and y = cos. of the ☽'s ascensional diff. in lat. 51° 32′; $s =$ sine, and c = cos. of the diff. between the ascen. diff. = 1°; then will $cx + sy$ = sine of ascens. diff. in lat. 52° 20′. Now, by spheric trigon. as $1 : t :: x : tx =$ tang. ☽'s declination. Again, $1 : T :: cx + sy : Tcx + Tsy$; hence $tx = Tcx + Tsy$; i. e. $y \div x = (t - Tc) \div Ts =$ 1·6808489 the cotang. of ☽'s ascens. diff. (in lat. 51° 32′) = 30° 45′. Hence her declin. is found = 22° 6½′. Then, to right ascen. ⊙ 341° 12′ add 250° 30′ (= time ☽'s southing) the sum rejecting 360°, will be 231° 42′ = right ascen. ☽ at southing. By the rule of proportion, the diff. of right ascen. from ☽'s rising to southing = 1° 58½′; consequently her right ascen. at rising, in lat 51° 32′, is = 229° 43½′. Thus, having her right ascen. and declin. I find her true place to be ♏, 23° 7′ 11″, lat 3° 38′ 27″ S.

Questions proposed in 1752, and answered in 1753.

I. QUESTION 347, *by Mr.* T. Cowper, *of Wellingborough.*

One side of a triangular field being 34 perches and the opposite angle 55°: What is the area, the angle formed by two lines meeting in the middle of the perpendicular upon the given side from the opposite angle being 92°?

Answered by Mr. Terey, *of Portsmouth.*

Let ABC be the triangle; then on the base AB (per 33 Euc. III.) describe two segments of a circle, ACB and AHB, containing the given angles; through the centres D, E draw DG; also draw DH, EC to the

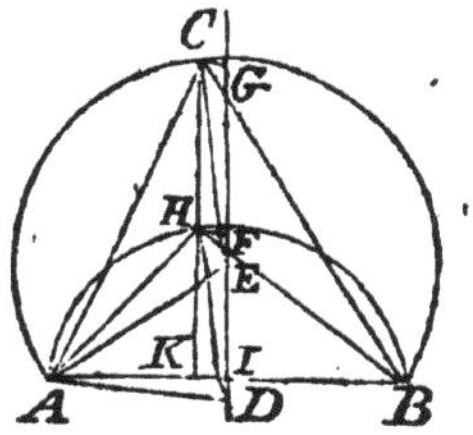

$\perp$ CK; and let fall CG and HF $\perp$s to GD. By trigonometry, AD $=$ DH $= 17{\cdot}0104 = a$; AE $=$ EC $= 20{\cdot}7531 = b$; ID $= {\cdot}593651 = c$; IE $= 11{\cdot}9035 = d$. Put $x =$ CK, then (by 47 Euc. I.) $aa - cc - cx - \frac{1}{4}xx = bb - dd + 2dx - xx =$ CG $=$ HF; hence $x = \frac{1}{3}(2c + 4d) \pm \frac{1}{3}\sqrt{3}\,(b - a)(b + a) + 2c + d^2 = 32{\cdot}5340$ $=$ CK; whence KI $= 2{\cdot}2524$; AC $= 35{\cdot}7204$; CB $= 37{\cdot}8036$; and the area $=$ 3 a. 1 r. 33·078 p. required.

The same answered by Nichol Dixon, *of Blackwell.*

Put $t =$ tangent $\angle$ACB $= 55°$, and T $=$ tangent $\angle$AHB $= 92°$, and $2b =$ AB $= 34$ poles, $y =$ HK $=$ HC, $x =$ KI, the distance from the $\perp$CK to the middle of the base. Then, by trigonometry, as $y : 1 :: b + x : (b + x) \div y =$ tangent $\angle$KHB; and $y : 1 :: b - x : (b - x) \div y =$ tangent angle AHK: Now, by prob. 8, p. 21, of Mr. Emerson's excellent Trigonometry, As $1 - \frac{b^2 - x^2}{y^2} : 1 :: \frac{2b}{y} : \frac{2by}{y^2 - b^2 + x^2} =$ T; and, by the same reasoning, $\frac{4by}{4y^2 - b^2 + x^2} = t$; from which equations $y = \frac{4b}{3t} - \frac{2b}{3\text{T}} = \left(\frac{2}{t} - \frac{1}{\text{T}}\right)\frac{2}{3}b$: But rad. divided by the tangent $=$ co-tangent: Therefore $y = (2 \times$ co-tangent $55° -$ co-tangent $92°)\ \frac{2}{3}b = 16{\cdot}2666$; hence AC $= 35{\cdot}718$, CB $= 37{\cdot}804$, and area $=$ 3 a. 1 r. 33·06 p. &c.

Mr. *Joseph Orchard, of Gosport,* putting b and x as above, tang. $\angle$AHB $= 92° = -v$, tang. $\angle$ACB $= t$, rad. 1, makes the tangents of the respective angles $\frac{2by}{yy - bb + xx} = -v$, and $\frac{4by}{4yy - bb + xx} = t$, the equations brought out of fractions, the former multiplied by t and the latter by v; and taking the sum of both, we get $(2v + t)\ 2b = 3vty$, whence $y = 2b(2v + t) \pm 3vt = 16{\cdot}2665$, whence the area as before.

Additional Solution.

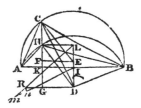

Analysis. On the given base AB let two segments of circles be described to contain the given angles; then we have only to draw CK perpendicular to AB so that CH may be $=$ HK. Let E and D be the centres of the circles, and draw EF, DG, and HL parallel to AB, and join CE and HD. Then the line ED will bisect the base perpendicularly in I; consequently ED will be parallel to CK, and GD $=$ KI $=$ FE $=$ HL. Also FK $=$ EI, KG $=$ ID, HK $=$ LI, and GH $=$ DL. But 47 Euc. I. $\text{CF}^2 = \text{CE}^2 - \text{FE}^2$ and $\text{HG}^2 = \text{HD}^2 - \text{GD}^2$

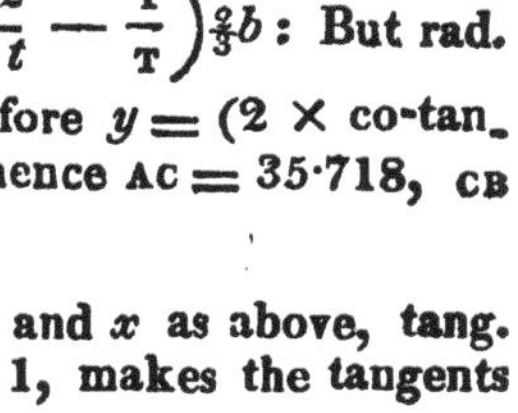

(FE^2), therefore, $CF^2 - DL^2 = CE^2 - HD^2$, a given space. Make DR^2 = this space, draw LR and produce it till Rm be = KF, join mI, and draw Rn parallel to LD meeting mI in n. Then because $RL^2 - LD^2 = DR^2 = CF^2 - LD^2$, RL is = CF, and since Rm = KF, mL is = CK; but CK is = twice LI, therefore mL is = twice LI, and therefore by similar triangles mR = twice Rn: Whence this construction. On a line drawn through D parallel to AB, take DR such that its square may be equal to the difference of the squares of the radii of the circles, make Rn parallel to ID and = to half IE, draw In, and from R apply Rm = EI, produce mR to meet DE in L, and draw LH parallel to AB to meet the circle in H; through H draw the perpendicular CK and the thing is done.

The demonstration is evident from the analysis. J. L.

II. QUESTION 348, *by Mr.* James Terey, *of Portsmouth.*

Required the contents of the greatest spheroid and parabolic conoid that can be inscribed in a cone whose altitude is 30 inches and the diameter of the base 35 inches?

Answered by Mr. J. Orchard, *Writing-master and Teacher of the Mathematics at Gosport.*

Let GC $= a =$ 30 inches, AE $= 2b =$ 35 inches, $m = \cdot 7854$, and KC $= x$. Then KG $= a - x$; and per sim. $\triangle$s GCA, GKP, $a : b :: a - x : (a-x)b \div a$ = PK; per conics, PK $\times$ AC = TN^2 $= bb(a-x) \div a$. Then $8mbb(a - 3x) \div a =$ the solidity of the spheroid, which, or $ax - xx$, is a maximum. In fluxions $a\dot{x} - 2x\dot{x} = 0$, whence $x = \frac{1}{2}a$. Again, let GL = LI $= x$; per sim. $\triangle$s GAC and GHI, $a : b :: 2x : 2bx \div a =$ HI; per conics $x : 4bbxx \div aa :: a - x$ (= LC) $: 4bb(ax - xx) \div aa =$ BC^2; then $8mbbx(a-x)^2 \div aa$ is the conoid's solidity, which, or $aax - 2axx + x^3$, is a maximum. In fluxions $aa\dot{x} - 4ax\dot{x} + 3x^2\dot{x} = 0$; whence $x = \frac{1}{3}a$. These values of x substituted in the maximums, and dimensions, by proper theorems for the segments and frustums, the content of each solid, with the ratio they are in, are exhibited in the following table, by Mr. Joseph Orchard.

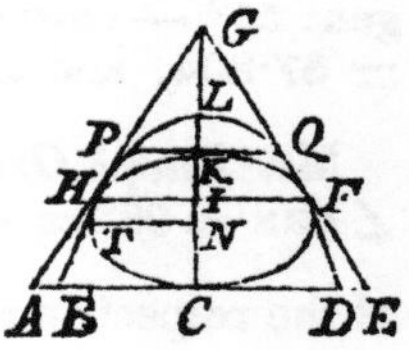

Generating lines from the cone's axis.	Ratio of each solid to the cone.	Content of each solid.
GHL	4	712·677
LHK	5	890·847
KHC spheroid.	27 : 54	4810·575
CHBC	16	2850·712
AHB	2	356·338
sum = whole cone.	54	9621·15

Mr. Terey's *Solution.*

Put AE $= b = 35$, CG $= a = 30$, and $z = \cdot 78539$, &c. whence the cone's solidity $= \frac{1}{3}bbaz = s = 9621\cdot0274$.

1. Then put CK $= x$: As $a : b :: a - x : (a - x)b \div a =$ PQ; but PQ × AE = □ diam. spheroid; whence its content $= (bbax - bbxx) \times \frac{2}{3}z \div a$, is a max. when in fluxions, $a\dot{x} - 2x\dot{x} = 0$, and $x = \frac{1}{2}a$, by writing which value in the above expression, its content $= \frac{1}{2}$ of s $= \frac{1}{2}$ content of the cone.

2. *To find* CI. (N is the centre of the spheroid) NG : NK :: NK : NI; *i. e.* $\frac{3}{4}a : \frac{1}{4}a :: \frac{1}{4}a : \frac{1}{12}a =$ NI. Whence CI $= \frac{1}{3}a$, IK $= \frac{1}{6}a$. By fluxions, content of the spheroidal segment $= \left(\frac{2ccxx}{t} - \frac{\frac{4}{3}ccx^3}{tt}\right)z$; for t put $\frac{1}{2}a$; for cc, $\frac{1}{2}bb$; and for x, $\frac{1}{6}a$; then the segment HKF $= \frac{7}{54}$ of s, and segment HCF $= \frac{10}{27}$ of s.

3. *To find the greatest parab. conoid* BLD. Let LC $= x$, then LG = LI $= a - x$; say, $a : b :: 2a - 2x : (2ab - 2bx) \div a =$ HF, and per known property, $a - x : 2b \times \frac{a - x}{a} \times 2b \times \frac{a - x}{a} :: x : \frac{4bbax - 4bbxx}{aa} =$ BD^2; whence the content BLD $= \frac{2abbxx - 2bbx^3}{aa}z$, a max. from whence, by fluxions, $x = \frac{2}{3}a =$ CL, and LG $= \frac{1}{3}a$; consequently, the contact, in this case, cuts off $\frac{1}{3}$ of the axis, viz. IC, the same as of the spheroid: For x put $\frac{2}{3}a$, and the conoid BHLFD $= \frac{8}{9}$ of s.

4. Conoid HLF $= \frac{4}{9}bb \times \frac{1}{6}az = \frac{2}{9}$ of s $= 2139\cdot006$.

5. (3 and 4) BHLFD — HLF = BHFD $= \frac{2}{3}$ of s $= 6414\cdot0182$.

6. Cone HGF $= \frac{8}{27}$ of s: Consequently, frustum AHFE $= \frac{19}{27}$ of s $= 6770\cdot3526$.

7. (4 and 6) cone HGF — conoid HLF = solid HGFL $= \frac{2}{27}$ of s $= 712\cdot6687$.

8. (2 and 4) conoid HLF — segment HKF = solid HKFL $= \frac{5}{54}$ of s $= 890\cdot8358$.

9. (2 and 5) frustum BHFD — segment HCF = solid HCFDB $= \frac{8}{27}$ of s $= 2850\cdot6748$.

10. (5 and 6) frustum AHFE — frustum BHFD = solid HABDEF $= \frac{1}{27}$ of s $= 356\cdot3343$.

III. QUESTION 349, *by Mr.* Obadiah Wittam, *of Whitby.*

On what two days of the year 1752 will the sun rise at the same instant of time at Petersburgh and Jerusalem?

Answered by Mr. T. Cowper, *Teacher of the Mathematics, at Wellingborough.*

By Dr. Halley's astronomical tables, lat. of Petersburgh $= 60°$, lat. of Jerusalem $= 31° 55'$, their diff. of long. $= 5°$. Put $a =$ tang. $60°$, $b =$ tang. $31° 55'$, and s and $c =$ sine and cos. $5°$, also $x =$ tang. sun's declination. Then, by spherical trigonometry, $1 : a :: x : ax =$ sine ascensional diff. at Petersburgh; cos. $= \sqrt{(1 - a^2x^2)}$; also $1 : b :: x : bx =$ sine ascensional diff. at Jerusalem. But $acx - \sqrt{(s^2 - a^2s^2x^2)} = bx$, or $(ac - b)\, x = s\sqrt{(1 - a^2x^2)}$. Therefore, $a^2c^2x^2 - 2abcx^2 + b^2x^2 = s^2 - a^2s^2x^2$, or (because $1 = s^2 + c^2$) $a^2x^2 - 2abcx^2 + b^2x^2 = s^2$; whence $x^2 = s^2 \div (a^2 + b^2 - 2abc)$; But writing $v =$ sine sun's declination, we have $v^2 \div (1 - v^2) = s^2 \div (a^2 + b^2 - 2abc)$, or $2v^2 = 2s^2 \div (a^2 + b^2 - 2abc + s^2) = \cdot 0121914$ $=$ the versed sine of $8° 57' 21'' =$ twice the sun's declination; consequently the sun's declin. $= 4° 28' 40\frac{1}{2}''$, corresponding to the 26th of February and the 20th of March, and the 31st of August and the 23d of September, O. S.

Mr. N. Dixon's *Answer.*

Let B be Petersburgh, I Jerusalem, P the pole, by Gordon's Geographical grammar, BP $= 38° 35'$; PI $= 57° 16'$; $\angle$BPI $= 5° 25'$, the diff. long. Let B☉ $= 90°$ be perpendicular to BI, then ☉ is the place of the sun at rising. In the triangle BPI, is given BP, PI, and $\angle$P, to find the $\angle$B $= 169° 54'$, from whence take $90°$, and you have $\angle$PB☉. Then in the quadrantal triangle BP☉ there is given BP $30° 35'$, $\angle$B $79° 54'$, to find P☉ the comp. of the sun's declin. $= 84° 53'$, whence the sun's declin. $= 5° 7'$. And the days correspondent thereto are the 22d March and the 28th of August, O. S. on which the sun will rise at both places nearly at the same time.

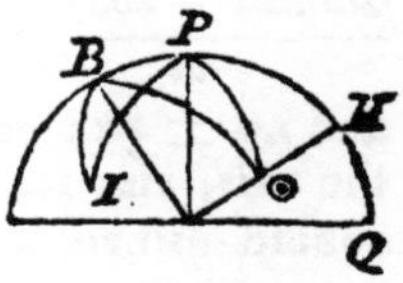

Mr. *James Hartley*, of *Yarum*, solves it thus: Lat. Jerusalem $= 32° 30'$, tang. $= t$; lat. Petersburgh $= 60° 4'$, tang. $=$ T; diff. of merid. $= 3° 30'$; let its sine $= s$, rad. $= 1$, and $x =$ tang. sun's declin. Then $1 : t :: x : tx$. Again, $tx + s : x ::$ T $: 1$, whence $s \div$ (T $- t$) $= x = 3° 10' 40''$, answering to the 1st and 17th of March, and the 3d and 19th of September, O. S.

IV. QUESTION 350, *by Mr.* William Honnor.

Required a theorem for determining the length of a lever of the first kind (supposed of no weight) capable of being divided into two brachias, y the greater, and x the lesser, so that $y^m - x^m = y^n \times x^n$; on whose ends two given weights being suspended, w the greater, and v the lesser, shall equipoise each other?

Answered by Mr. James Hartley, *of Yarum.*

Take $R = v \div w =$ the ratio of the given weights. Then, $Ry = x$; and the given equation will become $y^m - R^m y^m = R^n y^{2n}$, which being reduced, we have the following theorem.

When

m is greater than $2n$, $\left(\frac{R^n}{1 - R^m}\right)^{\frac{1}{m-2n}} = y$.

$m = 2n$, $R^m + R^n = 1$. Here R is limited, and y indeterminate.

m is less than $2n$, then $\left(\frac{1 - R^m}{R^n}\right)^{\frac{1}{2n-m}} = y$.

Mr. John Honey, *of Redruth, in Cornwall, solves it thus:*

Put $N =$ length of the lever, and $s = w + v$, then $x = \frac{Nv}{s}$, and $y = \frac{Nw}{s}$; also $\frac{N^m w^m - N^m v^m}{s^m} = \frac{N^{2n} w^n v^n}{s^{2n}}$ per question; reduced

$$N = \left(\frac{s^{2n-m} \times (w^m - v^m)}{w^n v^n}\right)^{\frac{1}{2n-m}} = \text{length required.}$$

Mr. *Charles Smith*'s theorem is

$$\left(\frac{w^n}{v^n} - \frac{v^{m-n}}{w^{m-n}}\right)^{\frac{1}{2n-m}} + \left(\frac{w^{m-n}}{v^{m-n}} - \frac{v^n}{w^n}\right)^{\frac{1}{2n-m}} = y + x.$$

Who says, if $m = 3$, and $n = 2$, that $y + x = \frac{w}{v}\left(1 + \frac{w}{v}\right) - \frac{v}{w}\left(1 + \frac{v}{w}\right)$.

V. QUESTION 351, *by* Taptinos.

In a right-angled triangle, there is given the distance from the angle at the base to the centre of an inscribed circle 4 chains; and if it be prolonged 2 chains further it will touch the cathetus: To find the sides?

Answered by Mr. Joseph Orchard, *of Gosport.*

Geometrical constructions of problems being always valuable, where they are to be had, we insert the construction first, as follows: Draw AD = 6, making AO = 4, and OD = 2; through O draw EF, at right angles to AD, making OE = OF = OD; draw AB and AC through F and E, and BC through D perpendicular to AB. Then ABC will be the triangle required, as is evident; which is the greatest of demonstration.

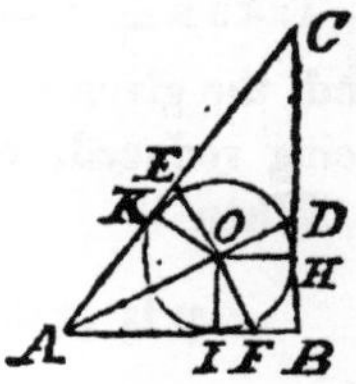

Calculation. In △AOF we have AO = 4, OF = 2, whence AF = $\sqrt{20} = 2\sqrt{5}$. Per sim. triangles, AF : AO :: AD : AB = $12 \div \sqrt{5}$. Per trigon. tang. ∠FAO = $\frac{1}{2}$, and (per schol. to prop. 2, of Mr. Emerson's Elements of Trigonometry) tang. ∠BAC = tang. 2∠FAO = $\frac{4}{3}$, ∴ 1 (rad.) : $12 \div \sqrt{5}$ (AB) :: $\frac{4}{3}$: BC = $16 \div \sqrt{5}$, and AC = $20 \div \sqrt{5}$; the sides are in arithmetical progression, and are AB = 5·366, BC = 7·155, and AC = 8·944.

Mr. Thomas Cowper, *Teacher of the Mathematics, at Wellingborough, computes it thus:*

Put x = sine, and y = cosine ∠DAB = ∠DOH; then 1 : 4 :: x : $4x$ = OI = OH. And 1 : 2 :: y : $2y$ = OH, ∴ $x \div y = 2 \div 4 = {\cdot}5$, the tangent of ∠OAI = 26° 33′ 54·2″; whence AB = 5·3665626, BC = 7·1554168, and AC = 8·944271.

VI. QUESTION 352, *by Mr.* Randles.

A gentleman has an orchard of fruit-trees, one-half of the trees bearing apples, one-fourth pears, one-sixth plumbs, and 50 of them bearing cherries: How many fruit trees in all grow in the said orchard?

Answered by Mr. Henry Watson, *of Gosberton School.*

Put x = the number of fruit trees unknown; then $\frac{1}{2}x + \frac{1}{4}x + \frac{1}{6}x + 50 = x$, and, by transposition, $x - \frac{1}{2}x - \frac{1}{4}x - \frac{1}{6}x = 50$; by division $x = \dfrac{50}{1 - \frac{1}{2} - \frac{1}{4} - \frac{1}{6}} = \dfrac{50}{1 - \frac{11}{12}} = \dfrac{50}{\frac{1}{12}} = 600$, the number of fruit trees required.

The same answered by Mr. John Fish, *of Dartford.*

$\frac{1}{2} + \frac{1}{4} + \frac{1}{6} = \frac{24}{48} = \frac{12}{48} + \frac{8}{48} = \frac{44}{48} = \frac{11}{12}$ wanting $\frac{1}{12}$ of the whole.

Therefore $\frac{1}{12}$ = 50 trees; consequently the whole = 600 trees. Now $\frac{1}{2}$ = 300 apple trees, $\frac{1}{4}$ = 150 pear trees, $\frac{1}{6}$ = 100 plumb trees, whose sum = 550, to which adding 50 trees, the sum total = 600 trees, the proof.

But Mr. *Thomas Huntley*, of *Burford*, putting $12x$ = trees; then $6x + 3x + 2x + 50 = 12x$; whence $x = 50$, and $12x = 600$, required.

VII. QUESTION 352, *by* Taptinos.

In a plain triangle there is given the rectangle of the sides 195, the rectangle of the segments of the base 45, and the perpendicular 12; to find the sides.

Answered by Mr. Henry Watson, *of Gosberton School.*

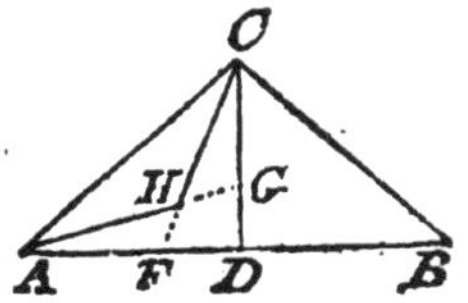

Put x = AD the greater, and y = DB the lesser segments; z = AC, and u = BC, the sides of the triangle; $a = 45$, $b = 195$, and $c = 12$. Then $xy = a$, $zu = b$, per quest. and $xx + cc = zz$, and $yy + cc = uu$, by 47 Euc. I. Whence $zz = \frac{bb}{uu} = xx + cc$, $uu = \frac{bb}{xx + cc} = yy + cc$, and $yy = \frac{bb}{xx + cc} - cc = \frac{aa}{xx}$; from which last $-x^4 + \frac{bbxx - aaxx}{cc} - ccxx = aa$: Put $d = \frac{bb - aa}{cc} - cc$, then $-x^4 + dxx = aa$, or $x^4 - dxx = -aa$; whence $xx = \frac{d}{2} \pm \sqrt{\left(\frac{dd}{4} - aa\right)} = 81$, and $x = 9$; whence $y = 5$, $z = 15$, and $u = 13$.

Another Solution by the Rev. Mr. Wildbore, *taken from the Appendix to Dr.* Hutton's *Edition of the Diaries.*

Having made two rectangles equal to the given ones, so that the given perpendicular may be the shortest side of the greater and the longest side of the less, upon the longest side of the greater as a diam. describe the circ. ABC, any where in which apply BE = the sum of the two sides of the less rectangle, on which take BD = the given perpendicular; through D draw AC ⊥ BE, cutting the circle in A and C; then, AB, BC being drawn, ABC is the required triangle.

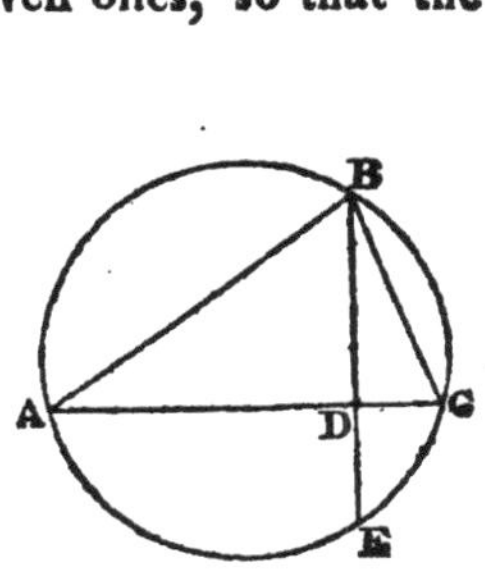

For, by Simpson's Geom. III. 25, calling

the diam. of the circle D, the rectangle $D \times DB = AB \times BC =$ the given one by construc. and, by III. 21, $AD \times DC = BD \times DE =$ the less given one by constr. and BD being the given perp. ABC must be the triangle required.

Corollary. Since $D \times DB = AB \times BC$, and $DE \times BD = AD \times DC$, the two given rectangles are in the ratio of D to DE; ∴ when, instead of the perpendicular, the vertical angle is given; AC being assumed at pleasure, and the circle described to contain thereon the given angle; and DE taken to D in the given ratio of the rectangles, and continued to B; ABC will be a triangle similar to the required one.

VIII. QUESTION 354, *by* Philotheoros.

Given the area of the greatest trapezium that can be inscribed in an apollonian parabola, whose abscissa and semi-ordinate are as 3 to 2, equal 256: Required the dimensions of the parabola and trapezium by a simple equation?

Answered by Mr. Tercy, *of Portsmouth.*

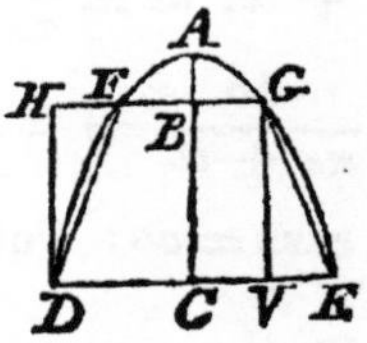

Let $AC = a$, $DC = b$, $BC = x$. Per property of the curve, $a : bb :: a - x : (a - x)\, bb \div a = FB^2$; and $FB = b\sqrt{(a - x)} \div a$. But $bx + bx\sqrt{(a - x)} \div a$ = area of the trapezium DFGE to be a maximum. In fluxions and reduced, $x = \frac{8}{9}a$ (let b be what it will). Now substituting this value of x in FB above, $FB = \frac{1}{3}b$, and also $\frac{2}{3}a$ for b (per quest.), $FB = \frac{2}{9}a$: But $(DC + FB)\, BC = (\frac{2}{3}a + \frac{2}{9}a)\, \frac{8}{9}a = \frac{8}{9}a \times \frac{8}{9}a = 256$; whence by extraction, $\frac{8}{9}a = 16$. And $a = 18 = AC$, and $DC = 12$, $BC = 16$, $FB = 4$; the area of the parabola $= \frac{4}{3}ab$; and of the trapezium $\frac{4}{3}b \times \frac{8}{9}a = \frac{32}{27}ab$. Hence every parabola is to the greatest inscribed trapezium, as $\frac{4}{3}$ to $\frac{32}{27}$, *i. e.* 9 to 8.

Mr. James Hartley *solves it thus:*

The ratio of the abscissa and semi-ordinate being as 3 to 2, the shortest side of the greatest inscribed trapezium will be the parameter, and its area will be equal to the square, whose side HG will be double the parameter. Put $2x = DC$, then $3x = AC$, and from the nature of the curve $DC^2 \div AC = FG = \frac{4}{3}x$. And $AB = \frac{1}{3}$; but $AC - AB = BC = \frac{8}{3}x$. And $DC + BG = DV = \frac{8}{3}x = \sqrt{256}$; whence $x = 6$: Consequently $DE = 24$, $FG = 8$, and $FD = GE = \sqrt{(256 + 64)} = 17{\cdot}8885$; and lastly, $AC = 18$.

IX. QUESTION 355, *by* Anagramensis Holy in Heart, Ebor.

There are two cities in the same parallel of latitude, whose difference of longitude is 144° 15′, and their distance in the arch of a great circle 6797·4 statute miles: Required their latitude, and what day of the year the sun rises to the one city exactly at the same time he sets to the other?

Answered by Mr. T. Cowper, *Teacher of the Mathematics.*

First, 6797·4 ÷ 69·5 = 97° 48′, the distances of the two places, in the arch of a great circle, half of which = 48° 54′. By spherics, As sine $\frac{1}{2}$ diff. long. = 72° 7$\frac{1}{2}$′ : sine $\frac{1}{2}$ the dist. = 48° 54′ :: rad. : sine 52° 21′ = comp. lat. Hence the lat. = 37° 39′. Again, as radius : co-tang. lat. (37° 39′) :: cos. $\frac{1}{2}$ diff. long. (72° 7$\frac{1}{2}$′) = sun's semi-diurnal arch : tang. sun's declination = 21° 42′ almost; answering to the 29th of November and the 12th of January, also the 29th of May and the 14th of July N. S.

Mr. F. Holden *solves it thus:*

6797·4 miles ÷ 69·5 = 97° 48′ 15″, the dist. of the cities in degrees. If a perpendicular be let fall from the pole, it will bisect that dist. and also the diff. long. Therefore, by opposite sides and angles, as sine $\frac{1}{2}$ diff. of long. 72° 7$\frac{1}{2}$′ : sine $\frac{1}{2}$ dist. 48° 54$\frac{1}{8}$′ :: rad. : cos. lat. 37° 38′ 38″. Now, the sun's semi-diurnal arch = $\frac{1}{2}$ diff. long. = 72° 7$\frac{1}{2}$′. Therefore by right-angled spheric triangles, as tang. lat. : rad. :: cos. 72° 7$\frac{1}{2}$′ : tang. sun's declin. = 21° 42′.

But if the apparent time of rising to one city and setting to the other be required, Say, as sine $\frac{1}{2}$ dist. cities 48° 54$\frac{1}{8}$′ : sine sun's refraction 33′ :: rad. : 43′ 48″, which being added to the above-found declin. gives 22° 25′ 48″ declin. at the time of rising and setting required.

Mr. *James Hartley*, of *Yarum*, solves it in the very same manner, and determines the situation of the cities upon the globe thus:

1758. { 29th of May and 14th of July, the sun rises at Japan when he sets in Spain.
12th of January and 30th of November, the sun sets at Japan when he rises in Spain. And vice versa.

Mr. *Cottam* determines the same latitude, and days of the year, very near; and properly observes, that, in the question, there should have been expressed, *on what days*, instead of *on what day* of the year, &c. But, we observe, as the declination can seldom, if ever, correspond with the true time of sun-rising or setting in the required latitude, there is no propriety in this sort of questions, which only admit of answer near the truth. And, for the same reason, the 349th question (where the days of sun-rising at Petersburgh and Jerusalem, at the same instant, are required) contains the same geometrical absurdities; since the required declination is hardly ever possible to hit the true time of sun-rising in both those different latitudes; the declination being still variable every moment of time. And, if the sun's declination be supposed the same for the space of twenty-four hours, in this sort of questions, still the declination and time of sun-rising, on a particular day of the month, cannot exactly correspond, according to computation of the sun's place for that day, at noon, or time of rising, by astronomical tables.

X. QUESTION 356, *by Mr.* John Williams, *of Mold, in Flintshire.*

A gentleman would make a corn mill to be turned by a current of water that runs a tun in a minute, and has 16 feet fall or perpendicular descent: It is required to know the diameter of the water-wheel, so that the issuing water may give the wheel the greatest power, or force possible?

Answered by Mr. William Cottam, *at his Grace the Duke of Norfolk's.*

When the water is conveyed by the trough AA = DC, then 'tis called an over-shot mill, and the diameter of the wheel is thus found: Let DP = 16 feet = a, the distance of the fall; CP = CA = x. By mechanics $nx = n\sqrt{(a-x)}$, where n = force of water at A; therefore $x = \sqrt{(a+\frac{1}{4})} - \frac{1}{2} = 3{\cdot}5313$, whence $7{\cdot}06226$ = diam. of the wheel required.

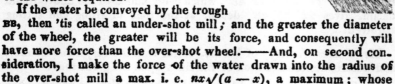

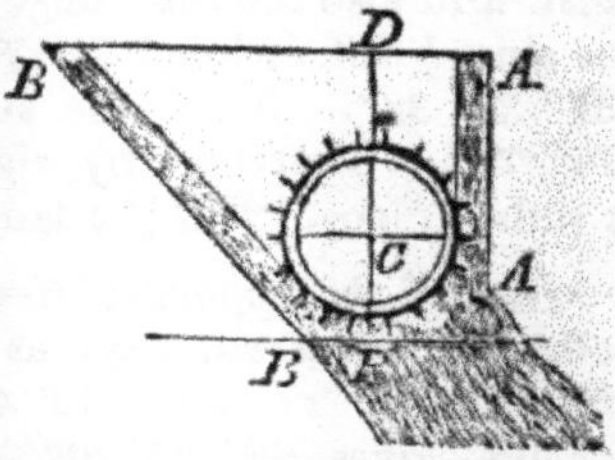

If the water be conveyed by the trough BB, then 'tis called an under-shot mill; and the greater the diameter of the wheel, the greater will be its force, and consequently will have more force than the over-shot wheel.——And, on second consideration, I make the force of the water drawn into the radius of the over-shot mill a max. i. e. $nx\sqrt{(a-x)}$, a maximum; whose fluxion $\frac{2a\dot{x} - 3x\dot{x}}{2\sqrt{(a-x)}} = 0$, being reduced $x = \frac{2}{3}a = 10\frac{2}{3}$, and therefore $21\frac{1}{3}$ feet = wheel's diameter.

But Mr. *James Hartley*, of *Yarum*, solves it thus: Let DP = 16 feet = a; $16\frac{1}{12}$ feet = c; x'' = time of descent of the water from D to C; then cx^2 = DC, and $2cx$ = velocity at C. But $\frac{1}{2}(a - cx^2)$ = semi-diam. of the wheel, which multiplied into the velocity is = $acx - c^2x^3$, and its fluxion made = 0, and reduced gives $\frac{1}{3}a = cx^2$ = DC = $5\frac{1}{3}$ feet, whence CP = $10\frac{2}{3}$; though I don't see the necessity of the querist's mentioning "*a tun a minute.*"

Mr. *John Rickerby*, of *Wooburn*, *Bucks*, says he has spent great part of his life among the best paper mills in the nation; and observes, that a swing wheel, which receives its force of water eight inches above its breast centre, exceeds an over-shot wheel; provided the current of water and fall are alike: And says, though mill-wrights differ in their opinions concerning the true pitch of the water-wheel, that this opinion of his own is true. He speaks of a pen, to give the discharge of water the greater force, at that part of the fall where the water-wheel receives its impetus, or depressing force; and computes the diameter of such a wheel to be 26 feet 8 inches; but on principles a little doubtful.

XI. QUESTION 357, *by the Rev. Mr.* Baker, *of Stickney*.

Let ABCE represent a compound barometer, filled with mercury from B to C, and with water from C to E: How then must the bores of the two tubes ABCF, and FEK be adjusted, or proportioned, so that the scale of variation in the lesser tube of this barometer may be to the common scale, as 10 to 1?

Answered by Mr. James Hartley, *of Yarum*.

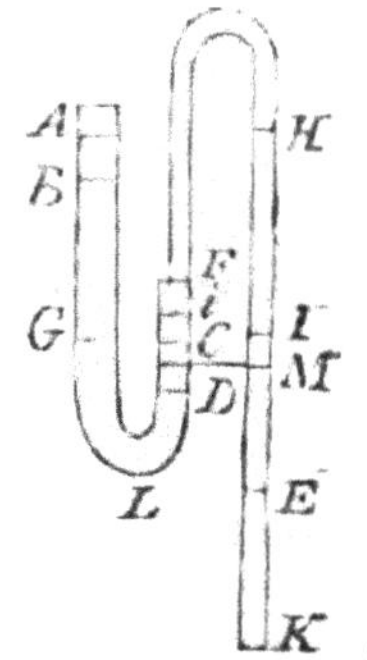

Let AB be the length of the whole scale in the common barometer $= s$; while the mercury descends from A towards B, it will equally ascend from D towards C, so that DC $= \frac{1}{2}s$. Let M be the place of the water and mercury at a middle state of the atmosphere; then per quest. MH $=$ MK $= 5s$. Put $d =$ diam. greater tube, $y =$ diam. lesser, and let $ci = x$; then $\left(\frac{s}{4} + x\right) d^2 = \left(\frac{21s}{4} + x\right) y^2$; and, as mercury is about fourteen times as heavy as water, $14xd^2 = \left(\frac{21s}{4} + x\right) y^2$; whence $s \div 52 = x$. Now, if instead of x, in the first equation, we substitute its value, and assume $d = 1$, we get, by reduction, $y^2 = \frac{7}{137}$, whence $y = \cdot 226$; if therefore d be taken at pleasure, it will be as $1 : \cdot 226 :: d : y$.

The Rev. Mr. *Baker*'s solution is thus: s to 1 the specific gravity of mercury to that of any fluid in a lesser tube; r to 1 the ratio of the tubes; v a given variation in the common barometer, and x the correspondent variation in the lesser tube. Then $r^2 : x :: 1^2 : x \div r^2 =$ variation at the upper surface C of the greater tube (being reciprocally as the squares of the tubes diameters); the whole variation of the lesser tube $= (r^2x + x) \div r^2$; the variation of the mercury's surface at C, in the greater tube, the same with that of the water in the same place, is $x \div r^2 =$ variation at B, of the same diameter. The whole variation of the greater tube BG $= 2x \div r^2$. The pressure of the mercury and water together upon the air at K is from the length of the tubes, the contained fluids in GLC and *cmi* being suspended in equilibrio; the variation in the pressure of the different columns depend on their weights $\frac{r^2x + x}{r^2}$ and $\frac{2x}{r^2}$: Say, $s : 1 :: \frac{r^2x + x}{r^2} : \frac{r^2x + x}{r^2 s} =$ weight of that column, in respect of a column of mercury of the same length; $\frac{2x}{r^2} - \frac{r^2x + x}{sr^2} = v$, the variation in the common barometer; whence $x = \frac{vr^2s}{2s - r^2 - 1}$, i. e. $x : v :: r^2s : 2s - r^2 - 1$; and per quest. $10 : 1 :: 14r^2 : 28 - r^2 - 1$, (here $s = 14$ nearly) whence $r = 3{\cdot}3541$, and the diameters of the tubes are as 3·3541 to 1.

Cor. I. If $s = 1, 2, 3, 4, 5, 6,$ &c. and $r = \sqrt{\frac{1}{2}}, \sqrt{\frac{3}{3}}, \sqrt{\frac{5}{4}}, \sqrt{\frac{7}{5}}, \sqrt{\frac{9}{6}}, \sqrt{\frac{11}{7}}$, &c. correspondent; or if $s = \frac{r^2 + 1}{2 - r^2}$, or $r = \sqrt{\frac{2s - 1}{s + 1}}$, the variations in this barometer will equal those in the common sort.

Cor. II. If $s = 1, 2, 3, 4,$ &c. and $r = \sqrt{1}, \sqrt{3}, \sqrt{5}, \sqrt{7}$, &c. or if $s = \frac{r^2 + 1}{2}$, or $r = \sqrt{(2s - 1)}$, the variation will be infinite. Hence,

Cor. III. The scale of variation in this barometer may have any assignable ratio to the variation in the common barometer.

Mr. *J. Williams* says, that Mr. Rowning (in his Compendious System, p. 112), determines the ratio of the variation of x, in the lesser tube, to the common scale by $x = \frac{vmd^2}{2m - s^2 - 1}$, whence $d = \sqrt{\frac{2mx - x}{vm + x}}$.

XII. QUESTION 358, *by* Upnorensis.

To determine the solidity and superficies of an elliptical ring*, of any dimensions (c the conjugate and t the transverse inches) the substance filling whose periphery is circular of p inches diameter?

*

Answered by ΦΙΛΟ-משקה.

t being the transverse, and c the conjugate diameters of the inner ellipsis of a solid elliptical ring, whose circumference is circular, of p inches diameter; then $t + p$ and $c + p$ will be the transverse and conjugate diameters of the ellipsis passing through the middle of the ring, whose circumference is in the centre of gravity; which circumference put $= a$; then, since the solid or surface generated is equal to the product of the generating plane, or circular line, respectively multiplied into the way made by the centre of gravity, therefore $\cdot 7854ppa$ is the solidity, and $3\cdot 1416pa$ the surface, of the elliptical ring required.

The same answered by Mr. John Hartley, *of Yarum.*

If d be taken $= (tt - cc) \div tt$, in the quest. and $m = 3\cdot 1416$, also A, B, C, &c. = the next preceeding term in the following series, then $1 - \frac{d}{2.2} - \frac{3d}{4.4}\text{A} - \frac{3.5d}{6.6}\text{B} - \frac{5.7d}{8.8}\text{C} - \frac{7.9d}{10.10}\text{D}$, &c.

$$\therefore \left\{ \begin{array}{l} m^2pt = \text{surface} \\ \frac{1}{4}m^2p^2t = \text{solidity} \end{array} \right\} \text{of the ring.}$$

N. B. The above series are taken from Mr. Emerson's excellent Doctrine of Fluxions, p. 174.

XIII. QUESTION 359, *by* Honorius.

Miss's apron grown short, she is full of complaint,
And to merit your pity she looks like a saint!
On the floor falls her tea; then her screams you may hear,
And fainting she sinks in a fit on the chair.
Mamma for the doctor immediately sends,
Who, in honour to miss, in his chariot attends;
He examines her pulse, and appearing so wise,
Descants on the languishing looks of her eyes—
But alas! neither spirits, nor letting Miss blood,
Specifics, nor preaching, are found to do good:
For a surgeon came in, who the cause did declare,
And the doctor's finesse, and his art made appear.
Mamma now was told Miss's hoop was too small,
Therein lay her grievance, disorder, and all;
The question was ask'd—Polly sighing reply'd,
A French hoop will cure me, and so will a bride.
A hoop of the fashion to cure her disease,
Extends from her centre quite round to her knees:
In the right and left wing a French placket * is made,
To her elbows advancing, and forms a parade.

* *Opens and shuts, forms a pair of bellows, and rises and falls by the means of strings or bowlings.*

Miss Polly to church now, or play can repair,
And wherever she goes is admir'd for her air!
At the sight of a beau, how her heart beats alarms!
While the winds swell her pride and her legs tell their charms;
Her hidden perfections she knows will invite,
Or ensnare the beholder, should chance give them sight.
By the pow'r of her hoop Polly steps into fame,
By out-priding the rest she conceals her own shame;
In the country she reigns o'er the 'squire and the clown,
O'er the lords and the fops she's triumphant in town.
Her hoop is the secret—and if you would know
What it holds with her petticoat, seek from below †.

† *Form of the hoop is the lower frustum of an ellipsoid, with its vertex next the head.*

Transverse } *diams.* { 42 *inches* } *above,* { 48 *inches* } *below.*
Conjugate } { 26 } { 29 }

Altitude of the frustum 12 *inches.*

From the lower part of the hoop's circumference to the bottom of the petticoat, the form is an elliptic cylinder, by the petticoat hanging nearly perpendicular from thence: the altitude of which elliptical cylinder is 18 *inches: Quere the content of the whole concavity in wine gallons?*

Answered by Mr. John Honey, *of Redruth, Cornwall.*

Put a and $b = 42$ and 26 inches, the transverse and conjugate diameters of the hoop above; and c and $d = 48$ and 29, the dimensions of those below; also $m = 12$, the frustum of the ellipsoid's altitude; and $n = 18$ inches, the elliptical cylinder's altitude: Then, by a known theorem, $(ab + cd + \sqrt{(abcd)})m \div 882{\cdot}36 = 50{\cdot}54$ wine gallons, the content of the hoop's concavity; and $cdn \div 294{\cdot}12 = 85{\cdot}18$ wine gallons, the content of the cylindrical concavity; whence the concavity of both $= 135{\cdot}72$ wine gallons.

Mr. *John Wigglesworth* answers it thus: Let $a = 42$ inches, $b = 26 =$ transverse and conjugate diameters above; $t = 48$, $c = 29 =$ transverse and conjugate diameters below; $h = 12$, $p = 18$, and $m = {\cdot}2618$, then the content of the whole concavity $= (mh\,(ct + \frac{1}{2}bt + ab + \frac{1}{2}ac) + 3mtcp) \div 231 = 1{\cdot}35{\cdot}7416$ wine gallons.

THE PRIZE QUESTION, *by* χρονονμονονπυβλικος.

Archimedes, the renowned mathematician of Sicily, once bathing himself, observed the water to rise so much higher on his going into the bath. It was from thence he first took the hint for measuring the solidities of all irregular bodies, not measurable by the known rules of art; and also for determining the different specific gravities of bodies. For, being transported with the discovery, he came out of the bath, forgetting he was naked, and ran home crying out, Ευρηκα, Ευρηκα, sigifying, *I have found it*; and, afterwards, discovered the quantity of silver mixed with the gold in King Hiero's crown, which the workman confessed.

It is proposed to determine by the best method, the nearest superficial content in inches of a modern mathematician, of a middle age, weighing 160 pounds avoirdupois, being naked, all his parts middle-sized, and meanly proportioned; and his muscles not rigidly swelled, nor yet quite unbraced?

N. B. The same rule will hold good for male or female mensuration; and man and woman being microcosms, expressions of many elegant and useful curves may thence be discovered; and several improvements made in the rectifications of curve lines, and quadrature of curvilinear spaces; besides cubation of several important solids; whose forms of fluxions, with their fluents, we shall insert in our new Harmonia Mensurarum.

Answered by Mr. F. Holden, *at Westhouse, near Settle, Yorkshire.*

Take a piece of wood or a stone, of a known superficies, and, dipping it into a vessel full of melted tallow, you may, by trying the weight of the tallow and dipping-vessel in a scale before and after dipping, know the quantity of tallow, in weight, taken up by the piece of wood or stone. Then take about 18 or 20 modern mathematicians, (the more the better) strip them stark-naked, and suspend them (like Absaloms) by the hair of their heads, as chandlers hang their candles, or else by soft bandages under the chin and behind the nape of the neck, so that they may be raised or let down by pullies without hurting; dip them also, one by one, in the same vessel where the wood or stone was lately dipped, and mark the tallow they all take up, by weighing the vessel and tallow, before and after they are all dipped, (keeping the tallow just melted and of an equal warmth): Then say, as the quantity of tallow, in weight, taken up by the wood or stone, is to the known superficies of either, so is the weight of tallow taken up by all the mathematicians to the superficies of all the mathematicians. But, by all means, take care that they are kept naked till they are shivering, and almost as cold as the wood or stone itself, before they are dipped, else this proportion will not hold good.

When they are all dipped, well scoured with soap, and cleansed from the tallow, let them be weighed, (or they may be all weighed before dipping) and say, as the weight of them all in pounds is to the late-found superficies of them all in square inches or square feet, so is 160 pounds weight to the superficies of the modern mathematician required to be known, (= $14\frac{1}{2}$ square feet, nearly, as we find by another method).

Questions proposed in 1753, *and answered in* 1754.

I. QUESTION 361, *by a Person of Honour.*

A water mill is to be built where there is a fall of water of 24 feet.—It is required to determine whether a wheel of 18 feet with 6 feet fall, or a wheel of 16 feet with 8 feet fall, will grind the most corn with least water?

Answered by Mr. J. M.

The spaces passed through by falling bodies being as the squares of the acquired velocities, the velocity of the water falling 8 feet is to its velocity acquired in falling 6 feet, as $\sqrt{8}$ to $\sqrt{6}$. Therefore by the property of the lever, the force of the same water to turn a wheel of 16 feet, with a fall of 8 feet, is to its force to turn a wheel of 18 feet, with a fall of 6 feet, as $16\sqrt{8}$ to $18\sqrt{6}$, or as $8\sqrt{4}$ to $9\sqrt{3}$; that is, as to $\sqrt{256}$ to $\sqrt{243}$. Whence, it is evident that the wheel of 16 feet diameter has the greatest advantage.

It was answered in like manner, and on the very same principles, by several other contributors. But these principles, though true in themselves, do not appear to us sufficient to give a right and full determination of the problem under consideration. To have the true quantity of the effect, not only the height of the fall, and the diameter of the wheel, but also the weight and force of the water in the wheel ought to be regarded; and consequently the different positions of the buckets with respect to the horizon.——As this is a subject of much importance, it is hoped our ingenious correspondents will think it worthy of a farther consideration, and communicate their thoughts thereon, for the benefit of the public, in our next Diary; to which we shall refer for a fuller discussion of this matter.

II. QUESTION 362, *by Mr.* John Fish, *of Dartford.*

A ball weighing 4 pounds upon the surface of the earth, to what height, in the air, must it be carried to weigh but 3 pounds, and how long would it be in falling to the ground?

Answered by Timothy Doodle.

Let CD be the earth's semi-diameter, and DA the required height from whence the ball must fall: Then 3 : 4 :: CD^2 : CA^2; and consequently CA = CD $\times \sqrt{\frac{4}{3}} = 3980\sqrt{\frac{4}{3}} = 4596$ miles. Whence DA is given = 616 miles, or 3252480 feet.

Now the distance descended in the first second of time being always as the force, it will here be $= \frac{3}{4}$ of $16\frac{1}{12}$ feet $= 12\frac{1}{16}$ feet: And consequently the time of descent through AD, with the same force uniformly continued $= \sqrt{3252480 \div 16\frac{1}{12}} = 519$ seconds. But, supposing a semi-circle ASC to be described upon AC, and DS perpendicular to AC, the true time of descent through AD will be in proportion to 519 seconds, the time just now found, as half the sum of the sine DS and the arch AS is to the chord AS (as is proved by the writers on fluxions). Now AC : CD (:: $\sqrt{4} : \sqrt{3}$) :: 2 (twice the radius of the tables : $\sqrt{3} = 1{\cdot}732 =$ the versed sine of the angle SOC $= 137° 4'$. Whose supplement AOS is therefore given $= 42° 56'$: The natural sine of which will be ·6811, and the measure of the angle itself = ·7494; the half sum of these is ·7152: But the chord of 42° 56′ (2 × sine 21° 28′) is ·7319. Hence it will be ·7319 : ·7152 :: 519 : 507 seconds, or $8^m 27^s$ the true time required.

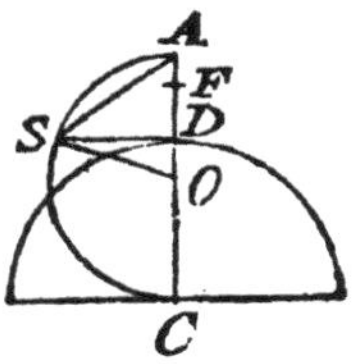

The same answered by Anthony Shallow, *Esq.*

If the earth's semi-diameter (CD) be denoted by a, it is plain that $a\sqrt{\frac{4}{3}} - a$ will express the required height AD from which the ball must fall. To determine therefore the time of the descent through AD, let the velocity of the ball, per second, acquired in falling through any distance AF ($= x$), be denoted by v; putting $c =$ AC; and $d (= \frac{3}{4} \times 16{\cdot}1) =$ the distance descended in the first second of time from A: Then, $2d$ being the measure of the velocity acquired in one second, with the force at A, it will be as $(c - x)^2 : c^2 :: 2d : \frac{2dcc}{(c-x)^2}$, the velocity generated, per second, by the force at F:

Therefore $v : \frac{2dcc}{(c-x)^2} :: \dot{x} : \dot{v}$; and consequently $\frac{v\dot{v}}{2d} = \frac{cc\dot{x}}{(c-x)^2}$.

Hence by taking the fluents, $\frac{vv}{4d} = \frac{cc}{c-x} - c = \frac{cx}{c-x}$. Therefore

$v = 2\sqrt{dc} \times \frac{\sqrt{x}}{\sqrt{(c-x)}}$, and consequently $\frac{\dot{x}}{v} = \frac{\dot{x}\sqrt{(c-x)}}{2\sqrt{cdx}} =$

$\frac{1}{2\sqrt{cd}} \times \frac{c\dot{x} - x\dot{x}}{\sqrt{(cx - xx)}} =$ the fluxion of the required time. Whose fluent is $= \frac{1}{4}\sqrt{(c \div d)}$ multiplied by the sum of the sine and the arch whereof the corresponding versed sine is $2x \div c$ (unity being the radius). But $\sqrt{(c \div d)}$, expressing the time of descent through AC, with an uniform force equal to that at D, is given = 1418 seconds.

And $2x \div c$ is $= 0{\cdot}26795$; answering to an arch of $42° 56'$; whose length is $= 0{\cdot}7494$; and that of its sine $= 0{\cdot}6811$. Hence we have $\frac{1}{4}(0{\cdot}6811 + 0{\cdot}7494) \times 1418 = 507$ seconds $= 8^m 27^s$.

III. QUESTION 363, *by Mr.* Nichol Dixon, *of Blackwell.*

In Craven, Yorkshire, lat. 54°, are three large stone obelisks, A, B, C, lying in the form of a right-angled triangle, of which the line joining A and B is the hypothenuse, and coincides with the meridian; and the line joining A, C is ten miles. At D, a point in AB, three miles distant from B, on the 28th of May, 1752, the sun was observed to rise directly behind C. Required the distance from D to C?

Answered by Mr. Rich. Gibbons.

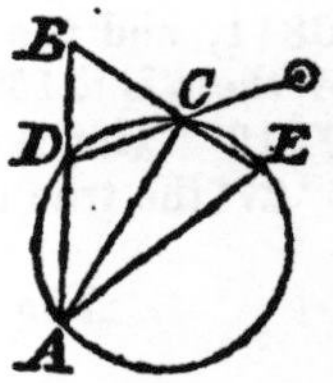

Construction. From the latitude of the place and the sun's declination, the sun's horizontal azimuth BDC is given $= 48° 35'$. Having therefore made the angle BD⊙ $= 48° 35'$ and DB $= 3$; let a square, whose side AC is 10, be so moved along the line D⊙, that the end A, of the side CA, may, at the same time, pass over the line BDA, till the other side of the square passes through the given point B. In which position draw BC and AC, and the thing is done, as is evident by inspection.

Algebraic Solution.

Put $b =$ AC $= 10$, $c =$ BD $= 3$, and $s =$ sine of BDC $=$ sine ADC; also put $d = s \div b = 0{\cdot}074992$, and $x =$ sine of ACD: Then is $\sqrt{(1 - xx)} =$ sine BCD; and we have $b : s :: x : dx =$ AD; also $\sqrt{(1 - xx)} : c :: s : cs \div \sqrt{(1 - xx)} =$ BC. And, by 47 Euc. 1, $b^4 + c^2s^2 \div (1 - x^2) = c^2 + 2cdx + d^2x^2$. Reduced, $x = 0{\cdot}5517$ $=$ the sine of $33° 29'$. From whence CD $= 3{\cdot}4738$ miles.

The same answered by Anthony Shallow, *Esq.*

Suppose the circumference of a circle to pass through A, D, and C, cutting BC produced in E. Then, AE being drawn, in the triangle ACE will be given (besides the right angle) the side AC and the angle E $=$ BDC $= 48° 34' =$ the sun's horizontal azimuth from the north: Whence CE is given $= 8{\cdot}8265$; which put $= a$, making AC $= b$, BD $= c$, and BC $= x$: Therefore AB $= \sqrt{(bb + xx)}$; and, by the property of the circle, $c\sqrt{(bb + xx)} = x(x + a)$. Hence $x =$ $2{\cdot}6966$; and consequently DC $= 3{\cdot}473$ miles.

IV. QUESTION 364, *by Mr.* T. Cowper, *of Wellingborough.*

On the 14th of last March, at half an hour after 11 in the forenoon, being in latitude 52° 22′, I observed 10 beats of my pulse between the time of a small cloud shading me, and that when the shadow thereof reached a tree, at the distance of 88 yards, easterly from me; immediately before, I likewise observed that the angle formed by the shadow of a stick perpendicular to the horizon, and a line drawn from the tree to the place of observation was 120°. Now, admitting 70 pulsations in a minute, the hourly velocity of the cloud, its direction, and what point of the compass the wind blew are thence required?

Answered by Mr. John Wigglesworth.

Let P represent the north pole, Z the zenith, ZS the shadow of the stick, and ZT the direction of the cloud; then the angle TZ⊙ = 60°; and, in the oblique spheric triangle ⊙ZP, is given PZ the complement of latitude, P⊙ the complement of the sun's declination, with the contained angle P; whence the angle ⊙ZB is found = 9° 42′ 30″; which added to TZ⊙, gives TZB = 69° 42′ 30″. Therefore the wind blew W. b. N. $\frac{3}{4}$N. nearly. Again, by allowing 70 pulsations to a minute, the shade of the cloud will move over a space = 88 yards in the time of 10 pulsations, or $8\frac{4}{7}$ seconds; which is at the rate of 21 miles per hour.

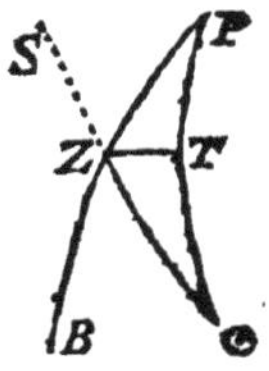

The same answered by Mr. James Robinson.

During 10 pulsations, we may suppose the sun does not sensibly change its place; and, considering the immense distance of the sun, when compared with that of a cloud, we may take the rays proceeding from the sun to the cloud, at both observations to be parallel Then, the sun's azimuth being (per spherics) 9° 26′ from the south, the direction of the cloud was E. b. S. $\frac{3}{4}$ S. fere; and the hourly velocity of the wind, blowing from the opposite point, 21 miles.

V. QUESTION 365, *by Mr.* John Williams.

What pounds principal, being put out at its equal value per cent. at simple interest, for an equal number of years, will raise an interest equal to half the principal?

Answered by Mr. J. Milbourn.

Let 100l. principal = a, and for the required principal, &c. put x; then $a : x :: x : xx \div a$, the interest for one year; and as 1 (year) is to x (years) so is $x^2 \div a$ to $x^3 \div a$, the interest for x years.

Whence (per quest.) $x^3 \div a = \frac{1}{2}x$; reduced, $x = \sqrt{\frac{1}{2}a} = 7{\cdot}07106$ pounds = 7l. 1s. 5d.

VI. QUESTION 366, *by the Rev. Mr.* Baker, *of Stickney.*

If the thickness of two microscopic glasses be three-eighths of their respective radii of convexity, and these be in the ratio of 10 to 3, how must those glasses be disposed, in a compound microscope, so that an object, eight inches distant from the eye, shall be thereby magnified a thousand times?

Answered by Κυβερνητης.

Let the radius of convexity of the lens O, next the object A, be put $= a$, thickness BC $= b$, the radius of the eye-glass S $= d$, and its thickness EF $= e$; and let the sine of incidence be to that of refraction, out of air into glass, as 1 to r; putting $q = 1 - r$, $t =$ the given linear amplification, and $x =$ the distance CD of the place of the image from the lens O.

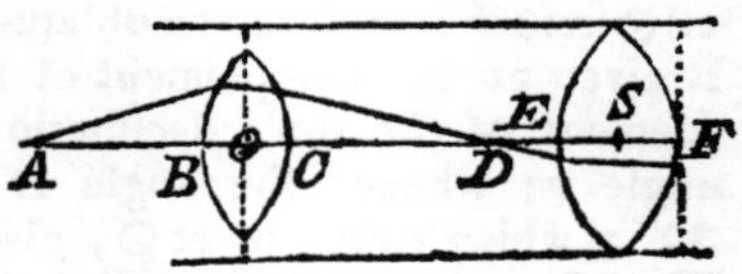

Then, A being the place of the image of an object at D, by a known theorem in optics, AB will be $= \frac{rAQ}{qQ - a}$; Q being put $= b + \frac{x}{r - qx \div a}$. And, by another known theorem (which is a corol. to the former) the principal focal distance ED of the lens S, will be $\frac{rd}{q} \times \frac{d - qe}{2d - qe}$.

But, by the question, $b = \frac{3a}{8}$, $e = \frac{3d}{8}$, and $d = \frac{10a}{3}$. Therefore, by the substitution of equal values, $Q = \frac{3a}{8} + \frac{x}{r - qx \div a}$, and DE $= \frac{10ra}{3q} \times \frac{1 - \frac{3}{8}q}{2 - \frac{3}{8}q} = a \times \frac{10r}{3q} \times \frac{8 - 3q}{16 - 3q}$. Whence DS (= DE + ES) $= a \times \frac{10r}{3q} \times \frac{8 - 3q}{16 - 3q} + \frac{5a}{8} = ca$, by putting $c = \frac{10r}{3q} \times \frac{8 - 3q}{16 - 3q} + \frac{5}{8}$. Again, by the quest. $\frac{OD}{OA} \times \frac{FA}{SD} = t$; that is, in species, $\frac{x + \frac{3a}{16}}{\frac{raQ}{qQ - 1} + \frac{3a}{16}} \times \frac{ca + x + \frac{3a}{8} + \frac{raQ}{qQ - a}}{ca} = t$. Put $\frac{x}{a}$

$=z$, and $\frac{3}{8}+\frac{z}{r-qz}=y$; then will Q $(=\frac{3a}{8}+\frac{az}{r-qz})=ay$; and our equation, by substitution, &c. will be reduced to $\frac{z+\frac{3}{16}}{\frac{ry}{qy-1}+\frac{3}{16}}$ $\times\frac{c+\frac{3}{8}+z+\frac{ry}{qy-1}}{c}=t$; or, $\left(z+\frac{3}{16}\right)\left(c+\frac{3}{8}+z+\frac{ry}{qy-1}\right)$ $=ct\left(\frac{3}{16}+\frac{ry}{qy-1}\right)$. From whence, and $\frac{3}{8}+\frac{z}{r-qz}=y$, the value of z, will be found by an equation of three dimensions: And then, the value of FA being given in terms of a, by putting that value $=8$ inches, a itself will be found, and from thence every thing else, required.

But as the finding of z this way (the terms being numerous) will be somewhat troublesome, the known method of approximation, by trial-and-error, may be here used with advantage. According to which, having assumed for the value of z, that of y $(=\frac{3}{8}+\frac{z}{r-qz})$ will be immediately found; and then, by substituting these two values in the other equation, the error will be determined, &c. &c.

VII. QUESTION 367, *by Mr.* T. Cowper, *of Wellingborough.*

On the 19th of September, 1751, at night, the vertical angle between Jupiter and the star Castor was observed to be 35° 48′, and that between Jupiter and the bright star in the whale's tail, 78° 29′; it is required to determine the latitude of the place and hour of the night, where and when those observations were made?

Answered by Mr. W. T—t.

Let P be the pole of the world, Z the zenith of the place, and B, I, C, the three stars: From the given longitudes and latitudes of which, or from their right ascensions and declinations, the distances BI and IC, and the angle BIC, may be found, by common trigonometry.

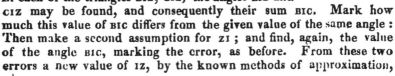

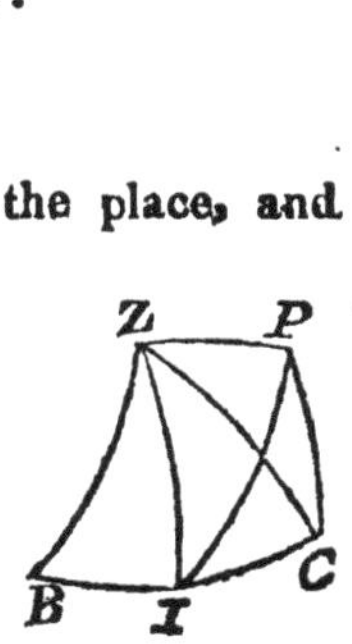

Assume the value of ZI as near as you can to its true value: Then, having two sides and one angle, in each of the triangles BIZ, CIZ, the angles BIZ and CIZ may be found, and consequently their sum BIC. Mark how much this value of BIC differs from the given value of the same angle: Then make a second assumption for ZI; and find, again, the value of the angle BIC, marking the error, as before. From these two errors a new value of IZ, by the known methods of approximation, may be found; and so on, till you arrive to what degree of exactness you please. Having thus determined ZI and ZIC, from the latter of these deduct PIC, the remainder gives the angle ZIP: From which, and the two given sides including it, both ZP and ZPI will become known.

VIII. QUESTION 368, *by Mr.* Christopher Mason, *of Eastburn, near Petworth, in Sussex.*

A constant quantity being put for a factor, in Mons. Ozanam's Mathematical Recreations, to be multiplied by variable factors, in order to produce 3 ones, 3 twos, 3 threes, &c. through all the digits, I desire to know both the constant and variable factors that will produce 6 ones, 6 twos, 6 threes, &c. also 9 ones, 9 twos, 9 threes, &c. through all the digits?

Answered by Mr. J. Robinson.

Let $999999 = a$ be the constant factor, in order to produce 6 ones, 6 twos, &c. The other factor call x, and the first product p; then $xa = p$, and consequently $x = p \div a$.

Universally, putting the constant factor (which is arbitrary) for the denominator, and the given product for the numerator, the fraction, or fractions, thence arising, will be the variable factor, or factors, required.

If any number of nines be taken for the denominator of a fraction, and the same number of any of the digits for a numerator, the fraction, when reduced to a decimal one, will have the very same figures as the numerator, repeated to infinity. Thus, for example, $\frac{1234}{9999}$ is $=$ $\cdot 1234123412341234$, &c. ad infinitum. Thus also $\frac{444444}{999999} = \cdot 44444$, &c. and consequently $999999 \times \cdot 4444$, &c. $= 444444$; and so of others. The variable factors, derived by this general method, are fractions; but there are particular answers to be had in whole numbers. Thus, because $\frac{111111}{3} = 37037$, and $\frac{111111111}{3} = 37037037$, it is evident that, if 3 be taken for the constant factor, the respective variable factors, to produce 6 and 9 ones, will be 37037 and 37037037. The multiples of which by 2, 3, 4, 5, &c. will consequently be the other variable factors required. In like manner, 37 being assumed for the constant factor, the variable ones will be 3003 and 3003003, together with their multiples.

IX. QUESTION 369, *by Mr.* Terey, *of Portsmouth.*

Required the superficial content of a scalenous cone, whose longest side is 12 inches, shortest 9, and base diameter 6 inches?

Answered by Κυβερνητης.

Let ABHDEC be the given cone, and AG its perpendicular height: Let EM be a tangent to the circular base BEDH at any point E; and supposing e to be another point in the curve indefinitely near to E, let EA and eA be drawn in the surface of the cone: And from the same points E, e, upon the diameter BD, passing through G, let fall the perpendiculars EF and ef: Draw the radius CE, and the line GK parallel thereto, meeting the tangent EM, at right angles, in K; to which point from the vertex of the cone draw KA, which will be perpendicular to the tangent EM; because (being equal to $\surd(AG^2 + GK^2)$) it will be the least possible, this position, where GK is the least possible.

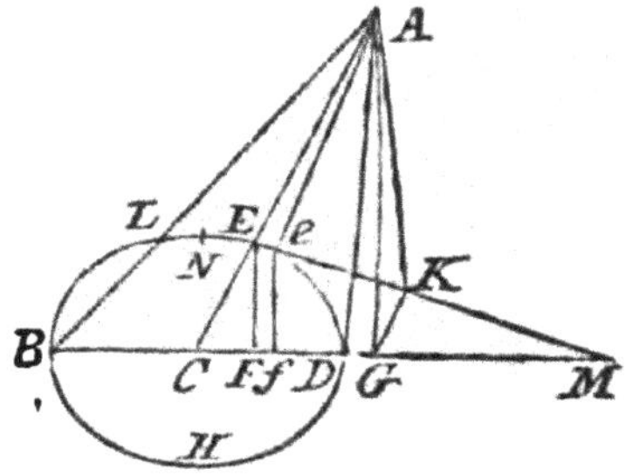

Put now CE $= a$, CG $= b$, AG $= c$, and CF $= x$, BE $= z$, and Ee $= \dot{z}$. Then, by the property of the circle, CM $= aa \div x$: And, by similar triangles, CM : CE :: MG $\left(\frac{aa}{x} - b\right)$: GK $= a - \frac{bx}{a}$. Whence AK $= \surd(AG^2 + GK^2) = \surd\left(c^2 + a^2 - 2bx + \frac{bbxx}{aa}\right)$: Which, multiplied by $\frac{1}{2}\dot{z}$, gives $\frac{1}{2}\dot{z}\surd\left(cc + aa - 2bx + \frac{bbxx}{aa}\right)$ for the area of the triangle EAe, or the fluxion of the required superficies. Which fluxion, because $\dot{z} = \frac{a\dot{x}}{\surd(aa - xx)}$, is also $=$

$$\frac{\frac{1}{2}a\dot{x}\surd\left(cc + aa - 2bx + \frac{bbxx}{aa}\right)}{\surd(aa - xx)}$$: Whose fluent will express the required superficies of the cone.

But as the finding of this fluent is extremely troublesome, and, when it is found, converges slowly (except where the cone is but little inclining) I shall therefore give the solution by a different, and more general method.

Let PT be a right line equal in length to the semi-circumference BNED; upon which, as a base, (or abscissa) suppose a curve $pqrst$ to be described, such that, taking PS always $=$ BE, the ordinate Ss shall be every where equal to half the corresponding perpendicular AK: Then it is plain, that the area PTtp of this curve, will be exactly equal to half the convex superficies of the cone. To approximate which, conceive the axis PT of the curve and the semi-circum-

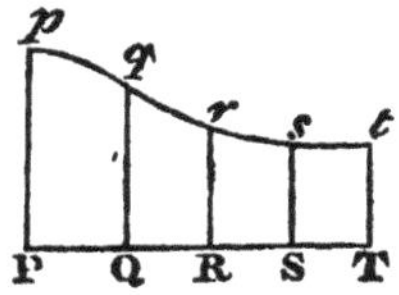

ference BNED of the circle, to be divided, each into four equal parts: and let the successive values of AK $\left\{=\sqrt{(c^2+(a-\frac{bx}{a})^2)}\right\}$ answering to the points of division B, L, N, E, D (or P, Q, R, S, T) be computed, and represented by d, c, f, g, and h, respectively. Then, these values being the doubles of the corresponding ordinates Pp, Qq, Rr, Ss, Tt, it is evident, by the method of equidistant ordinates, that $\frac{7(d+h)+32(e+g)+12f}{90}\times\frac{1}{2}$PT, will express the area of the curve, or half the superficies of the cone, very nearly.

Now, in the case proposed, AB being $=12$, AD $=9$, and BD$=6$, we have CE $=3=a$; DG $\left(=\frac{AB^2-AD^2-BD^2}{2BD}\right)=\frac{9}{4}$; CG $=5\frac{1}{4}=b$; and AG^2 $=75\cdot9375=c^2$. Here, therefore d (= AB) $=12$; $e(=\sqrt{cc+aa+\frac{1}{2}bb+ab\sqrt{2}}))=10\cdot9996$; $f(=\sqrt{(cc+aa)})=9\cdot2162$; $g(=\sqrt{(cc+aa+\frac{1}{2}bb-ab\sqrt{2})})=8\cdot7432$; and h(= AD) $=9$.

From whence $\left(\frac{7(d+h)+32(e+g)+12f}{90}\times\text{NED}\right)$ the content of the whole required superficies comes out 93·13 square inches. By taking a greater number of ordinates, the answer may be brought out to any degree of exactness desired, however great the inclination may be.

X. QUESTION 370, *by Mr.* Jos. Hilditch, *at Handel, near Shrewsbury.*

The three distances from an oak, growing in an open plain, to the three visible corners of a square field, lying at some distance, are known to be 78, 59·161, &c. and 78 poles, in successive order: Quere the field's dimensions, and the acres it contains?

Answered by Mr. J. Robinson.

Let $a=$ OD $=$ OB $=78$, $b=$ OA $=59\cdot161$, and $x=$ AF. Then (per Euc. I. 47.) $bb+2bx+2xx=aa$. Reduced, $x=\sqrt{\left(\frac{aa}{2}-\frac{bb}{4}-\frac{b}{2}\right)}$: From whence the side of the square will be found $=24$ poles, and the area of the field $=3$ acres, 2 roods, and 16 perches.

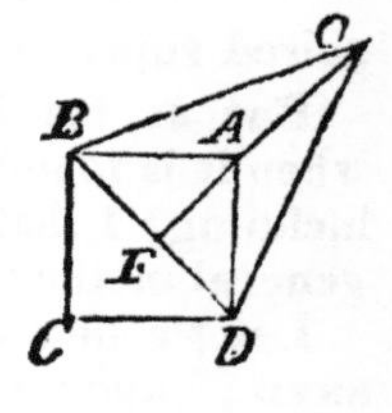

The same by Master John Birks, *a Tyro, at Gosberton School.*

The three visible corners of the field being represented by A, B, and D, it is evident, because the given distances of the two last from the oak, at O, are equal, that the line OAF bisects the angle BAD, and consequently that the angle BAO $= 135°$. From which, and the given sides AO and BO, the angle ABO is found (by trigonometry) $=$ $32° 26'$; AOB $= 12° 34'$; and the side of the square AB $= 24$ poles: Whence the area of the field $= 3$ acres, 2 roods, and 16 perches, which was required.

XI. QUESTION 371, *by Mr.* Tercy, *of Portsmouth.*

What is the content of a cask in ale gallons, whose staves are exactly circular, and dimensions of the head and bung diameters, and also length 24, 36, and 48 inches?

Answered by Mr. T. Coates, *Writing-Master, at Bristol.*

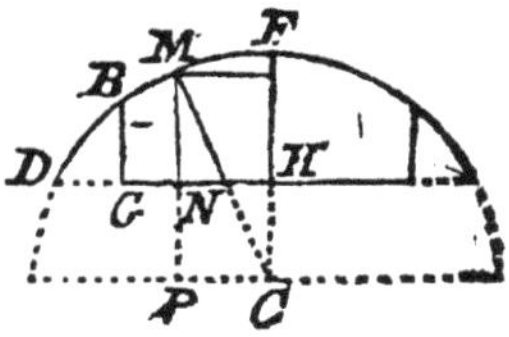

Put $r =$ the radius CF of the generating circle, which is found (by common properties) $= 51$; also put $c =$ CH $= 33$, DH $= b$, HN ($=$ CP) $= x$, and MN $= y$.

Then universally, $y = \sqrt{(rr - xx)} - c$: Therefore $yy = rr - xx + cc - 2c\sqrt{(rr - xx)}$ $= bb - xx - 2cy$; and $py^2\dot{x} = p(bb\dot{x} - xx\dot{x} - 2yc\dot{x}) = \dot{s}$; whose fluent is $p(bbx - \frac{1}{3}x^3 - 2c \times$ area MNHF$) = s$: which, when $x = 24$, gives the content 139·42 ale gallons.

The same answered by Mr. John Wigglesworth.

Let GH $=$ half the length of the cask $= 24 = a$, BG $= 12 = b$, HF $= 18 = c$; also put $r =$ rad. CM $=$ CF $= \dfrac{aa + (c - b)^2}{2c - 2b}$ (by the nature of the circle), $m =$ PN $=$ CH, $p = 3{\cdot}14159$, &c. $x =$ HN $=$ CP, and $y =$ MN: Then $y = \sqrt{(rr - xx)} - m$; and therefore $py^2\dot{x} = p\dot{x}(r^2 - x^2 + m^2) - 2mp\dot{x}\sqrt{(rr - xx)} =$ the fluxion of the solidity: Whose fluent $px(r^2 - \frac{1}{3}x^2 + m^2) - 2mp \times$ area CFMP $=$ the solid described by the rotation of HFMN about the line HN: And, if L be put for the length of the arc whose radius is r, and right sine x, then $\frac{1}{2}(r\text{L} + mx + xy) =$ area CFMP. Whence the solidity of the whole cask, when $x = a$, and $y = b$, is $2p(ar^2 - \frac{1}{3}a^3 - mr\text{L} - abm) =$ 39316·99 cubic inches, or 139·4219 ale gallons.

XII. QUESTION 372, *by Mr.* W. Bevil, *of Harpswell.*

Suppose the moon's diameter to be 2170, the earth's diameter 7970, and the distance of each other's surface 240000 miles, where must I view them on a line drawn betwixt their centres, to see the greatest quantity of surface of both bodies possible?

Answered by Timothy Doodle.

Let O and C be the centres of the earth and moon, and H the place required: Suppose HF and HB to touch the two surfaces in F and B, and let FG*f* and BD*b* be perpendicular to OC.

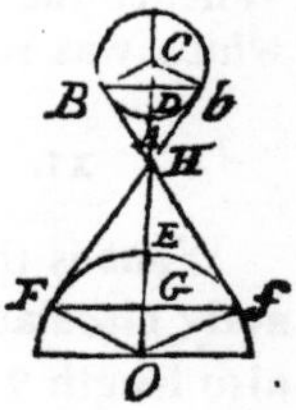

Put $a = \text{OE} = 3985$, $b = \text{CA} = 1085$, $c = \text{OC} = 245070$, and $x = \text{OH}$; and let $p = 2 \times 3\ 14159$, &c. So shall the circumference FE*f*, &c. $= pa$, and the circumference BA*b*, &c. $= pb$: And therefore the parts EF*f*, BA*b* of the two surfaces visible to an eye at H, are equal to $pa \times \text{EG}$ and $pb \times \text{AD}$, respectively.

But, by similar triangles, OH (x) : OF (a) :: OF (a) : $\text{OG} = aa \div x$: Whence $\text{EG} = a - aa \div x$: And, in the very same manner, $\text{AD} = b - bb \div (c - x)$. Therefore, by substitution, $\text{FE}f + \text{BA}b = pa \left(a - \frac{aa}{x}\right) + pb \left(b - \frac{bb}{c-x}\right)$: Which being a maximum, $\frac{a^3}{x} + \frac{b^3}{c-x}$ must be a minimum; and its fluxion $-\frac{a^3\dot{x}}{xx} + \frac{b^3\dot{x}}{(c-x)^2} = 0$. Hence $\frac{x^2}{a^3} = \frac{(c-x)^2}{b^3}$; or $(b \div a)^{\frac{3}{2}} x = c - x$, and consequently $x = \frac{c}{1 + (b \div a)^{\frac{3}{2}}} = 214585$. Therefore the place where friend *Bevil* must take his view is 210600 miles above the surface of the earth, if he can find his way up so high.

XIII. QUESTION 373, *by Mr.* J. Williams, *of Mould, Flintshire.*

Mathematicians take pains to describe curves and solids that never existed, yet say little or nothing to the properties of those things that are in nature; especially the sections, solidities, and curve superficies of the egg, which is one of nature's principal productions. If any of the problematic problemists would be pleased to give the solution of the quantity of curve superficies and solidity of the egg, when its axis is $2\frac{1}{2}$ inches, greatest ordinate $1\frac{1}{2}$, and the distance from that ordinate to the nearest end 1 inch, they would be intitled to a maximum of applause, instead of the minimum acquired, by puffing and cavilling about their superior dignity, who are odd fishes at foot-ball.

Answered by Anth. Shallow, *Esq.*

This question, in the form it is proposed, is indeterminate: The figure of the egg, as well as its principal dimensions, ought to have been given; since, of an infinity of curves that may be described through the same given points, experience is not sufficient to direct us which to choose; it not being known that ever two eggs were exactly of the same figure.

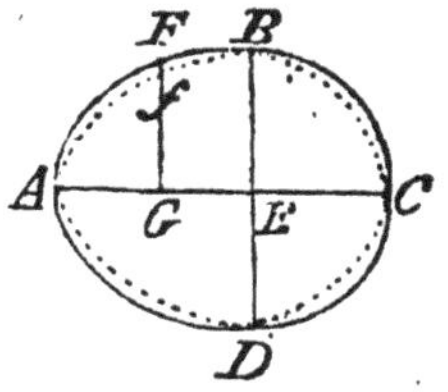

Let AFBCD be a section of the egg through its axis AC, and let BD be the given position of the greatest ordinate. It is visible that innumerable curves, AFBCD, AfBCD, &c. may be described through the given points A, B, C, and D, to cut AC and BD at right angles, conformable to the nature of the problem. But, the greatest ordinate BD dividing the axis AC unequally, no curve of a lower order than the second can possibly answer these conditions.

Let, therefore, AFBC be a curve of this order, whose equation is $yy = bx + cx^2 + dx^3$ (being the most simple the data will admit of): Also, let AC $= p$, AE $= q$, BE $= r$: Then, by making $x = p$, and $y = 0$, our general equation becomes $bp + cp^2 + dp^3 = 0$, or $b + cp + dp^2 = 0$.

Also, by making $x = q$, and $y = r$, we have $bq + cq^2 + dq^3 = r^2$, or $b + cq + dq^2 = r^2 \div q$.

Lastly, by making $b\dot{x} + 2cx\dot{x} + 3dx^2\dot{x}$ (the fluxion of $bx + cx^2 + dx^3$) $= 0$, and writing q in the room of x, we have $b + 2cq + 3dq^2 = 0$.

Now, from the three equations thus derived, d is found $= -\frac{rr(2q-p)}{q^2(p-q)^2}$; $c = \frac{rr(3qq-pp)}{q^2(p-q)^2}$; and $b = \frac{rrpq\times(2p-3q)}{q^2\times(p-q)^2}$.

Therefore the general equation, in known terms, is $yy =$

$$\frac{rr}{q^2(p-q)^2}\times((2p-3q)pqx + (3qq-pp)xx - (2q-p)x^3).$$

Whence, if a be put $= 3{\cdot}14159$, &c. we get $ayy\dot{x} = \frac{arr}{q^2\times(p-q)^2}$ $((2p-3q)pqx\dot{x} + (3qq-pp)x^2\dot{x} - (2q-p)x^3\dot{x})$ for the fluxion of the solidity. Whose fluent, when $x = p$, will be found $= ar^2p^3 \frac{6pq-6qq-pp}{12q^2\times(p-q)^2}$ expressing the true content of the whole solid. Which therefore is to ($arrp$) that of the circumscribing cylinder, in the proportion of $p^2\,\frac{6pq-6qq-pp}{12q^2\times(p-q)^2}$ to unity. This proportion, in the case proposed, (where $p = 2\frac{1}{2}$, $q = 1\frac{1}{2}$, and $r = \frac{3}{4}$) becomes as $\frac{275}{432}$ to unity. Therefore the solidity (according to the above assumption) comes out 2·8124 cubic inches.

As to the superficies, or shell of the egg, it may be also found from the same general equation; but it is hoped the facetious proposer will himself determine that, and accept it as a proper reward for his trouble and industry in promoting useful science.

Some correspondents consider the egg as formed of two unequal semi-spheroids; but this does not seem to agree well with the true figure, it being hard to conceive that the curvature shall be immediately changed by more than one-half, in passing from one side of the greatest ordinate to the other.

XIV. QUESTION 374, *by* Philotheoros.

My Lord Mayor's gold chain being 50 inches in measuring round it, at what distance must it be hung over two pins, horizontally fixed in a wall, covered with crimson damask, for a spectator to behold the most damask possible, within the circumference of the chain?

Answered by Mr. W. Bevil.

Let *acb* be a curve similar to that (ACB) formed by the chain, such that its ray of curvature (a) at the lowest point c may $= 1$. Then, the area of the semi-curvilineal space, *acd*, will be truly defined by $y\sqrt{(1+zz)} - z$, (as is proved by the writers on fluxions) z being $= ca$, and y ($= ad$) $=$ hyp. log. $(z + \sqrt{(1+zz)})$.

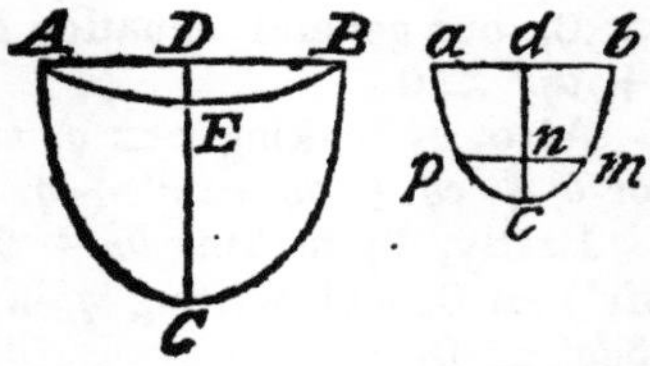

Hence, putting the length of the chain $= c$, we have (by the general property of similar figures) as

$$(z+y)^2 : \tfrac{1}{4}cc\,(\text{AC} + \text{AD})^2 :: y\sqrt{(1+zz)} - z : \tfrac{1}{4}\,cc \times \frac{y\sqrt{(1+zz)}-z}{(z+y)^2}$$

$=$ area ACD: Which being a maximum, let its fluxion be therefore taken and made $= 0$; whence, after proper reduction, there will come out $\tfrac{1}{2}yz \times \dfrac{z+y}{1+\sqrt{(1+zz)}} = y\sqrt{(1+zz)} - z$. From which equation (by the known methods of approximation) the values of z and y may be found. For, having assumed for z, y will be given from the equation $y =$ hyp. log. $(z + \sqrt{(1+zz)})$; and then, by substituting these values of z and y in the above equation, the error will be known; and from thence, by repeating the operation, &c. the true value of z; which comes out $= 5\cdot462$; and $y = 2\cdot399$. Then $7\cdot861$ $(ac+ad)$: $2\cdot399$ (ad) :: $\tfrac{1}{2}c(\text{AC}+\text{AD})$: $\text{AD} = 0\cdot15259 \times c$: Whence $\text{AB} = 0\cdot30518 \times c = 15\cdot259$ inches.

Anthony Shallow, Esq. solves this problem exactly in the same manner. But Mr. *Timothy Doodle*, and Mr. *O'Cavanah*, taking the meaning of the question in a different sense [supposing the arch ACB, and not ACB + AB, to be given = 50] bring out AB = 33·575 for the answer. In which case it appears that AD, CD, and AC, will be in the ratio of 0·6715, 0·6656, and 1, respectively; and that the area ACB is to the square of the arch ACB, as 0·1549 to unity.—— But there is yet another way in which the question may be taken, as it is not specified whether the chain is to be fastened to the pins, (in which case the area will be the greatest possible) or whether it is suffered to slide freely over them, till the two parts thereof (here represented by ACB and AEB) acquire an equilibrium. In this last sense the solution will be still more complex; but the best method will be to assume two arcs, *acb* (= 2z) and *pcm* (= 2z), of the same given catenaria, similar to the two parts ACB and AEB of the chain. From whence (by the properties of the catenaria, and the consideration of similar figures, and of the equal action of the two branches of the chain at B) there will be had $\frac{\sqrt{(1+\text{zz})}}{\sqrt{(1+zz)}} = \frac{\text{hyp. log. z} + \sqrt{(1+\text{zz})}}{\text{hyp. log. } z + \sqrt{(1+zz)}}$; and $\frac{z(1+\text{zz}) - \text{z}(1+zz)}{(z\sqrt{(1+\text{zz})} + \text{z}\sqrt{(1+zz)})^2}$, a maximum. From which the values of z and z may be determined.

PRIZE QUESTION, *by Mr.* James Hartley, *of Yarum.*

Suppose DAFE to be a vessel in the form of a conical frustum, whose top diameter DA = 20, bottom diameter = 10, and depth = 15 inches, suspended by an inflexible line C*a* = 100 inches (the line and vessel being supposed of no weight) and the vessel full of water to weigh 100 pounds weight, when the vessel is perpendicularly suspended at B. Let the vessel be drawn aside by a cord fastended at (*a*) the bottom of the vessel, while at the other end of the cord is a weight W suspended, the cord freely sliding over the pulley at P, placed at such a distance from C, in the horizontal line PC, as to require the least weight, W, possible, to equipoise the vessel when the tension of the cord P*a* is a maximum: Required the weight, W, and distance PC?

Answered by Mr. J. Ash.

'Tis plain that the least weight will equipoise the vessel in any degree of elevation when the cord P*a* is perpendicular to C*a*: Then, putting s = solidity of the whole frustum, c = ·7854, w = tension of the cord, n = the ratio of D*n* to *mn*, (A*m* being the horizontal surface of the water) and x^2 = diameter of the frustum at *m*, we have $\frac{1}{2}(a - x^2)$ = D*n*, $\frac{1}{2}(na - nx^2)$ = *mn*, $\frac{1}{2}(a + x^2)$ = A*n*, and, per theorem,

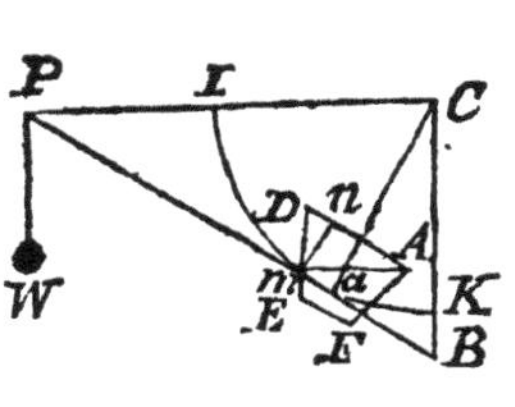

Diary 1744, quest. 240, $\frac{1}{2}(ca^3 - ca^{\frac{3}{2}}x^3) \times \frac{1}{3}n =$ (because $n = 3$) $\frac{1}{2}(ca^3 - ca^{\frac{3}{2}}x^3)$, the solidity of the hoof D*n*A*m*; therefore $\frac{1}{2}(2s - ca^3) + \frac{1}{2}(ca^{\frac{3}{2}}x^3)$ = the solidity of the remaining water, which let $= qx^3 - p$: Then (per Stone) C*a* : sine of IK*a* :: $qx^3 - p : w$; or which is the same thing, (because the triangles are similar) AM $\left(\sqrt{\frac{10a^2-16ax^2+10x^4}{4}}\right)$ $: mn\left(\frac{na-nx^2}{2}\right) :: qx^3-p : w = \frac{(na-nx)^2\times(qx^3-p)}{\sqrt{(10a^2-16ax^2+10x^4)}}$, the tension, a max. and consequently $\frac{(a-x)^2\times(qx^3-p)}{\sqrt{(a^4-1{\cdot}6ax^2+x^4)}}$ thrown into fluxions and properly reduced, gives $x^2 = 13{\cdot}6$ fere. Hence the weight required $= 24{\cdot}7054$, and PC $= 201{\cdot}57$.

The same answered by Mr. Patrick O'Cavanah.

Put $a = 15 =$ the depth of the frustum, $b = 20$ the greatest diameter AD, $c = 10 =$ the least diameter EF, $p = {\cdot}7854$, and $(2am) = x$. Then, supposing mn perpendicular to DA, it will be $b-c : b-x :: a : a(b-x)\div(b-c) = mn$: Whence, by a well-known theorem for the content of a conical ungula, the solidity of ADm is $= \frac{1}{3}pab\,\frac{bb-x\sqrt{bx}}{b-c}$. Which subtracted from $\frac{1}{3}pa(bb+bc+cc)$, the content of the whole frustum AFED, leaves $\frac{1}{3}pa\,\frac{bx\sqrt{bx}-c^3}{b-c}$ for the part AFEm, in which the water is contained. Now the tension of the rope, or the force of the weight W, acting at right angles to ca, so as to sustain the water in this position, is known to be in proportion to the weight of the water in the vessel, as the sine of the angle CPa (or DAM) to the radius; that is, as mn to Am, or in species, as $a\,\frac{b-x}{b-c}$ to $\sqrt{\left(\frac{a^2(b-x)^2}{(b-c)^2}+\frac{(b+x)^2}{4}\right)}$. Whence, by the question, $\frac{(b-x)(bx\sqrt{bx}-c^3)}{\sqrt{(a^2(b-x)^2+\frac{1}{4}(b-c)^2(b+x)^2)}}$ (which is in a given ratio to AFEm $\times\,\frac{mn}{\text{Am}}$) must be a minimum. This expression, by putting $yy = bx$, $d = \frac{b-c}{2}$, and $f = \frac{aa-dd}{aa+dd}2bb$, and dividing the denominator by $\sqrt{(aa+dd)}$, is reduced to $\frac{(bb-yy)(y^3-c^3)}{\sqrt{(b^4-f^2y^2+y^4)}}$. Which, in fluxions, &c. gives $\frac{3y}{y^3-c^3}-\frac{2}{bb-yy}+\frac{ff-2yy}{b^4-f^2y^2+y^4} = 0$. Hence $y = 16{\cdot}516$, $x = 13{\cdot}7578$, CP $= 202{\cdot}66$, and W $=$ 24·709 pounds.

Remark. As the taking of the fluxions of expressions compounded like that in the preceding solution, is somewhat troublesome, the following method may, in such cases, be of use.

Seeing $\frac{(y^3-c^3)(b^2-y^2)}{\sqrt{(b^4-f^2y^2+y^4)}}$ is to be a maximum, by the question, the logarithm thereof, or its equal, log. (y^3-c^3) + log. (b^2-y^2) — $\frac{1}{2}$ log. $(b^4-f^2y^2+y^4)$, must also be a maximum, and consequently its fluxion $\frac{3y^2\dot{y}}{y^3-c^3}-\frac{2y\dot{y}}{bb-yy}+\frac{ffy\dot{y}-2y^3\dot{y}}{b^4-f^2y^2+y^4}=0$: Whence, dividing by $y\dot{y}$, we have $\frac{3y}{y^3-c^3}-\frac{2}{bb-yy}+\frac{ff-2y^2}{b^4-f^2y^2+y^4}=0$, the same as before. Which equation stands in a much better form, for a solution, than that immediately resulting from the common method. In like sort, supposing $\frac{(a^2-x^2)^{\frac{1}{2}}(b^3+x^3)^{\frac{1}{3}}(c^4-x^4)^{\frac{1}{4}}}{(a-x)^2(a^2+2dx+x^2)}$ was to be a maximum, or minimum, we should have $\frac{1}{2}$ log. (a^2-x^2) + $\frac{1}{3}$ log. (b^3+x^3) + $\frac{1}{4}$ log. (c^4-x^4) — 2 log. $(a-x)$ — log. $(a^2+2dx+x^2)$ a max. or min. And consequently $-\frac{x}{a^2-x^2}+\frac{x^2}{b^3-x^3}-\frac{x^3}{c^4-x^4}+\frac{2}{a-x}-\frac{2d+2x}{a^2+2dx+x^2}=0$. And so of the others.

Questions proposed in 1754, *and answered in* 1755.

I. QUESTION 376, *by* Rusticus.

A person bought a horse for twenty-six shillings, and, after keeping it some time, sold it for three pounds, by which he lost one half of the prime cost, together with one-fourth of the keeping. What did the keep stand him in, and what did he lose?

Answered by Mr. Edward Gallyatt.

Put $4x$ = keeping, $2a = 26$, and $b = 60$: Then, per quest. $4x + 2a - b = a + x$; whence $x = \frac{1}{3}(b-a) = 15$s. 8d. Consequently $4x$ = 3l. 2s. 8d. = the charge of keeping, and $a + x$ = 1l. 8s. 8d. = the money lost.

The same answered by Mr. Phil. Williams.

Let $a = 26$, $b = 60$, and x = the keeping; then $a + x - b = \frac{a}{2} + \frac{x}{4}$ (per quest.) whence $x = \frac{4b-2a}{3} = 62\frac{2}{3}$.

II. QUESTION 377, *by Miss* Maria A—t—s—n.

There are three cities, A, B, and C, lying in the same road; whereof the first is 136 miles distant from the second, and the second 104 miles distant from the third: From A to B a courier travelled in two days; and from B to C in two days more, diminishing his distance every day alike, from the first to the last. What number of miles did he travel each particular day?

Answered by Mr. Samuel Kolt.

Let $2a = 136$, $2b = 104$, and $2x =$ the common difference of each day's journey; then $a + x$, $a - x$, $b + x$, and $b - x$ will be the respective distances travelled each day: But the first + the third = twice the second, that is $a + b + 2x = 2a - 2x$; whence $4x = a - b$, and $x = \frac{1}{4}(a - b) = 4$: Therefore 72, 64, 56, and 48, are the four distances required.

The same answered by Mr. W. Gawthorpe.

Put $x =$ the first day's journey, and $y =$ the common difference: Then $2x - y = 136$, and $2x - 5y = 104$ (per quest.); and by subtracting the latter equation from the former $4y = 32$: Whence $y = 8$, and $x = 72$.

III. QUESTION 378, *by Mr.* Charles Tate.

My wife's a scold, a niggard, and a slut,
And ev'ry day she's sure to pay my scott;
And yet for what, no mortal e'er can tell,
Unless her courage rise from living well:
The which to tame, that I may live in quiet;
I am resolv'd henceforth to stint her diet,
In quantity, to what it was before,
As e to a; which, gentlemen, explore,
From the equations * that you see subjoin'd:
Else come and take my place—if you've a mind.

* Given $(aa + ee) \times \frac{e}{a} = b = 83{\cdot}2$

$$(aa - ee) \times \frac{a}{e} = c = 1920.$$

Answered by J. Milbourn.

Multiplying the given equations crosswise into each other, we have $(aa - ee)\frac{ba}{e} = (aa + ee)\frac{ce}{a}$; whence $a^4 - \frac{b + c}{b}eeaa = \frac{ce^4}{b}$;

and, by completing the square, &c. $aa = ee \times \frac{b+c+\sqrt{(bb+6bc+cc)}}{2b}$ $= ee \times 25$; consequently, $a = 5e$: Whence, by substitution, &c. $e = 4$, and $a = 20$.

The same answered by Mr. Tho. Todd.

Put $\sqrt{xy} = a$, and $\sqrt{\frac{x}{y}} = e$; then $\frac{1}{y} = \frac{e}{a}$, and $y = \frac{a}{e}$; whence, by substitution, $\left(xy + \frac{x}{y}\right)\frac{1}{y} = b$, and $\left(xy - \frac{x}{y}\right) y = c$: From the first of which we get $yy = \frac{x}{b-x} = \frac{c+x}{x}$, by the second.

Hence $x = \sqrt{\left(\frac{bc}{2} + \frac{(c-b)^2}{16}\right)} + \frac{b-c}{4} = 4$. Therefore $y = 5$, $a = 20$, and $e = 4$.

The same answered by Mr. W. Enefer.

Let $re = a$; then, by substitution, $(rree + ee)\frac{1}{r} = b$, and $(rree - ee)\, r = c$: By multiplying these equations crosswise, we have $br^4 - br^2 - cr^2 = c$, or (putting $1 + \frac{c}{b} = 2m$, and $\frac{c}{b} = n$) $r^4 - 2mr^2 = n$; whence $r = \sqrt{(m + \sqrt{(n + mm)})} = 5$; and from thence $e \left(= \sqrt{\frac{br}{rr+1}}\right) = 4$, and $a (= re) = 20$.

The answer by Sylvius *is omitted being very nearly the same as that by Mr.* Enefer.

IV. QUESTION 379, *by Mr.* John Morland.

Two persons, A and B, having an equal claim to an annuity of 100l. to continue for 30 years, agree to share it between them in this manner, *viz.* A for his part is to enjoy the whole annuity for the first 10 years; B and his heirs being to have the entire reversion thereof for the remaining 20 years. The question is, To find the rate of interest allowed in this contract, with the present value of the annuity corresponding?

Answered by Mr. W. Kingston.

Let $x =$ the rate of interest, and $u = 100$: Then per Ward's theorem, we have $\frac{u}{x-1} - \frac{u}{x^{10} \times (x-1)}$ = present worth for 10 years, and $\frac{u}{x-1} - \frac{u}{x^{30} \times (x-1)}$ = present worth for 30 years; hence $\frac{x^{30}-1}{x^{30}} = 2 \times \frac{x^{10}-1}{x^{10}}$, or $2x^{20} - x^{30} = 1$; which, solved, gives $x = 1{\cdot}049298$; from which the required value of the annuity comes out 159*l*. 13*s*.

The same answered by Mr. Hugh Brown.

Let R be the amount of 1*l*. in one year; then, by the question and the doctrine of annuities, we have $2 \times \frac{1-R^{-10}}{R-1} = \frac{1-R^{-30}}{R-1}$; $\therefore R^{30} - 2R^{20} + 1 = 0$; which divided by $R^{10} - 1$, gives $R^{20} - R^{10} - 1 = 0$; Whence $R = \sqrt[10]{\frac{1+\sqrt{5}}{2}} = 1{\cdot}049297$, &c. and the present value sought = 159*l*. 13*s*. $0\frac{1}{2}$*d*.

V. QUESTION 380, *by* W. T—t.

The sum of the squares of the two diagonals, of any trapezium, together with the square of twice the line joining their middle points, is equal to the sum of the squares of all the four sides of the trapezium. A demonstration of this is required?

Answered by Mr. J. Randles, *Teacher of Mathematics, at Wem, in Shropshire.*

Let the two diagonals AC and DG be bisected in B and F: Then, by the 12th of 2d of Simpson's Geometry, $AD^2 + CD^2 = 2AB^2 + 2BD^2$, and $AG^2 + GC^2 = 2AB^2 + 2BG^2$, and consequently $AD^2 + CD^2 + AG^2 + GC^2 = 4AB^2 + 2BD^2 + 2BG^2$; but DG by hyp. is also bisected by BF, and so $BD^2 + BG^2 = 2DF^2 + 2BF^2$, or $2BD^2 + 2BG^2 = 4DF^2 + 4BF^2$; whence by equal substitution, $AD^2 + CD^2 + AG^2 + GC^2 (= 4AB^2 + 4DF^2 + 4BF^2) = AC^2 + DG^2 + 2BF^2$.

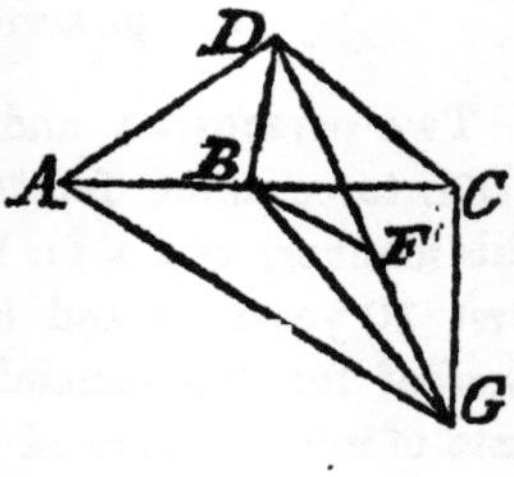

VI. QUESTION 381, *by* Bathonius.

Two ships sail, at the same time, from two ports under the same meridian, whose difference of latitude is 1° 25′. That from the southermost port runs due east at the rate of $4\frac{1}{2}$ miles per hour;

and that from the northermost E.S.E. at the rate of 7 miles per hour: I demand the distance sailed by each ship, when they are at their nearest distance from each other, and also what the distance will be?

Answered by Mr. J. Ash.

Let A and B represent the two ports, a the distance between them $=$ 85 miles, m and n the sine and cosine of the angle EBN, $q =$ the given ratio, and $x =$ BN $=$ the distance sailed by one of the ships; then $mx =$ EN, $nx =$ EB, and $qx =$ AM; whence $a - nx =$ AE $=$ hM, and $mx - qx$ (which let $= px$) $= h$N, and then $ppxx + aa - 2nax + nnxx =$ MN^2, a minimum; which, put into fluxions and reduced, gives $x = an \div (pp + nn) = 144 \cdot 3$; whence AM $= 92 \cdot 764$, and MN (the nearest distance) $= 50 \cdot 311$.

The same by Mr. R. Younge, *Teacher of Mathematics, Chester.*

Construction. Supposing the two courses to intersect in C, take CD to CB in the given ratio of the celerity in AC to that in BC; and having drawn BD, make AR perpendicular thereto; make also RN parallel to AC, and NM parallel to AR, so shall M and N represent the required places of the two ships when they are the nearest possible to each other.

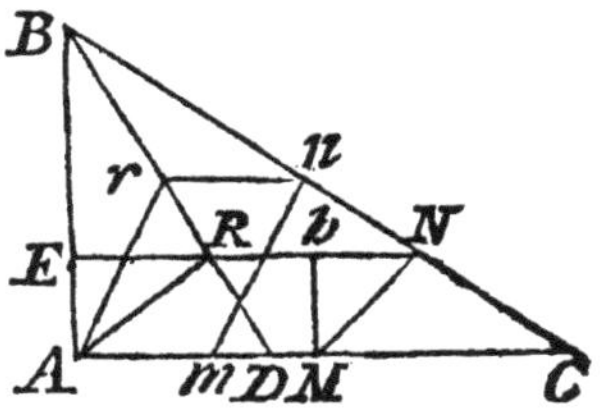

Demonstration. From any point r in BD, draw rA and rn, the latter parallel to AC; also draw nm parallel to Ar. By construction and similar triangles, Bn : nr (Am) :: BC : CD :: the velocity in BC : velocity in AC; whence it is evident, that n and m are always contemporary positions: But Ar, or its equal mn, is, evidently, the least possible when Ar coincides with AR, or when mn = MN = AR.

Calculation. As $7 + 4\frac{1}{2} : 7 - 4\frac{1}{2}$ :: co-tang. $\frac{1}{2}$C (11° 15′) : tang. $\frac{1}{2}$(CDB — CBD) $=$ 47° 32′; whence CBD $=$ 31° 13′, ABR $=$ 36° 17′, AR ($=$ MN) $= 50 \cdot 302$, RN ($=$ AM) $= 92 \cdot 79$, and BN $= 144 \cdot 34$.

VII. QUESTION 382, *by Mr* Thomas Moss.

To determine the least equilateral triangle that can be circumscribed about a given triangle, whereof the three sides are 8, 10, and 12 inches?

Answered by the Proposer, Mr. Moss.

It is evident, that one side AE of the required triangle AED must

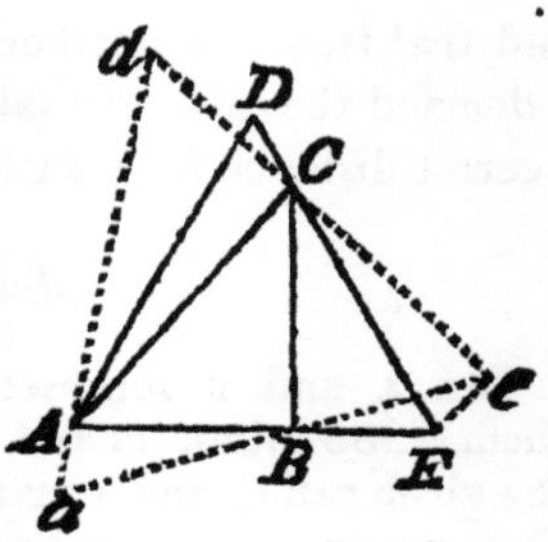

fall upon, or coincide with, one side AB of the given triangle ABC; for, if another equilateral triangle *aed* be described about ABC, near to the former, the side thereof (*ae*) will be greater than the side AE of the former, because both the angles *a*AB and BE*e* being obtuse, B*a* will be greater than BA and B*e* greater than BE, and consequently *ae* (B*a* + B*e*) greater than AE (BA + BE).

Now, all the three sides of the triangle ABC being given, the angle BAC may be found: Then it will be as the S. E (60°) : AC (12) :: S. ACE (120° — BAC) : AE; which quantity (as the two first terms are given) will be the least when 120° — BAC is the least, that is, when BAC is that angle of the given triangle which is nearest (below) the angle of the equilateral one. In the present case, the angle BAC (opposite to the mean side) being 56° 46′, AE will be found 12·48, and the area of the triangle 67·43 square inches.

Note. An equilateral triangle constituted on the longest of the three given lines as a base, will be less than that above determined: but cannot be said to be described about the given triangle (as the question requires) since all the angles of the latter are not situate in the sides of the former.

VIII. QUESTION 383, *by* Anthony Shallow, *Esq.*

To draw a right line parallel to a given line, which may cut three other lines given by position, in such sort, that the rectangle under the two parts thereof, intercepted by those lines, may be given in magnitude?

Answered by Mr. H. Watson.

Let B, A, and C be the points of intersection of the three given lines QBQ, RCR, and PAP; and let *cc* be the line to which the required line LMN is to be parallel; moreover, let ST be the side of a square, equal in area to the rectangle which is to be contained under the parts LM, MN, of the said line.

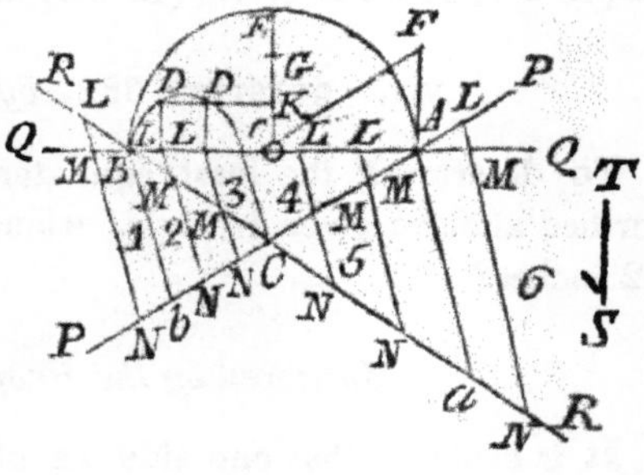

It is evident, that the problem (unless it be restrained by giving ST too large) will admit of six answers, or different positions of the line LN: But the determination of all these depends upon two cases only.

Construction of Case 1. (No. 1.) Upon AB let a semi-circle be described, and at c erect the perpendicular cE, meeting the circumference in E; make AF also perpendicular to BA and equal to a fourth proportional to Cc, cE, and ST; and, from F to the centre O, draw OF; take OM $=$ OF, and through M draw LN parallel to Cc, and the thing is done.

Demonstration. It is evident, (Euc. 6. 2.) that AM $\times$ BM ($=$ OM^2 — OA^2 $=$ OF^2 — OA^2) $=$ AF^2: And, by similar triangles, ML : BM :: Cc : Bc and MN : AM :: Cc : Ac; therefore ML $\times$ MN : AM $\times$ BM (AF^2) :: Cc^2 : Bc $\times$ Ac (cE^2). But (by construction) Cc^2 : cE^2 :: ST^2 : AF^2; whence, by equality, ML $\times$ MN : AF^2 :: ST^2 : AF^2, and consequently ML $\times$ MN $=$ ST^2.

Construction of Case 2. (No. 2 and 3.) Draw Bb parallel to Cc, meeting PAP in b; upon Bc let a semi-circle be described; and in cE, perpendicular to Bc, take cG a mean proportional between Bb and Cc, and take cK a fourth proportional to cG, Bc, and ST; draw KD parallel to BA, cutting the circumference of the circle in D, D; also draw DL perpendicular to BA, and then draw LMN parallel to Cc.

Demonstration. By similar triangles, LM : BL :: Cc : Bc and MN : cL :: Bb : Bc; therefore LM $\times$ MN : BL $\times$ cL (LD^2) :: Cc $\times$ Bb (cG^2) : Bc^2. But (by constr.) cG^2 : Bc^2 :: ST^2 : cK^2 (LD^2); whence, by equality, LM $\times$ MN : LD^2 :: ST^2 : LD^2; and consequently LM $\times$ MN $=$ ST^2.

The sixth position of LN has the very same construction with the first; and the fourth and fifth are determined like the second and third: But these four, when ST exceeds a certain limit, are impossible.——Besides the two cases above, there are other particular ones; such as, when all the three lines meet in a point, or when two of them are parallel, &c. But these, being much easier than the above, need not, I presume, be insisted on.

*** *In the fig. for* 6E *read* cE.

IX. QUESTION 384, *by Mr.* Tho. Moss.

Sailing due north, at the rate of 4 knots, in a current, a certain small island bore E.N.E. from us, at the distance of 40 miles: After running 12 miles (by the log) it bore due east; and having run 16 miles more, upon the same course, its bearing was then found to be S.E. To determine, from these observations, the direction and velocity of the current?

Answered by Mr. Rich. Gibbons.

Construction. By the first bearing and distance lay down the right-angled triangle SMK, so that M may represent the island, and S the ship, which, at the second observation, must be in the east and west line KM: Produce SK to R, so that KR = KM; and draw MR, in which the ship must be at the last observation. Moreover, as the given distances are in the ratio of 3 to 4, make KN = $\frac{4}{3}$ of SK; and draw NO parallel to KM, meeting RM in O, which is the ship's true place. From this construction (taking SL = 28) the distance LO, run by the current in seven hours, is found to be 18·25, and its direction (NLO) = 64° 58′.

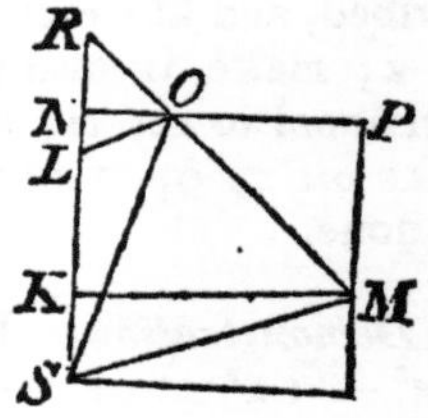

The same answered by Mr. S. Bamfield.

Let S be the place of the ship at the first observation, SN the meridian, M the island, and MK a perpendicular to SN, &c. Then (per trig.) MK (= NP) = 36 95, SK = 15·31; which, per log is but 12; say therefore 12 : 15·31 :: 16 : 20·413 = KN = MP) = PO (because the angle POM = 45°). Hence SN = 35·723, LN (= 35·723 — 28) = 7·723, and ON (= NP — OP) = 16·51. Hence, per trig. SO = 39·36 = the true distance sailed, NLO = 64° 59′ = the current's direction, and LO = 18·258 = the distance run by the current in seven hours, being at the rate of 2·608 miles per hour.

X. QUESTION 385, *by* W. T—t.

The vertical angle of a triangle being = 70°, and the sum of the two including sides = 100 feet; to determine the triangle itself, when the perpendicular is a mean proportional between the whole base and one of its two segments?

Answered by Mr. Harland Widd, *of Whitby.*

Let c = tang. 70° = ACB, y = AB, and x = AD; then, by the question, $\sqrt{xy}$ = CD; and (per trig.) $\sqrt{xy}$: 1 :: x : $\frac{x}{\sqrt{xy}}$ = tangent ACD; also $\sqrt{xy}$: 1 :: $y - x$: $\frac{y-x}{\sqrt{xy}}$ = tang. BCD: Whence, by the known theor. for the tangent of the sum of two angles, we have $\frac{yy}{x\sqrt{xy}} = c$; and therefore $y = x\sqrt[3]{cc} = rx$. Again, (per Euc. 47. I.) $\sqrt{(xx + rxx)} + \sqrt{(xx - rxx + rrxx)}$ (= AC + BC) =

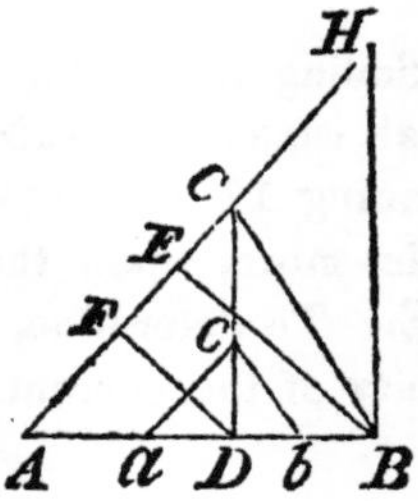

100; whence $x = \dfrac{100}{\sqrt{(1+r)} + \sqrt{(1-r+rr)}} = 29{\cdot}24$, and y $(= rx) = 57{\cdot}36 =$ AB.

The same answered by Sylvius.

Let the tang. ACB (70°) $= t$; and in acb, similar to ACB, assume CD $= 1$, and aD $= x$; then will $ab = \dfrac{1}{x}$ (per quest.) and consequently D$b = \dfrac{1}{x} - x = \dfrac{1-xx}{x}$: But $\dfrac{ab}{1 - a\text{D} \times \text{D}b} = t$ (= tang. acb), that is, $\dfrac{1}{x^3} = t$; whence $x = 0{\cdot}713984 =$ tang. 35° 31′ 34″ $= a$CD: From which the angle bCD $=$ 34° 28′ 20″; and it will be $ca + cb$ (the sum of the secants of these two angles) : AC + BC : : ac : AC $= 50{\cdot}3215$; whence the rest are easily found.

The same answered by Mr. Henry Watson.

Upon the side AC, of the required triangle ABC, conceive the perpendiculars BE and DF let fall; then, per similar triangles, AC : AD : : AB : AE and AC : CD : : CD : CF. From whence, as the rectangles under the means of both proportions are equal (per quest.) we have AE = CF, and consequently CE = AF. But AF (CE) : DF : : AD : CD, and DF : BE : : AD : AB; Therefore, by composition, CE : BE : : AD^2 : AB × CD : : AD^3 : AD × AB × CD $(= \text{CD}^3)$; whence $\dfrac{\text{BE}}{\text{CE}} = \dfrac{\text{CD}^3}{\text{AD}^3}$, and consequently $\frac{1}{3}$ log. tang. ECB ($\frac{1}{3}$ log. $\dfrac{\text{BE}}{\text{CE}} =$ log. $\dfrac{\text{CD}}{\text{AD}}$) $=$ log. tang. of A $=$ 54° 28′ 25″. Now, if in AC produced, there be taken CH = BC, and B, H be joined, in the triangle ABH will be given the side AH and all the angles; whence AB $= 57{\cdot}36$, BC $= 49{\cdot}67$, and AC $= 50{\cdot}32$.

XI. QUESTION 386, *by Mr.* Timothy Doodle.

Within a rectangular garden, containing just an acre of ground, I have a circular fountain, whose circumference is 28, 40, 52, and 60 yards distant from the four angles of the garden. From these dimensions the length and breadth of the garden, and likewise the diameter of the fountain, are required?

Answered by Mr. William Cottam, *at his Grace the Duke of Norfolk's.*

Let Bm, Ep, Dq, and An, (= 60, 52, 28, and 40) be denoted by m, p, q, and n, respectively; and let the radius of the fountain $= x$: Then

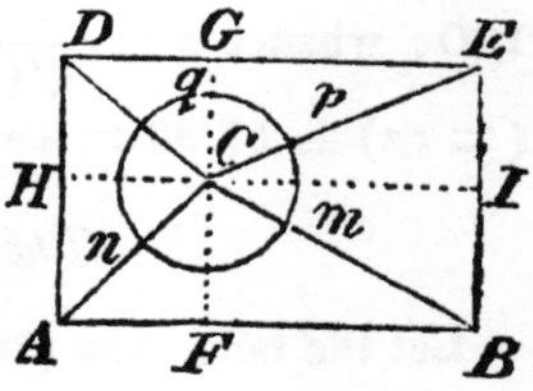

(per figure) $(m+x)^2 - (p+x)^2$ (= $BI^2 - EI^2 = AH^2 - DH^2$) $= (n+x)^2 - (q+x)^2$; whence $x = \frac{1}{2} \times \frac{m^2 + q^2 - p^2 - n^2}{p + n - m - q} = 10$. Now let m, p, q, n, (= 70, 62, 38, and 50) = BC, EC, DC, and CA, respectively, and let A = 4840 = the area of an acre in yards; also let x, now = GC; and then we have $(\sqrt{(pp - xx)} + \sqrt{(qq - xx)})\,(x + \sqrt{(nn - qq + xx)}) = A$; which, solved, gives $x = 15{\cdot}11145$: Hence DE = AB = 94·99631, and AD = BE = 50·94936.

The same answered by Mr. Hugh Brown.

It is manifest that the radius of the pond must be 10; because $BC^2 + DC^2 = EC^2 + AC^2$; consequently BC = 70 (= a), EC = 62 (= b), AC = 50 (= c), DC = 38 (= d). If through C, the centre of the pond, HI and FG be drawn parallel to the sides of the rectangle, and there be put AB = x, and AD = y; then will AF $= \frac{x}{2} - \frac{aa - cc}{2x}$, and AH $= \frac{y}{2} + \frac{cc - dd}{2y}$; from whence, and the question, we shall (because $AF^2 + AH^2 = AC^2$) have the two following equations $\frac{xx}{4} + \frac{yy}{4} + \frac{(aa - cc)^2}{4xx} + \frac{(cc - dd)^2}{4yy} = \frac{aa}{2} + \frac{dd}{2}$, and $xy = 4840 = A$. Let, now, the first equation be multiplied by 4, and $\frac{A}{x}$ substituted therein for y; so shall $xx + \frac{A^2}{xx} + \frac{(aa - cc)^2}{xx} + \frac{(cc - dd)^2}{A^2} xx = 2aa + 2dd$; and consequently $\left(1 + \frac{(cc - dd)^2}{A^2}\right) x^4 - 2(aa + dd)\,xx = -A^2 - (aa - cc)^2$: Put $1 + \frac{(cc - dd)^2}{A^2} = g$, $aa + dd = h$, $A^2 + (aa - cc)^2 = k$; and the equation will stand thus, $gx^4 - 2hx^2 = -k$; whence $x = \sqrt{\frac{h + \sqrt{(hh - gk)}}{g}} = 94{\cdot}9961$, and $y = \frac{A}{x} = 50{\cdot}9494$.

Additional Solution, taken from the London Magazine Improved, for 1784, where this question was proposed and answered, as follows, by Mr. Isaac Dalby.

QUESTION. Having given the area of a rectangle, and the lengths of four right lines drawn from its angles to a point within it; to determine the rectangle by construction.

Solution.

Analysis. Suppose ACEG to be the rectangle, P the point, and PA, PC, PE, PG the given lines; draw HD parallel to AC, and FB parallel to GA; join HB, BD, DF, FH. Then because the diagonals of parallelograms are equal, HB, BD, DE, FH will be respectively equal to the given lines; and, consequently, the trapezium HBDF is = half the rectangle ACEG: the problem is, therefore,

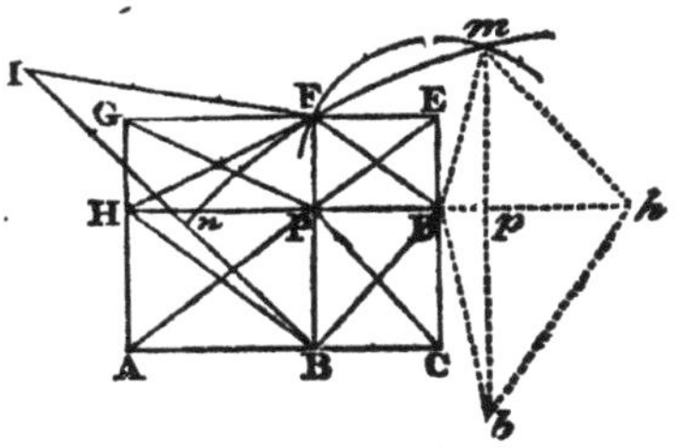

To make a trapezium of a given magnitude, with the sides given so, that the diagonals may intersect each other at right angles.

Perpendicular to one of the sides BD draw BI, and make the angle BFI = HBD; then, because the angle PBD is the comp. of each of the angles PDB, PBI to a right one, they must, therefore be equal; and, by construction, the angle BFI = HBD, hence BD : DH :: BF : BI, and BI × BD = DH × BF; that is, BI × BD is = to the area of the rectangle GC: but BD is given; consequently BI is given. And, because HB, BD are given, the ratio IF, FB is given, and hence we have this

Construction. Make the rectangle BI × BD = the given rectangle, and divide BI, in *n*, in the given ratio of HB to BD. Then, by the lemma at page 336, Simpson's Algebra, describe the arc *nm* so that lines from B and I, to meet in that arc, may be in that ratio. From D, as a centre, with the radius DF (= PE) describe an arc F*m*, cutting the former arc in F and *m*; join DF, D*m*, FB; and draw *mb* parallel, and HD*h* perpendicular to FB; take also D*b* = DB; then, if BH, FH, *bh*, *mh* be made equal to the other sides of the trapezium, DH, FB and D*h*, *mb* will be the sides of two rectangles answering the conditions of the problem. For it is well known that if the diagonals of a trapezium intersect at right angles, the sum of the squares of the opposite sides are equal, and the contrary; therefore, howsoever the angles are varied, if the sides are connected in the same order, the diagonals will intersect at right angles; whence the construction is manifest.

If the arcs *nm*, F*m* touch, instead of intersect, the problem evidently admits of but one answer; and, in that case, the area of the rectangle will be a *maximum*, and a circle will circumscribe the trapezium; which circle, and consequently the rectangle may be determined thus: *Make either of the two opposite sides of the trapezium the legs of a right-angled triangle, then a circle described about that triangle will be the circle required.*

XII. QUESTION 387, *by Mr.* Patrick O'Cavanah, *of Dublin.*

In the latitude of 51° 32′ north, stands two pillars S.W. and N.E. of one another, at the distance of 200 feet: The height of the southermost pillar is 60 feet, and that of the northermost 40 feet. At what time of the day, on June 20, do the shadows of their summits approach the nearest to each other?

Answered by Mr. W. Bevil.

Let P and W represent the places of the two pillars, whose given heights (40 and 60) let be denoted by a and b, respectively; then, supposing $v =$ the co-tangent of the sun's altitude, the lengths of their respective shades, DP and WG, will be av and bv: Draw DH parallel to PW, and WE perpendicular to the two meridian lines WA and BPE, putting $c =$ DH ($=$ PW) $= 200$, $f = b - a$, $s =$ sine $45° =$ EWP $=$ AWP, $z =$ sine of AWG ($=$ the sun's azimuth from the south), and $u =$ its cosine; then GH ($= bv - av$) $= fv$, and $sz + su =$ cosine of GWP $=$ GHD: Therefore $HD^2 + HG^2 - GH \times HD \times 2 \cos. \angle H = GD^2 = c^2 + f^2v^2 - 2csfvz - 2scfvu$, a minimum: Put $p = f \div 2cs$, $q = c \div 2fs$, then $q + pv^2 - vz - vu$ a minimum; let e and $d =$ the sine and cosine of the latitude, $n =$ the sine of the sun's declination, and x and y the sine and cosine of his altitude, then we have $q + \frac{py^2}{x^2} - \frac{y}{x}\sqrt{(1 - \left(\frac{ex-n}{dy}\right)^2)}$ $- \frac{y}{x} \times \frac{ex-n}{dy}$; which, by reduction, (putting $-dp + dq - e = r$, $ee + dd = 1$, and $dd - nn = h$) becomes $\frac{dp + rx^2 + nx - x\sqrt{(h - x^2 + 2enx)}}{dx^2}$. This, thrown into fluxions and reduced, gives $(2dp + nx)\sqrt{(h - x^2 + 2enx)} = hx + enx^2$; whence $x = \cdot 39359048 =$ sine of 23° 10′ 41″, the sun's altitude; and from thence the time of the day is found $5^h\ 27^m\ 3^s$ in the afternoon.

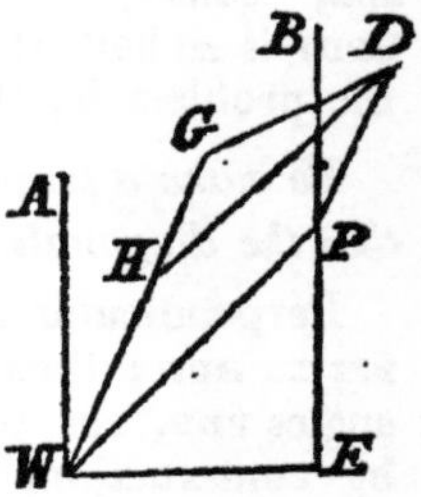

The same answered by Mr. T. Moss.

Let A be the place of the southermost pillar, and B that of the northermost; and suppose AE and BF to be any contemporary positions of the two shadows, taken as parallel: Then, if BG be made parallel to FE, it will be equal to FE, and GE likewise $=$ BF. But AE is to BF as the height of the pillar A, is to that of the pillar B; whence AG must be to AE in the constant ratio of the difference of the said heights to the altitude of the pillar at A: So that the path HG of the point G will be exactly similar to that of the point E, and is moreover the very same as would be described by the shadow of an object at A, whose height is the excess of the height of the pillar at A above the pillar at B: Which path being an hyperbola, suppose O to be its centre, and upon AOC produced let fall the perpendiculars GM and BI, producing BG to meet IA in N. Then, AB being equal 200, and the angle IAB = IBA, we have AI (= BI) = 141·42: Moreover, from the sun's meridian altitude, the length of the meridional shadow AH (supposing the height of the object to be 20) is given = 10·656: And (by art. 467, of Simpson's Fluxions) it will be as, the rectangle of the sines of the sun's altitude at noon and depression at midnight, is to the rectangle of the radius and the sine of twice the sun's declination, so is (20) the height of the object projecting the shadow, to (64·072) the transverse axis of the described hyperbola: Also, as the square root of the former of the said rectangles, is to the cosine of the sun's declination, so is the height of the object, to (38·402) the semi-conjugate axis of the hyperbola. Put now a (= 32·036) = OH, b = 38·402, c (= 98·728) = OI, and d (= 141·42) = BI; and let OM = az.

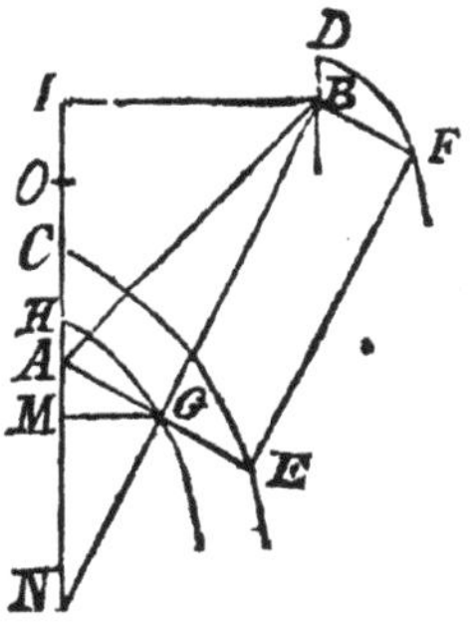

Then, by the property of the hyperbola, $MG = \frac{b}{a}\sqrt{(aazz - aa)} = b\sqrt{(zz-1)}$, and $MN = \frac{bb}{aa} \times az = \frac{bbz}{a}$ (since it is evident that BG, to be the shortest possible, must fall upon the curve at right angles): Hence, because of the sim. $\triangle$s, we have $\frac{MN}{MG} = \frac{NI}{BI}$; that is,

$$\frac{bz}{a\sqrt{(zz-1)}} = \frac{c + \left(a + \frac{bb}{a}\right)z}{d}, \text{ or } \frac{z}{\sqrt{(zz-1)}} = \frac{ac}{bd} + \left(\frac{aa}{bd} + \frac{b}{d}\right)z;$$

which, in numb. becomes $\frac{z}{\sqrt{(zz-1)}} = 0·58237 + 0·46052z$; whence $z = 1·56204$; OM (= az) = 50·04; AM = 7·348; MG (= b × tang. whose secant is z) = 46·083: and the angle MAG (= the sun's azimuth from the north) = 80° 56′ 27″; from which the required time is found to be 5h 26m 58s afternoon.

XIII. QUESTION 388, *by Mr.* Timothy Doodle.

Supposing p, q, r, s, t, &c. to represent the tangents of any number of arcs P, Q, R, S, T, &c. equal, or unequal: To determine a

general expression for the tangent of the sum (P + Q + R + S + T + &c.) of all those arcs; the common radius being unity?

Answered by Mr. **E. Rollinson.**

In order to give a general solution to this problem, it will be proper to premise the following

Lemma. **If** $\frac{\dot{p}}{1-mpp} + \frac{\dot{q}}{1-mqq} + \frac{\dot{r}}{1-mrr} + \frac{\dot{s}}{1-mss}$**, &c.** $= \frac{\dot{x}}{1-mxx}$; ***wherein* m *is constant, and* p, q, r, s, *&c. variable; then, if the sum of all the quantities* p, q, r, s, *&c. be denoted by* A, *the sum of all their rectangles by* B, *the sum of all their solids by* C, *&c. I say, that***
$x = \frac{A + mC + m^2E + m^3G + m^4I, \&c.}{1 + mB + m^2D + m^3F + m^4H, \&c.}$. **For, by taking the fluent, we have hyp. log.** $\frac{1+m^{\frac{1}{2}}p}{1-m^{\frac{1}{2}}p}$ **+ hyp. log.** $\frac{1+m^{\frac{1}{2}}q}{1-m^{\frac{1}{2}}q}$ **+ &c. = hyp. log.** $\frac{1+m^{\frac{1}{2}}x}{1-m^{\frac{1}{2}}x}$**; and consequently** $\frac{1+m^{\frac{1}{2}}p}{1-m^{\frac{1}{2}}p} \times \frac{1+m^{\frac{1}{2}}q}{1-m^{\frac{1}{2}}q}$**, &c.** $= \frac{1+m^{\frac{1}{2}}x}{1-m^{\frac{1}{2}}x}$**. Put this value of** $\frac{1+m^{\frac{1}{2}}x}{1-m^{\frac{1}{2}}x} = Q$**; whence** x **will be found** $= \frac{Q-1}{m^{\frac{1}{2}} \times (Q+1)} =$

$$\frac{(1+m^{\frac{1}{2}}p)(1+m^{\frac{1}{2}}q)(1+m^{\frac{1}{2}}r),\&c. - (1-m^{\frac{1}{2}}p)(1-m^{\frac{1}{2}}q)(1-m^{\frac{1}{2}}r),\&c.}{m^{\frac{1}{2}}(1+m^{\frac{1}{2}}p)(1+m^{\frac{1}{2}}q)(1+m^{\frac{1}{2}}r),\&c. + m^{\frac{1}{2}}(1-m^{\frac{1}{2}}p)(1-m^{\frac{1}{2}}q)(1-m^{\frac{1}{2}}r),\&c.}$$

(by substituting the value of Q): But, by multiplication, $(1+m^{\frac{1}{2}}p)(1+m^{\frac{1}{2}}q)(1+m^{\frac{1}{2}}r)$**, &c.** $= 1 + m^{\frac{1}{2}} \times (p+q+r)$**, &c.** $+ m \times (pq + pr)$**, &c.** $= 1 + m^{\frac{1}{2}}A + mB + m^{\frac{3}{2}}C$**, &c. &c. Hence our equation becomes** $x = \frac{A + mC + m^2E + m^3G, \&c.}{1 + mB + m^2D + m^3F \&c.}$. ***Q. E. D.***

If $m = -1$**, the given equation will become** $\frac{\dot{p}}{1+pp} + \frac{\dot{q}}{1+qq} + \frac{\dot{r}}{1+rr}$**, &c.** $= \frac{\dot{x}}{1+xx}$**; and the value of** $x = \frac{A - C + E - G, \&c.}{1 - B + D - F, \&c.}$.

Now, to apply this to the question proposed, let the arcs P, Q, R, &c. and their tangents p, q, r, **&c. be considered as in a flowing**

state; and let x be the required tangent of P + Q + R, &c. Then, it being known that $\dot{P} = \frac{\dot{p}}{1+pp}$, $\dot{Q} = \frac{\dot{q}}{1+qq}$, &c. we thence have $\frac{\dot{p}}{1+pp} + \frac{\dot{q}}{1+qq} + \frac{\dot{r}}{1+rr}$, &c. $= \frac{\dot{x}}{1+xx}$; and consequently $x = \frac{A - C + E - G + I, \&c.}{1 - B + D - F + H, \&c.}$, by the preceding lemma; A being the sum of all the tangents p, q, r, s, &c. B the sum of all their rectangles, C the sum of all their solids, &c. &c.

Corollary. If all the arcs P, Q, R, &c. are equal, and their number be denoted by n; then will $A = np$, $B = n \cdot \frac{n-1}{2}p^2$, $C = n \cdot \frac{n-1}{2} \cdot \frac{n-2}{3}p^3$, &c. and therefore $x =$

$$\frac{np - n \cdot \frac{n-1}{2} \cdot \frac{n-2}{3}p^3 + n \cdot \frac{n-1}{2} \cdot \frac{n-2}{3} \cdot \frac{n-3}{4} \cdot \frac{n-4}{5}p^5, \&c.}{1 - n \cdot \frac{n-1}{2}p^2 + n \cdot \frac{n-1}{2} \cdot \frac{n-2}{3} \cdot \frac{n-3}{4}p^4, \&c.}$$

The same answered by Mr. W. *Bevil.*

It is proved, by the writers on trigonometry, that *the tangent of the sum of two arcs (the radius being unity) is equal to the sum of the tangents of those arcs divided by the excess of the square of the radius above their rectangle or product.*

Hence the tang. of P + Q will be $\frac{p+q}{1-pq}$; and, if P + Q be considered as one arc, then the tangent of (P + Q) + R will, by the same rule, be $= \frac{\frac{p+q}{1-pq} + r}{1 - \frac{pr+qr}{1-pq}} = \frac{p+q+r-pqr}{1-pq-pr-qr}$: After the same manner the tangent of P + Q + R + S is found to be $= \frac{p+q+r+s-pqr-pqs-prs-qrs}{1-pq-pr-ps-qr-qs-rs+pqrs}$: And thus, by carrying on the process a step or two farther, the law of continuation will appear manifest; being such, that, if the sum of all the given tangents be denoted by A, the sum of all their rectangles by B, the sum of all their solids by C, &c. then will the tangent of the sum of all the arcs be $\frac{A - C + E - G, \&c.}{1 - B + D - F, \&c.}$.

XIV. QUESTION 389, *by Mr.* E. R——n.

To determine the ratio of the densities of the sun and earth, independent of the sun's parallax?

Answered by the Proposer, Mr. E. Rollinson.

Let R and r be the semi-diameters of the orbits of the earth and moon, P and p the periodic times in those orbits, S and s the sun's mean apparent semi-diameter and moon's mean horizontal parallax, and N and n any two numbers in the required ratio of the densities of the sun and earth respectively. Then, the real semi-diameters of the sun and earth being in the ratio of RS to rs, their masses will be as $R^3S^3 \times N : r^3s^3 \times n$; and consequently their forces, at the distances R and r, as $R^3S^3N \div R^2 : r^3s^3n \div r^2$, or as $RS^3N : rs^3n$. But these (by the laws of central forces) are also as $R \div PP : r \div pp$; therefore, by dividing the antecedents of these equal ratios by RS^3, and the consequents by rs^3, we have as $N : n :: \frac{1}{P^2S^3} : \frac{1}{p^2s^3} :: 1 : \frac{P^2}{p^2} \times \frac{S^3}{s^3}$; which, in numbers, (taking $P = 365^d\ 5^h\ 49^m$, $p = 27^d\ 7^h\ 43^m$, $S = 16^m\ 5\frac{1}{4}^s$, and $s = 57^m\ 17\frac{1}{2}^s$) will come out as 1 to 3·957, for the ratio of the density of the sun to that of the earth.

XV. QUESTION 390, *by* Anthony Shallow, *Esq.*

Having given any three computed visible latitudes of the moon, in a solar eclipse, together with the corresponding differences of longitude of the sun and moon: To shew the manner of finding, from thence, the true time of the greatest obscuration, and likewise the nearest approach of the two centres?

Answered by Mr. J. Morland.

Construction. In any right line AI set off SA, SB, and SC, equal to the three given longitudes of the moon from the sun; and make AD, BE, and CF perpendicular to AI, so as to express the given latitudes corresponding: Then, through the three points D, E, and F, let the circumference of a circle be described; and from O the centre thereof, through S, draw the radius OP, and upon AI let fall

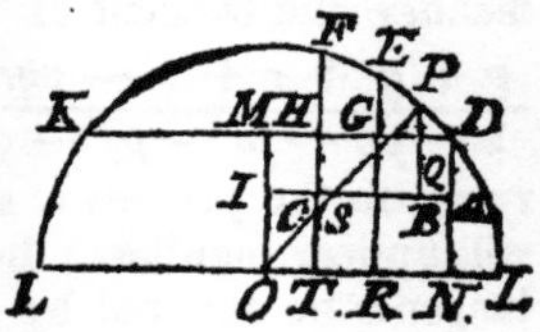

the perpendicular PQ: So shall SP be the distance of the two centres, at the time of the greatest obscuration, and SQ the required difference of longitudes at that time. For, since the circumference of the circle thus described, coincides with the real curve (whatever it is) in three points (D, E, F) which are but at a small distance from one another, it must necessarily have nearly the same degree of curvature, and therefore likewise coincide with it in the intermediate spaces, very near. To derive the numerical solution from this construction, let the chord DK be parallel to AI, and let OIM be perpendicular thereto; also let FHC and EGB be produced to meet the diameter LL, at right angles, in T and R.

Put DG (AB) $= a$, DH (AC) $= b$, EG (BE — AD) $= c$, FH (CF — AD) $= d$, OM $= x$, and MK (MD) $= y$: Then, by the property of the circle, GD $\times$ GK $=$ GE $\times$ (2GR + GE), and HD $\times$ HK $=$ HF $\times$ (2HT + HF); that is, $a(2y - a) = c(2x + c)$, and $b(2y - b) = d(2x + d)$: Whence x is found $= \frac{a(bb + dd) - b(aa + cc)}{2bc - 2ad}$, and $y = \frac{c(bb + dd) - d(aa + cc)}{2bc - 2ad}$: From which values, those of OK (OP), OI, OS, PS, SQ, and AQ will all become known.—As to the time answering to this (or any other) given value of AQ, it is best determined from the common method of interpolating by differences: According to which, the two given intervals corresponding to AB and BC being denoted by p and q, the required interval, between the position Q and the first position A, will be represented by $\frac{(pbb - qaa)\,AQ + (qa - pb)\,AQ^2}{ab(b + a)}$. See M. de Caille's Astron. p. 60.

The proposer resolves this problem by means of a parabolic curve described through three given points: And observes, 'That, if the equation $yy = g + hx + kx^2$ were to be assumed for the general relation of the latitude (y) to the difference of longitude (x), the result would come out more neat and simple than from any curve of the parabolic kind:' But adds, 'that this last method is not general, being only applicable when the moon has a considerable latitude during the whole time of the eclipse; since the assumed equation (which answers to an ellipsis or hyperbola) becomes impossible on the moon's passing from one side of the ecliptic to the other.' He observes farther, 'That the conclusions, according to either of the above methods, will seldom be found to differ by more than one minute in time from those arising from the common way of computation.' Which last he therefore thinks may be used as sufficiently near, till the theory of the moon's motion is known to a greater degree of exactness.

THE PRIZE QUESTION, *by* Anthony Shallow, *Esq.*

To determine the figure which the piers (or the starlings) of a bridge ought to have, so that the length, and greatest breadth of each, and their distances from one another, being given, the water in its passage through the bridge shall suffer the least resistance possible?

Answered by the Proposer.

Supposing cc and Ll to be the lengths, and AD and BP the semi-breadths of two adjacent piers (or starlings), let Ee, parallel to cc, bisect AB at right angles, in F; and let IH, parallel to Ee, be the direction of a particle of water impinging upon the surface CD in H; also let HM be a tangent at H, intersecting AC produced in M.

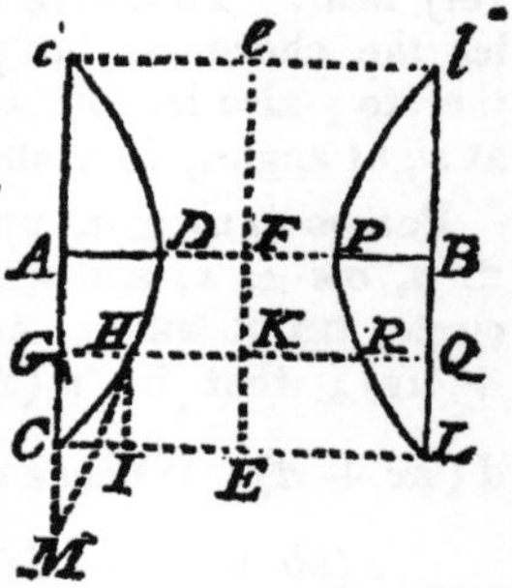

Call AC, a; AD, b; CE, c: CG, x; and GH, y; and let the tangents of the angles ECD and ADC (to the radius 1) be denoted by p and q respectively. The celerity wherewith the stream passes any section HR being inversely as the breadth, the velocity of the particle acting at H will therefore be as $\frac{1}{c-y}$,

and its force (by mechanics) as $\frac{1}{c-y} \times \frac{GH^2}{MH^2} = \frac{1}{c-y} \times \frac{\dot{y}\dot{y}}{\dot{x}\dot{x}+\dot{y}\dot{y}}$; which, drawn into $\dot{y}$, gives $\frac{1}{c-y} \times \frac{\dot{y}^3}{\dot{x}\dot{x}+\dot{y}\dot{y}}$ for the fluxion of the resistance upon GH. But (by art. 408, of Simpson's Fluxions) it appears that, if s be assumed to denote any quantity expressed in terms of y and given coefficients, the fluent of $s \times \frac{(\dot{x}\dot{x}+\dot{y}\dot{y})^n}{\dot{y}^{2n-1}}$ (corresponding to any given value of x) will be a max. or min. when the relation of x and y is such that the value of $\frac{s\dot{x} \times (\dot{x}\dot{x}+\dot{y}\dot{y})^{n-1}}{\dot{y}^{2n-1}}$ is every where the same. Therefore, by transforming our fluxion $\frac{1}{c-y} \times \frac{\dot{y}^3}{\dot{x}\dot{x}+\dot{y}\dot{y}}$, to $\frac{1}{c-y} \times \frac{(\dot{x}\dot{x}+\dot{y}\dot{y})^{-1}}{\dot{y}^{-3}}$, and then comparing it with that above, we shall get $\frac{1}{c-y} \times \frac{\dot{x}\dot{y}^3}{(\dot{x}\dot{x}+\dot{y}\dot{y})^2} = d$, (a constant quantity). In order to the resolution of this equation, put $\dot{x} = v\dot{y}$ (v being the tang. of the angle MHC), then, by substitution, &c. $\frac{1}{c-y} \times \frac{v}{(vv+1)^2} = d$; and consequently $d(c-y) = \frac{v}{(1+vv)^2}$: Which, in fluxions, gives $-d\dot{y} = \frac{\dot{v}-3v^2\dot{v}}{(1+vv)^3}$, and $dv\dot{y}\ (=d\dot{x}) = \frac{2v^3\dot{v}-v\dot{v}}{(1+vv)^3}$; whereof the corrected fluent will be $dx = \frac{1+3pp}{2(1+pp)^2} - \frac{1+3vv}{2(1+vv)^2}$,

To find d, take $y = 0$; in which circumstance, v being $= p$, the equation $\frac{1}{c-y} \times \frac{v}{(vv+1)^2} = d$ becomes $\frac{p}{c(1+pp)^2} = d$; which value being substituted for d, our two equations, after proper reduction, will become $y = c - \frac{cv(1+pp)^2}{p(1+vv)^2}$, and $x = c \times \frac{1+3pp}{2p} - c \times \frac{(1+3vv)(1+pp)^2}{2p(1+vv)^2}$.

These equations give the general relation of x, y, and v; but to apply them to any particular case proposed, something further remains to be done, since the value of p (the tangent of the angle ECH) is not given, but must be found from the known values of CE, CA, and AD. In order to this, suppose H to coincide with D; then, x becoming $= a$, $y = b$, $v = q$, if these values be substituted in the aforesaid equations, we shall, after due reduction, have $\frac{(1+pp)^2}{(1+qq)^2} = \frac{c-b}{c} \times \frac{p}{q}$, and $a = c \times \frac{1+3pp}{2p} - (c-b)\frac{1+3qq}{2q}$.

Put $r = \frac{2a}{c} + \frac{c-b}{c} \times \frac{1+3qq}{q}$, and then $p = \frac{\sqrt{(rr-12)}+r}{6}$; from which equations the values of p and q may be found. Thus, for example, if a be supposed $= 12$, $b = 5$, and $c = 8$; then p will come out $= 2$, and $q = 3$: So that the angles ECH and ADC are here 63° 26′ and 71° 34′, respectively.

Corollary. If c be supposed exceeding great, or, which comes to the same, if every particle of the fluid impinges with the same velocity, then $\frac{y}{c}$ will vanish, and the equation $\frac{y}{c} = 1 - \frac{v(1+pp)^2}{p(1+vv)^2}$ will become $\frac{v(1+pp)^2}{p(1+vv)^2} = 1$, and consequently $v = p$; therefore the angle CHI being every where the same, CD will, in this case, become a right line: From whence it appears, that the less DF is in respect of CE, the greater must be the curvature of the surface upon which the water acts.

Questions proposed in 1755, and answered in 1756.

I. QUESTION 391, *by Miss* Maria Atkinson.

(Addressed to Mr. E. P. who took the liberty to ask her age.)

Five times seven and seven times three
Add to my age, the sum will be

As many above six nines and four
As twice my years exceed a score:
From hence, sweet Sir, my age explore.

Mr. B. Lydal, *addressing himself to the Proposer, answers it thus:*

Let x denote your age, then will $56 + x - 58 = 2x - 20$; and consequently $x = 18$. A very good age for matrimony, Miss.

II. QUESTION 392, *by* Sylvius.

A ball descending by the force of gravity from the top of a tower, was observed to fall half the way in the last second of time: Required the tower's height, and the whole time of descent.

Answered by Mr. W. Stoker, *of Fatfield Staiths.*

The square roots of the distances being as the times, we have (per quest.) as $\sqrt{1}$ to $\sqrt{2}$, so is the time of falling through the first half, to the time of falling through the whole required height: And therefore as $\sqrt{2}-1$ is to $\sqrt{2}$, so is 1 second (the time of descent through the latter half) to $\sqrt{2} \div (\sqrt{2}-1) (= 2 + \sqrt{2}) = 3{\cdot}414$ the time of descent through the whole height: Whence the height itself is found $= 187{\cdot}48$ feet.

The same answered by Mr. J. Nichols.

Let $t =$ the whole time of descent, so will $t - 1 =$ the time of descent through the first half of the tower's height; and therefore (the spaces descended being always as the squares of the times) we have $tt : (t-1)^2 :: 2 : 1$; whence $tt - 2t + 1 = \frac{1}{2}tt$: From which $t = 2 + \sqrt{2} = 3{\cdot}414$, and the tower's height $= 187{\cdot}48$ feet.

The same answered by Mr. J. Boston.

Let $a = 16\frac{1}{12}$ feet (the ball's descent in the first second of time) and $x =$ the seconds the ball was falling; then (by the question and the law of the descent of heavy bodies) the tower's height will be $= ax^2$, and the half of that height $= a \times (x-1)^2$; hence we have $\frac{1}{2}ax^2 = a \times (x-1)^2$: consequently $x = 2 + \sqrt{2} = 3{\cdot}41421$; and $ax^2 = 187{\cdot}4806$ feet $=$ the tower's height.

III. QUESTION 393, *by Mr.* G. Brownbridge.
Addressed to the ingenious Mr. C. T——te.

In vain you hope to live in quiet
By stinting of your vixen's diet;
'Tis five to one that scheme wo'n't do;
But I'll exchange a wife with you,
And, if you please a daughter too,

If you their ages will make known
From what below is fairly shown.

Given $\left\{ \begin{array}{l} x^3z + xz^3 = 546560 \\ x^4 + z^4 = 1086992 \end{array} \right\}$ x and z being the two ages req.

Answered by Mr. Richard Gibbons.

Put $p = xz$, and $s = xx + zz$, $a = 546560$, and $e = 1086992$: Then, by the question $sp = a$, and $s^2 = e + 2p^2 = a^2 \div p^2$. Hence, by completing the square and reduction, p is found $= 448$, and $s (= a \div p) = 1220$. Consequently $x + z = \surd(s + 2p) = 46$, and $x - z = \surd(s - 2p) = 18$, and therefore $x = 32$, and $z = 14$.

The same answered by Mr. W. Phipps.

Putting $a = 546560$, and $b = 1086992$, we have, by the first equation, $x^2 + z^2 = a \div xz$; from the square of which take the second, so shall $2x^2z^2 = a^2 \div x^2z^2 - b$: whence $x^4z^4 + \frac{1}{2}bx^2z^2 = \frac{1}{2}aa$, and consequently $x^2z^2 = \frac{1}{4}\surd(8a^2 + b^2) - \frac{1}{4}b = 200704$, which call m^2, and then $xz = m = 448$; whence, by substitution, $x^2 + z^2 = a \div m$: From which and $xz = m$, we have

$$\left.\begin{array}{l} x \\ z \end{array}\right\} = \tfrac{1}{2}\sqrt{\left(\frac{a}{m} + 2m\right)} \pm \tfrac{1}{2}\sqrt{\left(\frac{a}{m} - 2m\right)} = \left\{ \begin{array}{l} 32 \text{ the wife's age.} \\ 14 \text{ the daughter's age.} \end{array} \right.$$

Mr. *W. Bacon* and some others, solve it by substituting for the sum and difference of the two unknown quantities. Thus, making $v + y = x$, $v - y = z$, $546560 = 2a$, and $1086992 = 2b$, the given equations are reduced to $v^4 - y^4 = a$, and $v^4 + y^4 + 6v^2y^2 = b$; and, by subtracting the former from the latter, and putting $b - a = 2c$, there comes out $3y^2v^2 = c - y^4$ or $9y^4 \times v^4$ $(= 9y^4 \times (y^4 - a)) = (c - y^4)^2$, or $8y^8 + (9a + 2c) \times y^4 = c^2$: Which equation, solved, gives $y = 9$; whence $v = 23$, $x = 32$, and $z = 14$.

IV. QUESTION 394, *by Mr.* W. Kingston.

The distance of the centres of two circles, whose diameters are 50 each, being given $= 30$; 'tis required to find the side of the square inscribed in the intersection, or space common to both circles.

Answered by Mr. Chas. Tate.

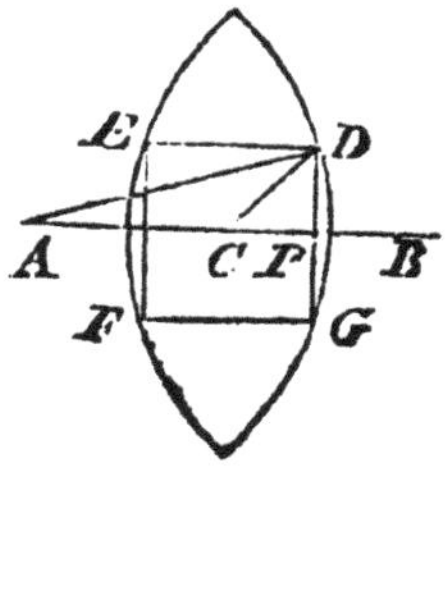

Construction. Let the given distance AB of the two centres, be bisected in C; and make the $\angle$BCD $= \frac{1}{2}$ a right angle; then a perpendicular DP, let fall from the intersection D, will evidently be half the side of the inscribed square DEFG.

Calculation. The radius AD being drawn, in the $\triangle$ACD will be given AD $= 25$; AC $= 15$, and the $\angle$ACD $= 135°$; whence DAC is found

$=19° 54'$, and from thence DP $=8\cdot5078$, the double of which 17·0156, is the side of the square.

The same algebraically by Mr. W. Harrison, *of Wigan.*

Putting the radius AD $= a = 25$, AC($=$BC) $= b = 15$, and CP ($=$ PD) $= x$, we have (per I. Euc. 47.) $(b+x)^2 + x^2 = a^2$; whence $x^2 + bx = \frac{1}{2}a^2 - \frac{1}{2}b^2$ and consequently $x = \sqrt{(\frac{1}{2}a^2 - \frac{1}{4}b^2)} - \frac{1}{2}b = 8\cdot5078$: Therefore ED ($=$ DG) $= 17\cdot0156$.

V. QUESTION 395, *by Mr.* Walter Trott.

Sailing on a certain course, we observed a head-land to bear due West from us; four hours after which, it was seen at W. S. W. and six hours after this (we still continuing to run at the same rate) its bearing was found to be S. S. W. What was our course at that time?

Answered by Mr. Philip George.

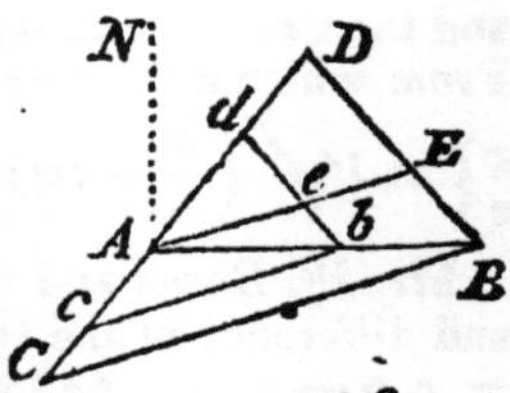

Construction. Draw the meridian AN; and supposing A to be the place of the head-land, draw AB, AE, and AD to represent the given bearings of the ship therefrom; in DAC take AC to Ad in the given ratio of the times, or as 4 to 6, and draw cb parallel to AE, meeting AB in b; then draw bed, to which the ship's course BED (let AB be what it will) must be parallel: Because (by VI. Euc. 4.) $be : ed$ (:: AC : Ad :: 4 : 6) : BE : ED.

Calculation. In the triangle Abc are given all the angles and the side AC ($=4$), to find A$b = 7\cdot391$; and then from Ab, Ad ($=6$) and the contained angle dAb, we get Abd ($=$ ABD) $= 47° 25'$, which is the complement of the course.

The same answered by Mr. J. Shipman, *of Hull.*

Let A represent the head-land, and AN the meridian; and through A draw the given bearings AB, AE, and AD; make AC $= 4$, AD $= 6$, and draw CB parallel to AE; then from B draw BD, so will ABD be the ship's course from the west. For, because of the parallel lines AC : AD :: EB : ED. From hence the course is found, by calculation, to be N. 42° 35′ westerly.

The same answered by Mr. J. Milbourn.

Let A represent the head-land, and B, E, and D the places of the ship at the three observations. Put the sine of BAE (22° 30′) $= a$, and that of DAE (45°) $= b$; then will AB $=$ (BE $\times$ sin. E) $\div a$, and AD $=$ (DE $\times$ sin. E) $\div b$ (by trig.): whence we have AB : AD (:: BE $\div a$:

DE $\div b$: : $4 \div a : 6 \div b$) : : $4b : 6a$. From which known ratio of the sides, and the included angle BAD, the course ABD, from the west, is found to be 47° 25′.

VI. QUESTION 396, *by Mr.* Thos. Moss.

In the three sides of an equi-angular field stand three trees, at the distances of 10, 12, and 16 chains from one another; To find the content of the field, it being the greatest the data will admit of?

Answered by Mr. W. Bevil.

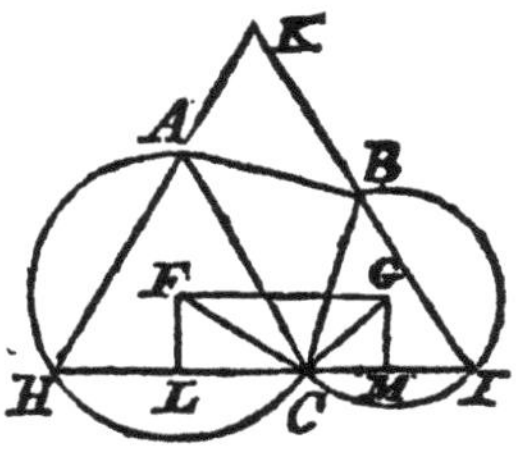

Construction. The given points A, B, C being joined, upon AC and BC let two segments of circles be described, each to contain an angle of 60°; join their centres by the line FG, and parallel thereto draw HCI cutting the two circles in H and I; then through A and B draw HK and IK, so shall HIK be the triangle required.

For, supposing FL and GM to be perpendicular to HI, and HI (always terminated by the circles) to revolve about the point C, it is plain that, when HI is a maximum, LM, being the half thereof must also be a maximum; and this will evidently be when LM is parallel to FG.

Calculation. In the triangle ABC are given all the sides, to find the angle C = 38° 38′; then in the triangle FCG are given the angle C = 98° 38′, and the ratio of the containing sides, as AC to BC (or 16 to 12); whence is found the angle F (= HCF) = 33° 41′, and the angle G (= ICG) = 47° 4′: therefore HCA = 63° 41′, ICB = 77° 4′, CAH = 56° 19′, CBI = 42° 56′, HI (= HC + CI) = 24·7, and the area = 26·42 = 26 acres, 1 rood, 27 perches.

Algebraic Solution to the same by Mr. W. Spicer.

Put $a =$ BC $= 12$, $b =$ BA $= 10$, $s =$ sine of 60°, and p and $-q$ equal to the sine and cosine of $\angle$BCI + $\angle$BAK = 152° 51′ 58″, and x and y those of $\angle$BAK: Then will $py + qx =$ sine $\angle$BCI; and (per trig.) $s : b :: x : bx \div s =$ BK; and $s : a :: py + qx : (apy + aqx) \div s =$ BI: Therefore $(bx + aqx + apy) \div s =$ IK, a max. In fluxions, $b\dot{x} + aq\dot{x} + ap\dot{y} = 0$: But $\dot{y} = -x\dot{x} \div y$; whence $b\dot{x} + aq\dot{x} - apx\dot{x} \div y = 0$: Solved, $x \div y$ (= tang. $\angle$BAK) $= (b + aq) \div ap$. Hence $(1 \div s)\sqrt{(aa + 2abq + bb)} =$ IK $= 24{\cdot}7002$; and $\sqrt{\frac{1}{3}} \times (aa + 2abq + bb) = 26{\cdot}42$ the area required.

Many contributors have solved this question upon a supposition that one side of the required triangle is parallel to the longest side of the given one; which though very near, is not strictly true; the said sides, when the area is the greatest possible, being inclined to each other in an angle of 3° 41′.

VII. QUESTION 397, *by Mr.* J. Ash.

Standing at the end of a visto, at the distance of 50 yards from each of the parallel sides, the opening of the said sides, at a sylvan statue before me, appeared to be $\frac{3}{2}$ of the opening of the further end; the part, or length, on this side the statue being to the part on the other side in the ratio of 1000 to 1501; From hence I would know the distance of the statue from the point of observation, and the whole length of the visto?

Answered by Mr. Abr. Botham.

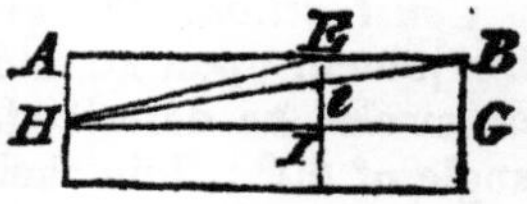

Supposing HG to be the length of the visto, and I the place of the statue; let t express the tang. of $\frac{1}{3}$ of the $\angle$EHI, which (by the quest.) is $\frac{1}{2}$ of $\angle$BHG; also let $a =$ 1501, and $b = 1000$: Then the tangents of EHI and BHG, being expressed by $\frac{3t - t^3}{1 - 3tt}$ and $\frac{2t}{1 - tt}$, (vide theor. p. 41, Diary 1755) and these tangents being to each other in the given proportion of a to b, we therefore have $\frac{3bt - bt^3}{1 - 3tt} = \frac{2at}{1 - tt}$; which, properly reduced, becomes $t^4 + \left(\frac{6a}{b} - 4\right) t^2 = \frac{2a}{b} - 3$; and this, solved, gives $t =$ ·019987: From whence the length of the visto is found = 1250 yards.

Some very ingenious gentlemen have taken a good deal of pains, in order to convince us, that this question was not properly limited; and others have hinted the same thing; not considering, that, from the ratio of any two angles (EHI, eHI) and the ratio of their tangents (which ratios are here given) both the angles and tangents may always be determined. The smallness of the angles in the case proposed (where the tangents are nearly in the same ratio with them) is, we presume, what embarrassed these gentlemen: The method of trial-and-error (as it is called) whereby most of them attempted the solution, being of little or no use here.

VIII. QUESTION 398, *by Mr.* Richard Gibbons.

Being on a journey, I took a guide at Modbury for Dartmouth; with whom having travelled 66 minutes, I asked him how far we were come? who replied, Just half so far as we are now from Totness. Having jogged on together seven miles further, I asked him how far we had now to travel? whose answer was still the same, Just half as far as we are from Totness: These indirect answers I did not like, though

I found when we were arrived at Dartmouth, which we reached in 55 minutes more, that all the fellow had said was strictly true, and that the two roads leading from Totness to Dartmouth and Modbury formed a right angle. From hence I would know the true distances of the three towns, and the rate at which we travelled which was uniform?

Answered by Mr. G. Brownbridge.

The place of the three towns being represented by M, D, and T, let the three parts of MD (as specified in the problem) be denoted by $6x$, a, and $5x$, respectively; then $a : 22x :: 2x : \frac{44xx}{a} = AC - BC$; $\therefore$ $AC = \frac{a}{2} + \frac{22xx}{a}$, $BC = \frac{a}{2} - \frac{22xx}{a}$, $MC = \frac{a}{2} + 6x + \frac{22xx}{a}$, and $DC = \frac{a}{2} + 5x - \frac{22xx}{a}$. Now $MC \times DC \; (= CT^2) = AT^2 - AC^2$; whence we have $44x^3 + 184ax^2 - 11a^2x = a^3$ or $44x^3 + 1288x^2 - 539x = 343$; which solved, gives $x = \cdot75284$. Therefore $MD = 15\cdot281$, $MT = 12\cdot236$, and $DT = 9\cdot153$.

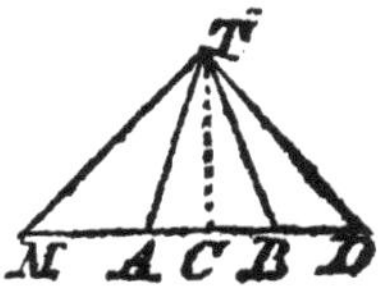

The same answered by Mr. R. Butler.

Supposing M, D, and T to represent the three towns, let $MA = 6x$, $TA = 12x$, $BD = 5x$, $BT = 10x$, and $AB \; (=7) = a$; Then (by trig.) $a : 22x :: 2x : \frac{44x^2}{a}$ = diff. segments AC and BC; whence $CA = \frac{a}{2} + \frac{22x^2}{a}$, and $CB = \frac{a}{2} - \frac{22x^2}{a}$. But (by II. Euc. 12.) $MA^2 + AT^2 + 2MA \times AC + DB^2 + BT^2 + 2BD \times BC \; (= MT^2 + DT^2) = MD^2$ (by I. Euc. 47.); that is, in species, $305x^2 + 11ax + 44x^3 \div a = (a + 11x)^2$, or $44x^3 + 184ax^2 - 11a^2x = a^3$. Whence x is found $= \cdot752846$: Therefore $MD = 15\cdot2813$, $MT = 12\cdot2365$, $TD = 9\cdot1534$, and the rate at which he travelled was $4\cdot1064$ miles per hour.

IX. QUESTION 399, *by Mr.* H. Watson.

The latitudes and longitudes of three places on the earth's surface, suppose London, Moscow, and Constantinople, being given, as below; required the latitude and longitude of that place which is equidistant from the former three?

The latitude of London is 51° 30′ the latitude and longitude of Moscow 55° 45′ and 38° 00′, and those of Constantinople 41° 30′ and 29° 15′, respectively.

Answered by Mr. H. Watson, *the Proposer.*

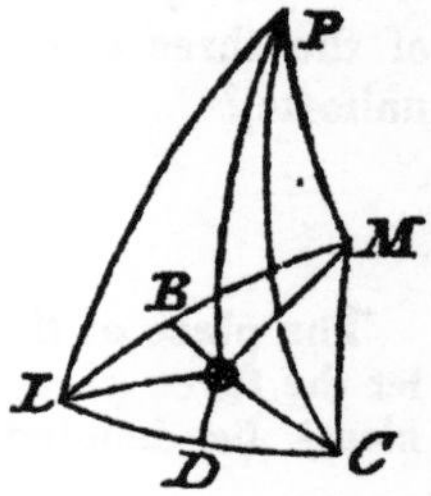

Let P represent the pole of the earth, and L, C, and M the three places proposed: From the latitudes and longitudes of which the distances LM, LC, and the contained angle CLM, will (by common proportions) be found 22° 38′, 22° 17′, and 40° 55′ respectively. Then supposing OB and OD perpendicular to LM and LC (O being the required place) we have given LB $(= \frac{1}{2}\text{LM}) = 11° 19'$, LD $(= \frac{1}{2}\text{LC}) = 11° 8\frac{1}{2}'$, with the sum of the angles BLO, DLO, at the bases (the hypothenuse LO being common); whence it will be (per spherics) sin. (LB + LD) : sin. (LB — LD) :: co-tang $\frac{1}{2}$(DLO + BLO) (20° $27\frac{1}{2}'$) : tang. of $\frac{1}{2}$(DLO — BLO) $= 1° 13\frac{1}{2}'$; whence BLO = 19° 14′; from which, and LB, we have LO (= MO = CO) = 11° 58′, and then, in the triangle LOP, will be given LO, LP, and the angle OLP (= 83° 27′ = PLM + BLO); whence OP = 38° 43′, and ∠LPO = 19° 13′, being the complement of the latitude, and the longitude from London, respectively.

Algebraic Solution to the same by Mr. R. Young.

Having found (by common spherics) LB, LD, and the angle DLB, put the tang. of LB $= a$, that of LD $= b$, the sine and co-sine of DLB equal to m and n, and those of BLO equal to x and y, respectively: Then, the co-sine of DLO being expressed by $mx + ny$, it will be, per spherics, $y : a :: 1$ (rad.) : tang. LO; $mx + ny : b :: 1$: tang. LO; whence $\frac{a}{y} = \frac{b}{mx + ny}$; and consequently $\frac{x}{y} = \frac{b - na}{ma}$ = tang. BLO = 19° 14′: From which the latitude and longitude of the place are found to be 51° 17′, and 19° 13′, respectively.

Mr. *P. George*, by producing the perpendicular DO to meet the side LM, solves it by common spherics: And Messrs. *Peart* and *Rollinson*, after finding LM, LC, and the angle CLM (as above) compute the difference of the other two angles LMC, LCM, which (because OMC = OCM) will also be the difference of OML and OCL, or of their equals OLB and OLD; whence both these angles are known, and from thence LO.

X. QUESTION 400, *by Mr.* Hugh Brown.

A lends B 1000*l.* for which B repays him as follows, viz. at the end of three months 180*l.* of five months 150*l.* of six months 140*l.* of eight months 100*l.* of nine months 90*l.* of ten months 120*l.* and at the year's end 250*l.* The rate of interest is required?

Answered by Mr. Hugh Brown, *the Proposer.*

Putting $z = \frac{1}{12}$ of the interest of 1*l.* for one year, we shall (by discounting at simple interest) have $\frac{250}{1+12z} + \frac{120}{1+10z} + \frac{90}{1+9z} + \frac{100}{1+8z} + \frac{140}{1+6z} + \frac{150}{1+5z} + \frac{180}{1+3z} = 1000$. The fractions reduced into series, and properly ordered, gives $z - \frac{3605z^2}{397} + \frac{36133z^3}{397}$, &c. $= \frac{3}{794}$; whence $z = \cdot0037783 + \cdot0001296 + \cdot0000040 = \cdot003912$; Therefore the required rate is 4·694, or 4*l.* 13*s.* 10½*d.* per cent.

The same answered by Mr. John Honey.

Let x^{12} = rate per annum; then x (according to compound interest) will be the rate for one month, x^2 for two months, &c. whence the present value of 180*l.* to be received at the end of 3 months will be $\frac{180}{x^3}$, &c. Hence we have $\frac{180}{x^3} + \frac{150}{x^5} + \frac{140}{x^6} + \frac{100}{x^8} + \frac{90}{x^9} + \frac{120}{x^{10}} + \frac{250}{x^{12}} = 1000$ (per quest.); and therefore $x^{12} - \cdot18x^9 - \cdot15x^7 - \cdot14x^6 - \cdot1x^4 - \cdot09x^3 - \cdot12x^2 = \cdot25$: Solved, $x = 1\cdot003852$; and $x^{12} = 1\cdot047216$; and consequently 4*l.* 14*s.* 5*d.* the rate per cent. required.

Mr. R. Flitcon *answers it thus:*

Let x, $2x$, and $3x$ be the interest of one pound for 1, 2, and 3 months, respectively; then the interest of 1000*l.* for three months will be $3000x$, and its amount $1000 + 3000x$, from which subtracting 180*l.* then paid, the remainder $820 + 3000x$ will be a new principal; which, in 2 months more, amounts to $820 + 4640x + 6000x^2$; from whence subtracting 150*l.* then paid, the remainder $670 + 4640x + 6000x^2$ will be the principal (or debt) at the end of 5 months: And by proceeding in the same manner with all the rest of the payments, the equation resulting (according to the question) will, in its least numbers, be $-3 + 758x + 4871x^2 + 14503x^3 + 24016x^4 + 22692x^5 + 11456x^6 + 2400x^7 = 0$. Hence $x = \cdot0038609$, and the rate, per cent. per annum, 4*l.* 12*s.* 7¼*d.*

XI. QUESTION 401, *by Mr.* E. Rollinson.

To investigate the value of an annuity, on a life of a given age, according to any table of observations on the degrees of mortality of mankind, by dividing the whole extent of life into differents periods, during which the decrements, or numbers dying off yearly, may be esteemed equal; without having any other series to sum than a common geometrical progression.

Answered by the Proposer Mr. Rollinson.

Let a denote the number of the living corresponding, in the table, to the given age; and let the succeeding decrements, or the numbers that die off yearly (for as long as they continue equal) be represented, each, by b; also let the next succeeding decrements (for as long as they continue equal) be denoted, each, by c; and the next after those by d; and so on: Putting r to represent the rate, and P the value of the perpetuity corresponding: Moreover, suppose that, after the decease of the proposed life A, the estate, or annuity, is to go to another person B, and his heirs, for ever. Then the probability that the life A fails the first year being $\frac{b}{a}$, the value of B's expectation on that contingency will therefore be $P \times \frac{b}{a}$: And the probability of A's dropping the second year being also $\frac{b}{a}$, the expectation of B thereon will be $\frac{P}{r} \times \frac{b}{a}$, ($\frac{P}{r}$ being the value of the perpetuity discounted for one year). In the same manner the expectation of B on the third year will appear to be $= \frac{P}{r^2} \times \frac{b}{a}$, and on the fourth year $= \frac{P}{r^3} \times \frac{b}{a}$, &c. Whence it is evident, that B's whole expectancy, on the contingency of the life's failing in the first interval, (m), during which the decrements are equal, each, to b, is truly defined by $rP \times \frac{b}{a} \times : \frac{1}{r} + \frac{1}{r^2} + \frac{1}{r^3} \ldots + \frac{1}{r^m}$, or its equal $rP \times \frac{b}{a} \times \frac{1 - r^{-m}}{r - 1}$, which is also $= rP^2 \times \frac{b - br^{-m}}{a}$, because $P = \frac{1}{r-1}$. Again, the decrements during the second interval (n) being each $= c$, the probability of the life's failing in any assigned year of this interval will therefore be denoted by $\frac{c}{a}$; and the value of B's expectation, on the whole interval, by $\frac{P}{r^m} \times \frac{c}{a} + \frac{P}{r^{m+1}} \times \frac{c}{a} + \frac{P}{r^{m+2}} \times \frac{c}{a}$, &c. $= \frac{rP}{r^m} \times \frac{c}{a} \times : \frac{1}{r} + \frac{1}{r^2} + \frac{1}{r^3} \ldots + \frac{1}{r^n} = rP^2 \times \frac{c - cr^{-n}}{ar^m}$. In the very same manner the expectancy on the third interval (p) appears to be $= r\text{P}^2 \times \frac{d - dr^{-p}}{ar^{m+n}}$; and on the fourth ($q$) $= r\text{P}^2 \times \frac{e - er^{-q}}{ar^{m+n+p}}$, &c. &c. Therefore by collecting all these values together, the whole expectation of B and his heirs comes out $= \frac{r\text{P}^2}{a} \times : b + \frac{c-b}{r^m} + \frac{d-c}{r^{m+n}} + \frac{e-d}{r^{m+n+p}} +$, &c. which, subducted from P, the value of the perpetuity leaves $\text{P} - \frac{r\text{P}^2}{a} \times : b + \frac{c-b}{r^m} + \frac{d-c}{r^{m+n}} + \frac{e-d}{r^{m+n+p}}$, &c. for the true value of the life A: But $r\text{P}$ is $= \text{P} + 1$, and $\frac{1}{r^m}$, $\frac{1}{r^{m+n}}$, &c. are the present values of 1*l.* to be received at the end of m, $m+n$, &c. years; which being found (from the tables) and represented by M, N, Q, &c. respectively, the value of the annuity will be $\text{P} - \frac{\text{PP}+1}{a} \times : b + (c-b) \times \text{M} + (d-c) \times \text{N} + (e-d) \times \text{Q} + (f-e) \times \text{R}$, &c. where the series is to be continued till it terminates, and where the co-efficient of the last term will be the last of the quantities b, c, d, &c. with a negative sign, the next letter in order being equal to nothing.

XII. QUESTION 402, *by Mr.* W. Bevil.

Suppose the ends of a thread, ten feet long, be fastened to two tacks, in the same horizontal line, at the distance of six feet: I would know where two weights, the one three and the other five ounces, must be fixed to the thread, so as to hang at rest in the same horizontal line at the distance of three feet from the level of the tacks?

Answered by Mr. S. Bamfield.

Supposing CE and DF to be perpendicular to the horizontal line AB, it is evident, from mechanics, that AE : BF :: 3 : 5. If therefore EC ($=$ FD) $= b = 3$, AB $= a = 6$, AC + CD + DB $= c = 10$, and AE $= 3x$, then will BF $= 5x$; EF ($=$ CD) $= a - 8x$; and $\sqrt{(bb + 9xx)}$ (AC) $+ \sqrt{(bb + 25xx)}$ (BD) $+ a - 8x = c$. From the resolution of which equation x is found $= \cdot 31786$; and from thence AC $= 3\cdot 1479$, BD $= 3\cdot 3950$, and CD $= 3\cdot 3951$.

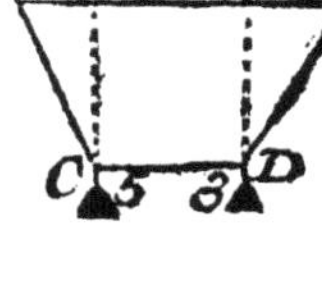

The same answered by Bircloverensis.

Put AB $(=6)=l$, CE $(=3)=b$, AC + CD + DB $(=10)=s$, and AC $=x$; then, AE $=\sqrt{(xx-bb)}$, and, by the property of the centre of gravity BF $=\frac{5}{3}\sqrt{(xx-bb)}$: Consequently CD (= EF) $=l-\frac{8}{3}\sqrt{(xx-bb)}$, and BD $=\frac{1}{3}\sqrt{(25xx-16bb)}$. Whence we have $x+l-\frac{8}{3}\sqrt{(xx-bb)}+\frac{1}{3}\sqrt{(25xx-16bb)}=s$. This equation, properly reduced, will give $x^3-3{\cdot}35x^2-10{\cdot}8x+36=0$; where x is found $=3{\cdot}1479$.

XIII. QUESTION 403, *by Mr.* Thomas Moss.

Suppose that, from the top of a mountain, in form of a paraboloid, whose perpendicular height is 600 yards and its base diameter three miles, a cannon ball is to be discharged with a quantity of powder sufficient to carry it to the height of 700 yards in a vertical direction; I would know the elevation of the piece so that the ball may fall at the greatest distance possible from the place of projection.

Answered by Mr. Thomas Peart.

From the given height of the mountain AB, and its semi-diameter BC, the parameter of the parabola will be found = 11616 yards; from whence the curve AEC may be constructed. Take AK = twice the height of the perpendicular shot, and draw KM perpendicular thereto; in KM take the point M, so that a perpendicular let fall from thence, meeting the curve in E, shall be equal to the straight line AE; then join A, M, which will be the line of direction.

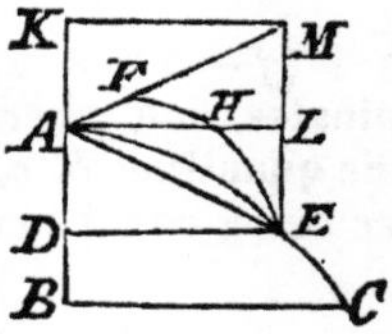

Calculation. Let d = the height of the perpendicular shot, and a = 11616 = the parameter, AD $=x=$ LE; then ME (= AD + AK) $=2d+x$, DE $=\sqrt{ax}$, and AE $=\sqrt{(ax+xx)}=2d+x$ (= ME); therefore $x=4dd\div(a-4d)=222{\cdot}323$ yards, and AE $(2d+x)=1622{\cdot}323$; from whence the angle AED is found = 7° 52′ 36″; the half of which being taken from 45°, there remains 41° 3′ 42″ for the angle of elevation LAM.

The same answered by Mr. Walter Trott.

Put the height AB of the mountain (= 600) $=a$, the semi-diameter BC of the base (= 2640) $=b$, and the double impetus (= 1400) $=c$; Then, suppose E to be the place where the ball (whose path is AHE, and first direction AF) impinges on the surface, let AE and ED be drawn, the latter parallel to CB, and let AD $=x$; and then, by the property of the parabola, DE$^2=b^2x\div a$, and therefore AE $=\sqrt{(bbx\div a+x^2)}$. Hence, because the dist. is to be a max. we have (by Simpson's Exercises, p. 199) $\sqrt{(bbx\div a+xx)}-x$

(=AE − AD)=c; from which $x = acc \div (bb - 2ac) = 222{\cdot}323 =$ AD. Whence the angle EAD=82° 8′; the half of which (=41° 4′) is the true angle of elevation (HAF).

XIV. QUESTION 404, *by Mr.* H. Watson.

To determine the centre of attraction of a semi-spherical body, or that point in the axis where a corpuscle may be placed to remain in equilibrio by the equal and contrary action of the matter of the hemipshere surrounding it.

Answered by Mr. Lionel Charlton, *of Whitby.*

Let BAD be the given hemisphere, P the required point, NPM and GHF sections of the solid perpendicular to the axis AC; let PF, CF, and PD be drawn; putting AC $= r$, PC $= a$, AP $= e$, PM $= f$, PD $= g$, PH $= x$, and PF $= z$. Then the effect of the atraction of all the particles in the plane of the circle GHF, on a corpuscle at P, will, it is well known, be as $1 - \frac{x}{z}$; and consequently the fluxions of the attraction of the segment NGFM as $\dot{x} - \frac{x\dot{x}}{z}$:

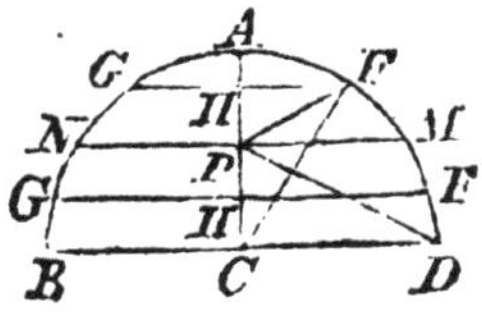

Which (because $a^2 + z^2 + 2ax = r^2$, and therefore $x = \frac{ff - zz}{2a}$, and $\dot{x} = -\frac{z\dot{z}}{a}$) will be reduced to $\dot{x} + \frac{f^2 - z^2}{2a} \times \frac{\dot{z}}{a}$; whereof the corrected fluent is $x + \frac{3f^2z - z^3 - 2f^3}{6a^2}$: Which, when $x = e = z$, becomes $e + \frac{3ef^2 - e^3 - 2f^3}{6aa}$ for the force of the upper segment NAM in the direction PA.

In the same manner, taking GF below NM, the fluxion of the attraction of the part NGFM will be, still, expressed by $\dot{x} + \frac{f^2 - z^2}{2a} \times \frac{\dot{z}}{a}$, and the corrected fluent by $x + \frac{3f^2z - z^3 - 2f^3}{6aa}$, which, by taking $x = a$(=PC) and $z = g$ (=PD) gives $a + \frac{3f^2g - g^3 - 2f^3}{6aa}$ for the whole force of the lower segment NBDM in the opposite direction PC. Hence we have $e + \frac{3ef^2 - e^3 - 2f^3}{6aa} = a + \frac{3f^2g - g^3 - 2f^3}{6aa}$, or $6a^2(e - a) + 3ef^2 - e^3 = g(3ff - gg)$. But $e = r - a$, $ff = rr - aa$, and $g = \sqrt{(rr + aa)}$ so that, by substitution, $2r^3 - 8a^3 = \sqrt{(rr + aa)} \times (2r^2 - 4a^2)$ which, ordered, gives $12a^4 - 8r^3a + 3r^4 = 0$; where taking $r = 1$, we have $a = 0{\cdot}4230428$. Therefore the ratio of AP to CP will be that of 0·5769572 to 0·4230428, or as 4 to 3, nearly.

The author of a very neat solution to this problem (whom we should be sorry to disoblige by an improper step) will, we hope, candidly excuse our giving preference to the above; as room would not possibly admit of both, and as the fluents, here, may be comprehended by common readers, for whose improvement we are solicitous.

XV. QUESTION 405, *by Mr.* Patrick O'Cavanah.

Given $\dot{x}\dot{y} - x\ddot{y} - a\ddot{y} - \frac{x\dot{y}^2}{b} = 0$, to find the general relation of the fluents x and y.

Answered by Mr. T. Moss.

Divide the given equation by $\frac{\dot{y}\dot{y}}{x}$, and you will have $\frac{x^2\dot{y} - xx\ddot{y}}{\dot{y}\dot{y}} - \frac{ax\ddot{y}}{\dot{y}\dot{y}} - \frac{x\dot{x}}{b} = 0$; which, by taking the fluent, becomes $\frac{x\dot{x}}{\dot{y}} - \frac{a\dot{x}}{\dot{y}} - \frac{xx}{2b} = d$ (a constant quantity) whence $\dot{y} = 2b \times \frac{a\dot{x} + x\dot{x}}{2bd + x^2}$, and consequently $y = a\sqrt{\frac{2b}{d}} \times$ arc whose tangent is $\frac{x}{\sqrt{2bd}}$ (to radius 1) $+ b \times$ hyp. log. $\frac{2bd + x^2}{e}$; where d and e may denote any constant quantities at pleasure.

The same answered by Mr. Patrick O'Cavanah, *the Proposer.*

In the given equation ($\dot{x}\dot{y} - x\ddot{y} - a\ddot{y} - \frac{x\dot{y}^2}{b} = 0$) let $-\frac{\dot{y}\ddot{x}}{\dot{x}}$ * be wrote in the place of $\ddot{y}$ (so that $\dot{y}$ may become the quantity flowing uniformly) then will $\dot{x}\dot{y} + x \times \frac{\dot{y}\ddot{x}}{\dot{x}} + a \times \frac{\dot{y}\ddot{x}}{\dot{x}} - \frac{x\dot{y}^2}{b} = 0$, or $\dot{x}\dot{x} + x\ddot{x} + a\ddot{x} - \frac{x\dot{x}\dot{y}}{b} = 0$; whose fluent, it is evident, will be $x\dot{x} + a\dot{x} - \frac{x^2\dot{y}}{2b} =$ some constant quantity, let it be $c\dot{y}$, so shall $\dot{y} = 2b \times \frac{a\dot{x} + x\dot{x}}{2bc + xx}$; and consequently $y = a\sqrt{\frac{2b}{c}} \times A + b \times L$; A being = the arch whose radius is 1, and tangent $\frac{x}{\sqrt{2bc}}$; L $=$ the hyp. log. of $\frac{2bc + xx}{dd}$; and c and d any constant quantities.

Mr. *R. W.* (whom for particular reasons, we should be glad to oblige) will not, we hope take it amiss that his solution to this problem was not inserted, as his reasoning upon the correction of the fluent (tho' very ingenious) is nevertheless, defective.

* That the substitution of $-\frac{\dot{y}\ddot{x}}{\dot{x}}$ for $\ddot{y}$, in any fluxionary equation of the second (or any higher) degree, causes no difference in the equation of the fluents, may be thus demonstrated. Suppose the relation of the fluents to be expressed by the general equation $y = Ax^m + Bx^n + Cx^p$, &c. then it is evident that $\frac{\dot{y}}{\dot{x}} = mAx^{m-1} + nBx^{n-1} + pCx^{p-1}$ &c. And, if $\dot{x}$ be made constant, we shall, by taking the fluxion a second time, have $\frac{\ddot{y}}{\dot{x}} = m(m-1)Ax^{m-2}\dot{x} + n(n-1)Bx^{n-2}\dot{x}$, &c. But if $\dot{y}$ be made constant, we shall then have $-\frac{\dot{y}\ddot{x}}{\dot{x}\dot{x}} = m(m-1)Ax^{m-2} \times \dot{x} + n(n-1)Bx^{n-2}\dot{x}$, &c. Therefore, seeing $\frac{\ddot{y}}{\dot{x}}$ and $-\frac{\dot{y}\ddot{x}}{\dot{x}^2}$ are both equal to one and the same quantity ($m(m-1)Ax^{m-2}$, &c.) they must necessarily be equal each other, and consequently $\ddot{y} = -\frac{\dot{y}\ddot{x}}{\dot{x}}$.

In like manner it will appear that $\dddot{y} = \frac{3\dot{y}\ddot{x}^2}{\dot{x}\dot{x}} - \frac{\dot{y}\dddot{x}}{\dot{x}}$. For, by taking again, the fluxions of the preceding equations, $\frac{\ddot{y}}{\dot{x}\dot{x}} = m(m-1)Ax^{m-2} + n(n-1)Bx^{n-2}$, &c. and $-\frac{\dot{y}\ddot{x}}{\dot{x}^3} = m(m-1)Ax^{m-2} + n(n-1)Bx^{n-2}$, &c. (making $\dot{x}$ constant in the former and $\dot{y}$ in the latter) there arises $\frac{\dddot{y}}{\dot{x}^4}$ $(= m(m-1)(m-2)Ax^{m-3}\dot{x}$, &c.) $= -\frac{\dot{y}\dddot{x}}{\dot{x}^3} + \frac{3\dot{y}\ddot{x}^2}{\dot{x}^4}$; and from thence $\dddot{y} = \frac{3\dot{y}\ddot{x}^2}{\dot{x}\dot{x}} - \frac{\dot{y}\dddot{x}}{\dot{x}}$.

THE PRIZE QUESTION, *by Mr.* E. Rollinson.

Three ships, A, B, C, sail at the same time from three different ports: The ship A, from the southermost port, runs due east, at the rate of five knots: The ship B sails S. E. by S. at the rate of three knots, her port lying N.N.E. from that of A, and at the distance of 12 miles: The other ship, C, always, bears down upon (or steers directly towards) the two former, which she keeps constantly in a line. I would know how far she must run before she comes up with B, together with the distance of her port from the other two, and the path she describes?

Answered by Κυβερνητης.

Let P, Q, and R represent the three ports, and A, B, and C the ships sailing from thence, the former two in the right lines PS, QS, and the latter C in a curve line RCE whose nature must be such that a right line ABC passing through A and B, may always touch the curve in that very point where the ship C then is. Let E be the place of the ship B when A arrives at S, the intersection of the courses, and C is come up with B: Let also SI be to SE in the ratio of the given celerities, or as PS to QE: Draw GIEM, and make AH parallel thereto, and CK and RM parallel to QES.

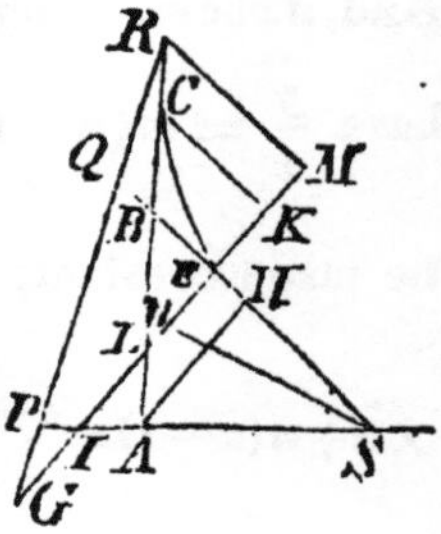

It is evident, that SH and EB will be always equal to each other, being each to AS in the constant ratio of SE to SI, or of the celerity in BS to that in AS; whence BH and ES must likewise be equal: But, by similar triangles, AH : BH (ES) :: EL : BE, and EI : AH :: ES : SH (BE): From the composition of which propositions, we have $BE^2 = (ES^2 \div EI)\,EL = En \times EL$ (sn being drawn to make the $\angle ESn = EIS$) which is a known property of the parabola: For in the parabola (supposing p to denote the parameter of any diameter EM it is well known that $CK^2 = p \times EK$, and that $EL = EK$, and consequently $EB = \frac{1}{2}CK$; so that $BE^2\ (= \frac{1}{4}CK^2) = \frac{1}{4}p \times EL$; whence, it not only appears that the curve is a parabola, but that En is $\frac{1}{4}$ of the parameter thereof, corresponding to the diameter EKM.

To determine now the length of the arch ECR, &c. we have, in the $\triangle$ESI, the ratio of SI to SE (as 5 to 3), and the contained $\angle S\ (= 56^\circ\ 15')$: Whence the $\angle$SEI (or MKC) which the ordinate makes with the diameter EM, is found $= 86^\circ\ 56\frac{1}{2}'$. Again, since $\angle S = \angle Q$, we have $PS = PQ = 12$, and $QE\ (= \frac{3}{5}PS) = 7{\cdot}2$; therefore $RM\ (= 2QE) = 14{\cdot}4$, and $QR\ (= GQ = QE\ (\text{sin. } E \div \text{sin. } G)) = 14{\cdot}086$ = the required distance of the port R from the port Q. Again, (by trigonometry) $QS = 13{\cdot}3337$, $SE\ (= QS - QE) = 6{\cdot}1337$, and $En = 4{\cdot}42$; which last put $= \frac{1}{2}b$, and let the cosine of the angle MKC, $(86^\circ\ 56\frac{1}{2}')$

$= m$: Then (supposing EK $= x$, and CK $= y$) we have (by the property of the parabola) $2bx = y^2$; whence $\dot{x} = y\dot{y} \div b$, and consequently $\sqrt{(\dot{y}^2 + 2m\dot{y}\dot{x} + \dot{x}^2)}$ the flux. of the arch EC $= \frac{\dot{y}}{b}\sqrt{(b^2 + 2bmy + y^2)}$; which by putting $v = mb + y$ and $h = b\sqrt{(1 - mm)}$, will be transformed to $\frac{\dot{v}}{b}\sqrt{(h^2 + v^2)}$; whereof the (corrected) fluent will be $\frac{v}{2b}\sqrt{(hh+vv)} - \frac{1}{2}mb + \frac{h^2}{2b} \times$ hyp. log. $\frac{v + \sqrt{(hh+vv)}}{mb+b}$: From whence (by taking $y =$ RM $= 14{\cdot}4$) the distance RE run by the ship C is found $= 19{\cdot}77$.

Questions proposed in 1756, *and answered in* 1757.

I. QUESTION 406, *by* Rusticus.

John and Hodge met together at market, John had brought sheep, and Hodge pigs and geese; whilst taking a friendly pot together they agreed to an exchange of goods, viz. John to give his sheep for Hodge's pigs and geese. The value of every sheep to be that of a pig and goose, and each goose 2s. Now the number of the pigs and geese together exceeded the number of sheep by 16; and the number of geese exceeded that of the pigs by 10: What was the number of each, the pigs and geese together being worth 5 pounds?

Answered by Mr. T. Barker, *of Westhall, in Sussex.*

Let x denote the number of pigs, and y the price of a pig; then will $x + 10$ be the number of geese, and $2x - 6$ the number of sheep; and therefore, by the question, $xy + (x + 10) \times 2 = 100$, and $(2x - 6) \times (y + 2) = 100$. From the first equation $y = 80 \div x - 2$; and from the second $y = 100 \div (2x - 6) - 2$; $\therefore 80 \times (2x - 6) = 100x$: whence $x = 8$, and $y = 8$. Therefore there were 8 pigs, 10 sheep, and 18 geese.

The same answered by Mr. Tho. Wilkin.

Let $x =$ number of pigs; then $x + 10 =$ number of geese, and $2x - 6 =$ number of sheep; whence $2x + 20 =$ price of all the geese; $100 - 2x - 20$ $(= 80 - 2x) =$ price of all the pigs; and $(80 - 2x) \div x$ or $80 \div x - 2$ the price of one pig; and consequently

$80 \div x$ (= the price of one sheep) which multiplied by the number of sheep gives $(80 \div x)(2x - 6) = 100$: Hence $160x - 480 = 100x$, and $x = 8$. Therefore there were 8 pigs at 8 shillings each, 18 geese at 2 shillings each, and 10 sheep at 10 shillings each.

II. QUESTION 407, *by* Philo-Pesos.

In a given triangle, to inscribe a rhombus, having one of its angular points coincident with a given point in the base.

Answered by Mr. Abr. Botham.

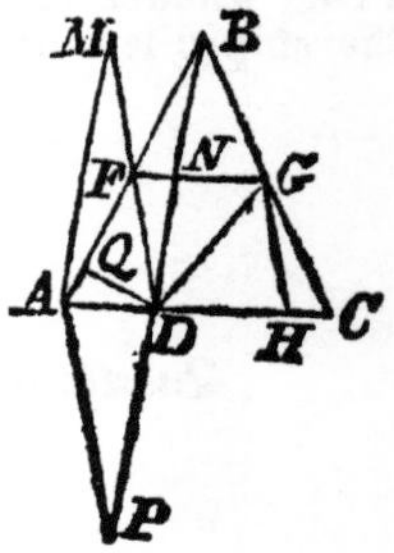

Construction. From the vertex B, of the triangle, through the given point D, draw BDP; to which, from A, apply AP equal to the base AC; draw DF, FG, GH parallel to AP, AC, and FD, respectively; and the thing is done.

For, by sim. $\triangle$s, AP : FD (:: BA : BF) :: AC : FG; but AP = AC (by construction); whence FD = FG.

The same answered by Mr. T. Peart.

Draw BD from the vertex to the given point D; and make AM parallel thereto, and equal to the base AC of the triangle; draw MD cutting AB in F; then draw FG and GH parallel to AC and DF, for the other side of the rhombus.

Demonstration. Let BD cut FG in N. The triangles ADM and DFN (because of the parallel lines) will be similar; whence AM (AC) : DF :: AD : FN :: AC : FG; and consequently DF = FG.

Method of Calculation. From AB, AD, and the $\angle$DAB, the $\angle$BDC (= MAD) will be known; from which and the given sides AD and AM, the $\angle$ADF will also be known, and consequently the side DF.

Algebraic Solution, by Mr. W. Smith, *of Irthlingborough.*

Put AB $= b$, AC $= c$, DQ (perp. to AB) $= d$, BQ $= e$, and DF (= FG) $= x$: Then $c : b :: x$ (FG) : BF $= bx \div c$; whence FQ $= e - bx \div c$, and consequently $(e - bx \div c)^2 + d^2 = x^2$. Which solved gives $x = \frac{bce}{cc - bb} + \sqrt{(c^2 \times \frac{dd + ee}{cc - vv} + (\frac{bce}{cc - bb})^2)}$.

III. QUESTION 408, *by Mr.* R. Young, *Writing-master, Chester.*

The sides of a quadrangular field are known to be 9, 10, 11, and 12 chains, respectively: And, if a diagonal be therein drawn, the part included by it and the two shortest sides will be to the remaining part in the given proportion of 3 to 5: From hence the content of the field is required?

Answered by Mr. Wm. Kingston, *of Bath.*

Let DFGH be the quadrangular field (see last figure); then, since by a well-known theorem, the area of the triangle DFG is $\frac{1}{4}\sqrt{\{((DF+FG)^2 - DG^2)\times(DG^2 - (DF-FG)^2)\}}$, if DG be put $= x$, we shall have (because FG is given $= 9$, and DF $= 10$) DFG $= \frac{1}{4}\sqrt{((361 - x^2)\times(x^2 - 1))}$. And in the very same manner, the area of DHG will be $= \frac{1}{4}\sqrt{((529 - x^2)\times(x^2 - 1))}$: which being to that of DFG as 5 to 3 (by the quest.) it is evident that $25 : 9 :: 529 - x^2 : 361 - x^2$; whence $xx = 266\cdot5$; and the area DFG $= \frac{1}{4}\sqrt{(94\cdot5 \times 265\cdot5)} = 39\cdot5994$: and consequently, $\frac{8}{3}\times 39\cdot5994 = 105\cdot5984$ ($=$ 10 ac. 2 r. $9\frac{1}{2}$ p.) the content of the field.

IV. QUESTION 409, *by Mr.* H. Watson.

To determine a point in a given triangle (whose sides are 16, 20, and 24 inches) from whence perpendiculars being let fall on all the sides, the solid (or continual product) contained under them, shall be the greatest possible?

Answered by Birchoverensis.

When FG $\times$ FH $\times$ FI is a maximum, the product thereof by the constant quantity $\frac{1}{2}$AB $\times$ $\frac{1}{2}$BC $\times$ $\frac{1}{2}$AC, will also be a maximum; that is, the product of the three parts ABF, BCF, and ACF of the given triangle, will be a maximum, and consequently those parts equal among themselves; since it is evident (from Euc. 5. II.) that the continual product of the parts of any given quantity (whatever their number is) will be the greatest when the parts are all equal: Therefore, ABF being $= \frac{1}{3}$ABC, it is evident that FG will be $= \frac{1}{3}$ of the perpendicular falling from C upon AB, &c. and consequently that F will be the centre of gravity of the given triangle ABC. Hence the three required perpendiculars are found to be 5·2915, 4·4095, and 6·6143, and their continual product $= 154\cdot33$.

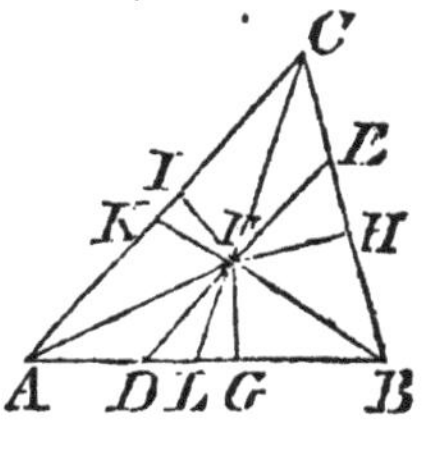

The same answered by Mr. Walter Trott.

From any point F, in DE $\parallel$ to one side AC of the triangle, conceive $\perp$s FG, FH, and FI to be let fall; then FG being in a constant ratio to DF, and FH to FE, it is evident, that, when DF $\times$ FE is a maximum, FG $\times$ FH or FG $\times$ FH $\times$ FI (because FI is supposed to continue the same) will likewise be a maximum; which therefore is known to be when FD $=$ FE, or when AK $=$ CK, supposing the right line BFK drawn meeting AC in K. Hence it appears that (let the distance between DE and AC be what it will) the quantity FG $\times$ FH $\times$ FI cannot be a maximum, unless AK $=$ CK: Neither can it be a maximum (by the very same argument) unless AL (supposing CFL drawn) is $=$ BL; therefore it must be so when AK $=$ CK and AL $=$ BL: Whence the construction is manifest; the point F required being the centre of gravity of the triangle, and the three required perpendiculars equal to $\frac{5}{3}\sqrt{7}$, $2\sqrt{7}$, and $\frac{5}{2}\sqrt{7}$, or 4·4096, 5·2915, and 6·6144, and the content of the solid contained under them $= \frac{175}{3}\sqrt{7} = 154{\cdot}33$ cubic inches.

A Fluxionary Solution to the same, by Mr. W. Allen.

Put AB $= a$, BC $= b$, AC $= c$, FG $= x$, FH $= y$, FI $= z$, and the area ABC ($= 158{\cdot}745$) $= d$; then $ax + by + cz = 2d$: and xyz a max. which last, when z is exterminated by means of the former, will give $2dxy - ax^2y - bxy^2$, a max. From which, supposing y invariable, we have $2dy\dot{x} - 2ayx\dot{x} - by^2\dot{x} = 0$; whence $y = (2d - 2ax) \div b$. Again, by making x invariable, we get (after reduction) $y = \frac{2d - ax}{2b}$. Therefore $\frac{2d - 2ax}{b} = \frac{2d - ax}{2b}$; whence $x = \frac{2d}{3a} = 4{\cdot}40958$; also $y = \frac{2d}{3b} = 6{\cdot}61437$; $z = \frac{2d}{3c} = 5{\cdot}2915$; and $xyz = 154{\cdot}33$.

V. QUESTION 410, *by Mr.* Walter Trott.

The distances of the three corners of a right-angled triangular field, from a watering place within the field, are known to be 38, 50, and 62 perches, respectively: To find, from thence, the content of the field, it being the greatest the data will admit of?

Answered by Mr. Pat. O'Cavanah.

Let ABC represent the field, and D the watering place. Let DE be $=$ and $\parallel$ to AB, intersecting CB in F, and let BE and CE be drawn: Then will the trapezium BDCE ($=$ DE $\times$ $\frac{1}{2}$BC) and the $\triangle$ABC ($=$ AB $\times$ $\frac{1}{2}$BC) be mutually equal to each other. Moreover, since CE^2 $-$ CD^2 ($=$ EF^2 $-$ DF^2) $=$ BE^2 $-$ BD^2, it is evident that CE^2 $=$ BE^2 (AD^2) $-$ BD^2 $+$ CD^2; from which CE is given ($= 70$). Therefore, all the four sides of the trapezium DCEB being given, the area will be a maximum when the figure is inscribed in a circle. But the rectangle of the two diagonals of any trapezium inscribed in a circle is equal to the sum of the rectangles of the opposite sides: Therefore DE $\times$ BC $=$ DC

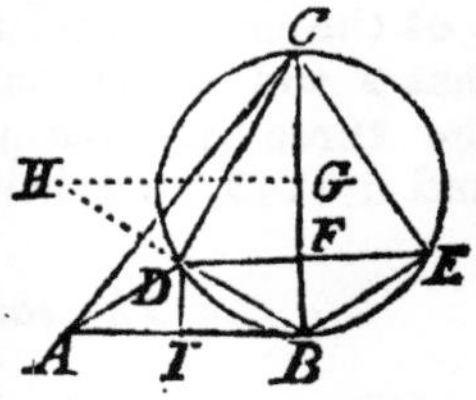

$\times$ BE + DB $\times$ CE; the half of which is (= 2880 perches, or 18 acres) the true area sought. The geometrical construction from hence is very easy; for (by sim. $\triangle$'s) DF is to BF in the given ratio of CD to BE (or AD). Therefore, having first drawn two $\perp$ lines at pleasure, in the one of them take BG = AD, and $\parallel$ to the other draw GH = CD; in BH (when drawn) take BD of the given length; and from D, to BA and BC, apply DA and DC also of the given lengths; from whence the points A and C, and consequently the $\triangle$ ABC itself, will be determined.

A Fluxionary Solution to the same, by Mr. J. Honey.

Let AD = 62 = a, BD = 38 = b, CD = 50 = c, and AB = x; then $x : a + b :: a - b : (aa - bb) \div x$ = diff. segments of the base; whence, putting $aa - bb = n$, we get $(xx - n) \div 2x$ = IB = DF; and (p. Euc. 47. I.) BF = $\frac{1}{2x}\sqrt{(4b^2x^2 - x^4 + 2nx^2 - n^2)}$; also CF $= \frac{1}{2x}\sqrt{(4c^2x^2 - x^4 + 2nx^2 - n^2)}$; and consequently BC = $\frac{1}{2x}\sqrt{(rx^2 - x^4 - n^2)} + \frac{1}{2x}\sqrt{(mx^2 - x^4 - n^2)}$, by putting $r = 4bb + 2n$, and $m = 4cc + 2n$. Hence $\sqrt{(rx^2 - x^4 - n^2)} + \sqrt{(mx^2 - x^4 - n^2)}$ = 4 times the area ABC; whose fluxion being taken and made = 0, we thence get $x^4 - \frac{mr + 4nn}{m + r} \times x^2 = -n^2$; solved, $x = 78{\cdot}3435$ perches. Whence AC $\doteq 73{\cdot}5223$, and the area ABC = 18 acres.

Mr. *Peart*, (who gives the solution without fluxions) says the angles BAD and BCD must be equal to each other; the truth of which is evident from the foregoing construction: For BED, which is = DAD, must necessarily be = BCD, standing on the same arch BD.

VI. QUESTION 411, *by Mr.* J. Ash.

Being on a journey, in lat. 52° 30′ N. the 3d of April, 1755, in the afternoon, I observed the two intersections of the inner part of the interior rain-bow with the horizon, to form an angle, at the place where I stood, of 60°: From which the time of observation is required?

Answered by Mr. L. Charlton.

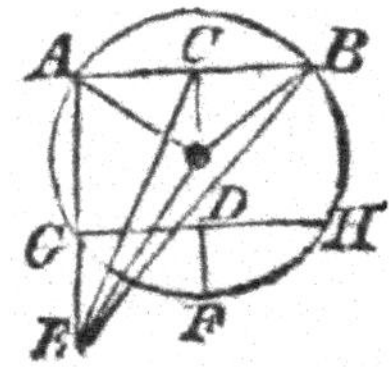

Let O be the centre of the Iris, E the eye of the spectator, and EACB the plane of the horizon: Then, if AE be taken as the radius, it is evident (from the writers on optics) that OA will be the sine of an $\angle$(AEO) of 40° 17′ = ·6465; from whence and AC, which is given = sin. 30° = ·5, OC is found = ·4099. Then in the $\triangle$ EOC (right-angled at o) it will be, as EO (= ·7628 = cos. 40° 17′) is to OC, so is rad. (1) to the tang. of OEC = 28° 15′ = the depression of the centre of the bow, or the sun's altitude: From which the time is found 3^h 15^m afternoon.

The same answered by Mr. Samuel Bamfield.

The arch (AO or BO) of a great circle of the sphere, drawn from the interior side of the bow ABF to the centre thereof, is (according to the writers on optics) = 40° 17′. Therefore in the isosceles spherical △AOB (where AB represents the horizon) we have given AO = OB = 40° 17′, and AB = 60°; whence the perpendicular OC (= the sun's altitude) is found = 28° 15′; and from thence the time of observation 3^h 15^m afternoon.

VII. QUESTION 412, *by Mr.* Chr. Mason.

The human stage is three-score years and ten;
So short the span! yet seldom reach'd by men.
As man's quietas ought to be his care,
'Tis now high time for Mason to prepare.
The rolling years that o'er my head have flown,
If you would know, they're in the margin * shown.

* Viz. $v + x + y + z = 1734$, $v^2 + x^2 + y^2 + z^2 = 2850372$, $vxyz = 3240960$, and $x = z$; v being the year of my birth, x the month, y the day, and z the hour P.M. all which are required?

Answered by Mr. W. Allen, *and several others.*

Making $a = 1734$, $b = 2850372$, and $c = 3240960$, we have $v + y = a - 2x$, $vv + yy = b - 2xx$, and $vy = c \div xx$. From the square of the first of which equations subtract the double of the last, so shall $v^2 + y^2 = a^2 - 4ax + 4x^2 - 2c \div xx = b - 2x^2$; and consequently $x^4 - \frac{2}{3}ax^3 + \frac{1}{6}(aa - b)x^2 = \frac{1}{3}c$; that is, $x^4 - 1156x^3 + 26064x^2 = 1080320$: Whence $x = 8$, $v = 1688$, $y = 30$, and the time of the proposer's birth Oct. 30, 1688, 8 hours P.M.

VIII. QUESTION 413, *by Mr.* Pat. O'Cavanah, *of Dublin.*

(*Addressed to the ingenious Authors of the 378th and 393d Questions.*

Ah! much, my friends, I mourn your lot,
Who have such woeful help-mates got:
Both sluts and scolds! oh, dreadful curse?
Bad daughters too! What can be worse?
Yet hear me, and to ease your grief,
I'll teach you what will yield relief; *

But like rare nostrums, little known,
In mystic terms it must be shown—
For brother Philomaths alone.

This powerful specific is denoted by a word consisting of five letters, and these have their places in the alphabet expressed by the values of u, w, x, y, and z, in the subjoined equations : + By means whereof the important mystery may be disclosed, and that, without having the root of any equation to extract, higher than a quadratic.

+ Viz. $w^2 + z^2 = 89$; $wz + w + z = 53$; $\frac{xx + yy}{x + y} = 18\frac{1}{10}$; $\frac{x^3 + y^3}{x^4 + y^4} = 18\frac{172}{181}$; $\frac{u^5 + x^5}{ux \times (u+x)^3} = 2\frac{1}{20}$.

Answered.

Messrs. *Barker, Bevil, Birchoverensis, Botham, Juvenis, Honey, Peart, Penovius, Smith, and Trott,* solve this problem by substituting for the half sum, and the half difference of the quantities sought in the several given equations : Thus, let $w = s + d$ and $z = s - d$; then the two first equations ($w^2 + z^2 = 89$ and $wz + w + z = 53$) will become $2ss + 2dd = 89$, and $ss - dd + 2s = 53$; whence $ss + s = 48\frac{3}{4}$, and $s = 6\frac{1}{2}$: Therefore $d (= \sqrt{(44\frac{1}{2} - ss)}) = 1\frac{1}{2}$, $w = 8$, and $z = 5$.

Again, by making $x = s + d$, and $y = s - d$, the 3d and 4th equations ($xx + yy = a\ (x + y)$, and $x^3 + y^3 = b\ (xx + yy)$) will be $2ss + 2dd = 2as$, and $2s^3 = 6sdd = b\ (2ss + 2dd) = b \times 2as$ (a being $= 18\frac{1}{10}$, and $b = 18\frac{172}{181}$): Whence $dd = as - ss = \frac{1}{3}ab - \frac{1}{3}ss$, and $ss - \frac{3}{2}as = -\frac{1}{2}ab$. From which $s = \frac{3}{4}a \pm \frac{1}{4}\sqrt{(9a^2 - 8ab)} = 10$, and $d (= \pm\sqrt{(as - ss)}) = \pm 9$; therefore $x = 10 \pm 9$, and $y = 10 \mp 9$; that is, the greatest number will be 19, and the lesser 1; but which of these x must be, depends on the other given equation $\frac{u^5 + x^5}{ux \times (u + x)^3} = 2\frac{1}{20}$; which equation, by substituting in like manner ($u = s + d$, and $x = s - d$) becomes $= \frac{s^4 + 10ssdd + 5d^4}{(ss - dd) \times 4ss} = 2\frac{1}{20}$; whence $d^4 + 3{\cdot}64ssdd = 1{\cdot}44s^4$. Now, by completing the square, and taking the root, $dd = 0{\cdot}36ss$, and $d = \frac{6}{10}s$: Therefore $x (= s - d) = s - \frac{6}{10}s$; whence $s = \frac{10}{4}x$; also $d\ (\frac{6}{10}s) = \frac{6}{4}x$; and consequently $u\ (s + d) = \frac{16}{4}x = 4x = 4$ or 76. But, if u be taken as the lesser of the two numbers, it is evident that, then, $x = 4u$, or $u = \frac{1}{4}x$; that is, $u = \frac{1}{4}$, or $\frac{19}{4}$; but of these four different values of u, the first only can fulfil the conditions of the problem : So that the five numbers required will be 4, 5, 1, 19, and 8; and the letters corresponding D, E, A, T, H.

Another Answer to the same, by Mr. H. Watson.

1. By adding the double of the second equation to the first, we have $(w+z)^2 + 2 \times (w+z) = 195$; whence $w + z = \sqrt{196} - 1 = 13$: From which, and $w^2 + z^2 = 89$, the greater number is found $= 8$, and the lesser $= 5$.

2. By multiplying together the third and fourth equations, we have $\frac{x^3+y^3}{x+y} = ab$; that is, $xx - xy + yy = ab$ (a being $= 18\frac{1}{10}$, and $b = 18\frac{172}{181}$): The double of which, taken from the triple of the third equation, $xx + yy = a \times (x+y)$, gives $xx + 2xy + yy = 3a \times (x+y) - 2ab$, or $(x+y)^2 - 3a \times (x+y) = -2ab$; whence $x + y = \frac{3a}{2} \pm \frac{1}{2}\sqrt{(9aa - 8ab)} = 20$: from which, and $xx + yy = 18\frac{1}{10} \times 20$ ($= 362$), the greater quantity is found $= 19$, and the lesser $= 1$.

3. It is plain that $(u+x) \times (u^4 - u^3x + uuxx - ux^3 + x^4) = u^5 + x^5$; whence $\frac{u^5 + x^5}{ux \times (u+x)^3} = \frac{u^4 - u^3x + uuxx - ux^3 + x^4}{ux \times (u+x)^2} = 2\frac{1}{20}$; and consequently $u^4 - u^3x + uuxx - ux^3 + x^4 = 2{\cdot}05ux \times (uu + 2ux + xx)$; that is, $(uu + xx)^2 - ux \times (uu + xx) - uuxx = 2{\cdot}05ux \times (uu + xx) + 4{\cdot}1uuxx$, or $(uu + xx)^2 - 3{\cdot}05ux \times (uu + xx) = 5{\cdot}1uuxx$; whence, by completing the square, and extracting the root, $uu + xx - 1{\cdot}525ux = 2{\cdot}725ux$: and therefore $uu - 4{\cdot}25xu = -xx$; and, by completing the square again, $u = 2{\cdot}125x \pm 1{\cdot}875x = 4x$ or $\frac{1}{4}x$; that is, $u = 4$, 76, $\frac{1}{4}$, or $\frac{19}{4}$; but the first of these values must be the required one; and the letters, answering the conditions of the problem, D, E, A, T, H.

Oh! cruel case! how fix'd the grief!
When DEATH alone can yield relief.

IX. QUESTION 414, *by Mr.* E. Rollinson.

A beam BC is to be supported in a given position, by means of a prop DE, of a given length, insisting on the horizontal beam AB: 'Tis proposed to determine the position of the prop, so that the beam AB whereon it stands may be the least subject to break, or so that the force, whereby it actually tends to break, may be to the whole force it can sustain at E in the least ratio possible, the thickness of the beam being every-where the same.

Answered by the Proposer, Mr. Rollinson.

Since the stress or pressure upon the prop ED is (by mechanics) as

$\frac{1}{BF}$ (supposing BF perpendicular to ED), the force in the perpendicular direction, whereby the beam tends to break at E, will be as $\frac{\cdot}{BE}$ (it being to the absolute force, in the direction DE, as BF to BE). But the strength of (or the whole force necessary to break) the beam at E, is known to be

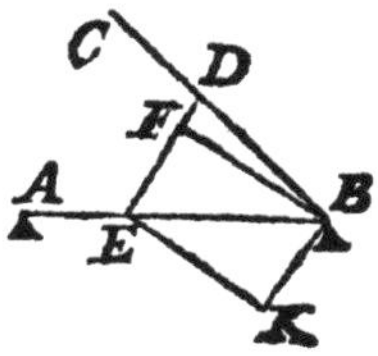

as $\frac{1}{AE \times BE}$; which is to $\frac{1}{BE}$ the force above-mentioned in the proportion of $\frac{1}{AE \times BE}$ to $\frac{AE}{AE \times BE}$, or of 1 to AE; whence (by the question) AE is to be a minimum, and consequently BE a maximum: But BE is to the sine of the $\angle$D in the given ratio of ED to the sine of B; whence it is evident that BE will be a maximum when the sine of D is so, or when D itself is a right angle.——Therefore having made BK $\perp$ BC and = the given length of DE, draw KE $\parallel$ BC, meeting AB in E, so shall E be the place where the end of the prop must stand.

X. QUESTION 415, *by Mr.* Hugh Brown.

An usurer lends 2000l. at 5 per cent. per annum compound interest, which the borrower is to clear off by quarterly payments, viz. 1l. at the end of the first quarter, 2l. at the end of the second, 3l. at the end of the third, and so on. Quere, at what time will the debt be the greatest possible?

Answered by Mr. W. Bevil.

Let $s = 2000 =$ the sum proposed, $r\ (1{\cdot}05^{\frac{1}{4}}) =$ the amount of 1l. in one quarter, and $n =$ the number of quarters required; then the amount of the sum s will be sr^n, and the amount of all the quarterly payments (exclusive of that due at the end of n quarters) $=$ $r^{n-1} + 2r^{n-2} + 3r^{n-3} + 4r^{n-4}$, &c. continued to $n - 1$ terms; the sum of all which will be found $= \frac{r}{(r-1)^2} \times (r^n - rn + n - 1)$*; and consequently the money then owing $= sr^n - \frac{r}{(r-1)^2} \times (r^n - rn + n - 1)$; whereof the interest for one quarter is $(r-1) \times (sr^n - \frac{r}{(r-1)^2} \times (r^n - rn + n - 1))$; and

* See Turner's Mathematical Exercises No. 2.

this, by the nature of the question, must be equal to rn (the sum r paid at the end of n quarters, and its interest for the same quarter): Hence we have $(r-1)^2 \times sr^{n-1} - r^n + 1 = 0$, and $n = \frac{\log. a}{\log. r}$ $= 28{\cdot}99$ (a being put equal to $\frac{r}{r - s\,(r-1)^2} = 1{\cdot}42361$): Therefore at the end of 29 quarters the debt will be the greatest possible (the moment before the 29th quarter is paid).

The same answered by Mr. Hugh Brown.

Put $a = 2000$, $r = 1{\cdot}05^{\frac{1}{4}} = 1{\cdot}0122722$, &c. and z the number of quarters at the end of which the debt will be the greatest; then (by the question, and known principles) $ar^z - r^{z-1} - 2r^{z-2} - 3r^{z-3} \ldots\ldots - (z-1) \times r = ar^z + \frac{rz}{r-1} - \frac{r^{z+1} - r}{(r-1)^2}$ a max. Therefore $a\dot{z}r^z n + \frac{r\dot{z}}{r-1} - \dot{z}r^z n \times \frac{r}{(r-1)^2} = 0$; n being $=$ ($\cdot0121975$) the hyperbolic log. of r: Whence $r^z = \frac{r}{n} \times \frac{r-1}{r - a \times (r-1)^2} = 1{\cdot}43233$, &c. and $z = \frac{\log. 1{\cdot}43233, \text{ \&c.}}{\frac{1}{4}\log. 1{\cdot}05} = \frac{\cdot156042}{\cdot005297} = 29{\cdot}45$, &c. Hence, as the answer by the nature of the question is restrained to a whole number, it is manifest, that at the end of 29 quarters (the moment before the payment then due is made) the debt will be the greatest possible.

XI. QUESTION 416, *by Mr.* Tho. Moss.

Suppose 12 half-pence to be thrown up, and those that come up heads to be taken away, and the remaining ones to be thrown up again, and so on, in the same manner, till all the half-pence have been thrown up heads: 'Tis proposed to find in what number of throws, according to an equality of chances, this may be effected?

Answered by Mr. E. Rollinson.

The probability of missing a head x times together with a single halfpenny being $\frac{1}{2}^x$, the probability of throwing any halfpenny a head, in x trials, will therefore be expressed by $1 - \frac{1}{2}^x$; and consequently that of throwing all 12 assigned heads, in x trials, $= (1 - \frac{1}{2}^x)^{12}$ (for the throwing all the 12 heads may be considered as 12 independent events; since the happening of some of these, sooner or later, no ways influences the happening of the others). Hence, by the conditions of the problem, we have $(1 - \frac{1}{2}^{x})^{12} = \frac{1}{2}$; and consequently $x = \frac{\log.(1-\frac{1}{2})^{\frac{1}{12}}}{\log \frac{1}{2}} = 4{\cdot}155$, the number of throws required; which, not being an integer, shews there can be no exact equality of chance, in the case proposed; 4 being a small matter too little, and 5 considerably too great.

Corollary. It appears from hence that the exact odds of bringing up all the heads of any number (n) of half-pence, in x throws, will be as $(1 - \frac{1}{2}^{x})^{n}$ to $1 - (1 - \frac{1}{2}^{x})^{n}$, universally.

XII. QUESTION 417, *Mr.* W. Bevil.

Suppose that a chain is to be suspended, at its extremes, by two tacks, in the same horizontal line, at the distance of 10 feet: To find the length of the chain, such that the stress or force thereof, upon the tacks, may be a minimum?

Answered by Mr. Lionel Charlton.

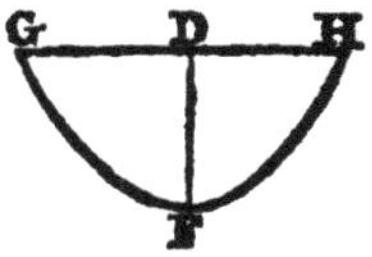

If half the length of the chain or curve (GF or HF) be denoted by z, then will the abscissa DF $= \sqrt{(aa + zz)} - a$, and the given semi-ordinate GD $(= c) = a \times$ hyp. log. $\frac{z + \sqrt{(aa + zz)}}{a}$ (see Landen's Mathematical Lucubrations, p. 34). Moreover, by mechanics, the stress or force upon both the pins, at G and H, will be to the weight of the chain, as the radius is to the sine of the $\angle$G or H, that is, as $\dot{z}$ to $\dot{x}$ $\left(\frac{z\dot{z}}{\sqrt{(aa + zz)}}\right)$, or as $2\sqrt{(aa+zz)}$ to $2z$. Therefore the weight of the chain being as the length $2z$, the stress upon the pins will consequently be as $2\sqrt{(aa + zz)}$; which quantity is to be a minimum, by the question: From whence, and the above equation, $c = a \times$ hyp. log. $\frac{z + \sqrt{(aa + zz)}}{a}$, both the values of a and z will be found for the hyp. log. $\frac{z + \sqrt{(aa + zz)}}{a}$ being $= \frac{c}{a}$, supposing N to denote the number whose hyp. log. $= 1$, it is evident that $N^{\frac{c}{a}} = \frac{z + \sqrt{(aa + zz)}}{a}$; whence, by reduction, z is found $= \frac{1}{2}aN^{\frac{c}{a}} -$

$\frac{1}{2}a\text{N}^{-\frac{c}{a}}$, and from thence $2\sqrt{(aa+zz)} = a\text{N}^{\frac{c}{a}} + a\text{N}^{-\frac{c}{a}}$: This, in fluxions, &c. gives $\dot{a}\text{N}^{\frac{c}{a}} - \frac{c\dot{a}}{a}\text{N}^{\frac{c}{a}} + \dot{a}\text{N}^{-\frac{c}{a}} + \frac{c\dot{a}}{a}\text{N}^{-\frac{c}{a}} = 0$; whence $\text{N}^{\frac{2c}{a}} = \frac{c+a}{c-a}$, or $\text{N}^{2v} = \frac{v+1}{v-1}$ (by putting $a = \frac{c}{v}$); which (in logarithms) becomes $2v =$ hyp. log. $\frac{v+1}{v-1}$; whence $v =$ 1·1996; and from thence $a = c \times {\cdot}8336 = 4{\cdot}168$, and $2z$ (GFH) $=$ $2c \times 1{\cdot}2578 = 12{\cdot}578$.

The same answered by Penovius.

Let (GF) half the length of the chain $= z$, GD (HD) $= b$, and DF $=$ x; then will $b = a \times$ hyp. log. $\frac{z + \sqrt{(ua+zz)}}{a}$, by the property of the curve. Also as $\dot{x}\left(\frac{z\dot{z}}{\sqrt{(aa+zz)}}\right) : \dot{z} :: z$ (half the chain) : $\sqrt{(aa+zz)}$ = the stress on each pin, which put $= n$; then, by exterminating z, we have $\frac{b}{a} =$ hyp. log. $\frac{n+\sqrt{(nn-aa)}}{a} =$ hyp. log. $(n+\sqrt{(nn-aa)})$ − hyp. log. a; whereof the fluxion, when n is a minimum, will be $-\frac{b\dot{a}}{aa} = \frac{-a\dot{a}}{\sqrt{(nn-ua)} \times (n+\sqrt{(nn-aa)})} - \frac{\dot{a}}{a}$; whence by reduction, $n = \frac{ba}{\sqrt{(bb-aa)}}$; Therefore $\sqrt{(nn-aa)} =$ $\frac{aa}{\sqrt{(bb-aa)}}$, and $n + \sqrt{(nn-aa)} = \frac{ba+aa}{\sqrt{(bb-aa)}} = a\sqrt{\frac{b+a}{b-a}}$; which value substituted above gives $\frac{b}{a} =$ hyp. log. $\sqrt{\frac{b+a}{b-a}}$, or a $\times$ (hyp. log. $(b+a)$ − hyp. log. $(b-a)) = 2b$: Whence $a = 4{\cdot}1677$; and from thence $2z$ (GFH) $= b \times 1{\cdot}2578 = 12{\cdot}578$.

Corollary. Hence the length of the chain GFH (when the stress is a minimum) is to the given distance of the tacks G, H, as 1·2578 to 1; and the stress on the tacks is to the weight of the chain, as 1·1996 to 1, or as 6 to 5, nearly.

XIII. QUESTION 418, *by Mr.* Tho. Peart.

Suppose a mountain, formed by the rotation of the catenaria, whose superficial content is equal to the square of its base diameter;

and suppose two equal pendulums, one at the foot, and the other at the vertex of the mountain, to be put in motion the same instant; and at the end of 24 hours, measured by the former, a cannon ball is there discharged in such a direction, as to fall the farthest possible upon the mountain, and arrives at the vertex at the same instant the pendulum there has measured 24 hours: It is required to find the mountain's height, and the velocity with which the ball is discharged?

Answered by Mr. Peart, *the Proposer.*

From the equation $zz = 2ax + xx$ of the generating curve, the surface of the mountain AEBC is easily found to be $4c\,(yz - ax)$ (where $c = \frac{3\cdot1416}{2}$), which being given $= 4yy$, we thence have (by completing the square) $y = \frac{1}{2}cz + \sqrt{(\frac{1}{4}c^2z^2 - cax)} = \frac{1}{2}c\sqrt{(2ax + xx)} + \sqrt{(\frac{1}{4}c^2 \times (2ax + xx) - cax)}$; but y is also $= a \times$ hyp. log. $\frac{a + x + \sqrt{(2ax + xx)}}{a}$ (by property of the curve): From which equal values, by substituting $ux = a$, we have $\frac{1}{2}c\sqrt{(2u + 1)} + \sqrt{(\frac{1}{4}c^2 \times (2u + 1) - cu)} =$ hyp. log. $\frac{u + 1 + \sqrt{(2u + 1)}}{u}$; whence u is found $= 1\cdot779 +$, and from thence $y = 1\cdot8078x$. Having, therefore, made BC to AB in the given proportion of 1·8078 to 1, and taken AD = the right line AC $= 2\cdot066x$, let CD be drawn, which (because the range is a *maximum*) will be the line of direction.

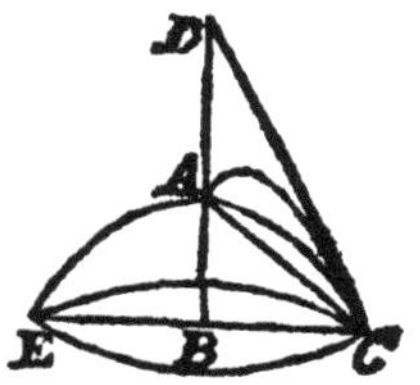

Now let $r =$ the earth's semi-diameter, in feet; $b = 16\frac{1}{12}$ feet; and $s =$ the number of seconds in 24 hours: Then it will be as $r : r + x :: s : s + sx \div r =$ the seconds taken up in performing an equal number of vibrations by the pendulum on the top of the mountain: Therefore $sx \div r$ is the time of the ball's flight; and consequently $s^2x^2 \div r^2 \times b = 2\cdot066x$ (= AD); whence x is given $= 2\cdot066rr \div bss = 7598$ feet, the mountain's height; and the time of flight $(sx \div r) = 31\cdot24$ seconds: By which number dividing CD (= 27046), the quotient $865\frac{3}{4}$ feet will be the velocity, *per second*, with which the ball is discharged.

In this solution I have supposed the gravitation to be proportional to the square of the distance from the earth's centre, inversely, without having regard to the attraction of the mountain: But if *this last* (which is to the former as the height of the mountain to the semidiameter of the earth, nearly) be also taken into the consideration, the time lost by the pendulum on the top of the mountain will then be only the half of $(sx \div r)$ what it is above found to be: Therefore the height of the mountain will come out four times as great here, as when the mountain's attraction is neglected. *

XIV. QUESTION 419, *by Mr.* R. Weston, *Discip. Landenii.*

From the equation $a^3\ddot{y} = a^2\dot{y}^2 - y^2\dot{x}^2$ it is proposed to find x in terms of y, without first finding y in terms of x, and then reverting the series.

Answered by Mr. R. Weston, *the Proposer.*

Since it appears, by what is done in the last Diary, that $-\frac{\dot{y}\ddot{x}}{\dot{x}}$ may be substituted for $\ddot{y}$, if such substitution be made in the given equation, the required value of x may then be readily obtained by the common method of finding fluents by infinite series: Or that value may be found in the following manner:

Assume $x = b + fy + py^2 + qy^3 + ry^4$, &c. supposing that when y is $= 0$, x is $= b$, and $\frac{\dot{x}}{\dot{y}} = f$: Then the given equation ($a^3\ddot{y} = a^2\dot{y}^2 - y^2\dot{x}^2$) being multiplied by the invariable quantity $\dot{x}$, we have $a^3\dot{x}\ddot{y} = a^2\dot{x}\dot{y}^2 - y^2\dot{x}^3$; whence substituting for $\dot{x}^3$ its value found by the assumed equation, we have, after dividing by $\dot{y}^2$, $\frac{a^3\dot{x}\ddot{y}}{\dot{y}\dot{y}} = a^2\dot{x} - f^3y^2\dot{y} - 6f^2py^3\dot{y}$, &c. Hence, by taking the fluents, we find $-\frac{a^3\dot{x}}{\dot{y}} = a^2x - \frac{f^3y^3}{3} - \frac{6f^2py^4}{4} - \frac{9f^2q + 12fp^2}{5}y^5$, &c. But, x being $= b$, and $\frac{\dot{x}}{\dot{y}} = f$, when y is $= 0$, the correct equation of the fluents is $-\frac{a^3\dot{x}}{\dot{y}} = -a^2b - a^3f + a^2x - \frac{f^3y^3}{3} - \frac{6f^2py^4}{4}$ &c. which multiplying by $-\frac{\dot{y}}{a^3}$ and substituting for x its assumed value, becomes $\dot{x} = f\dot{y} - \frac{fy\dot{y}}{a} - \frac{py^2\dot{y}}{a} + \left(\frac{f^3}{3a^3} - \frac{q}{a}\right) \times y^3\dot{y}$, &c. whence, taking the correct fluents, we get

$$x = b + fy - \frac{fy^2}{2a} - \frac{py^3}{3a} + \frac{f^3 - 3a^2q}{3.4a^3}y^4 + \frac{6f^2p - 4a^2r}{4.5a^3}y^5, \text{ \&c.}$$

Consequently, by comparing the two values of x; p, q, r, &c. will be known.

---

* This may be explained thus: Suppose the earth's attraction to be 1, and that it is to that of the mountain as r to x; then $x \div r$ will be the attraction of the mountain, and $1 + x \div r$ the sum of both; and, because the times of *vibration* are as the roots of the forces reciprocally, $\sqrt{1} : \sqrt{(1 + \frac{x}{r})} = 1 + \frac{x}{2r}$ nearly $:: s :$ $s + \frac{sx}{2r}$ nearly, and $\frac{sx}{2r}$ = time lost nearly. M.

The same answered by Mr. Henry Watson.

Put $\dot{y} = z\dot{x}$, then $\ddot{y} = \dot{z}\dot{x}$; which values being substituted in the given equation, $a^3\ddot{y} = a^2\dot{y}^2 - y^2\dot{x}^2$, and the whole divided by $\dot{x}$, we have $a^3\dot{z} = a^2z^2\dot{x} - y^2\dot{x} = a^2z^2 \times \frac{\dot{y}}{z} - y^2 \times \frac{\dot{y}}{z}$ (because $\dot{x} = \frac{\dot{y}}{z}$) and consequently $a^3z\dot{z} = a^2z^2\dot{y} - y^2\dot{y} = (a^2z^2 - y^2) \times \dot{y}$: Put $v = a^2z^2 - y^2$; then $zz = \frac{v + yy}{aa}$ and $z\dot{z} = \frac{\frac{1}{2}\dot{v} + y\dot{y}}{aa}$; and so, by substitution, $a \times (\frac{1}{2}\dot{v} + y\dot{y}) = v\dot{y}$, or $\frac{1}{2}a\dot{v} = (v - ay) \times \dot{y}$: Put now $w = v - ay$ ($= a^2z^2 - y^2 - ay$) then $v = w + ay$, and $\dot{v} = \dot{w} + a\dot{y}$; whence again, by substitution, $\frac{1}{2}a \times \dot{w} + a\dot{y} = w\dot{y}$, or $\frac{1}{2}a\dot{w} = (w - \frac{1}{2}aa) \times \dot{y}$, and therefore $\dot{y} = \frac{\frac{1}{2}a\dot{w}}{w - \frac{1}{2}aa}$; whereof the fluent is $y = \frac{1}{2}a \times \text{hyp. log.} \frac{w - \frac{1}{2}aa}{d}$ (d being any constant quantity at pleasure) $= \frac{1}{2}a \times \text{hyp. log.} \frac{a^2z^2 - y^2 - ay - \frac{1}{2}aa}{d}$. Hence, putting M = the number whose hyp. log. = 1, we have $M^{\frac{2y}{a}} = \frac{a^2z^2 - y^2 - ay - \frac{1}{2}aa}{d}$, and consequently $z^2 = \frac{y^2 + ay + \frac{1}{2}aa + dM^{\frac{2y}{a}}}{aa}$; from whence $\dot{x}\ (= \frac{\dot{y}}{z}) = \frac{a\dot{y}}{\sqrt{(y^2 + ay + \frac{1}{2}aa + dM^{\frac{2y}{a}})}}$: From which, when $d = 0$, x will be found $= a \times \text{hyp. log.} \frac{y + \frac{1}{2}a + \sqrt{(y^2 + ay + \frac{1}{2}aa)}}{c}$, wherein c may be any constant quantity at pleasure.

XV. QUESTION 420, *by Mr.* Patrick O'Cavanah, *of Dublin.*

To find the sum of the infinite series $1 - \frac{z^3}{2.3} + \frac{z^6}{2.3.4.5.6} - \frac{z^9}{2.3.4.5.6.7.8.9}$, &c. by means of circular arcs and logarithms.

Answered by Κυβερνητης.

Conceive the given series $1 - \frac{z^3}{2.3} + \frac{z^6}{2.3.4.5.6}$, &c. to be composed of three others,

$$\text{A} \times \left(1 + pz + \frac{p^2z^2}{2} + \frac{p^3z^3}{2.3} + \frac{p^4z^4}{2.3.4}\right), \text{ \&c.}$$
$$\text{B} \times \left(1 + qz + \frac{q^2z^2}{2} + \frac{q^3z^3}{2.3} + \frac{q^4z^4}{2.3.4}\right), \text{ \&c.}$$
$$\text{C} \times \left(1 + rz + \frac{r^2z^2}{2} + \frac{r^3z^3}{2.3} + \frac{r^4z^4}{2.3.4}\right), \text{ \&c.}$$

Then, by taking A, B, C each $= \frac{1}{3}$, and equating the homologous terms, we shall have $p + q + r = 0$, $pp + qq + rr = 0$, $p^3 + q^3 + r^3 = -3$, $p^4 + q^4 + r^4 = 0$, &c. &c. Make now $p^3 = -1$, $q^3 = -1$, and $r^3 = -1$; that is, let p, q, and r be the three roots, $(-1, \frac{1}{2} + \sqrt{(-\frac{3}{4})}, \frac{1}{2} - \sqrt{(-\frac{3}{4})})$ of the cubic equation $x^3 = -1$, or $x^3 + 1 = 0$; then, as both the second and third terms of this equation are wanting, not only the sum of all the roots $(p + q + r)$ but the sum of all their squares $(pp + qq + rr)$ will vanish, or be equal to nothing, as they ought, to fulfil the conditions of the two first of the preceding equations. Moreover, because $p^3 = -1$, $q^3 = -1$, and $r^3 = -1$, it is likewise evident that $p^4 + q^4 + r^4 = -p - q - r = 0$, $p^5 + q^5 + r^5 = -p^2 - q^2 - r^2 = 0$, and $p^6 + q^6 + r^6 = -p^3 - q^3 - r^3 = 3$; which equations being nothing more than the three first repeated, the values of p, q, r, above determined, will equally fulfil the conditions of these also: So that the series arising from the addition of the three assumed ones will agree, in every term, with that propounded. But $1 + pz + \frac{p^2z^2}{2} + \frac{p^3z^3}{2.3}$, &c. the first of those series is known to express the number whose hyperbolic logarithm is pz: Therefore if N be taken to denote the number (2·71828, &c.) whose hyp. log. is unity, then will $1 + pz + \frac{ppzz}{2}$, $= \text{N}^{pz}$: And in the same manner $1 + qz + \frac{qqzz}{2}$, &c. $= \text{N}^{qz}$, &c. and consequently $1 - \frac{z^3}{2.3} + \frac{z^6}{2.3.4.5.6}$, &c. $= \frac{1}{3}$ into $\text{N}^{pz} + \text{N}^{qz} + \text{N}^{rz} = \frac{1}{3}\text{N}^{-z} + \frac{1}{3}\text{N}^{\frac{1}{2}z} \times (\text{N}^{nz\sqrt{-1}} + \text{N}^{-nz\sqrt{-1}})$ (making $n = \sqrt{\frac{3}{4}}$). But $\text{N}^{nz\sqrt{-1}} + \text{N}^{-nz\sqrt{-1}}$ is known to express the double of the cosine of the arch nz (the radius being 1); which cosine let be denoted by S, and let the number whose hyp. log. is z be represented by T; then we shall have $1 - \frac{z^3}{2.3} + \frac{z^6}{2.3.4.5.6}$, &c. $= \frac{1}{3\text{T}} + \frac{2\text{ST}^{\frac{1}{2}}}{3}$.

From the same method, and the known roots of the equation $x^n \pm 1 = 0$, the series $1 \pm \frac{z^n}{1.2.3\,(n)} + \frac{z^{2n}}{1.2.3\,(2n)}$, &c. may be summed, n being any whole positive number.

THE PRIZE QUESTION, *by Mr.* H. Watson.

A person being to pass from one place A to another B, at the distance of three miles, between which places there lies a morass, is desirous to know the path he must describe, and also how far he must travel, to perform his journey in the shortest time possible; the way, by reason of the morass, being rendered so bad, that he can only move with a celerity every where proportional to his distance from the centre thereof, lying one mile distance from A, and two from B.

Answered by Mr. O'Cavanah.

From C, the centre of the morass, let a semi-circle AMN be described; and from any two points D, d in the required curve ADB, indefinitely near to each other, conceive DC and dc to be drawn, intersecting the semi-circle in M and m: Calling CA, 1; CB, b; AM, z; CD, y; Mm, $\dot{z}$; and de, $\dot{y}$. Then CM (1) : Mm ($\dot{z}$) :: CD (y) : De $= y\dot{z}$. Whence Dd ($=\sqrt{(\text{D}e^2 + de^2)}$) $= \sqrt{(y^2\dot{z}^2 + \dot{y}^2)}$. This, divided by y, gives $\sqrt{(\dot{z}^2 + \dot{y}\dot{y} \div yy)}$; which (by the question) is the time of describing Dd, and which, by making $\dot{x} = \dot{y} \div y$ (or $x =$ hyp. log. y), will be reduced to $\sqrt{(\dot{z}\dot{z} + \dot{x}\dot{x})}$.

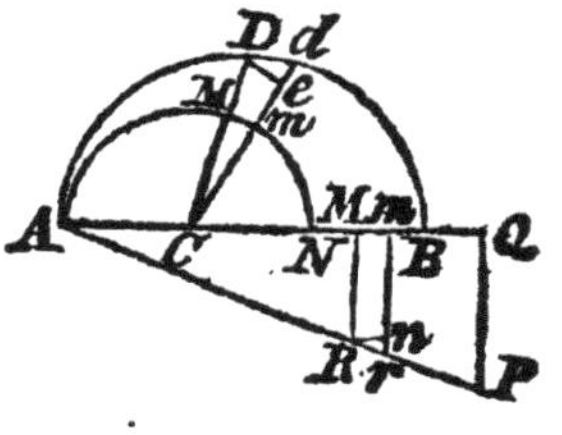

Make now the right line AQ = the semi-circumference AMN, and QP (perpendicular thereto) = hyp. log. b (= the last value of x, when $y = b =$ CB): And let the line ARP be of such a nature, that, taking the abscissa AM = the arch AM the corresponding ordinate MR shall be every-where equal to x: Then Rr (the fluxion of AR) being also expressed by $\sqrt{(\dot{z}\dot{z} + \dot{x}\dot{x})}$, it is evident that AR will truly express the time of describing the arch AD; and consequently, AP the whole time of describing the arch ADb; which will evidently be the shortest possible, when AP is a straight line (as being the shortest that can possibly be drawn between the two given points A and P).

Hence, putting AQ (= 3·1416) = p, and QP (= hyp. log. b, or of CB ÷ CA = ·693147) = q, we have, by similar triangles, $p : q :: \dot{z} : \dot{x} = q\dot{z} \div p = \dot{y} \div y$ (per above); and consequently De $= y\dot{z} = p\dot{y} \div q$; which being to de ($\dot{y}$) in the constant ratio of p to q, the angle Dde must be every-where the same, and therefore the curve ADB, the proportional, or logarithmical spiral; wherein Dd being to ed in the constant ratio of $\sqrt{(pp + qq)}$ to q, we shall therefore have $q : \sqrt{(pp + qq)}$ :: NB (the whole increase of the distance CD) : $\sqrt{(pp \div qq + 1)} \times$ NB = 4·641444 the true length of the spiral arch ABD.

Corollary. It appears from hence that the time of describing the spiral ADB will be to the time of uniformly describing the arch of the semi-circle AMN, with the first velocity at A, in the given ratio of AP to AQ, or of $\sqrt{(pp+qq)}$ to p.

The answer by *Plus-Minus* (though a small mistake is therein committed) sufficiently discovers the author to be a man of genius.

Questions proposed in 1757, *and answered in* 1758.

I. QUESTION 421, *by* Juvenis.

Three men to share a stock agree, of fifteen hundred pound:
The part of A to that of B, as four to three was found;
But C's exceeding that of A by pounds just ten times seven.
What each man shar'd, pray, ladies, say, from what above is given.

Answered by Mr. W. Bacon, *of Ipswich.*

Let the share of A be $= x$; then that of B will be $= \frac{3}{4}x$, and that of C $= x + 70$ (by the quest.); whence $2x + \frac{3}{4}x + 70 = 1500$ (by the quest.) and consequently $x = 520$. Therefore the share of A was 520l. that of B 390l. and that of C 590l.

The same answered by Mr. R. Flitcon.

Let $4x$, $3x$, and $4x + 70$ denote the shares of A, B, and C, respectively; then will $11x + 70 = 1500$ (per question): Whence $x = 130$; and the three shares 520, 390, and 590 pounds.

II. QUESTION 422, *by Mr.* William Spry, *Engineer.*

Suppose that a cannon ball is discharged to hit a target (or other obstacle) at the distance of 500 yards: How far must I stand from the target, in a perpendicular to the line drawn between it and the cannon, so as to hear the report of the shot and the explosion of the piece at the same instant of time, allowing the velocity of the ball to be to that of sound in the proportion of 3 to 2?

Answered by Mr. J. Hampson, *of Leigh.*

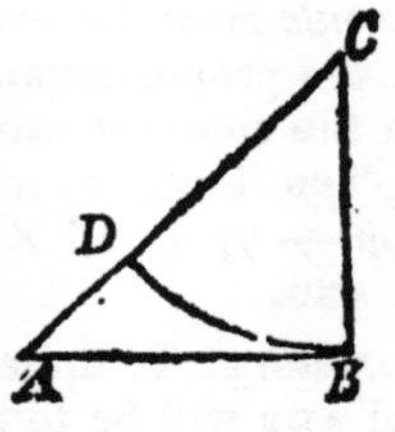

Let A be the place of the cannon, B that of the target, and C that of the spectator; and in CA take CD = CB: Then the sound of the cannon will be at D, on the ball's striking the target at B; and so AD will be to AB in the given proportion of 2 to 3 (by the question). Putting, therefore, $AB = 3a$, $AD = 2a$, and BC ($=$ CD) $= x$, we have $AC = x + 2a$; and from thence (by Euc. 47. I.) $xx + 4ax + 4aa = xx + 9aa$: From which $x = \frac{5}{4}a = 208\frac{1}{3}$ yards.

III. QUESTION 423, *by* Birchoverensis.

To determine the position of a point with respect to four given points, so that lines being drawn from thence to the given points, the sum of the four squares formed upon them shall be the least possible?

Answered by Mr. Will. Kingston, *of Bath.*

Let A, B, C, and D be the four given points: Bisect AB in E, and and CD in F; then assuming P for the required point, and putting $AE = BE = a$, $DF = CF = b$, $EP = x$, and $FP = z$, we shall (by the 12th of the 2d of Simpson's Geometry) have $AP^2 + BP^2 = 2aa + 2xx$, and $DP^2 + CP^2 = 2bb + 2zz$: Hence $AP^2 + BP^2 + DP^2 + CP^2 = 2aa + 2bb + 2xx + 2zz$, a minimum; And consequently $xx + zz$ ($EP^2 + FP^2$) a minimum also; which will evidently be when $EP + FP$ is a min. that is, when EP and FP make a right line: and EP will then be $= FP$; because (by Euc. 9. II.) the sum of the squares of the two parts of a line equally divided is always less than the sum of the squares of the parts, when the line is divided unequally.

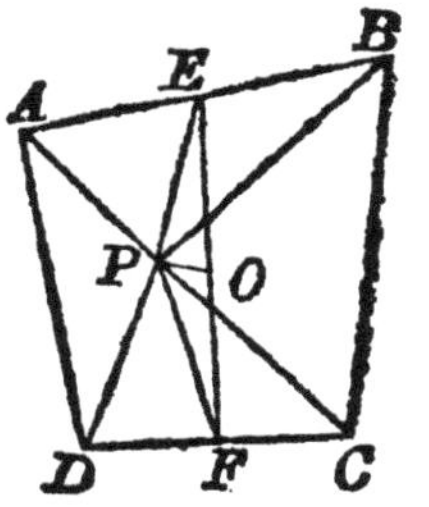

The same answered by Mr. W. Lucas.

Let A, B, C, and D represent the four given points, and P any point taken at pleasure; let PE and PF be drawn to bisect AB and DC in E and F; then (by a theorem in geometry) $AP^2 + BP^2 = 2AE^2 + 2EP^2$, and $DP^2 + CP^2 = 2DF^2 + 2FP^2$; therefore $AP^2 + BP^2 + DP^2 + CP^2 = 2 \times (AE^2 + DF^2 + EP^2 + FP^2)$, a minimum (per quest.); and consequently $EP^2 + FP^2$ a minimum also: But, drawing PO to bisect EF, we get $EP^2 + FP^2 = 2EO^2 + 2PO^2$: Therefore PO^2 a minimum, and so PO being $= 0$, O must be the point required.

IV. QUESTION 424, *by Mr.* T. Baxtonden, *of Liverpool.*

At a station due south of a tower, I observed the altitude of the top of the tower to be 30°, and that of its base 12°: Proceeding from thence 100 yards, north-east, down a path making an angle of 5° 30′ with the plane of the horizon, I again took the altitude of the tower's summit, which I then found to be 38° 30′. From whence I desire to know the height of the tower, and the distance thereof from my first station?

Answered by Mr. Tho. Baxtonden, *the Proposer.*

Let the ∠s BAC and BAR be the given elevations of the top and

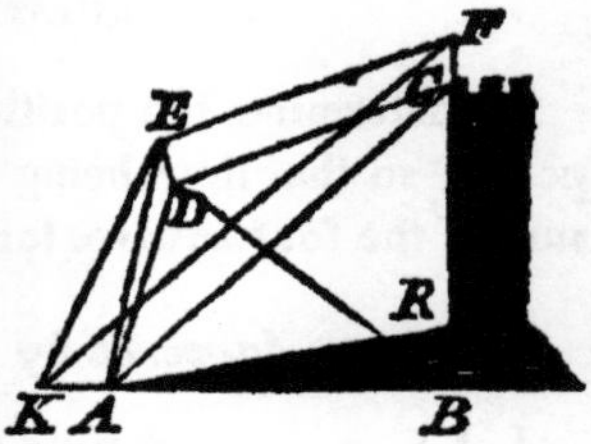

base of the tower above the plane of the horizon AB, at the first station A; let D be the second station, and DE its perpendicular distance below the plane ABE of the first. In BC produced, suppose CF to be taken = DE; let FK ∥ CA, and let EF, EB, FK, and AD be drawn. Then, because DE is = and ∥ FC, EF will also be = and ∥ DC, and consequently the ∠EFB (= DCB) = the complement of the given elevation at the second station D. Now, from AD (100) and the ∠DAE (5° 30′), we have given AE = 99·5396, and DE (FC) = 9·5846: and it will be as rad. : tang. BCA (60°) (:: BC : BA) :: FC : KA = 16·601: from which, together with AE and the contained ∠KAE (135°), the ∠AKE will be found = 38° 59′ Moreover the ∠s FBE and FBK being both right ones, it will be tang. BFE (51° 30′) : tang. BFK (60°) :: (BE : BK) :: sin. BKE (38° 59′) : sin. BEK = 60° 5′ : whence ABE is also given = 80° 56′, AEB = 54° 4′ : from which, and AE (= 99·5396) will be found AB = 81·61; and from thence BC = 47·12, BR = 17·34, and RC = 29·77 = the tower's true height.

An Algebraic Solution to the same by Mr. Rich. Mallock.

I have some reason to believe that this question is wrong printed, viz. north for south, or north-east for south-east; which being corrected, I solve it thus:

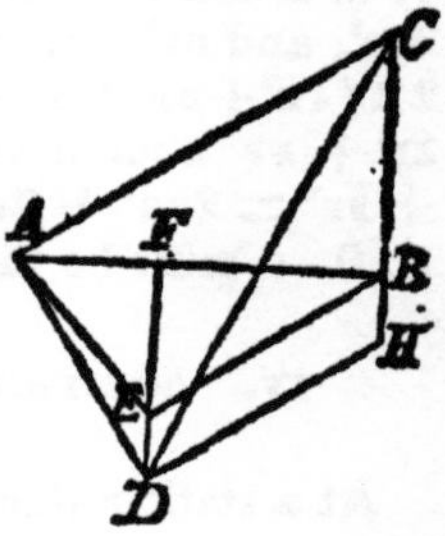

Let DE be the perpendicular distance of the second station D, below the horizontal plane (ABE) of the first station A; in CB (the tower's height) produced, take BH = ED; join D, H, and upon AB let fall the perpendicular EF.

By means of AD (= 100) and the given ∠DAE (= 5° 30′) I find BF (= BH) = 9·58458, and AE = 99·53955; from which last AF (EF) = 70·3853. Now put BH = b, AF (= EF) = c, and BC = x; and let m and n denote the tangents of the complements (BCA and HCD) of the tower's elevation at the two stations A and D: Then (by trigonometry) BA = mx, and HD = $n \times (x+b)$ = BE (because BH being ∥ and = ED, BE will also be ∥ and = HD). Hence BF = mx — c; and consequently $(mx - c)^2 + cc = (nx+nb)^2$ (Euc. 47. I.) : which ordered, gives $xx - \frac{mc + nnb}{mm - nn} \times 2x = -\frac{2cc - nnbb}{mm - nn}$ in numbers, $xx - 193{\cdot}11x = -6877{\cdot}8$; whence $x = 96{\cdot}555 \pm 49{\cdot}44$; that is, the tower's height above A is either 47·11 yards, or 146 yards; from the former whereof the required distance AB is found = 81·5 yards.

V. QUESTION 425, *by Mr.* Ja. Beresford.

The four sides of a field, whose diagonals are equal, are known to be 25, 35, 31, and 19 perches, in a successive order; from whence the content of the field is required?

Answered by Mr. O'Cavanah.

In the case proposed, where the sums of the squares of the opposite sides of the trapezium are equal (and where the diagonals do, therefore, cut each other at right angles) the problem may be thus constructed.

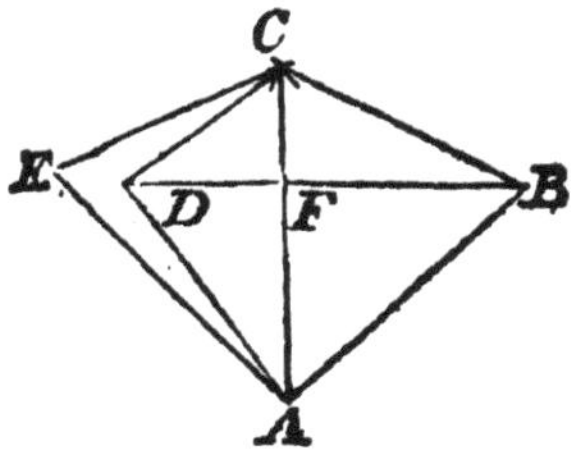

Having made AB and AE perpendicular to each other, and equal each to any one of the given sides (as 35) from the centres B and E with radii equal to the two opposite sides (25 and 31) let two arcs be described, intersecting in C; draw AC, and also BD ⊥ and = thereto; so shall ABCD (when AD and CD are drawn) be the trapezium sought.

For the ∠s ABD and EAC are equal, being both complements of CAB, to a right angle (by construction), and the sides AB, BD, AE, AC, containing them, are also respectively equal (by construction): Whence the remaining sides AD and EC must necessarily be equal. Also $AB^2 - AD^2$ ($= BF^2 - DF^2$) $= BC^2 - DC^2$, and therefore $AB^2 + DC^2 = AD^2 + BC^2$, that is, in the present case, $35^2 + DC^2 = 31^2 + 25^2$, and consequently $DC = 19$, as it ought to be. The numeral solution is from hence very easy, whereby the area comes out = 4 a. 1 r. 38¼ p.*

Mr. *W. Harrison*, Mr. *J. Honey*, Mr. *R. Mallock*, Mr. *J. Pearce*, Mr. *W. Stoker*, and Mr. *W. Terrill* (whose solutions agree in almost every step) give the following analytical investigation of this problem.

'Since the sums of the squares of the opposite sides of the trapezium ABCD are equal (by the question), the diagonals will therefore cut each other at right angles in F: so that, putting BC ($= 25$) $= a$, BA ($= 35$) $= b$, DA ($= 31$) $= c$, CD ($= 19$) $= d$, and AC ($=$ BD) $= x$, we shall have $x : b + a :: b - a :$ AF − CF $= (bb - aa) \div x$;

and $x : b + c :: b - c :$ BF − DF $= (bb - cc) \div x$; whence AF $= \frac{x}{2}$

$+ \frac{bb - aa}{2x}$, and BF $= \frac{x}{2} + \frac{bb - cc}{2x}$; and consequently $\left(\frac{x}{2} + \right.$

---

* The numerical calculation at large may be seen at p. 129, Hutton's Mensuration.

$\left.\frac{bb-aa}{2x}\right)^2 + \left(\frac{x}{2}+\frac{bb-cc}{2x}\right)^2 = bb$; from which equation $2x^4 - (aa+cc) \times 2xx + (bb-aa)^2 + (bb-cc)^2 = 0$. This solved, gives $x = \sqrt{\left\{\frac{aa+cc}{2} \pm \sqrt{(bbdd - \frac{1}{4} \times (cc-aa)^2)}\right\}} = 37{\cdot}9$; and the area sought $(\frac{1}{2}xx) = 4$ a. 1 r. 38 p.'

A General Solution of the same Problem by Birchoverensis.

Let the given sides DA, AB, BC, CD of the trapezium be denoted by a, b, c, and d respectively: and let each of the equal diagonals AC, BD, be called x; Then (supposing DE and CF to be $\perp$ AB, and DG $\parallel$ AB) we shall have $b : x + a :: x - a :$ BE $-$ AE $= \frac{xx-aa}{b}$; whence AE $= \frac{bb+aa-xx}{2b}$, and DE $= \sqrt{\frac{-x^4+(aa+bb)\times 2xx-(bb-aa)^2}{4bb}}$.

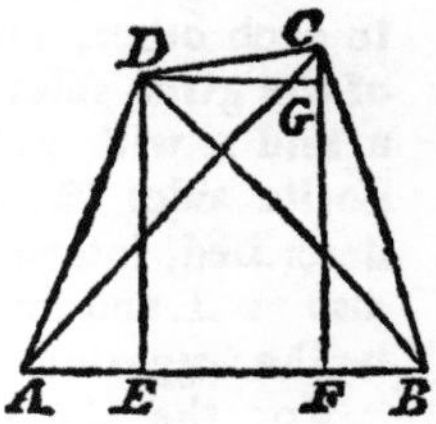

In the same manner BF $= \frac{bb+cc-xx}{2b}$, and CF $= \sqrt{\frac{-x^4+(cc+bb)\times 2xx-(bb-cc)^2}{4bb}}$; and consequently DG (EF $=$ AB $-$ AE $-$ BF) $= \frac{2xx-aa-cc}{2b}$; From which (because $\sqrt{(\text{DC}^2 - \text{DG}^2)} = \text{GC} \doteq \text{CF} - \text{DE}$) we get $\sqrt{(4bbdd - 4x^4 + (aa+cc) \times 4xx - (aa+cc)^2)} = \sqrt{(-x^4 + (cc+bb) \times 2xx - (bb-cc)^2)} - \sqrt{(-x^4 + (aa+bb) \times 2xx - (bb-aa)^2)}$; whence, by involution and proper reduction, there results $2x^6 - (aa+bb+cc+dd) \times x^4 - (2aacc + 2bbdd - (aa+cc) \times (bb+dd)) \times xx + (aacc - bbdd) \times (aa+cc-bb-dd) = 0$; from which x may be found, let the values of a, b, c, and d be what they will: But, in the case proposed, $aa+cc$ being $= bb+dd$, the last term will vanish, and we shall have $2x^4 - (2aa+2cc) \times xx - 2aacc - bbdd + (aa+cc)^2 = 0$; which solved, gives $x = \sqrt{\left\{\frac{aa+cc}{2} \pm \sqrt{(bbdd - \frac{1}{4} \times (cc-aa)^2)}\right\}} = 37{\cdot}90025$. Whence the area ABCD $(= \frac{1}{2}xx) = 718{\cdot}2145 = 4$ a. 1 r. 38·2145 p.

VI. QUESTION 426, *by Mr.* W. Spicer.

A man laid out 60 pounds in sheep, of three different sorts; for the first sort he paid 9 shillings a-piece, for the second 12, and for the third 15: And the number he bought of each sort was such, that the sum of their three squares was less than it could possibly have been, had he bought more of any one sort and less of another. What number of sheep did he buy?

Answered by Mr. L. Charlton.

Let x, y, and z represent the sheep bought of each sort, and a, b, and c the given prices paid for each sort per head, respectively; then will $ax + by + cz = 1200\ (= d)$, and $xx + yy + zz$ a minimum (by the quest). Now, seeing any two of the quantities x, y, z, may be here varied independently of the other, we shall (by making x and y to flow, while z remains constant) have $a\dot{x} + b\dot{y} = 0$, and $2x\dot{x} + 2y\dot{y} = 0$; from whence $\dot{x} = -\frac{b\dot{y}}{a} = -\frac{y\dot{y}}{x}$; and therefore $y = \frac{bx}{a}$. In the very same manner $z = \frac{cx}{a}$: Whence, by substitution, $ax + \frac{bbx}{a} + \frac{ccx}{a} = d$; and consequently $x = \frac{ad}{aa + bb + cc} = 24$; whence y is also given $= 32$, and $z = 40$.

VII. QUESTION 427, *by Mr.* T. Moss.

The rectangle under the two diagonals of any trapezium drawn into twice the cosine of the angle contained by them (the radius being 1) is equal to the difference of the aggregates of the squares of the opposite sides of the trapezium: and the area of the trapezium is equal to one-fourth of the same difference drawn into the tangent of the said angle. A demonstration of this is required.

Answered by Mr. J. Thompson, *of Whitherly Bridge.*

In the trapezium ABCD, put AG $= m$, BG $= n$, CG $= p$, DG $= q$, area ABCD $=$ A, x and y for the sine and cosine of the $\angle$AGD; then, by a known theorem,

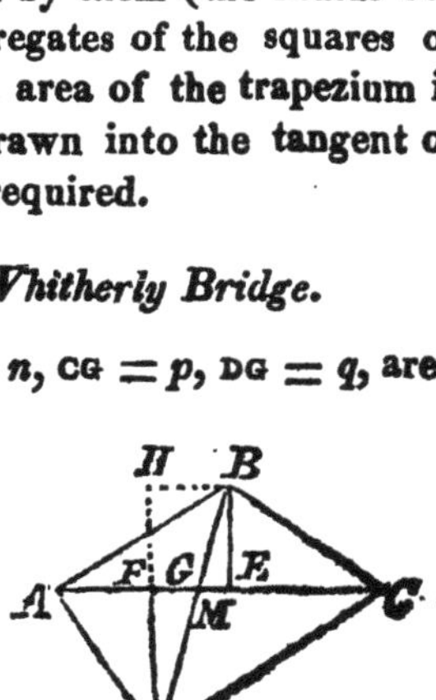

$$\text{AB}^2 = mm + nn + 2mny$$
$$\text{DC}^2 = pp + qq + 2pqy$$
$$\text{AD}^2 = mm + qq - 2mqy$$
$$\text{BC}^2 = nn + pp - 2npy$$

Whence, by equal subtraction, $\text{AB}^2 + \text{DC}^2 - \text{AD}^2 - \text{BC}^2 = (mn + pq + mq + np) \times 2y = (m + p) \times (n + q) \times 2y = \text{AC} \times \text{BD} \times 2y$. Again, by a known theorem, $mnx + npx + pqx + qnx = 2\text{A}$, or $(m + p) \times (n + q) \times x^* = 2\text{A}$; hence $\frac{2\text{A}}{x} = \frac{\text{AB}^2 + \text{DC}^2 - \text{AD}^2 - \text{BC}^2}{2y}$, and $(\text{AB}^2 + \text{DC}^2 - \text{AD}^2 - \text{BC}^2) \times \frac{x}{4y} = \text{A}$.

Corollary. From the equation * the area of any trapezium is equal to half the rectangle under the two diagonals drawn into the sine of the angle contained by them.

The same answered by Mr. T. Moss, *the Proposer.*

Suppose the diagonal AC to be bisected in M; and having let fall the ⊥s BE and DF, draw BH ∥ AC, meeting DF, produced, in H. It is well known, that $AB^2 - BC^2 = 2AC \times ME$, and $CD^2 - AD^2 = 2AC \times MF$: Whence, by equal addition, $AB^2 + CD^2 - BC^2 - AD^2 = 2AC \times (ME + MF)$ $= 2AC \times BH$: But $BD = \frac{BH \times \text{rad}}{\cos AGD}$; whence $AC \times BD = AC \times BH \times \frac{\text{rad.}}{\cos G} = (AB^2 + CD^2 - BC^2 - AD^2) \times \frac{\frac{1}{2}\text{ rad.}}{\cos G}$. Again, $DH = BH \times \frac{\text{tang. } G}{\text{rad.}}$; whence $\frac{1}{2}AC \times DH$ (= area ABCD) $= \frac{1}{2}AC \times BH \times \frac{\text{tang. } G.}{\text{rad.}} = (AB^2 + CD^2 - BC^2 - AD^2) \times \frac{\frac{1}{4}\text{ tang. } G}{\text{rad.}}$. Q. E. D.

VIII. QUESTION 428, *by Mr.* Lionel Charlton.

The sum (200) of the two extremes, and the sum (300) of the four means of six numbers in continued geometrical proportion being given; to find the numbers themselves, by an equation not exceeding a quadratic.

This question, Mr. Charlton observes, was proposed to him by a Gentleman at the Mathematical College at Edinburgh.

Answered by Mr. Robert Butler.

Let a = the first term, r = common ratio; then $a + ar + ar^2 + ar^3 + ar^4 + ar^5$ $\left(= a \times \frac{r^6 - 1}{r - 1}\right) = 500$, and $a \times (r^5 + 1) = 200$ (per question). Multiply these two equations cross-wise; whence $2 \times (r^6 - 1) = 5 \times (r^5 + 1) \times (r - 1)$. Now put $2x = r + 1$, and $2z = r - 1$; then will $x + z = r$, and $x - z = 1$; therefore, by substitution, $(x + z^6) - (x - z^6) = 5z \times ((x + z^5) + (x - z^5))$; which, being expanded, gives $19z^4 = x^4 - 30xxzz$; whence $xx = (15 + \sqrt{244}) \times zz$; therefore $z = \frac{1}{\sqrt{(15 + \sqrt{244})} - 1} = \cdot 22057$. Hence $r = 1\cdot 4412$, and $a = 27\cdot 527$: Therefore the six numbers answering the question are 27·527, 39·94, 57·56, 82·95, 119·55, and 172·473.

* *The same answered by Messrs.* Ja. Bank and J. Wilson.

Let x and z represent the two middle terms, and put $a = 200$, and $c=300$: Then, by the question, $x+z+\frac{xx}{z}+\frac{zz}{x}=c$, and $\frac{x^3}{z^2}+\frac{z^3}{x^2}=a$, or (by reduction) $(x+z)\times xz+x^3+z^3=cxz$, and $x^5+z^5=ax^2z^2$; which, by putting $s=x+z$ and $p=xz$, will become $s^3-2ps=cp$, and $s^5-5ps^3+5p^2s=ap^2$; from the former whereof we have $p=\frac{s^3}{2s+c}$; which value, substituted in the latter, gives $s^5-\frac{5s^6}{2s+c}+\frac{5s^7}{(2s+c)^2}=\frac{as^6}{(2s+c)^2}$: Whence $s^2+(a+c)\times s=c^2$, and consequently $s=\frac{\sqrt{(4c^2+(a+c)^2)}-a-c}{2}=50\sqrt{61}-250=140\cdot51248$. Hence $p\left(=\frac{s^3}{2s+c}\right)=\frac{2590000-320000\sqrt{61}}{19}$, and $x-z\ (=\sqrt{(ss-4p)})=\frac{\sqrt{(805000\sqrt{61}-6275000)}}{19}=25\cdot39269$; from whence the numbers are found to be 27·714, 39·939, 57·559, 82·952, 119·55, and 172·286.

The same answered by Mr. William Davies.

Put $2a=400$ the sum of the two extremes, $3a=300$ the sum of the four means, $a-ax=$ the first term, $a+ax=$ the last term, and $y=$ the common ratio: Then $(a-ax)\times y^5=a+ax$, and therefore $y^5=\frac{1+x}{1-x}$; also $(a-ax)\times(y+y^2+y^3+y^4)$, or $(a-ax)\times\frac{y-y^5}{1-y}=3a$; whence $y^5=\frac{4y-3-xy}{1-x}=\frac{1+x}{1-x}$; and consequently $y=\frac{4+x}{4-x}=\frac{c+x}{c-x}$ (putting $c=4$.) Hence we have $\frac{(c+x)^5}{(c-x)^5}=\frac{1+x}{1-x}$; from which, by reduction, $(5c-1)\times x^4+(10c^3-10c^2)\times x^2=5c^4-c^5$, or $19x^4+480x^2=256$; solved $x=\sqrt{\frac{32\sqrt{61}-240}{19}}$; whence the rest may be found.

The same by Mr. T. Barker *of Westhall, in Suffolk.*

Let $x-y$ and $x+y$ denote the third and fourth numbers (putting $a=200$ and $b=300$). Then (by the question) we shall have $\frac{(x-y)^3}{(x+y)^2}+\frac{(x+y)^3}{(x-y)^2}=a$, and $2x+\frac{(x-y)^2}{x+y}+\frac{(x+y)^2}{x-y}=b$; which, cleared of fractions, give $2x^5+20x^3y^2+10xy^4=a\times(x^4-2x^2y^2+y^4)$, and $4x^3+4xy^2=b\times(xx-yy)$; from the last whereof we obtain $y = \frac{\frac{1}{4}bx^2 - x^3}{\frac{1}{4}b + x}$. Now substitute the value of yy in the other equation; from whence, after proper reduction, there will come out $xx + \frac{a+b}{2} \times x = \frac{bb}{4}$; which, solved, gives $x = \frac{\sqrt{(aa + 2ab + 5bb)} - a - b}{4} = 70{\cdot}256$; therefore $y = 12{\cdot}696$; and the six numbers sought are 27·72, 39·964, 57·564, 82·952, 119·44, and 172·28.

The same answered by Mr. Hugh Brown.

Let the two extremes be denoted by $a - x$ and $a + x$; then, putting the given sum of the four means $= b$, the sum of the 1st, 2d, 3d, 4th, and 5th terms will be $= b + a - x$, and that of the 2d, 3d, 4th, 5th, and 6th terms $= b + a + x$: Whence (by the nature of proportionals) it is manifest, that the first term is to the second as $b + a - x$ to $b + a + x$, or as 1 to $\frac{b+a+x}{b+a-x}$. But the ratio of the first term to the sixth is the quintuplicate of that of the first to the second, that is, $\frac{a+x}{a-x} = \left(\frac{b+a+x}{b+a-x}\right)^5$. From whence, putting $c = b + a = 400$, we get $x\,((c+x)^5 + (c-x)^5) = a \times ((c+x)^5 - (c-x)^5)$, or $x \times (2c^5 + 20c^3x^2 + 10cx^4) = a \times (10c^4x + 20c^2x^3 + 2x^5)$; and consequently $x^4 + \frac{c-a}{5c-a} \times 10c^2x^2 = \frac{5a-c}{5c-a} \times c^4$: Whence x will be found, and, from thence, all the other quantities sought.

IX. QUESTION 429, *by Mr.* Walter Trott.

Two ships, A and B, sail from a certain port, in north latitude, to two other ports lying under the equinoctial line, at the distance of 666⅔ leagues from each other. The ship A, steering full south (which was her direct course), made her port in 15 days; but B, though she steered the shortest course possible, and run at the same rate as A, did not arrive at her port till the end of 25 days. Now I demand the latitude of the port sailed from, and the true distance run by each ship?

Answered by Mr. W. Spicer.

Let A be the port sailed from, and B and C the ports to which the two ships are bound. Then, in the right-angled spherical triangle ABC, we have given the base BC, and the ratio AB to AC, as 5 to 3: Therefore, putting $a =$ cosine BC (33° 20′) and $x =$ cosine $\frac{1}{3}$AC (or $\frac{1}{5}$AB) we shall (by a well-known theorem) have $4x^3 - 3x =$ cosine of AC, and $16x^5 - 20x^3 + 5x =$ cosine of AB; and therefore (per spherics) 1 (rad.) : a :: $4x^3 - 3x : 16x^5 - 20x^3 + 5x$: and consequently $16x^4 - 20x^2 + 5 = 4ax^2 - 3a$;

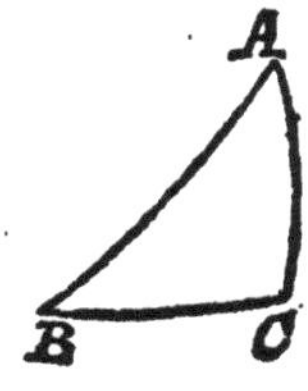

which solved, gives $x = \sqrt{\frac{5+a+\sqrt{(24+(1-a)^2)}}{8}}$ $= 0{\cdot}990091$; answering to an arch of 8° 4′ 21″. Therefore AC = 24° 13′ 3″, or 1453·05 miles; and AB = 40° 21′ 45″, or 2421·75 miles.

X. QUESTION 430, *by Mr.* Henry Watson.

In a given triangle (whose three sides are 40, 50, and 60) to describe the greatest ellipsis possible; and to determine the area, and the principal diameters thereof.

This problem, or one like it, was printed in the Ladies' Diary for 1739, but never was answered in any succeeding Diary, or elsewhere, that I have been able to discover. Your proposing of it, at this time, will oblige many of your readers, and particularly your humble servant, HENRY WATSON.

Answered by Penovius.

Draw the tangent GH, and the conjugate diameter EF, parallel to BC. Put BD $=$ DC $= 20 = b$, AD $= 51{\cdot}478 = a$, the sine of the angle at D (74° 30′ 30″) $= s$, and DI $= x$; then will AI $= a - x$, GI $=$ IH $= (b \div a) \times (a-x)$, by similar triangles; and, by the property of the ellipsis, EO $=$ OF $= b\sqrt{((a-x) \div a}$ (whence $bx\sqrt{((a-x) \div a)}$ is a maximum; in fluxions $2ax\dot{x} - 3x^2\dot{x} = 0$, and $x = \frac{2}{3}a$; whence EF $= 2b \div \sqrt{3}$, and the area of the ellipsis $= 3{\cdot}1416sab \div 3\sqrt{3} = 599{\cdot}8582$. Now put the transverse axis $= y$, and the conjugate $= z$; then, by conics, $yz = 4ab \div 3\sqrt{3} = 2n$, and $yy + zz = \frac{4}{9}aa + \frac{4}{3}bb = 4m$; from which $y + z = 2\sqrt{(m+n)}$, and $y - z = 2\sqrt{(m-n)}$; whence $\sqrt{(m+n)} \pm \sqrt{(m-n)} = \begin{cases} y = 35{\cdot}229236 \\ z = 21{\cdot}67976 \end{cases}$, the principal diameters required.

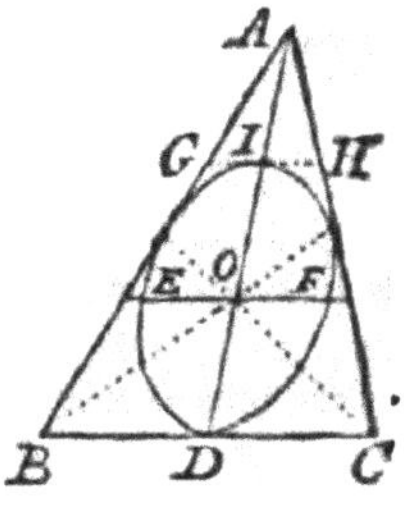

Corollary 1. All the sides of the triangle are bisected by the points of contact; and the centre of the ellipsis coincides with the centre of gravity of the triangle.

Corollary 2. The area △ : the area of the ellipse :: $3\sqrt{3} : 3{\cdot}1416$. Hence the areas of all triangles circumscribing the same ellipsis, having their sides bisected by the points of contact, are equal: When $s = 1$, the triangle is isosceles; and equilateral, when the ellipsis becomes a circle.

Mr. *O'Cavanah*, after demonstrating geometrically, that the least triangle that can be described about any oval figure, will have all its sides bisected by the points of contact, derives, by a different method, the very same conclusions above exhibited; and farther adds, by way of note, that the greatest parabola that can be described in a given triangle, may be determined by the same method; the area thereof being to that of the triangle, as $\sqrt{3}$ is to 2; which proportion is general, whatever the species of the triangle is, or which ever of its sides the base of the parabola supposed to stand upon. Those persons (continues he) are therefore mistaken, who make the area to be greater or less, according as the base is made to coincide with this, or that side of the triangle.——The axis of the parabola will not be perpendicular to the base of the triangle, but parallel to a line drawn from the vertex to bisect the base.

XI. QUESTION 431, *by Mr.* W. Bevil.

Suppose a round post one foot in diameter, and fixed perpendicular to the horizon, on which are wound 100 rounds of Manchester binding (one upon another) whose thickness is one twenty-eighth of an inch. Now, if a person takes hold of the extremity of the outward end, and moves round the said post until he has unwrapt it all, how many yards will he have travelled when he arrives at the end of his journey, always keeping as far from the post as the binding will admit him?

Answered by the Proposer, Mr. Bevil.

Let a = diameter of the post, b = twice the thickness of the binding, n = the number of rounds, $f = a + nb$, and $c = 3{\cdot}14159$, &c. then will f, $f - b$, $f - 2b$, $f - 3b$, &c. be the diameter of the post and binding, after 0, 1, 2, 3 &c. rounds are disengaged; and cf, $c \times (2f - b)$, $c \times (3f - 3b)$, &c. will be the length disengaged in 1, 2, 3, &c. rounds respectively. But, if d be taken to denote the diameter, and s the length of the part disengaged, at the end of any number (r) of rounds; then (the length unwrapped at the end of the next round after, being $= s + cd$) it appears by p. 163 of Simpson's Fluxions) that the distance moved over in unwrapping that round, will be $= \dfrac{(s + cd)^2 - ss}{d} = 2cs + ccd$. Now, when $r = 0$, then $d = f$, and $s = 0$; therefore, in the first round, $2cs + ccd = ccf$. When $r = 1$, then $d = f - b$, $s = cf$, and consequently $2cs + ccd = 2ccf + ccf - ccb = cc \times (3f - b)$. When $r = 2$, then $d = f - 2b$, $s = c \times (2f - b)$; and therefore $2cs + ccd = cc \times (5f - 4b)$:

In the same manner, when $r = 3$, then $2cs + ccd = cc \times (7f - 9b)$, &c. Now, by collecting all these values of $2cs + ccd$ together, we get the two following series, viz. $ccf \times : 1 + 3 + 5 + 7 +$ &c. and $- ccb \times : 0 + 1 + 4 + 9$, &c. where each series is to be continued to n terms: These series being summed, we have $ccf \times nn - ccb \times \frac{1}{6}n \, . \, (n - 1) \, . \, (2n - 1) = ncc \times (nf - \frac{1}{6}b \, . \, (n - 1) \, . \, (2n - 1)) =$ 26 m. 1 f. = the distance sought.

The same answered by Mr. H. Watson.

Let a = the semi-diameter of the post, b = the thickness of the binding wrapped thereon, n = the whole number of rounds at first, x = the number of rounds unwrapped at the end of any time, and u = the length of the part disengaged at that time: And supposing N and M to be two positions of the end of the line indefinitely near to each other, to the points of contact n and m, from the centre C, let Cn and Cm be drawn.

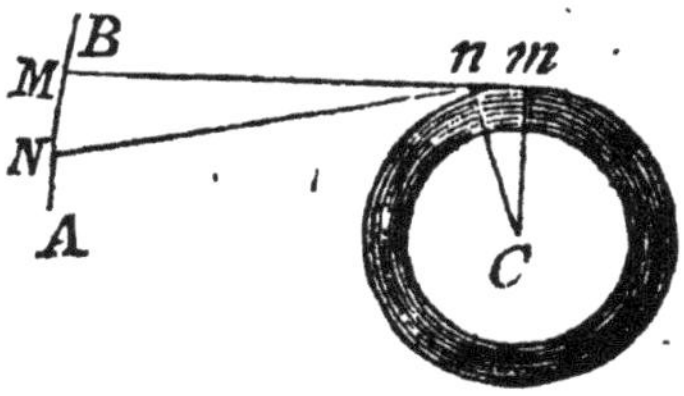

It is evident, that $a + nb$ will be the semi-diameter of the post and binding together, and consequently that the length of the first round (computed at the middle of the binding's thickness) will be $= p \times (a + nb - \frac{1}{2}b)$ (p being put $= 2 \times 3{\cdot}1416$). In the same manner the length of the last of x rounds will be $= p \times (a + nb - (x - \frac{1}{2}) . b)$. Therefore we have here the first and last terms of an arithmetical progression, whereof the number of terms is x: whereby the sum of the progression is found $= p \times (2a + (2n - x) . b) \times \frac{1}{2}x = u$.

Now, since the angles NnC and MmC are both right ones, the angle NmM will be $= n$Cm; and therefore the figures NmM and nCm being similar, we have as Cn $(a + (n - x) . b)$: nm $(\dot{u})$ :: Nm (u) : NM $= \dfrac{u\dot{u}}{a + (n - x) . b}$: But $\dot{u} = p\dot{x} \times (a + (n - x) . b)$; and consequently NM $= p\dot{x} \times u = pp \times ((a + n) . x\dot{x} - \frac{1}{2}bx^2\dot{x})$; whose fluent $(pp \times ((a + nb) . \frac{1}{2}xx - \frac{1}{6}bx^3))$ when $x = n$, will be $= nnpp \times (\frac{1}{2}a + \frac{1}{3}nb) =$ 1654320 inches (= 26 m. 1 f. nearly) = the distance sought.

N. B. If the curve AMB be considered as the involute of *Archimedes's* (or any other kind of) spiral, the length thereof will be always found, by multiplying Nn and its fluxion together, dividing the product by the radius of curvature of the evolute (at n), and then taking the fluent of the quotient.

XII. QUESTION 432, *by Mr.* Edw. Rollinson.

From the equation $m\dot{x}^2\dot{z}^2 - nx^2\dot{z}^2 = (p + zz)^2 \times q\dot{x}^2$ to determine the general relation of x and z; and also to find in what circumstances of the coefficients m, n, p, q, that relation can be expressed in finite terms.

"Dr. *Brook Taylor*, in his *Increments*, after having given a solution of that case where $m = 4$, $n = 4$, $p = 1$, and $q = 1$, (in which there seems to be a mistake) says, that if the coefficients be changed, it does not appear to him that x can be determined, in terms of z, by a finite equation."

Answered by Mr. O'Cavanah.

By division, and extracting the square root on both sides of the proposed equation, we have $\sqrt{\frac{q}{n}} \times \frac{\dot{x}}{x\sqrt{\left(\frac{mx}{n} - 1\right)}} = \frac{1}{p} \times \frac{\dot{z}}{1 + \frac{zz}{p}}$; which, by substituting $u = \frac{z}{\sqrt{p}}$, and $y = \sqrt{\left(\frac{mx}{n} - 1\right)}$ is further transformed to $\frac{\dot{y}}{1 + yy} = \frac{1}{2}\sqrt{\frac{n}{pq}} \times \frac{\dot{u}}{1 + uu}$: whence, supposing A and B to denote the two circular arcs whose tangents are $y \left(\sqrt{\left(\frac{mx}{n} - 1\right)}\right)$ and $u\left(\frac{z}{\sqrt{p}}\right)$, we shall, by taking the fluent, have $A + C = \frac{1}{2}B\sqrt{\frac{n}{pq}}$; C being a constant arch, serving to correct the fluent. From this equation, when either of the quantities z or x is given, the other may be determined (from a table of tangents) in all cases, except when imaginary quantities enter into the consideration: Thus, z being supposed given, the tangent $u\left(\frac{z}{\sqrt{p}}\right)$ will also be given; and from thence (by the table) the arch B corresponding; whereby the arch A $\left(= \frac{1}{2}B\sqrt{\frac{n}{pq}} - C\right)$, and its tangent y, will be known, and consequently $x \left(= \frac{n}{m} \times (1 + yy)\right)$.

When n, p, and q are such, that $\frac{1}{2}\sqrt{\frac{n}{pq}}$ is a rational quantity, then may the relation of x and z be algebraically expressed by a finite equation: For, here the fraction $\frac{r}{s}$, where r and s are integers, may be assumed $= \frac{1}{2}\sqrt{\frac{n}{pq}}$; and we shall have $s \times (A + C) = rB$:

But, if c be taken to denote the tangent of the arch C, that of A + C will be $= \frac{c+y}{1-cy}$. Whence (by the theorem in the solution of quest. 388) the tangent of $s \times (A + C)$ will be had $= \left(s \cdot \frac{c+y}{1-cy} - \frac{s}{1} \cdot \frac{s-1}{2} \cdot \frac{s-2}{3} \cdot \frac{(c+y)^3}{1-cy} + \&c.\right) \div \left(1 - \frac{s}{1} \cdot \frac{s-1}{2} \cdot \frac{(c+y)^2}{1-cy} + \&c.\right)$ which must necessarily be equal to the tangent of $rB = \left(ru - \frac{r}{1} \cdot \frac{r-1}{2} \cdot \frac{r-2}{3} \cdot u^3 \&c.\right) \div \left(1 - \frac{r}{1} \cdot \frac{r-1}{2} \cdot u^2 \&c.\right)$ where r and s being integers, the serieses will terminate, and the relation of $y \left(= \sqrt{\left(\frac{mx}{n} - 1\right)}\right)$ and $u \left(\frac{z}{\sqrt{p}}\right)$ will therefore be expressed in finite terms.

XIII. QUESTION 433, *by Mr.* R. Weston, *Discip. Landenii.*

If a straight, uniform, slender rod, or bar, of heavy metal, of a given length, be left to descend after being set leaning, in a given position, with its lower end (n) on the immoveable horizontal plane AB, and its upper end (m) full against the immoveable vertical plane AC (the lower end being at liberty to slide freely along the first-mentioned plane, while the upper end is descending), what will be the position of the rod when it shall cease to touch the said vertical plane? how long will it then have been in motion? and how far from the point A will the end (m) strike the horizontal plane?

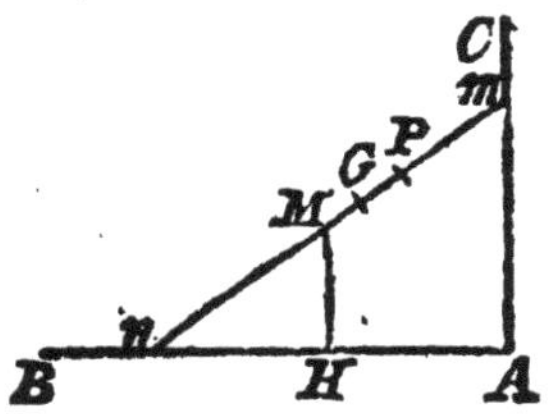

Answered by Mr. Peter Walton.

Let A′ denote the force which, acting at the end n, at right angles to the rod, would accelerate the velocity of that end about (G) the centre of gravity of the rod, at the same rate as the said velocity is accelerated by the action of the rod against AB; and let A″ denote the force which, acting at the end m, at right angles to the rod, would retard the velocity of that end about G, at the same rate as the said velocity is retarded by the action of the rod against AC: Also let u be the velocity of m, or n, about G; v, the velocity of G towards the horizon; w, its velocity in a direction parallel to the horizon; t, the time the rod has been in motion, while it touches AC; T, the time of the rod's ceasing to touch AC; d, the distance of G from AB, at the commencement of the motion; x, the distance of G from AB, at any time after the motion has commenced; $2a$, the length of the rod; $s = 16\frac{1}{12}$ feet; and $y = \sqrt{(aa - xx)}$.

Then will the absolute weight of the rod, the pressure against AB, and the pressure against AC, be as $2s$, $\frac{aA'}{3y}$, and $\frac{aA''}{3x}$ respectively: And $\frac{aA''}{3x} \times \frac{-\dot{x}}{v}$ will be $= \dot{w}$, $(A' - A'') \times \frac{-\dot{x}}{v} = \dot{u}$, $\left(2s - \frac{aA'}{3y}\right) \times \frac{-\dot{x}}{v} = \dot{v}$, $av = uy$, and $aw = ux$.

By means of which equations, we find $3w\dot{w} + u = -6s\dot{x} - 3v\dot{v}$, while the rod continues to touch AC.

Hence, by taking the correct fluents and expunging u and w, we have $v = \frac{\sqrt{(3ds - 3sx)} \times y}{a} = \frac{\sqrt{(3ds - 3sx)} \times \sqrt{(aa - xx)}}{a}$.

Therefore w is $= \frac{\sqrt{(3ds - 3sx)} \times x}{a}$; whose fluxion is $= 0$; and consequently $x = \frac{2d}{3}$, when $A'' = 0$, i. e. when the rod ceases to touch AC. Moreover $\dot{t}$ is $= \frac{-a\dot{x}}{\sqrt{(3ds - 3sx)} \times \sqrt{(aa - xx)}}$; whose correct fluent, when x is $\frac{2d}{3}$, is the time required in the question.

After the rod has ceased to touch AC, (A'' being $= 0$, and w invariable) we shall have $A' \times \frac{-\dot{x}}{v} = \dot{u}$, $\left(2s - \frac{aA'}{3y}\right) \times \frac{-\dot{x}}{v} = \dot{v}$, and $av = uy$. From which equation we get $u\dot{u} = -6s\dot{x} - 3v\dot{v}$: Hence, by taking the correct fluents, expunging u, and putting $d - \frac{d^3}{9aa} = h$, we find $v = \frac{2y\sqrt{(3sh - 3sx)}}{\sqrt{(aa + 3yy)}} = \frac{2\sqrt{(aa-xx)} \times \sqrt{(3sh-3sx)}}{\sqrt{(4aa - 3xx)}}$.

Consequently $\dot{T}$ is $= \frac{\dot{x}\sqrt{(4aa - 3xx)}}{2\sqrt{(aa - xx)} \times \sqrt{(3sh - 3sx)}}$; whereby the time from the rod's ceasing to touch AC, to its coincidence with AB will be found; from whence and the given (invariable) velocity $\frac{2d\sqrt{ds}}{3a}$, of the centre of gravity from the plane AC, during that time, the required distance of the end of the rod from AC will be obtained.

Mr. *O'Cavanah* has also obliged us with a solution to this very difficult problem, which agrees in every particular with that exhibited above. This gentleman at the end of his solution, subjoins the following remark: "If the figure of the rod should be such that the centre of gravity (G) is not in the middle point M; then, putting h = NG, and k (= NP) = the distance of the centre of oscillation (P) from the end *n* (considered as the point of suspension), it will appear, in like manner, that the celerity with which the said end recedes from the plane AC, will be $2y\sqrt{\frac{hf \times (c-y)}{\frac{1}{4}dhk-(h-d)\times yy)}}$.* When this quantity is a maximum, the end (*m*) will quit the plane AC; after which the relative celerity wherewith *n* recedes from MH, will be $= 2y \times \sqrt{\frac{df \times (r-y)}{ddk-hyy}}$: wherein *r* is a constant quantity, to be determined (like *m* above) from the value of *y*, when the end *m* ceases to touch the plane AC."

* *Note*, That, in this author's solution, $d =$ MN, $y =$ MH, $c =$ the first value of y, and $f = 16\frac{1}{12}$ feet.

THE PRIZE QUESTION, *by Mr.* O'Cavanah, *of Dublin.*

A pert young exciseman, who boasted his knowledge
In gauging of vessels, and taking an ullage,
A wager would lay, his skill to make good;
And the case we propos'd for his trial thus stood:
'Eighteen inches, five-tenths, a cask's length is given:
'The heads, which are equal, are each thirty-seven*.
'It likewise is known, that such is the make,
'The cask it is formed the least † wood to take.'
To find the content quite baffled his art;
But he hopes you the method next year will impart.

* Each head diam. is 37 inches. † The superficies is the least possible.

Answered by Mr. E. Rollinson.

Let ABCD represent the cask or vessel; put the given semi-length EI $= c$, and the given semi-diameter (AE) at the end $= b$; put also IL $= x$, and LK $= y$, supposing LK to be variable, and ∥ IG: Then, the fluxion of the curve surface being $3{\cdot}1416 \times 2y\sqrt{(\dot{y}\dot{y}+\dot{x}\dot{x})}$, it follows from the conditions of the problem, that the fluent of $\sqrt{(\dot{y}\dot{y}+\dot{x}\dot{x})}$ ought to be a minimum, when that of $\dot{x}$ becomes equal to the given quantity IE. Hence, by the general rule for the resolution of isoperimetrical problems (vide p. 101, of Simpson's Miscellaneous Tracts) it appears that

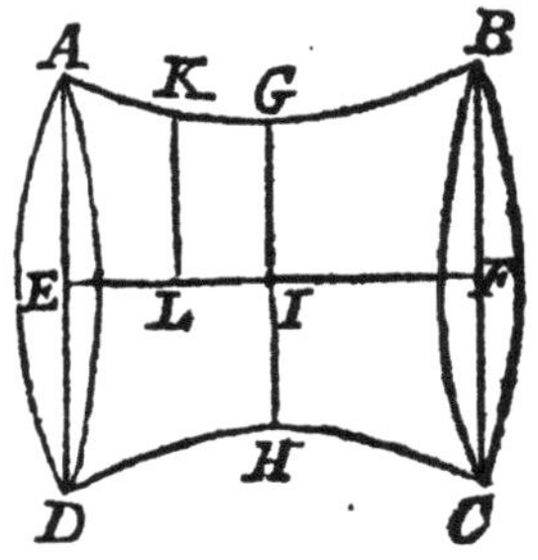

$\frac{y\dot{x}}{\sqrt{(\dot{y}\dot{y}+\dot{x}\dot{x})}} - a = 0$; whence $\dot{x}$ is found $= \frac{a\dot{y}}{\sqrt{(yy-aa)}}$; and

consequently $x = a \times$ hyp. log. $\frac{y + \sqrt{(yy - aa)}}{a}$, answering to the catenaria; in which $a = 1G$ (because when $y = a$, x will be $= 0$).

The equation of the curve being thus known, the fluxion of the generated solid, $3{\cdot}1416 \times yy\dot{x}$, will be found $= 3{\cdot}1416a \times \frac{y^2\dot{y}}{\sqrt{(yy-aa)}}$, and consequently the solid itself $= 3{\cdot}1416a \times \frac{1}{2}y\sqrt{(yy - aa)} + \frac{1}{2}ax$; the double of which, when $x = c$, and $y = b$, will be the whole content of the vessel.

But now to find a (which is yet unknown), the general equation, when $y = b$, and $x = c$, will give $c = a \times$ hyp. log. $\frac{b + \sqrt{(bb - aa)}}{a}$, or $\frac{c}{a} =$ hyp. log. $\frac{b}{a} + \sqrt{\left(\frac{bb}{a^2} - 1\right)}$; which, by putting $\frac{b}{a} = v$, will become hyp. log. $v + \sqrt{(vv - 1)} = cv \div b = {\cdot}5v$ (in the present case); whence v is found $= 1{\cdot}1788$; and from thence $a = 31{\cdot}388$. Hence the content will come out 16121 cubic inches, or 69·79 wine gallons. And the curve surface will be less than that of the circumscribing cylinder, in the proportion of 95·54 to 100.

Questions proposed in 1758, *and answered in* 1759.

I. QUESTION 431, *by Miss* T. S—e.

Addressed to Mr. U. T—r, who took the liberty to ask her the following Questions, viz. What age? What fortune? And what height she was?

My height, Sir, in inches, is three times my years;
  My fortune their squares will both shew;
Put all these together, there then, Sir, appears
  The number exposed to your view.* (*4494.
From which, Sir, determine the things you required;
  And then, if more favours you want,
As lovers of science I always admired,
  Those favors, perhaps, I may grant.

Answered by Mr. Tho. Baker.

Let x represent the lady's age, then her height (in inches) will be $3x$, and her fortune (in pounds) $= 10xx$, by the conditions of the question; from whence we have also given $10xx + 3x + x = 4494$; therefore $xx + 0{\cdot}4x = 449{\cdot}4$, and consequently $x = \sqrt{(449{\cdot}44)} - {\cdot}2\sqrt{(449{\cdot}44)} - {\cdot}2 = 21$. Hence the lady's age appears to be 21, her height five feet three inches, and her fortune 4410 pounds.

II. QUESTION 435, *by Mr.* P. O'Cavanah.

A tortoise once (or Æsop lies)
Run with a hare, and won the prize;
The hare, in the first minute's space,
Four furlongs run o'th' destin'd race,
While far behind her foe crept on,
And only crawl'd the fourth of one:
Puss, looking back, observes the case,
Disdains her foe, and bates her pace:
In the next minute she run o'er
But half the ground she run before;
And in the third was only reckon'd,
One half the space she went the second,
Decreasing still (as artists' call)
In ratio geometrical:
Mean time the tortoise still crept on,
At the same rate as he began:
Now deign to shew, ye learned fair,
In what time he o'ertook the hare.

Answered by Mr. B. Lydal.

Since the space run by the hare, in the several succeeding minutes, are expressed by 4, 2, 1, $\frac{1}{2}$, $\frac{1}{4}$, &c. respectively, it is plain, from addition only, that the whole space gone over from the beginning, will be always less than 8 furlongs by a space equal to that passed over in the last minute. Therefore since the space passed over in the last minute (after a continual decrease of one half every minute) must be extremely small, the whole space gone over by the hare, when the tortoise overtakes her, will amount to 8 furlongs, extremely near: which at the rate of $\frac{1}{4}$ of a furlong in a minute, will take the tortoise 32 minutes to crawl over, in order to come up with the hare.

The same answered by Mr. W. Kingston, *of Bath.*

If the number of minutes be denoted by x, the space gone over by the tortoise will be $= \frac{1}{4}x$, and that by the hare $= 4 + 2 + 1 + \frac{1}{2} + \frac{1}{4}$, &c. $= 8 \times (\frac{1}{2} + \frac{1}{4} + \frac{1}{8}$, &c.) continued to x terms, whereof the last is evidently $= 8 \times \frac{1}{2^x} \left(= \frac{8}{2^x}\right)$. Whence the sum of the whole progression (being = the double of the first term *minus* the last, in all cases where the ratio is $\frac{1}{2}$) is had $= 8 - \frac{8}{2^x}$. Hence we have $8 - \frac{8}{2^x} = \frac{x}{4}$, or $32 - \frac{1}{2^{x-5}} = x$; where, if $\frac{1}{2^{x-5}}$ be rejected on account of its extreme smallness, we shall have $x = 32$. To correct this value, let $\frac{1}{2^{x-5}} = \frac{1}{134217728}$, be now subtracted from 32; whence $x = 31\frac{134217727}{134217728}$.

The same answered by Penovius.

Supposing that the hare moved uniformly in each minute, the distance run by her was $= 16 + 8 + 4 + 2 + 1 + \frac{1}{2} + (\frac{1}{2})^2 + (\frac{1}{2})^3 + \&c. \ldots + (\frac{1}{2})^{27} \times x = 31 + \frac{67108863}{67108864} + \frac{x}{134217728}$, and by the tortoise $= 1 + 1 + 1 + \&c. \ldots + x = 31 + x$: Therefore $x = \frac{67108863}{67108864} + \frac{x}{134217728}$; whence $x = \frac{134217726}{134217727}$; and therefore $31\frac{134217726}{134217727}$ minutes is the true time required.

III. QUESTION 436, *by Mr.* John Brickland, *Teacher of the Mathematics, in Oxford.*

What is the side of that equilateral triangle, whose area cost as much paving at 8d. a foot, as the pallisadoing the three sides did at a guinea a yard?

Answered.

If the length, in feet, of each side of the triangular garden be denoted by x, then the pallisadoing of the three sides, at 21s. per yard, will amount to $21x$ shillings: Moreover, the perpendicular being $= \surd(xx - \frac{1}{4}xx) = \frac{1}{2}x\surd 3$, the area of the garden, in square feet, will be $= \frac{1}{4}xx\surd 3$; whereof the paving, at $\frac{2}{3}$ of a shilling per foot, will come to $\frac{1}{6}xx\surd 3$ shillings. Hence, by the question, we have $\frac{1}{6}xx\surd 3 = 21x$; and therefore $x = 126 \div \surd 3 = 42\surd 3 = 72{\cdot}74615$, the required length of each side.*

Thus the problem is re-solved by Mr. *J. Brickland* (the proposer) and by several other persons.

IV. QUESTION 437, *by Mr.* Thomas Moss.

From the vertical angle of a triangle (whose base is 70, and its two sides 40 and 50) to draw a line, terminating in the base, so as to be a mean proportional between the two segments of the base made thereby.

---

* See this Question solved without Algebra, p. 118, Hutton's Mensuration.

Answered by Mr. Rob. Shirtcliffe, *Author of the Theory and Practice of Gauging.*

About the given triangle ABC let a circle be circumscribed, and in the side CB, produced, take BM = CB, drawing MNN parallel to AB, intersecting the circle in N, N'; then draw CN, cutting AB in P; so shall CP be a mean proportional between AP and BP. For, MN being parallel to AB (by construct.) and BM = BC, it is evident that PN is also = PC (II. Euc. 6.); and consequently $CP^2 = CP \times PN = AP \times BP$ (III. Euc. 35.) or AP : CP :: CP : BP. (XVI. Euc. 6.).

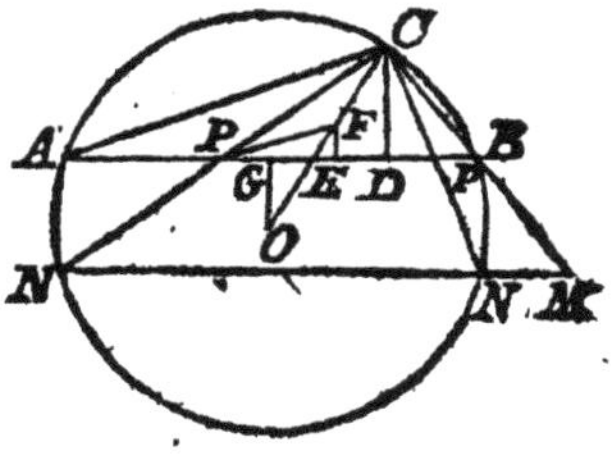

The same answered otherwise by Mr. W. Davies.

Let a circle be described about the given △ ABC; and, having bisected the radius OC in F, from this point to AB apply FP = FC (or FO); then will CP, when drawn, be the required mean proportional between AP and BP.

Calculation. There being given BC = 40, AC = 50, and AG = BG = 35, the perpendicular CD will be found $= \frac{80\sqrt{6}}{7}$, GD $= \frac{45}{7}$, and GE = ED $= \frac{45}{14}$: Then $\frac{AC \times BC}{2CD} = OC = \frac{175}{2\sqrt{6}}$, $\frac{1}{2}CD - \frac{1}{2}OG =$ FE $= \frac{40\sqrt{6}}{7} - \frac{35}{4\sqrt{6}}$, and $(FP^2 - FE^2)^{\frac{1}{2}} = EP = \frac{5\sqrt{1649}}{14} = 14{\cdot}5$: Whence AP = 23·712, and BP = 46·288.

An Algebraical Solution to the same by Mr. Ed. Johnson.

All the sides of the triangle being given, the segments of the base are readily obtained, viz. AD $= 41\frac{3}{7}$, and BD $= 28\frac{4}{7}$. Putting therefore AC $= a$, AD $= b$, BD $= c$, and DP $= x$, we shall have AP $= b - x$, and BP $= c + x$; whence CP^2 (= AP × BP per quest.) $= (b - x) \times (c + x)$: But CP^2 is also $= CD^2 + DP^2 = aa - bb + xx$; consequently $aa - bb + xx = (b - x) \times (c + x)$; whence $xx - \frac{1}{2}(b - c)x = \frac{1}{2}(bb + bc - aa)$, and $x = \frac{1}{4}(b - c \pm \sqrt{((3b + c)^2 - 8aa)})$ = 17·71708, or − 11·28851. From which AP = 23·71148, or 52·71708.

V. QUESTION 538, *by Mr.* Peter Walton.

To place three circles (whose radii are 10, 15, and 20) so that three right lines may be drawn each, to touch all the three circles.

Answered by Mr. Henry Watson.

Let GPM, QHK, and NLI be the three right lines which each circle, is to touch, (forming by their intersections the △ EDF): and from the centres A, B, C, to the points of contact, let radii be drawn. Because EG = EH, and DI = DH, it is evident that the sum of the two (equal) tangents FG and FI is equal to the sum of all the three sides of the triangle EDF. And by the same argument, the sum of the tangents, EM + EK, or DQ + DN, must also be equal to the very same quantity, and consequently all these tangents equal among themselves. Now from FG = EM, let EF (common) be taken away, and there remains EG = FM; but EH = EG, and FL = FM; consequently EH = FL. In the very same manner, DH = PF, and DL = EP. Now per sim. figures,

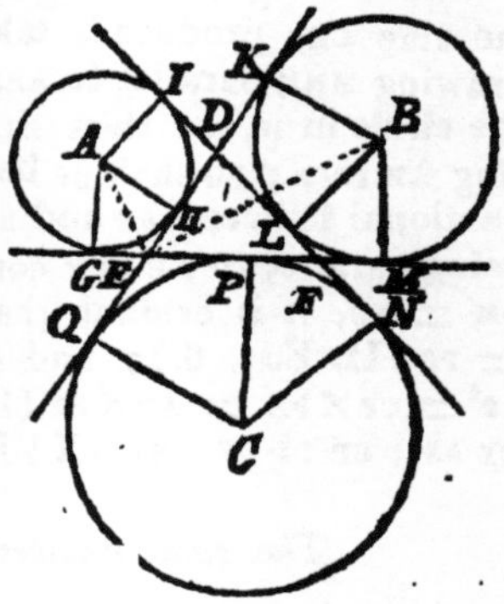

$$AH : EH :: CP : EP = EH \times CP \div AH:$$

$$BL : FL\ (EH) :: CP : FP = EH \times CP \div BL.$$

Hence it appears, that the three sides EF, ED, DF, of the triangle DEF, are equal to $\frac{EH \times CP}{AH} + \frac{EH \times CP}{BL}$, EH + $\frac{EH \times CP}{BL}$, and EH + $\frac{EH \times CP}{AH}$, respectively; which, therefore, are to each other in the given ratio of CP + $\frac{AH \times CP}{BL}$, AH + $\frac{AH \times CP}{BL}$, and AH + CP: From whence the problem may be very easily constructed, and the angles, &c. all found by common trigonometry.

The same answered algebraically by Mr. Patrick O'Cavanah.

This gentleman, after proving EG and FM to be equal, (as in the solution above given by Mr. *Watson*) puts AG = a, BM = b, CP = c, and EM = x: Then (from the similarity of the triangles EBM, AEG it will be EM (x) : BM (b) :: AG (a) : EG = $ab \div x$. Again (by sim. fig.) AG (a) : GE ($ab \div x$) :: CP (c) : EP = $bc \div x$; and BM (b) : FM (= EG = $ab \div x$) :: CP (c) : FP = $ac \div x$: Therefore $\frac{bc}{x} + \frac{ac}{x} + \frac{ab}{x}$ (= EP + PF + FM = EM) = x: and consequently $x = \sqrt{(bc + ac + ab)} = 25{\cdot}495098$: Whence every thing else is easily determined.

Messrs. *L. Charlton*, *R. Pitches*, and *Penopius*, have also given new and very neat algebraical solutions to this problem. The two latter have determined the distances of the centres A, B, C, in numbers, viz. AB $= 31\cdot774$, AC $= 34\cdot807$, and BC $= 37\cdot596$.

Penovius making $m = \surd(ab + ac + bc)$ (= EM), proves also, that the sides (DE, DF, EF) of the $\triangle$ DEF will be $= a \times \frac{b+c}{m}$, $b \times \frac{a+c}{m}$, and $c \times \frac{a+b}{m}$, respectively; and the radius of the circle inscribed therein $= \frac{abc}{ab + ac + bc}$.

VI. QUESTION 439, *by Mr.* W. Bevil.

The base of an isosceles triangle being given = 100, and each of the equal sides = 60; so to draw a right line from the vertex to the base, that the solid (or continual product) under it and the two segments of the base shall be the greatest possible.

Answered by Messrs. J. *and* R. Hudson.

Let CD be perpendicular to AB, and put AC (= BC = 60) $= a$, AD (= BD = 50) $= b$, and $x =$ CE, the line required: Then $aa - bb =$ CD^2, and $\surd(xx + bb - aa) =$ DE (Euc. 47. I.); whence AE $= b - \surd(xx + bb - aa)$, and BE $= b + \surd(xx + bb - aa)$: Therefore, AE $\times$ BE $\times$ CE $= (aa - xx) \times x = a^2x - x^3$; which being a maximum, we have, in fluxions, $a^2\dot{x} - 3x^2\dot{x} = 0$; whence $x = a \div \surd 3$: Consequently AE = 40, and BE = 60.

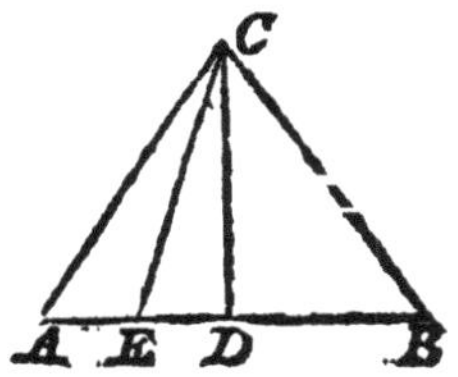

The same answered by Mr. T. Hopkinson.

Making CD $\perp$ AB, and putting AC $= a$, and EC $= x$, we have (by Euc. 5. II.) AE $\times$ BE $=$ AD^2 $-$ ED^2 $=$ AC^2 $-$ EC^2 $= aa - xx$; and consequently AE $\times$ BE $\times$ CE $= aax - x^3$: Which in fluxions, gives $aa\dot{x} - 3x^2\dot{x} = 0$; whence $x = \surd\frac{1}{3}aa$; therefore DE $(= \surd(\text{AD}^2 - \frac{2}{3}\text{AC}^2))$ = 10, AE = 40, BE = 60.

Geometrical Solution by β. Cygni, *taken from the Monthly Magazine, April,* 1796.

Analysis. Let ABC be the isosceles triangle, ACBF a circle described round it, and CD a perpendicular on the base. Suppose the line CE drawn as required in the question, so that CE $\times$ AE $\times$ BE is a maximum; and produce CE to meet the circle in F. Then, as AE $\times$ BE is $= CE \times EF$, $CE^2 \times EF$ is a maximum, but $EF = CF - CE$, therefore $CE((CE \times CF) - CE^2)$ is likewise a maximum; and as by the property of the circle AC^2 is $= CE \times CF$, $CE(AC^2 - CE^2)$ is a maximum. Upon AC describe a semi-circle, and apply the chord CE′ equal to CE; join AE′, and draw E′H perpendicular to AC. Then AE'^2 being $= AC^2 - CE'^2$, $CE' \times AE'^2$, is a maximum, and consequently by Simpson's Gemometry, page 208, AH is $= 2CH$. Hence the construction is manifest.

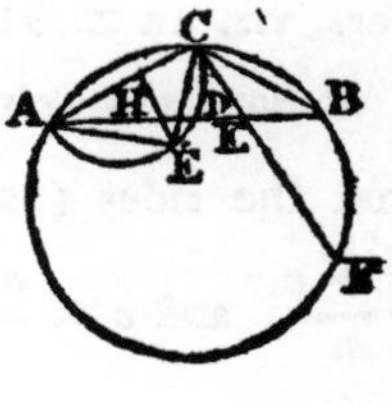

Cor. 1. The square of AC is equal $3CE^2$.

Cor. 2. When the square of AC is to the square of AB as 3 to 8, or in any greater ratio, the line required is the perpendicular on AB.

Cor. 3. If the equal sides be constant, and the base vary, the locus of the point E will be a circle, whose centre is C; also the solid under AE, BE, and CE, will be constant.

VII. QUESTION 440, *by Mr.* J. Fellows, *of H. M. S. Captain.*

Suppose two ports bear N.E. and S.W of each other, distant 50 leagues; and suppose a S. E. current runs between them at the rate of 3 knots an hour: How must a vessel, from the northern port, shape her course, so that running at the uniform rate of 5 knots an hour she may reach the other port in the least time?

Answered by Mr. J. Milbourn.

Let AS be the given direction of the current at the southermost of the two ports N, S: Take $sa = 3$, expressing the rate of the current's motion) and, from a to SN, apply $an = 5$, the rate of the ship's motion. Then draw NA parallel to na, which will give the course and distance required; because, by the composition of motions, the ship will then really move in the right line NS, directly towards her port.

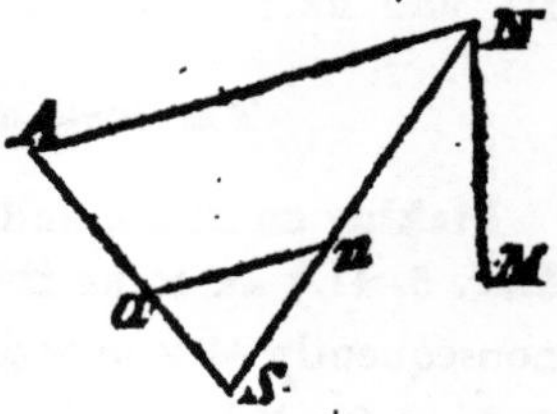

Calculation. In the right-angled triangle sna, we have given $sa \div na = 0{\cdot}6 =$ nat. sin. of sna (or SNA) $= 36° 52'$, whence the angle of the course ANM $= 81° 52'$; also NA (the dist. run by the log) $=$ NS $\times \frac{5}{4} = 62{\cdot}5$; and consequently the time $= 37\frac{1}{2}$ hours.

The same answered by Mr. Richard Terry.

Let NM represent the meridian of the place sailed from, N and S the two points, AS the direction of the current, and NA the course steered by the compass; then, by the composition of motions, the ship's real direction will be in the line NS. Put now NS $= a$, SA $= 3x$, and NA $= 5x$: Then (by Euc. 47. I.) $25xx - 9xx$ $(= NA^2 - SA^2 = NS^2) = aa$; whence $x = \frac{1}{4}a = 37 \cdot 5$ miles: From which the course is found to be W. by S. 3° 7′ S. and the time required 37^h 30^m.

VIII. QUESTION 441, *by Mr.* Lionel Charlton.

It being a common practice among ship-chandlers to make their log-lines 42 feet long, and their half-minute glasses 28 vibrations of a pendulum, whose length is 38 inches and a half: Quere, how far a ship's reckoning will vary from the truth, that uses only these, and sails by Mercator on one rhumb from the Lizard to the Capes of Virginia?

Answered by Mr. Rich. Mallock, *of Lyme Regis.*

By the known rules for pendulums, $\sqrt{39 \cdot 2} : \sqrt{38 \cdot 5} :: 28 : 27 \cdot 75$ = the number of seconds that the glass is running, which let be denoted by n; then it will be as 3600 (the number of seconds in an hour) is to n, so is 6120 (the feet in a nautical mile) to $1 \cdot 7n =$ the distance which the knots ought to be asunder, in order to measure truly the ship's way; which distance, in the present case, coming out 47·175 feet, (instead of 42 feet) we therefore have, as 42 to 47·175, so is 3100 miles (the whole given distance of the Lizard from the capes of Virginia) to 3482 miles, the distance run, by the ship's reckoning, when she arrives at her port; exceeding the truth by 382 miles.

IX. QUESTION 442, *by Mr.* J. Wilson.

Given $\left\{ \begin{array}{l} x^2z + z^2x = a \\ x^2z^7 + z^2x^7 = b \end{array} \right\}$ To find x and z, without having the root of any adfected equation to extract higher than a quadratic.

Answered.

In the given equations $xz \times (x + z) = a$, and $xxzz \times (z^6 + x^6) = b$, let there be assumed $x + z = s$, and $xz = p$; then will $z^5 + x^5 = s^5 - 5s^3p + 5sp^2$; and so by substitution, there will be had $ps = a$, and $p^2 \times (s^5 - 5s^3p + 5sp^2) = b$: In the latter of which equations let $a \div s$, the value of p (as given by the former) be substituted; then will $\frac{aa}{ss} \times (s^5 - 5s^3 \frac{a}{s} + 5s \frac{aa}{ss}) = b$; whence $s^6 - (5a + \frac{b}{aa})s^3 = -5aa$; and consequently $s = \sqrt[3]{\left\{\frac{5a}{2} + \frac{b}{2aa} \pm \sqrt{\left(\left(\frac{5a}{2} +\right.\right.}\right.}$

$\left.\frac{b}{2aa}\right)^2 - 5aa\Big\}$: From which p $(= a \div s)$ will also be known; whence x and z are easily obtained.

Thus the answer is given by Mr. *L. Charlton*, and several others.

The same answered otherwise.

Put $y+v=x$, and $y-v=z$; then the two given equations will become $(yy-vv)\,2y = a$, and $(yy-vv)^2 \times (2y^5 + 20y^3v^2 + 10yv^4) = b$; from the former whereof $vv = yy - a \div y$; which value substituted in the latter, gives (after proper reduction) $y^6 - \frac{5a^3+b}{8aa}y^3 = -\frac{5aa}{64}$; whence $y^3 = \frac{5a^3+b \pm \sqrt{(5a^6+10a^3b+bb)}}{16aa}$.

In this manner the solution is given by Mr. *R. Butler*, and some others.

Mr. W. M. *of Plymouth answers it thus:*

Let $c = \frac{1}{2}a$, $c - y = xxz$, and $c + y = zzx$; which values being wrote in the second equation, it becomes $\frac{(c-y)^4}{c+y} + \frac{(c+y)^4}{c-y} = b$. whence, by reduction, $y^4 + 2ccyy + \frac{byy}{10c} = \frac{bc-2c^4}{10}$; and consequently (by putting $d = 2cc + \frac{b}{10c}$) $y = \left(\sqrt{\left(\frac{dd}{4} + \frac{bc-2c^4}{10}\right)} - \frac{d}{2}\right)^{\frac{1}{2}}$: But $\frac{(c-y)^2}{c+y} = x^3$, and $\frac{(c+y)^2}{c-y} = z^3$; whence x and z are also known.

X. QUESTION 443, *by Mr.* H. Watson.

From the equation $44000xx + 1 = zz$, to find both x and z in whole numbers.

Answered by Birchoverensis.

In the given equation $44000xx + 1 = zz$, let z be assumed $= 210x - b$; then will $100xx - 1 = 420xb - bb$: Here x being greater than $4b$, and less than $5b$, let $4b+c=x$, then will $79bb+1 = 380bc + 100cc$. Here $b \sqsubset 5c$, and $\sqsupset 6c$; let therefore $5c + d = b$; then will $25cc - 1 = 410cd + 79dd$. Here c being $\sqsubset 16d$, and $\sqsupset 17d$, make $17d - e = c$; then will $176dd - 1 = 440de - 25ee$. From whence, by proceeding on in this manner, (making $2e + f = d$, $2f + g = e$, $4g - h = f$, $5b - i = g$, $17i - k = h$, $105k - l = i$, $17l - m = k$, $5m - n = l$, $4n + o = m$, $2o + p = n$, $2p -$

$q = o$, $17q + r = p$, and $5r + s = q$) you will, at last, come to the equation $100rr - 1 = 380sr + 79ss$; where $4s = r$, and $s = 1 = ss$; whence $r = 4$, $q = 21$, $p = 361$, $o = 701$, $n = 1763$, &c. x coming out $= 40482981221781$, and $z = 8491781781142001$: From which two numbers, answering the conditions of the problem, an infinity of others may be found.

The same otherwise answered by Mr. Peter Walton, *in a new method.*

First let $y \div 20$ be substituted, in the proposed equation ($44000xx + 1 = zz$), instead of x; then will $100yy + 1$ be $= zz$. Now, z being greater than $10y$, suppose $z = 10y + b$, and you will by substitution and reduction, have $10yy - 20by = bb - 1$. In this last equation y is greater than $2b$; therefore suppose $y = 2b + c$, and you will find $bb - 20cb = 10cc + 1$; from whence it is found $b = 10c + \sqrt{(110cc + 1)}$.

Suppose now $c = 0$; then will you have $b = 1$, $y = 2$, $z = 21$, and $x = 2 \div 20$; which last not being an integer, we must seek further. To that end, take $c = 2$, (the value of y just now found); then will $\sqrt{(110cc + 1)}$ be $= 21$ (the above found value of z), and b will be $= 41$, $y = 84$, $z = 881$, and $x = 84 \div 20$; which not being an integer, we must proceed farther, by taking $c = 84$ (the last value of y); then will $\sqrt{(110cc + 1)}$ be $= 881$ (the last value of z); and b will be $= 1721$, $y = 3526$, $z = 36981$, and $x = 3526 \div 20$; which not being an integer, we must proceed yet farther, by taking $c = 3526$, &c. At length, by proceeding in that manner, we find $x = 40482981221781$, and $z = 8491781781142001$.

A third method of Solution, by Mr. Patrick O'Cavanah, *of Dublin.*

From the given equation ($44000xx + 1 = zz$) it is evident that z is nearly $= 210x$. Putting therefore $(210 - y) \times x = z$, we thence get $420y - yy - 100 + 1 \div xx = 0$; which, by making $v = \frac{1}{10}y$ (to reduce the coefficients still lower) will become $42v - vv - 1 + 1 \div 100xx = 0$. Now, it may be observed, that as the term $1 \div 100xx$ is very small, the true value of v, in this equation, must be an approximate value of v, in the equation $42v - vv - 1 = 0$ (where the said term $1 \div 100xx$ is omitted). In order therefore to find such an approximate value of v, let a series of numbers, 1, 42, 1763, 74004, 3106405, &c. be so formed that each new term may be equal to 42 times the last, *minus* the last but one (42 and -1 being the coefficients of v and vv in the equation here to be resolved): Then (by what is demonstrated at page 174 of Simpson's Algebra, 2d. edit.) it will appear, that $\frac{1}{42}$, $\frac{42}{1763}$, $\frac{1763}{74004}$, $\frac{74004}{3106405}$, &c. are so many successive approximations to the value of v, each more exact than the preceding one. It will also appear (and might be easily demonstrated in a general manner, if room would permit) that the error, or quantity arising by substituting any one of these fractions for v, in the given equation $42v - vv$

L 2

$-1+\frac{1}{100xx}=0$, will be always expressed by $-\frac{1}{D^2}+\frac{1}{100xx}$, D denoting the denominator of the fraction so substituted. Therefore, in order that the error may entirely vanish, and a solution in integers be obtained, we have nothing to do but to continue the preceding series of numbers to five terms further, so that the last of the these (D) may be a multiple of 10, or so that x $(=\frac{1}{10}d)$ may be an integer; by which means D is found $=40482981221781 0$, $x=40482981221781$, and $z=84917817811420 01$.

XI. QUESTION 444, *by* F. Bell.

A gentleman having in his garden an elliptical fountain, whose greater diameter is 30, and the lesser 24 feet, orders a free-stone walk to be made round it, to be every where of an equal breadth, and to take up just the same quantity of ground as the fountain itself: Quere, what must the breadth of the walk be?

Answered by Mr. C. Wildbore.

Let dab represent half the elliptic fountain, foe its evolute, ro and so two radii of curvature indefinitely near each other, and let $Aa=Rr=Ss=Bb$ be every where perpendicular both to the ellipsis and the curve ARSB, and then it is evident that this will represent the breadth of the required walk. Put $ar=z$, $AR=v$, $Rr=b$, and $ro=r$; then $ro:\dot{z}::RO:\dot{v}=\frac{(r+b)\times\dot{z}}{r}$, which multiplied by $\frac{1}{2}RO$, gives $\frac{(r+b)^2\times\dot{z}}{2r}$ for the area ROS; and in the same manner the area ros is found to be $\frac{r\dot{z}}{2}$; the difference of which area is $b\dot{z}+\frac{1}{2}b^2\times\frac{\dot{z}}{r}$ the fluxion of the walk AaRr; the fluent of which is $bz+\frac{1}{2}b^2\times\angle ron$ (to rad. unity): This, when AR becomes = AB, is $bz+\frac{1}{2}b^2\times 1\cdot5708$; which expression is general, let the curve arb be what it will, and in the present case, must be equal to the area of the quarter of the ellipsis $abc=a$; this reduced (putting $1\cdot5708=c$) gives $b=\frac{\sqrt{(2ac+zz)}-z}{c}=\frac{\sqrt{(282\cdot744\times1\cdot5708+21\cdot25\times21\cdot25)}-21\cdot25}{1\cdot5708}$ $=5\cdot5247$ feet, the breadth of the walk required.

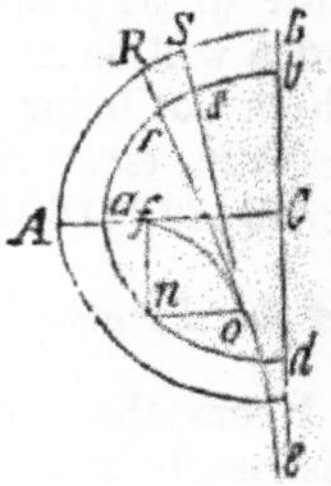

The same answered by Mr. E. Rollinson.

Let ACDE be the fountain, and PQ (pq, Aa, cc, or Dd) the breadth of the walk surrounding it: Let pq be supposed indefinitely near to PQ, and pm parallel to PQ; moreover, supposing OFRG to be a circle whose radius is $=$ PQ, let OR be conceived parallel to PQ, and or to pq. Then, the area mpq being $=$ Ror, and Qppm $=$ PQ $\times$ Pp $=$ Aa $\times$ Pp, it is manifest that (QPpq) the whole increment of the area APQa, is every where equal to a rectangle under Aa, and the corresponding increment (Pp) of the arch AP, together with the increment (Ror) of the circular sector ROF; and consequently, that the area APQa $=$ Aa $\times$ AP $+$ RO $\times \frac{1}{2}$FR $=$ Aa $\times$ (AP $+ \frac{1}{2}$FR). Hence the area of the whole walk is evidently $=$ Aa $\times$ (ACDEA $+ \frac{1}{2}$FRGHF); which is general, let the curve ACDEA be what it will. But, in the present case, ACDEA being an ellipse (whose two axes are 30 and 24) the length thereof, by the known series for the periphery of an ellipsis, will be found $= 27\cdot0834 \times p$ (p being $= 3\cdot14159$, &c.). Whence, if Aa be denoted by x, the area of the walk will here come out $x \times (27\cdot0834p + px) = 30 \times 24 \times \frac{1}{2}p$ (by the quest.). From which $xx + 27\cdot0834x = 180$; and consequently $x = 5\cdot520766$, the required breadth of the walk.

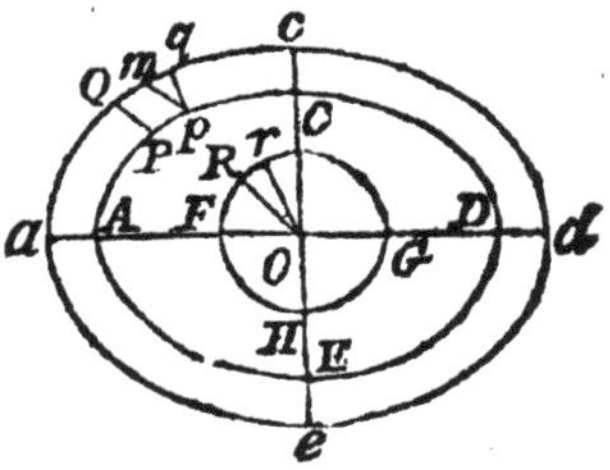

The same answered by Mr. T. Barker, *of Westhall, Suffolk.*

Although the external boundary of the walk is not accurately an ellipsis, yet the deviation therefrom is so very little, that the considering of it as such, will be sufficient to give the solution very near the truth. Let therefore AO ($= 16$) $= a$, OC ($= 12$) $= b$, and Aa ($=$ Cc) $= x$; then we shall have $(a + x) \times (b + x)$ (O$a \times$ Oc) $= 2ab$ (2OA $\times$ OC) by the question; whence $xx + (a + b) \times x = ab$, and consequently $x = \sqrt{\left(ab + \frac{(a+b)^2}{4}\right)} - \frac{a+b}{2} = 5\cdot5328 =$ the required breadth of the walk, very near.

XII. QUESTION 445, *by* Plus-Minus.

One morning last summer, being in a gentleman's garden, I saw a very good horizontal sun-dial, which was not fixed down, but neatly let into the post on which it stood: Taking hold of the gnomon. I pulled it towards me, thereby elevating the south side of the dial, still keeping the gnomon in the plane of the meridian: Upon this, the shadow came forward from the hour-line of 7, till it marked 7h. 40m. and no nearer to the meridian, or 12 o'clock, would it come; for, if the dial was further elevated, it turned back again: Now the day of the month (which I have forgot, and for private reasons should be glad to recollect) is here required?

Answered by Mr. O'Cavanah.

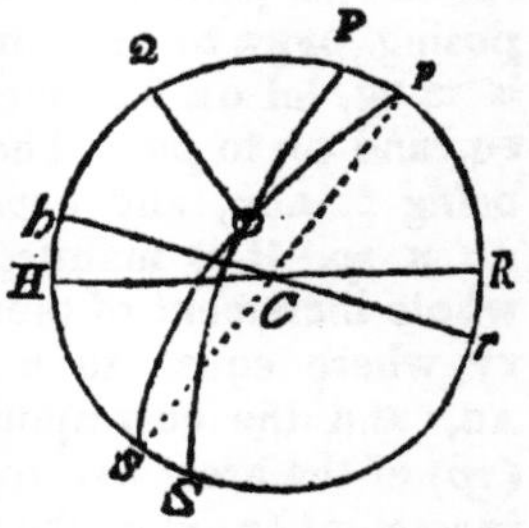

Let HPRS be the meridian (or plane of the style), P the north pole, HCR the horizon ⊙PH the given hour angle from noon, P⊙ the sun's distance from the north pole, and ⊙Q a great circle perpendicular to the meridian: Moreover, let *hcr* be supposed to represent the position of the plane of the dial, and *cp* that of the gnomon, when the south side of the dial *hc* is elevated (as by the question) above its proper horizontal position HC.

It is evident, from the general construction of all dials, that the hour from 12 marked out by the shadow (converted into degrees) will, in every possible position of any dial, be the true measure of the angle made by the plane of the style and another (imaginary) plane passing by the edge of the style through the sun's centre: Whence it appears that the ∠Q*p*⊙, made (in the present case) by the plane of the style HPRS and the plane *p*⊙*s*C passing by the edge of the style in the position *p*C, will be truly expressed by 65°, answering to $4^h\ 20^m$ the time from 12, given by the question. Now, the ∠Q*p*⊙ (here given) being (by the quest.) less than it can, otherwise, be in any other position of *p* (or *cp*) it is plain that, let ⊙Q be what it will, Q*p* and ⊙*p* must be arcs of 90 degrees each; since in this case, the measure of the ∠Q*p*⊙, or Q*s*⊙, (which in all other positions of *p* is greater) will become = ⊙Q. Hence we have (per spherics) as sin. ⊙PQ (75°) : sin. ⊙Q (65°) : : rad. : sin. P⊙ = 69° 45′ 52″; whose complement 20° 14′ 8″, is the declination sought: The time corresponding being May the 21st, or July the 22d.

Corollary. In all latitudes, the shadow, while the plane of the dial is elevating, will proceed to the same hour before it returns back: Because the true time (or hour) being supposed the same in all latitudes, ⊙Q as well as PQ will remain invariable, therefore P*p* (the complement of PQ) is invariable likewise.

XIII. QUESTION 446, *by Mr.* Patrick O'Cavanah, *of Dublin.*

How must I inclose an acre of land into a garden, with a fence of 80 poles in circumference, so as to be able to form therein the longest (straight) walk possible, and what will that length be, supposing the breadth of the walk to be 10 feet?

Answered by Mr. L. Charlton.

Supposing AMBCKD to represent the garden, and ABCD the walk; it is evident, seeing the length AMB is given, that AB will be a maximum, when the excess of AMB above AB is the least quantity possible (the area being given, or supposed to remain the same) which will therefore be when AMB is an arch of a circle; because, the circle being the most capacious figure, any other curve drawn from A to B, to contain with AB an area equal to AMBHA, must necessarily differ more from AB than AMB does, in this supposition. To determine, now, the radius of the circle (and from thence the length of the walk) put AM (= 325 feet) = a, AD (= 10) = b, the area AMBCKD (= 43560) = c, and AO (= MO = BO) = r. Then, the area AMH being = $\frac{1}{2}$AM × OM — $\frac{1}{2}$AH × OH = $\frac{1}{2}ar$ — $\frac{1}{2}$AH × OH, and the area AHnD = b × AH, we thence get $2ar$ — 2AH × OH + $2b$ × AH = c. From whence, by series, or any of the known methods of approximation, the required length AB of the walk may be determined, and will come out = 642·312 feet, or 38·928 poles.

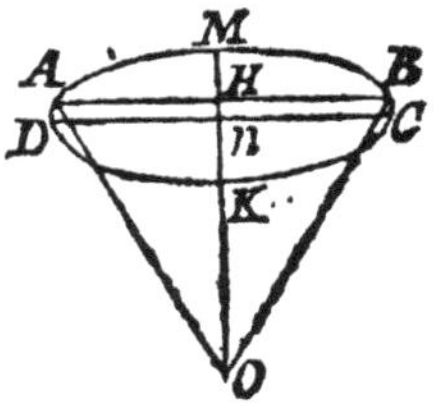

XIV. QUESTION 447, *by Mr.* Peter Walton, *Descip. Landenii.*

A chain, 10 yards long, consisting of indefinitely small equal links, being laid straight on an horizontal (perfectly polished) plane, except one part, a yard in length, which hangs down perpendicularly below the plane: In what time will the said chain (drawn by the gravity of the descending part) entirely quit the plane?

Answered by Mr. T. Allen, *of Spalding.*

Let the given length of the chain (= 30 feet) = a; the part hanging down at the commencement of motion (= 3 feet) = c; $s = 32\frac{1}{6}$ feet, the velocity generated at the earth's surface in one second; x the space moved over by a particle in any variable time t; and v = the corresponding velocity of the chain at that instant: Then it is evident that $c + x$ is as the motive force acting on the chain; and therefore, $\frac{\dot{x}}{v}$ being $= \dot{t}$, $s \times (c + x) \times \frac{\dot{x}}{v}$ will be $= a\dot{v}$, the fluxion of the quantity of motion generated in the time t; whence $s \times (c\dot{x} + x\dot{x}) = av\dot{v}$; and taking the fluents, &c. $v = \sqrt{((s \div a) \times (2cx + xx))}$, which substituted in the above value of $\dot{t}$, gives $\dot{t} =$ $\frac{\dot{x}}{\sqrt{((s \div a) \times (2cx + xx))}}$, whose correct fluent is $t = \sqrt{\frac{a}{s}} \times$ hyp. log. $\frac{c + x + \sqrt{(2cx + xx)}}{c}$, which, when x becomes $= a - c$, is = 2·890663 seconds, the time required.

The same answered by Mr. Robert Butler.

Let a = the chain's length (= 30 feet), $d = 32\frac{1}{6}$ feet, the velocity generated by the uniform force of gravity in one second, x the length of the part of the chain disengaged from the plane at the end of any time t, and v = the velocity acquired in that time. Then $a : x :: d : dx \div a$ the velocity that would be acquired in one second by the uniform force of x length: Therefore, as 1 (second) is to $\frac{\dot{x}}{v}(=\dot{t})$ $:: \frac{dx}{a} : \frac{dx\dot{x}}{av} = \dot{v}$ (the uniform increase of the velocity in the time $\dot{t}$).

Hence, $\frac{dx\dot{x}}{a} = v\dot{v}$; and therefore, by taking the correct fluents $\frac{dxx}{d}$ $- \frac{dbb}{a} = vv$, b being (= 3 feet) = the first or given value of x, when the motion commences. Now $\dot{t}\ (= \frac{\dot{x}}{v})$ being given from hence = $\sqrt{\frac{a}{d}} \times \frac{\dot{x}}{\sqrt{(xx - bb)}}$, by taking again the correct fluents, we get $t = \sqrt{\frac{a}{d}} \times$ hyp. log. $\frac{x + \sqrt{(xx - bb)}}{b} = \sqrt{\frac{a}{d}} \times$ hyp. log. $(10 + \sqrt{99})$ (when $x = a = 30$) = 2·89066 seconds; the time required.

THE PRIZE QUESTION, *by Mr.* G. Witchell.

To determine the nearest distance of the orbit of the expected comet from that of the earth, together with the longitude of the earth and comet in that situation, supposing, 1. That the orbit of the earth is a circle. 2. That the trajectory of the comet is a parabola, whose focus is the centre of the earth's orbit. 3. That the perihelion distance of the comet is to the radius of the circle, as 0·5868 to 1. 4. That the inclination of the planes is 17° 2′, the place of the perihelion ♒ 3° 30′, and that of the descending node ♍ 22° 13′.

Answered by Mr. G. Witchell, *the Proposer.*

Let OTR be the earth's orbit, PDC the comet's trajectory, ADB the orthographic projection of it on the plane of the ecliptic, D the descending node, and P the place of the perihelion. From any point C of the parabola let fall CE perpendicular to the plane of the ecliptic; through S and E draw ST and join C, T, it is evident that T is the nearest point in the circle to C. Now put SC = 1 − x, PS = $\frac{1}{4}a$, $n = \frac{1}{2}a - 1$, $m = a - \frac{1}{4}aa$, t = sin. of E☋C (the greatest inclin.) b = the sine and d = the cosine of PS☋; then, by the nature of the parabola, we shall have SF

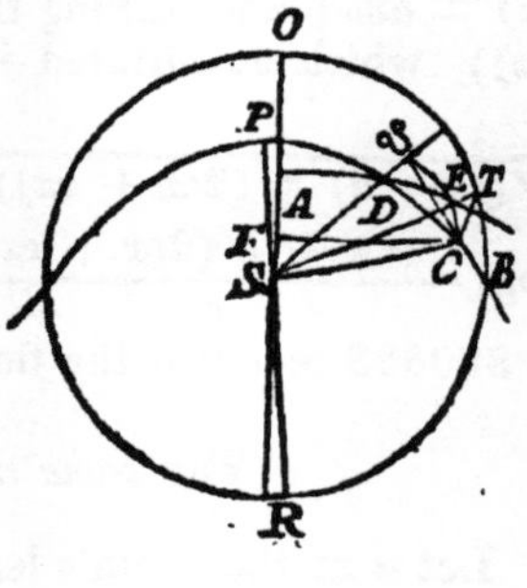

$= n+x$ and FC $= \sqrt{(m-ax)}$; therefore $1-x : 1 :: \sqrt{(m-ax)} : \frac{\sqrt{(m-ax)}}{1-x} :: n+x : \frac{n+x}{1-x} =$ sin. and cos. of PSC; but sin. PSC $-$ PS☊ $=$ sin. PSC $\times d -$ cos. PSC $\times b = \frac{d\sqrt{(m-ax)} - b(n+x)}{1-x}$ $=$ sin. ☊SC; then $1 : 1-x :: \frac{d\sqrt{(m-ax)}-b(n+x)}{1-x} : d\sqrt{(m-ax)} - b(n+x) =$ ☊C, and $1 : d\sqrt{(m-ax)} - b(n+x) :: t : td\sqrt{(m-ax)} - tb(n+x) =$ EC; which being squared and subtracted from $(1-x)^2 =$ SC^2, we have $(1-x)^2 - ttdd(m-ax) - ttbb(n+x)^2 + 2ttdb(n+x) \times \sqrt{(m-ax)} =$ SE^2: This being resolved into a series (and A, B, C, &c. subst. for the known coefficients) we shall get $A^{\frac{1}{2}} \times (1 - \frac{B}{2A}x - \frac{B^2+4AC}{8A^2}xx + \frac{B^3-4ABC+8A^2D}{16A^3}x^3$ &c.) $=$ SE: But $1 +$ SC$^2 - 2$SE $=$ CT^2; whence we have $2 - 2A^{\frac{1}{2}} - \left(2 + \frac{B}{A^{\frac{1}{2}}}\right)x + \left(1 + \frac{B^2-4AC}{4A^{\frac{3}{2}}}\right)xx - \frac{B^3-4ABC+8A^2D}{8A^{\frac{5}{2}}}x^3$ &c. $=$ CT^2; which, when CT is a minimum, will give $2 + \frac{B}{A^{\frac{1}{2}}} = \left(\frac{B^2-4AC}{2A^{\frac{3}{2}}} + 2\right)x - \frac{B^3-4ABC+8A^2D}{8A^{\frac{5}{2}}}3xx$ &c. Whence we get $x = 0{\cdot}0146$, and CT $= 0{\cdot}041974$; which (taking ST $= 80000000$) is $= 3357920$ miles. Moreover, the cos. of PSC being $= (n+x) \div (1-x) =$ cos. $78° 59'$, we shall have the long. of the comet in its orbit ♏ $14° 31'$, and its arg. of latitude $7° 42'$; from which and the given inclination, its longitude in the ecliptic will be found ♏ $14° 51'$; which place the earth transits, May 4th.

Hence it appears, that we need not be under any apprehensions from the return of this comet, notwithstanding the prediction of a certain experimental philosopher; for if it should return at the time he mentioned, it would be distant from the earth (by his own scheme) 9000000 miles.

Questions proposed in 1759, *and answered in* 1760.

I. QUESTION 448, *by* Richard *of the Vale.*

Teach me, fair artists, how to find,
When my dear Nancy will be kind;
The charming maid my love allows,
Yet is averse to crown my vows;

In mystic terms* prescribes her will,
Which to disclose, exceeds my skill.
T' unfold the knot, *ye fair, descend,
And serve, for once, an unlearn'd friend.

* $\left\{\begin{array}{l} xx + yy - 5x + 9y = 968 \\ 8yy + xy - 30x + 54y = 5808 \end{array}\right\}$; x being (she says) the months, and y the days, before she can consent to complete my happiness.

Answered by Mr. G. Witchell.

The first equation being multiplied by 6, and compared with the second, there results $yy + \frac{1}{2}xy = 3xx$; whence by completing the square, $y = \sqrt{(3xx + \frac{xx}{16})} - \frac{x}{4} = \frac{3x}{2}$; which value substituted in the first equation, gives $\frac{13xx}{4} + \frac{17x}{2} = 968$, whence $x = 16$ months, and $y = 24$ days.

II. QUESTION 449, *by Mr.* Paul Sharp.

The difference of the two legs of a right-angled triangle, whose area is double to that of its inscribed circle, being given ($= 6$); to find all the sides, and the radius of the circle.

Answered by Mr. T. Barker, *of Westhall, Suffolk.*

Put $\frac{1}{2}$ the sum of the two legs $= x$, and $\frac{1}{2}$ their difference $= a$; then $x - a =$ the less leg, and $x + a =$ the greater; and consequently $\sqrt{(2xx + 2aa)} =$ the hypothenuse. Also, by a well-known theorem, $2x - \sqrt{(2xx + 2aa)}$ ($=$ the sum of the legs minus the hyp.) $=$ the circle's diameter; whence its periphery (putting $n = 3{\cdot}1416$) will be $= 2nx - n\sqrt{(2xx + 2aa)}$; the double of which (by another well-known theorem, and the conditions of the question) must be equal to the perimeter of the triangle; that is, $4nx - 2n\sqrt{(2xx + 2aa)} = 2x + \sqrt{(2xx + 2aa)}$; whence x is found $= a\sqrt{\frac{(2n+1)^2}{2 \times (2n-1)^2 - (2n+1)^2}} = 13{\cdot}1$; therefore the sides are 10·1, 16·1, 19, and the diam. $= 7{\cdot}19$.

III. QUESTION 450, *by Mr.* Richard Gibbons.

Given the content of a cone $133333\frac{1}{3}$ cubic feet, and a heavy body will descend down the slant side in $2\frac{1}{2}$ seconds: Required the perpendicular height of the cone?

Answered by Mr. W. Bamfield, *and several others.*

Put the content of the cone ($= 133333\frac{1}{3}$) $= c$, and its altitude $=$

x, also put $p = 3{\cdot}1416$, and let $a =$ diameter of a sphere described about the cone; which (by the laws of descent of heavy bodies) is given $= 100\frac{25}{18} =$ the perp. descent in the given time in which the chord or side of the cone is described. Now, by the property of the circle, $x(a - x)$ is $=$ the square of the rad. of the cones's diam. at the base; and consequently $\frac{1}{3}pxx(a - x) = c$ ($=$ the cone's content): From which equation $x = 81{\cdot}219$, or $x = 50{\cdot}408$.

IV. QUESTION 451, *by Mr.* T. Barker, *Westhall, Suffolk.*

Two lines drawn to bisect and terminate in the opposite sides of any trapezium, will also bisect each other: A demonstration of this is required?

Answered by Mr. Allen, *and several others.*

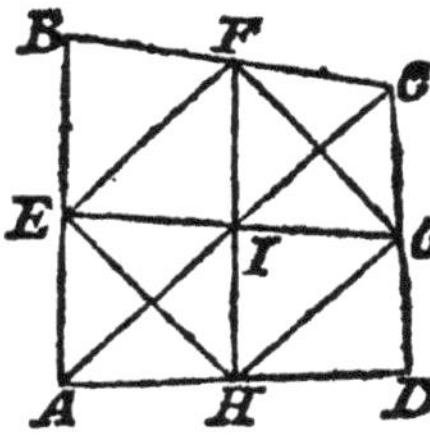

Let the bisecting points be joined by the lines EF, FG, GH, and HE, and let AC be drawn: Then, because the sides of the triangles ABC and ADC are divided proportionally, both EF and HG will be parallel to AC (Euc. II. 6.); and, therefore, parallel to each other; and in the very same manner is EH parallel to FG; and so, the triangles EIF and HIG being equiangular (Euc. 29. I.) and EF being $=$ HG, thence will EI $=$ GI, and FI $=$ HI.

V. QUESTION 452, *by Mr.* Christopher Mason.

A farmer proposes to have a bushel to contain three quarts above statute (being the customary measure of the neighbourhood) to be 1-5th of an inch in thickness, of hammered brass; the depth and diameter to be such as will require the least metal; he will likewise allow 16d. per lb. and desires to know the expence at that rate.

Answered by Mr. Barker, *and several others.*

Put $\cdot7854 = p$, the content of the bushel in cubic inches $= a$, and the internal diameter $= x$; then will the area of the base $= pxx$, the circumference of the base $= 4px$, the depth of the bushell $= a \div pxx$, the concave surface $= 4px \times a \div pxx = 4a \div x$, and the whole internal surface (including the base) $= 4a \div x + pxx$: Which being a minimum (by the quest.) we thence have $-4a\dot{x} \div xx + 2px\dot{x} = 0$: whence $x = \sqrt[3]{(2a \div p)} = 18{\cdot}1604$; and the depth ($a \div pxx = \frac{1}{2}x$) $= 9{\cdot}0802$: Hence $158{\cdot}85 =$ the quantity of the metal; which, at 16d. per lb. (supposing a cubic foot of hammered brass to weigh 8349 ounces, as by the tab. in Mis. Cur.) comes to 3l. 3s. 11d.

The same otherwise, by Mr. T. Baxtonden, *and others.*

Let x and y denote the internal diameter and depth, and a the thickness of the metal; then the external diameter and depth being $x + 2a$ and $y + a$, we shall have $pyxx = c$ (= the given content) and $p \times (y+a)(x+2a)^2 - c$ a minimum (by the question): Whence, in fluxions, $\dot{y}xx + 2yx\dot{x} = 0$, and $\dot{y} \times (x+2a)^2 + (y+a)(x+2a) \times 2\dot{x} = 0$; hence $\dot{y} = -\frac{2y\dot{x}}{x} = -\frac{(y+a)\times 2\dot{x}}{x+2a}$, $\frac{y}{x} = \frac{y+a}{x+2a}$, and $x = 2y$: Whence also $\frac{1}{2}px^3 (= px^2y) = c$; from which x is given $= \sqrt[3]{2c \div p} = 18\cdot1604$, and $y = 9\cdot0802$; hence the quantity of metal $(= p \times (y+a) \times (x+2a)^2 - p \times yx^2 = 4pa \times (3yy + 3ay + aa))$ comes out $= 158\cdot85$ cubic inches, and the value thereof 3l. 3s. 11d.

From this last method it appears, that, let the thickness of the metal be ever so great, the external diameter and depth (as well as the internal) will be, accurately, in the ratio of two to one.

VI. QUESTION 453, *by Mr.* J. Hudson, *of Louth.*

Having given the lengths (14 and 20) of two right lines drawn from the same point to make a given angle (60°) with each other; so to draw another right line through the point of concourse, that two perpendiculars being let fall thereon from the extremes of the first lines, the sum of the two triangles, thereby formed, shall be the greatest possible.

Answered by Mr. Thomas Moss.

Let BC, joining the extremes of the given lines AB and AC, be bisected by AF, upon which conceive a semicircle FGA to be described, and conceive FG to be perpendicular to ED, and GH to BC; Then, it is evident, that DE × FG, or BC × GH, will be = the area BCED; which (as BAC is given) must be a maximum (by the quest.), and therefore GH (as BC is invariable) must also be a maximum; and this, since G is always in the circumference of the circle, must necessarily happen when HG passes through the centre O; in which case the angle OFG $= \frac{1}{2}$FOH $= \frac{1}{2}$OFM (FM being supposed perpendicular to BC): Whence this

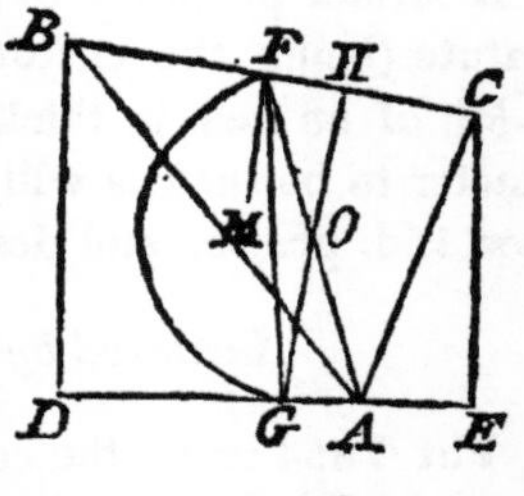

Construction. Having joined A, F, and made FM perpendicular to BC, draw FG to bisect the angle AFM; then, through A, draw DE perpendicular to FG, and the thing is done.——The numerical calculation is, from hence, very easy; whence the angle DAB comes out = 54° $2\frac{1}{2}'$, and the sum of the two triangles (ADB + ACE) = 131·53.

An algebraical Solution to the same, by Philarithmus.

Let AD be the line sought, and let AF bisect the given angle BAC;

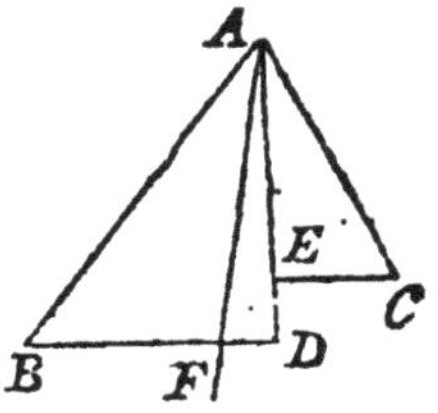

call 2BAF ($60°$) A; 2FAD, V; tab. rad. 1; AB (20), a; AC (14), b; sin. A, s; cos. A, c; sin. V, x; cos. V, y; then by trigonometry $2\text{BD} \times \text{DA} \div aa =$ sin. 2BAD, or sin. $(\text{A} + \text{V}) = sy + cx$, and $2\text{CE} \times \text{EA} \div bb =$ s. 2CAE, or s. $(\text{A} - \text{V}) = sy - cx$; therefore $\frac{1}{2}\text{BD} \times \text{DA}$, or the $\triangle$ABD $= (sy + cx) \times \frac{1}{4}aa$, and $\frac{1}{2}\text{CE} \times \text{EA}$, or the $\triangle$ACE $= (sy - cx) \times \frac{1}{4}bb$, and their sum $= sy \times \frac{1}{4}(aa + bb) + cx \times \frac{1}{4}(aa - bb)$, a minimum; therefore $s\dot{y} \times (aa + bb) + c\dot{x} \times (aa - bb) = 0$: whence substituting $- x\dot{x} \div y$ for $\dot{y}$, and dividing by $-s\dot{x}$, and transposing, $(x \div y)(aa + bb) = (c \div s)(aa - bb)$; or $aa + bb : aa - bb ::$ cot. A : tan. V; whence V is given $= 11°\ 10{\cdot}7'$, and FAD, its half, $= 5°\ 35{\cdot}35'$, and the greatest sum of the triangles $131{\cdot}532$.

VII. QUESTION 454, *by Mr.* William Toft.

A ship, plying to windward, then at N.N.E. sails, with her starboard tacks on board, 60 miles; she then tacks, and, after having run 45 miles farther, finds, by an observation, that her whole difference of latitude on both tacks, is just 40 miles: Hence you are desired to determine how near to the wind the ship made good her way?

Answered by Mr. O'Cavanah.

Draw AB and BC equal to the two given distances (60 and 45), so that the supplement (ABF) of the angle contained by them, shall be equal to ($45°$) double the given angle included between the wind and the meridian; join AC, and let a semi-circle be described thereon, in which apply CD ($= 40$) $=$ the given difference of latitudes; draw AN parallel thereto; then, if NAW be made $= 22°\ 30'$, I say that BAW will be the angle sought, which the ship makes with the wind.

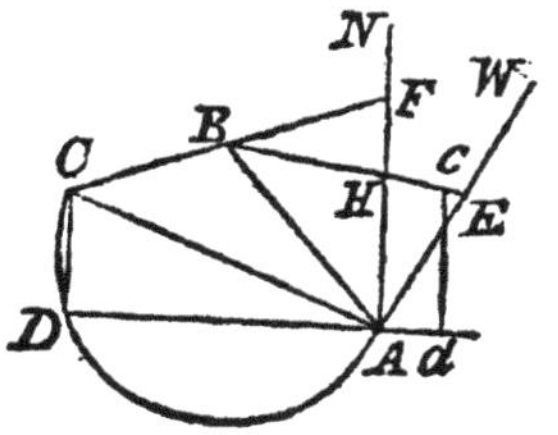

For, having produced CB to meet AN in F, let BHCE be so drawn, as to make the $\angle$BHA $=$ BCD $=$ BFN, in which take Bc $=$ BC, and let cd be parallel (and consequently equal) to CD; then AB, Bc, and cd, being (by construction) equal to the three quantities given, it only remains to prove, that the $\angle$BEA, which the ship makes with the wind, on her second course (Bc), is equal to the $\angle$BAE, on her first course (AB). Now (by construction) BFN $=$ BHA; but BFN $-$ FBA (2NAW) $=$ BAH; whence BFN $-$ NAW $=$ BAE: But BHA (BFN) $-$ NAW $=$ BEA; therefore BAE $=$ BEA.

Calculation. In the triangle ABC we have given AB ($= 60$), BC ($= 45$), and the angle ABC ($= 135°$), to find BAC ($= 19°\ 7'$), and

AC (= 97·178); from which last, and CD (= 40), CAD is found = 24° 18′: Whence BAD = 43° 25′, and BAW = 69° 5′.

VIII. QUESTION 455, *by Mr.* Thŏ. Moss.

To divide a given trapezium into two equal parts (geometrically) by a right line cutting off from opposite sides two segments, adjacent to the base, which shall also be equal to each other.

Answered by P. M. *of Durham.*

From the angle C draw CH to bisect the given trapezium ABCD (p. 4, 6. Simpson's Elem. Geom.); produce BC, AD indefinitely, cutting each other in G, from which point, on CG produced, take GK = GH, and, on DG produced, take GL = GB; draw two right lines bisecting BG, LA perpendicularly in *n* and *m*, and intersecting each other in O; round which point O as a centre, with the distance OC or OK, describe a circle cutting AD (produced both ways if needful) in F and I; from G take GE = GI; join E, F, and the thing is done.

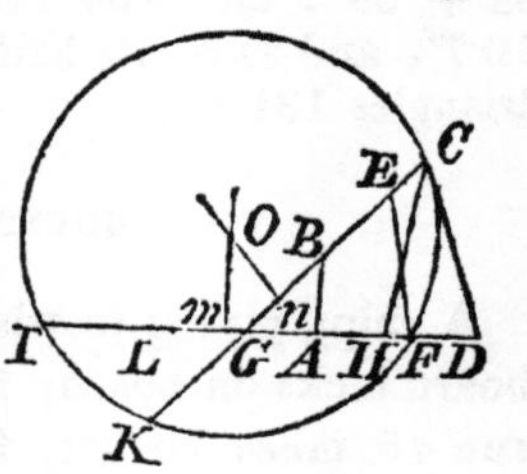

Demonstration. *m*L and *m*A are equal (by hypoth.) and also *m*I, and *m*F (3 Euc. III.); therefore the remainders IL, AF will be equal; but since GE, GI, and GB, GL are equal (by hypoth.) IL, BE, and consequently AF, BE will be equal. But I say moreover, that EF bisects the trapezium, for CG : GF : : GE (GI) : GH (GK) (35 Euc. III.) therefore the triangles CGH, EGF are equal (15 Euc. VI.); and, taking away the common triangle AGB, there will remain the quadrilateral BH = BF; therefore, because CH bisects the given trapezium BD (by hypoth.) EF will bisect it also.

The same constructed otherwise, by Mr. W. Davies.

Draw AH so as to make the triangle ABH equal to half the given trapezium ABCD (p. 4, 6. of Simps. Geom.); produce AD and BC to meet in E, and, in the former of them, take AM = EB, and EF = BH; then take EG a mean proportional between AE and EF; and from G, to the middle of EM, draw GO; make OK = OG, BI = AK, and draw KI, which will divide the given trapezium ABCD into two equal parts.

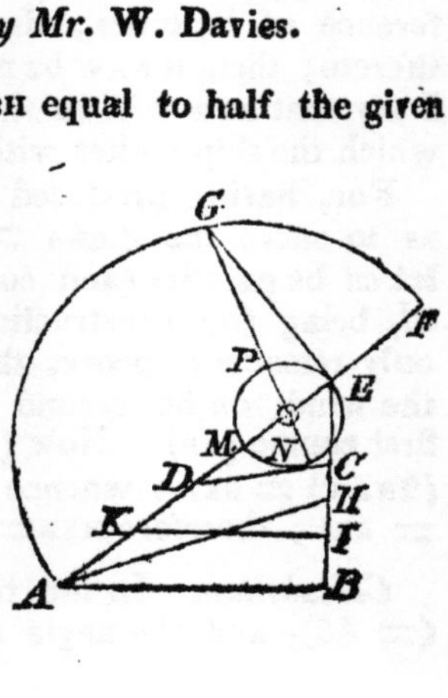

For if GO be conceived to cut, in P and N, a circle described on the diameter ME, then will KE × KM (= GN × GP) = GE² (36 Euc. III.) = AE × EF = AE

× EH; but, because AM = EB and AK = BI, thence is KM = EI; therefore KE × EI (= KE × KM) = AE × EH; and so, the triangle EKI being = EAH (15 Euc. VI.), the trapezium ABIK is also = ABH = $\frac{1}{2}$ABCD.

In the two solutions here exhibited, it will be observed, that the authors have considered different cases: For, of the two equal trapeziums, into which the given one is to be divided, the opposite sides of either the one or the other, may, by the question, be taken as equal; according as the intersection of the other opposite sides is supposed to fall above or below the base.

IX. QUESTION 456, *by Mr.* Mor. M'Roy.

In a right-angled triangular field there are three trees, viz. one in each fence: The distance of the tree in the base from that in the hypothenuse, and from the acute angle adjacent, and of the tree in the perpendicular from the right angle, are all equal, and given; and if lines be drawn from the tree in the base to the other two, those lines will form a right angle: The perpendicular of the proposed triangle is known to be the least of its kind (or that the data will admit of): Hence you are desired to find the sides of the triangle, and to construct the same geometrically.

Answered by Mr. H. Watson.

Let ABC represent the field, and D, E, F the places of the trees; and let AIG be parallel to DF: Then AIE being = EDF = a right angle, EAI (or CAG) is (= comp. AEI = comp. EAD) = C; whence CG = AG, and BC = BG + AG. From which, as many points as you please, in the curve bounding the perpendicular (BC) may be determined.

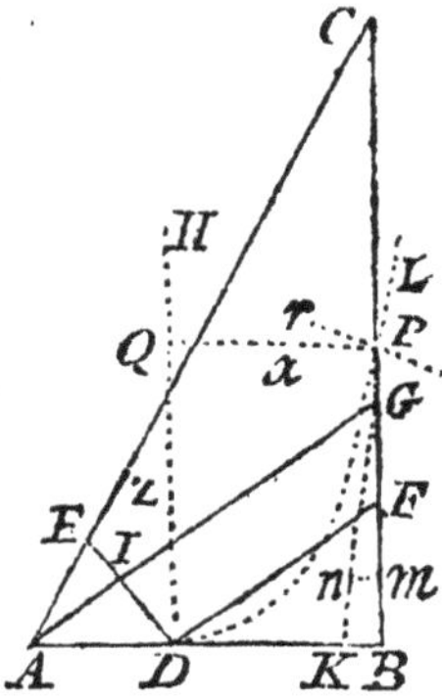

But to determine the least value of BC (calling BF, a; and DB, x), we have DB (x) : BF + DF ($a + \surd(xx + aa)$) :: AB ($a + x$) : BG + AG (= BC) = $(a + \surd(aa + xx)) \times (a + x)) \div x$: Whence, by taking the fluxion, &c. $x^4 = 2a^3x + a^4$. To construct this equation, put $az = xx$, and let the parabola DPL answering to $az = xx$ be described, having its axis DH perpendicular to AD; then if there be taken DK = a, Kn (parallel to DH) = $\frac{1}{2}a$, and from the centre n, with the radius $\frac{3}{2}a$, an arch rPr of a circle be described cutting the parabola in P; the perpendicular sought will pass through this point P; and, if in the perpendicular, so drawn, there be taken BF = AD, and the length of AG (parallel to DF) be set off from G to C, then shall ABC (when AC is drawn) be the triangle required to be constructed.

X. QUESTION 457, *by Mr.* G. Witchell.

The poet Hesiod tells us, That, in his time Arcturus rose at the setting of the sun 60 days after the winter solstice. It is required to find from hence, and the subjoined data,* collected from astronomical observations, at what time that author flourished.

* Lat. of the place 38° 54′ N.; long. of the star, at the beg. 1690, ≏ 10°53′52″, its lat. 30° 57′ N. annual precession of the equinox 50·3″.

Answered by Philarithmus.

Let E be the point of the ecliptic opposite to the sun's centre, and rising with Arcturus; $*l$ an arc of the circle of longitude passing through the star; then (by carrying back the perihelion) the sun's distance from the winter solsticial point in 60 days, about the year 850 A. C. (the middle of the century in which Hesiod is usually placed) will be 60° 10′; whence assuming ≏E (= its complement) 29° 50′, in the triangle ≏EC there are also given the angles at ≏ and C, whence ≏EC is given (= 109° 12′); therefore

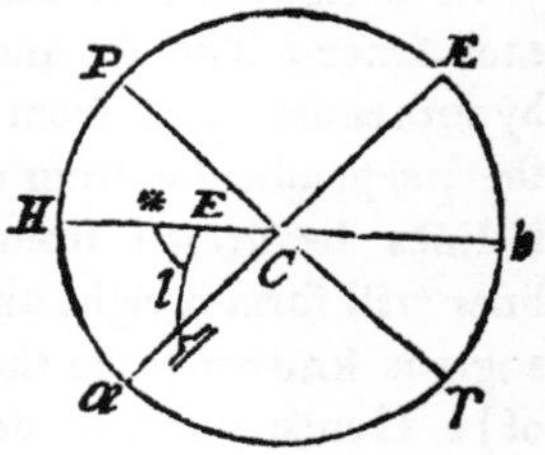

in the triangle $*l$E the angle at E being given, and $*l$ (= Arcturus's latitude), El is found = 12° 3′, and ≏l = 17° 47′, which, added to 19° 53′ 52″, gives the precession of the equinox from Hesiod's time to 1609, equal to 37° 30′ 52″, answering to the year 997 A. C. or rather 1002 A. C.

But the arc ≏E ought to be shortened at least 1° for the refractions of the sun and Arcturus, and the sun's semi-diameter: Now 1° in the arc ≏E (or an error of a day in Hesiod's account) lessens the arc ≏l, 50′, and consequently the time 60 years. So that Arcturus's image was in the horizon 60 days after the winter solstice, just when the sun disappeared about the year 940 A. C. But how far the data are to be depended upon as accurate, I am not able to say, considering that Hesiod lived before astronomy was cultivated among the Greeks as a science, and these observations of the stars were used to mark the returns of tho seasons in the room of an artificial year, which they wanted. There is one thing indeed might be alledged in favor of those who place him rather later than earlier, that supposing the observations made with all the accuracy they were capable of at first, yet it might have been done before Hesiod's time, and retained for the sake of a round number, whilst its deviations from truth were not very great, or the cause of the alteration known.

Mr. *G. Witchell*, the proposer of this question, proceeds in the solution of it exactly in the same manner. He makes the year sought to be ant. Chr. 1007: which he thinks is near a century and a half earlier than the truth; but observes, that, if the sun had been set about $8\frac{1}{2}'$ before the appearing of the star, the result would give ann. ant. Chr. 870, as Sir Isaac Newton has it in his chronology. Hence he infers, that chronologers ought not to lay any great stress, or too much rest their systems on such observations as these.

XI. QUESTION 458, *by Mr.* Edw. Johnson.

On May 12th, 1758, in latitude 53° 40′ N. a person sets out at six in the morning, and walks uniformly in the direction of his own shadow, till 50 minutes past ten; when he arrives at a place known to lie 12 miles west of the meridian of that from whence he set out: The question is to find how far, and at what rate he travelled in going from one place to the other?

Answered by Mr. G. Witchell.

Let P represent the north pole, Z the zenith, and ss a very small part of the sun's parallel of declination described in a given particle of time; supposing PS, Ps, ZS, Zs to be drawn, and also sn perpendicular to ZS. It is evident, that the celerity with which the man (walking uniformly in the direction of his own shadow) leaves the meridian, is always proportional to the sine of the sun's azimuth PZS; but the sine of PZS is to the sine of PSZ (or Ssn) in the constant ratio of the sine of PS to the sine of PZ: And the sine of Ssn is, again, to Sn (the increase of the sun's altitude) in the given ratio of the sine sns to Ss. Therefore the celerity aforesaid is every where in a constant ratio to that with which the sun's altitude increases: And consequently the whole increase of the sun's altitude from 6 to 50^{m} after 10 o'clock (found by calculation to be 37° 27′) must be to 12m. (the whole of the man's westing during that time) in that same constant proportion. But the increase of the sun's altitude when due east (at which time the man leaves the meridian with his absolute motion) is at the rate of $8°\ 53\frac{1}{4}'$ per hour, or at the rate of 42° 57′ in $4^{h}\ 50^{m}$ (the whole time of walking): Therefore it will be, as 37° 27′ : 42° 57′ : : 12 miles : 13·76 = whole distance travelled; being at the rate of 2·85 miles per hour.

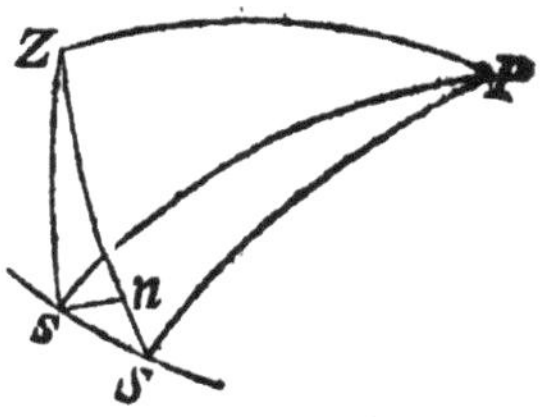

XII. QUESTION 459, *by Mr.* Hugh Brown.

Suppose a cistern whose length, breadth, and depth are 60, 28, and 32 inches, to be supplied with water by a cock, running uniformly into it, at the rate of 2 gallons per minute; and that the discharge of water by a cock at the bottom, when the cistern is full, is at the rate of 5 gallons per minute. Now I desire to know, supposing the vessel to be full, and both cocks opened together, in what time 150 gallons may be drawn off, at the evacuating cock.

Answered by Mr. Hugh Brown, *the Proposer.*

Put $a =$ the depth (AB) of the cistern, and $b = 190{\cdot}638 =$ the content, in ale gallons: Then, denoting the three given quantities 2, 5, and 150, by m, n, and p, and calling AM, x; Mm, $-\dot{x}$, and the time of descent through BM, t; we shall have $n\sqrt{(x \div a)} =$ the rate (per min.) at which the water runs out; and consequently $n\sqrt{(x \div a)} - m =$ the rate of the real decrease. Hence, as $1 : \dot{t} :: n\sqrt{(x \div a)} - m : -b\dot{x} \div a$ (= the decrease MNPpnm, in the time $\dot{t}$); whence $\dot{t} = -\frac{\dot{x}}{a} \times \frac{b}{n\sqrt{(x \div a)} - m}$: Also (by the quest.) we have $mt + b(a - x) \div a = p$. But to find the value of t from these equations, let y be assumed $= \frac{n\sqrt{(x \div a)} - m}{n - m}$ (= the ratio of the decrease at M to that at B) so shall $\frac{x}{a} = \frac{((n-m)y+m)^2}{nn}$, and $\dot{t} = -\frac{2b}{nn} \times \left((n-m)\dot{y} + \frac{m\dot{y}}{y}\right)$; and consequently $t = \frac{2b}{nn} \times ((n-m) \times (1-y) - m \times \text{h. log. } y)$. But by the other equation, t is $= \frac{p}{m} - \frac{b}{m} + \frac{b}{m} \times \frac{((n-m) \times y + m)^2}{nn}$: Which two values being compared together, we get $\left(\frac{n}{m} - 1\right) \times 2y + \left(\frac{n}{m} - 1\right)^2 \times \frac{yy}{2} + \text{h. log. } y = \left(\frac{n}{m} - 1\right) \times \left(\frac{n}{2m} + \frac{3}{2}\right) - \frac{nnp}{2mmb}$: Which, in the present case, becomes $3y + \frac{9yy}{8} + \text{h. log. } y = 1{\cdot}666$; whence y is found $= {\cdot}5954$; and from thence $t = 34{\cdot}33 = 34^{m}\ 20^{s}$, the time required.

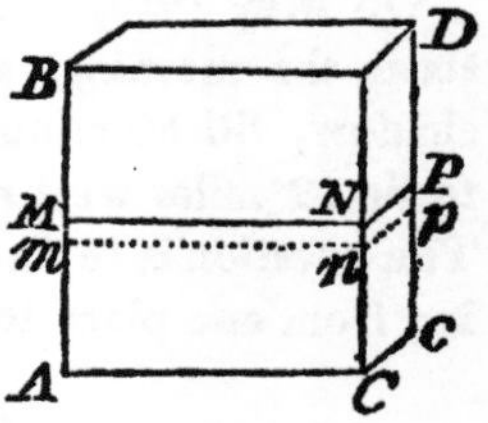

XIII. QUESTION 460, *by Mr.* Edw. Rollinson.

To determine the curve in which a body must move, so as to continue always at the same invariable distance from another body moving uniformly in a right line; the velocity of the former body being also uniform, and exceeding that of the latter, in any given ratio.

Answered by Mr. O'Cavanah, *of Dublin.*

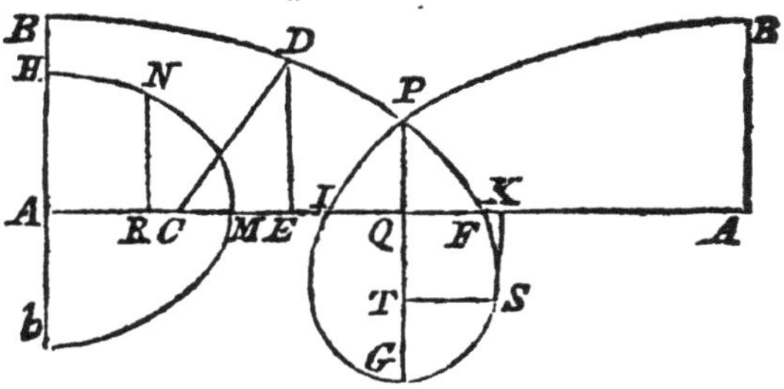

Let BDPFGIPB′ be the required curve, described (in a whole revolution) by the body D moving uniformly therein, whilst the other proposed body C moves uniformly over the right line AA′: Conceive CD, and also the ordinate DE, to be drawn; putting CD (= AB) = a, CE = z, AC = x, AE = y, and the given ratio of BD to AC, as n to 1. Then will DE = $\sqrt{(aa-zz)}$, AE = $x+z$, and BD = nx; whereof the fluxions being $\frac{-z\dot{z}}{\sqrt{(aa-zz)}}$, $\dot{x}+\dot{z}$, and $n\dot{x}$, we therefore have $\frac{z^2\dot{z}^2}{aa-zz}$ $+(\dot{x}+\dot{z})^2=n^2\dot{x}^2$; whence $\dot{x}=\frac{\dot{z}}{nn-1}+\frac{\dot{z}\sqrt{(nnaa-zz)}}{(nn-1)\sqrt{(aa-zz)}}$, and $\dot{y}\ (=\dot{x}+\dot{z})=\frac{nn\dot{z}}{nn-1}+\frac{\dot{z}\sqrt{(nnaa-zz)}}{(nn-1)\sqrt{(aa-zz)}}$. To assign the fluent hereof, let HMh be a semi-ellipsis, having its greater semi-axis AM = AB, and its lesser (AH) in proportion thereto, as $\sqrt{(nn-1)}$ is to n; then take AR = z (= CE) and make RN parallel to AB; so shall the arch HN = the fluent of $\frac{\dot{z}\sqrt{(nnaa-zz)}}{n\sqrt{(aa-zz)}}$; and we shall therefore have $y=\frac{nn\times \text{AR}+n\times \text{HN}}{nn-1}$; whence the ordinate DE is also known.

Corollary 1. When z becomes $=a$, then y is = AF = $\frac{nn\times \text{AM}+n\times \text{HNM}}{nn-1}$: But, when the two bodies are arrived at Q and G, and z is (again) = 0, then y = AQ = $\frac{n\times \text{HM}h}{nn-1}$; whence QF (= AF — AQ) is given = $\frac{nn\times \text{AM}-n\times \text{HM}}{nn-1}$.

Corollary 2. If $\dot{y}\ (=\frac{nn\dot{z}}{nn-1}+\frac{\dot{z}\sqrt{(nnaa-zz)}}{(nn-1)\sqrt{(aa-zz)}}$ be taken $=0$, then will $z=na\div\sqrt{(nn+1)}$; and the ordinate (KS) which touches the curve in (S) will be $=a\div\sqrt{(nn+1)}=\sqrt{(aa-zz)}$. And if the value of y corresponding hereunto be found (by the general equation) and from it AQ be subtracted, we shall thence get the semi-breadth QK (or ST) of the nodus PSGIP.——As to the value of PQ it may be also obtained from the same equation, by which $\frac{nn \times AR + n \times HN}{nn - 1}$ (= y) is given $= \frac{n \times HMh}{nn - 1}$ (= AQ): Where, if the value of n be such, that AR = AM, then will $n \times$ AM = HM, or $n : 1 :: $ HM : AM; in which case the intersection P will fall in the line AA′: But if the value of n be less than that determinable from hence (but still greater than unity) the point P will then fall below the line AA′, but never by a distance greater than $\frac{2894}{10000} \times$ AB; nor can ST be ever less than $\frac{266}{1000} \times$ AB.

When n is supposed less than unity, the radical quantity $\sqrt{(nnaa - zz)}$ will be no longer possible than till z becomes $= na$; after which time the conditions of the problem can no longer be fulfilled.

XIV. QUESTION 461, *by* Peter Walton, *Discip. Landenii.*

To assign the sum of the series $1 - \frac{1^2}{2^2} + \frac{1^2.3^2}{2^2.4^2} - \frac{1^2.3^2.5^2}{2^2.4^2.6^2} +$ &c. by means of circular and elliptic arcs.

Answered by the Proposer, P. Walton.

$\frac{\dot{x}}{\sqrt{(1 - x^4)}}$ being $= \frac{\dot{x}}{\sqrt{(1 - x^2)} \times \sqrt{(1 + x^2)}} = \frac{\dot{x}}{\sqrt{(1 - x^2)}} \times$ $(1 - \frac{1}{2}x^2 + \frac{1.3}{2.4}x^4 - \frac{1.3.5}{2.4.6}x^6 +$ &c.) it appears, by art. 286 of Mr. Simpson's Fluxions, that the whole fluent of $\frac{\dot{x}}{\sqrt{(1 - x^4)}}$ is $= f \times$ $(1 - \frac{1^2}{2^2} + \frac{1^2.3^2}{2^2.4^2} - \frac{1^2.3^2.5^2}{2^2.4^2.6^2} +$ &c.) f being the whole fluent of $\frac{\dot{x}}{\sqrt{(1 - x^2)}}$; which, it is well known, is $\frac{1}{4}$ of the periphery of a circle whose radius is 1.

Now, if x be supposed $= y^{\frac{1}{2}}$, $\frac{\dot{x}}{\sqrt{(1 - x^4)}}$ will be $= \frac{\frac{1}{2}y^{-\frac{1}{2}}\dot{y}}{\sqrt{(1 - y^2)}}$ of which fluxion (by Landen's Math. Lucubrations, p. 146) the whole flu. is $\frac{e + \sqrt{(ee - 2f)}}{2}$, e denoting $\frac{1}{4}$ of the periphery of an ellipsis, whose semi-axes are $\sqrt{2}$ and 1.

It follows therefore, that the series $1 - \frac{1^2}{2^2} + \frac{1^2.3^2}{2^2.4^2} - \frac{1^2.3^2.5^2}{2^2.4^2.6^2}$ + &c. is $= \frac{e + \sqrt{(ee - 2f)}}{2f}$.

The same answered by Mr. E. Rollinson.

It is well known that $Q \times : 1 - \frac{1}{2}e + \frac{1.3}{2.4}f - \frac{1.3.5}{2.4.6}g +$ &c. is the fluent of $\frac{\dot{x}}{\surd(1-x^4)} \times : 1 - ex^2 + fx^4 - gx^6 +$ &c. when $x = 1$; Q being the fluent of the first term, and $= \frac{1}{4}$ of the periphery of the circle whose radius is unity. Hence, if e be taken $= \frac{1}{2}$, $f = \frac{1.3}{2.4}$, $g = \frac{1.3.5}{2.4.6}$, &c. then $Q \times : 1 - \frac{1^2}{2^2} + \frac{1^2.3^2}{2^2.4^2} - \frac{1^2.3^2.5^2}{2^2.4^2.6^2} +$ &c. $=$ fluent $\frac{\dot{x}}{\surd(1-xx)} \times : 1 - \frac{1}{2}x + \frac{1.3}{2.4}x^2$ &c. $=$ fluent $\frac{\dot{x}}{\surd(1-xx)} \times (1 + xx)^{-\frac{1}{2}} =$ flu. $\frac{\dot{x}}{\surd(1-x^4)}$; and consequently $1 - \frac{1^2}{2^2} + \frac{1^2.3^2}{2^2.4^2} - \frac{1^2.3^2.5^2}{2^2.4^2.6^2} +$&c.$= \frac{1}{Q} \times$ flu. $\frac{\dot{x}}{\surd(1-x^4)}$ (when $x=1$). But $\frac{\dot{x}}{\surd(1-x^4)} + \frac{x^2\dot{x}}{\surd(1-x^4)}$ is known to express the fluxion of an elliptical arch (R), whose two semi-axes are 1 and $\surd 2$; and it may be easily demonstrated (by the method laid down at p. 76, in Simpson's Laws of Chance) that, if A and B be assumed to denote the fluents of $\frac{x^{pn-1}\dot{x}}{\surd(1-x^n)}$ and $\frac{x^{pn+\frac{1}{2}n-1}\dot{x}}{\surd(1-x^n)}$ (p and n being any positive numbers whatever), the product (AB of these fluents) will (when $x = 1$) be $= \frac{2Q}{pnn}$. Whence, by taking $n = 4$, and $p = \frac{1}{4}$, so that A and B may become the respective fluents of $\frac{\dot{x}}{\surd(1-x^4)}$ and $\frac{x^2\dot{x}}{\surd(1-x^4)}$ (whereof the sum is given above $=$ R) we have AB, in this case, $= \frac{1}{2}Q$: From whence, and the equation $A + B = R$, A is found $= \frac{R}{2} + \frac{1}{2}\surd(R^2 - 2Q)$; and $\frac{A}{Q} = \frac{R}{2Q} + \frac{1}{2}\sqrt{\left(\frac{RR}{QQ} - \frac{2}{Q}\right)} =$ the series proposed.

Some correspondents have objected against this question, as an unfair one; "because there can be no reasoning from the data, to arrive at a conclusion."——We shall not take upon us to decide this point; but shall not, however, scruple to allow, that problems of a very abstracted and intricate nature, are less eligible in a work like this, than some others of a different kind.

THE PRIZE QUESTION, *by Mr.* O'Cavanah, *of Dublin.*

To determine on what day of the year, in the lat. of London, the length of the afternoon exceeds that of the forenoon by the greatest difference possible, reckoning the day to begin at sun-rising, and to end at sun-setting.

Answered by Κυβερνητης.

Let *ss* represent the increase of the sun's longitude (♈*s*) in the space of six hours (or $\frac{1}{4}$ of a day); and *sg* be the increase of declination corresponding thereto: Putting $s =$ sin. ♈, $x =$ sin. ♈s, and $u = ss$: Then will $sx =$ sin. dec. BS, and $\sqrt{(1 - ssxx)} =$ its cos.

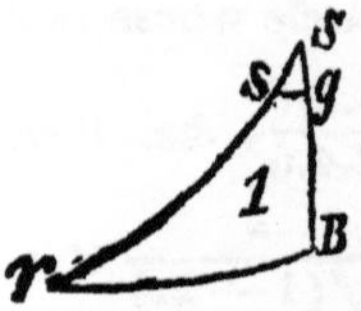

And so, again (per spherics) tang. ♈s $\left(\frac{x}{\sqrt{(1-xx)}}\right)$: tang. BS $\left(\frac{sx}{\sqrt{(1-ssxx)}}\right)$ :: rad. : cos. S :: ss (u) : sg = $\frac{su\sqrt{(1-xx)}}{\sqrt{(1-ssxx)}}$ = the increase of decl. in $\frac{1}{4}$ of a day. Therefore, if m be now put for $\frac{1}{4}$ of the whole periphery of the circle (the rad. being unity) and z for the arch, expressing the sun's ascensional difference, we shall have $m : m + z :: \frac{su\sqrt{(1-xx)}}{\sqrt{(1-ssxx)}} : \frac{su \times (m+z)\sqrt{(1-xx)}}{m\sqrt{(1-ssxx)}}$ = the alteration of decl. from noon to sun-setting. Let this (*in Fig.* 2.) be denoted by sk (supposing PS to be the sun's polar distance, and S his place in the horizon HON, at setting) and let b, c, and d be taken to express the sine, cosine, and tangent of PN the pole's elevation. Then, the sine of S being $= \frac{b}{\sqrt{(1-ssxx)}}$, its cosine will be $= \frac{\sqrt{(cc - ssxx)}}{\sqrt{(1-ssxx)}}$; whence $hk = \frac{b}{\sqrt{(cc-ssxx)}} \times (sk)$;

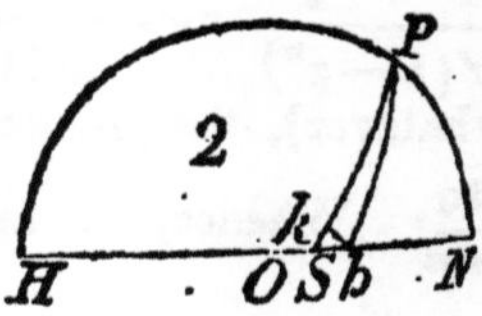

and consequently sin. kPh $\left(= \frac{\text{rad.}}{\text{sin. PS}} \times (hk)\right) =$

$$\frac{b \times (sk)}{\sqrt{(1-ssxx)} \times \sqrt{(cc-ssxx)}} = \frac{bsu \times (m+z)\sqrt{(1-xx)}}{m \times (1-ssxx) \times \sqrt{(cc-ssxx)}};$$

which, by the question, is to be a maximum; and consequently its log. ($=$ log. $\frac{bs}{m}$ + log. u + log. $(m+z)$ + $\frac{1}{2}$ log. $(1-xx)$ − log. $(1-ssxx)$ − $\frac{1}{2}$ log. $(cc - ssxx)$) a maximum also: Whence, in fluxions (putting $t = \frac{s}{c}$) we have $\frac{\dot{u}}{u} + \frac{\dot{z}}{m+z} - \frac{x\dot{x}}{1-xx} + \frac{2ssx\dot{x}}{1-ssxx} + \frac{ttx\dot{x}}{1-ttxx} = 0$; but the sine of z being $\frac{bsx}{c\sqrt{(1-ssxx)}}$,

we have $\dot{z} = \frac{bs\dot{x}}{(1 - ssxx)\sqrt{(cc - ssxx)}} = \frac{ds\dot{x}}{(1 - ssxx)\sqrt{(1 - ttxx)}}$.
And to determine u and $\dot{u}$ also, let e $(= \cdot 017)$ denote the eccentricity of the earth's orbit, and y the sine complement of the sun's (or earth's) true anomaly: Then $(1 - ey)^2$, as is well known, will express the rate of the increase of this anomaly (or of the sun's true longitude): Which quantity being therefore proportional to u, we have $\frac{\dot{u}}{u} = \frac{-2e\dot{y}}{1 - ey}$. But the difference of the two angles, whose sines are y and x, being given $(= 9° \ 14') =$ the distance of the aphelion from the solstitial point, we shall, by putting the sin. of $9° \ 14' = p$, and its cosine $= q$, have $y = qx - p\sqrt{(1 - xx)}$; and consequently $\frac{\dot{u}}{u}$ $\left(= \frac{-2e\dot{y}}{1 - ey}\right) = - \frac{2eq\dot{x}\sqrt{(1 - xx)} - 2epx\dot{x}}{\sqrt{(1 - xx)} \times (1 - eqx + ep\sqrt{(1 - xx)})} = -2e\dot{x}$, very near (because the terms both in the numerator and denominator, after the two first, may, on account of their smallness, be neglected). This value, therefore, with that of $\dot{z}$, being substituted in the general equation, we at length get $-2e + \frac{ds}{(1 - ssxx)\sqrt{(1 - ttxx)} \times (m + z)}$ $- \frac{x}{1 - xx} + \frac{2ssx}{1 - ssxx} + \frac{ttxx}{1 - ttxx} = 0$: From whence, (either by resolving the whole into a series, or by any other of the known methods of approximation) the value of x may be found, and will come out $= \cdot 493 =$ sin. of $29° \ 32' =$ sun's longitude sought; answering to the 19th day of April; the afternoon of that day exceeding the forenoon by $1' : 4\frac{1}{4}''$.

Questions proposed in 1760, *and answered in* 1761.

I. QUESTION 462, *by Miss* Ann Nichols.

Two partners, in a venture made,
Gain twice two hundred pounds in trade:
The stock of A, when they began,
Exceeded B's by eighty-one;
Twice ninety-five B gain'd in all.
Now for the stock of both I call.

Answered by Amaryllis, *and several others.*

Since, according to the question, A's stock exceeded B's by 81l.

and his gain was greater than B's by 20l. Say therefore, as 20 : 81 :: 210 (A's whole gain) : 850l. 10s. the stock of A; whence that of B is found = 769l. 10s.

II. QUESTION 463, *by Miss* S. T.

A drover bought in as many sheep, of different sorts, as cost him 48l. one-third of which he sold again at 20s. a-piece; one-fourth at 18s. and the rest at 16s. a-piece; and found his gain, upon the whole, to be 5l. 10s. What number of sheep did he buy and sell?

Answered by Mr. T. Bromhall, *and others.*

Let $12x$ = the number of sheep; then, by the question, $4x \times 20 + 3x \times 18 + 5x \times 16 = 1070$, or $214x = 1070$: Therefore $x = 5$, and $12x = 60$, the number of sheep required.

III. QUESTION 464, *by Mr.* Tho. Harris.

Near Ouse's verdant banks, in Bedfordshire,
Stands Carlton, blest with a sagacious fair,
In whom at once Minerva's wit is seen,
Diana's chastness, and the graces' mien.
Would you the name of this fair charmer know,
First solve th' equations which you'll find below.

$x + y + z + v = 56$, $xx + yy + zz + vv = 910$, $xv + 2yy - zz = 6$, $z = 2y$; in which the values of x, y, z, and v denote the places in the alphabet of the four letters that compose this amiable maiden's name.

Answered by Mr. Edw. Nott, *of Stamford.*

By writing $2y$ for z, the equations become $x + 3y + v (= 56) = a$, $xx + 5yy + vv (= 910) = b$, $vx - 2yy (= 6) = c$: To the second of which add twice the third, and we have $xx + 2vx + vv$, or $(x + v)^2 = b + 2c - yy = (a - 3y)^2$ (by the first); whence $10yy - 6ay = b + 2c - aa$; from which y will be found $= 9$, $x = 8$, $z = 18$, and $v = 21$: And the letters of the alphabet corresponding, whereby the young lady's name is formed, are, I, H, S, W.

IV. QUESTION 465, *by Mr.* Geo. Brown.

From two given points, to draw two lines to meet in the circumference of a circle given in position and magnitude, so that the sum of their squares shall be the least possible?

Answered by Mr. Davies, *of Newent, Gloucestershire.*

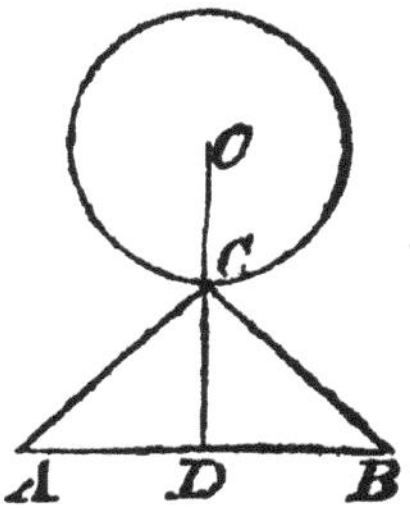

Let O be the centre of the given circle, and let A and B be the two given points; draw the line AB and bisect it in D; also draw DO cutting the circle in the point C. So shall AC and BC, when drawn, be the two lines required.

Demonstration. Because DC is a minimum, or the shortest line that can possibly be drawn from D to the circumference of the circle, $2DC^2$ must therefore be a minimum. But, by a known theorem, $AC^2 + BC^2 = 2AD^2 + 2DC^2$; therefore ($2AD^2$ being constant) $AC^2 + BC^2$ must be a minimum.

V. QUESTION 466, *by Mr.* W. Chapman.

There are three circles whose diameters are 30, 20, and 25 inches, having their centres all placed in the same right line; whereof the distance of the first from the second is 30, and of the second from the third 28 inches; now you are desired to determine the diameter and position of a fourth circle, that shall touch all the three given ones.

Answered by Mr. E. Batten.

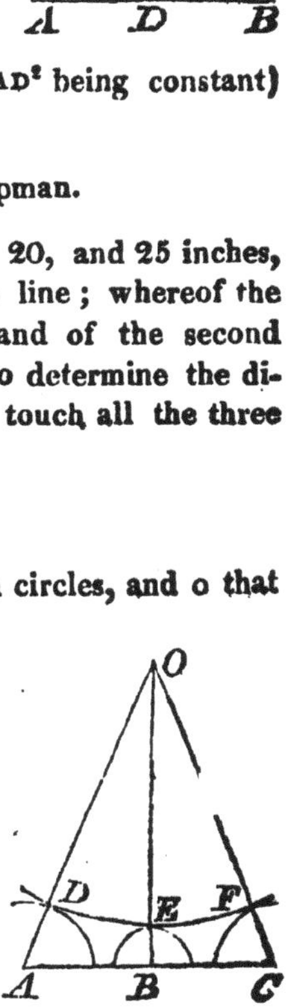

Let A, B, and C be the centres of the three given circles, and O that of the required one. Put $a = AB = 30$, $b = BC = 28$, $c = AC = 58$, $x = OB$, $x + m\ (x + 5) = OA$, $x + n\ (x + 2\frac{1}{2}) = OC$. Now (per lemma, p. 128, of Mr. Simpson's Select Exercises) $(x + m)^2 \times b + (x + n)^2 \times a = (xx + ab) \times c$; that is, $(b + a).\ xx + (2bm + 2an)\ x + bmm + ann = cxx + abc$; or, $(2bm + 2an)\ x + bmm + ann = abc$; whence $x = \dfrac{abc - bmm - ann}{2bm + 2an} = 111\frac{41}{172}$; and consequently the radius OE of the required circle $= 101\frac{41}{172} = 101{\cdot}238$, &c.

A *Construction* of this prob. may be seen in Lawson's Apollonius on Tangencies.

VI. QUESTION 467, *by Mr.* W. Spicer.

Surveying a field, I found the four sides thereof to be 10, 9, 7, and 6 chains, in a successive order; I likewise, at the two extremes of the longest side, took the bearings of the opposite angles, which were N.E. by E. and W.S.W. Hence the content of the field is required?

Answered.

The best solutions that have been received to this question are derived in virtue of a theorem in the Diary, for 1758, where it is demonstrated, that the quadruple of the area of any trapezium ABCD is equal to $(AB^2 + CD^2 - BC^2 - AD^2) \times \frac{\text{tang. E}}{\text{rad.}}$: From whence (seeing the angle E, as well as sides of the trapezium, are all given, by the question) the area will be found: And in this manner is actually determined by Mr. *W. Davies*, Mr. *D. Hastings*, Mr. *T. Harris*, Mr. *Malachy Hitchins*, Mr. *W. Kingston*, and *W. Spicer*; and is found to be $= 8 \times$ tang. $11° 15' = 1{\cdot}5912992$ square chains.—If the second bearing were to be W.N.W. instead of W.S.W. (as it was misprinted) the angle made by the two diagonals would then be 5 points (instead of one); and the area of the trapezium 11·9728 square chains.

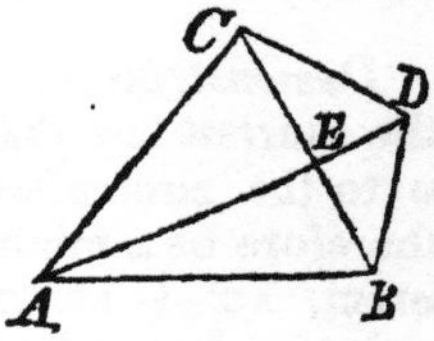

VII. QUESTION 468, *by Mr.* Tho. Barker.

Supposing two sides of a triangle to be given (equal to 25 and 16 feet), whereof the greater is parallel to the horizon; I desire to know what the length of the third side must be, so that the time of the descent of a heavy body along the same shall be a minimum?

Answered by Mr. T. Barker, *the Proposer.*

Construction. With a radius BE equal to the lesser of the two given sides, let a semi-circle EFG be described; and, having made BF perpendicular to the longer side AB, draw AF, cutting the circle in C, from whence draw CB: Then shall ABC be the triangle sought.

Demonstration. Make AH perpendicular to AB, and produce BC to meet it in O; and with the radius OA or OC (for they are equal, because BF and BC are equal) let the semi-circle ACH be described. Then the time of the descent along the chord CA will, it is well known, be equal to the time of descent along the diameter HA; which will be the shortest possible, because the semi-circle ACH only touches the given one ECFG, and has therefore its diameter less than any other semi-circle that can be described on AH produced, to cut the given semi-circle ECFG.

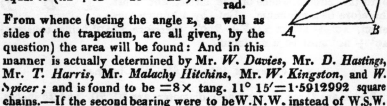

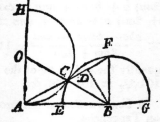

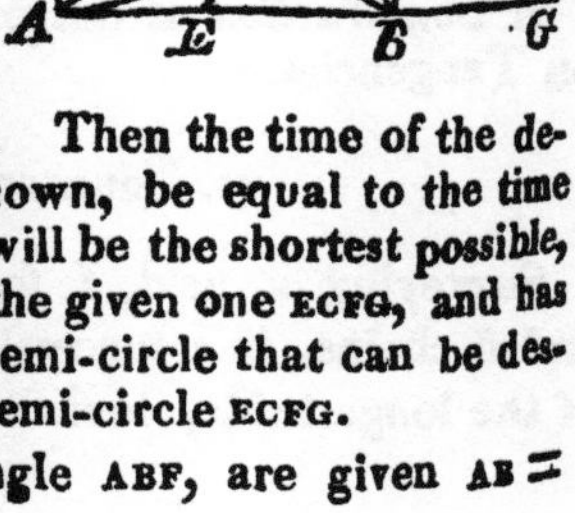

Calculation. In the right-angled triangle ABF, are given AB =

25, and $BF = 16$; whence the angle $BAF = 32° 27'$: Then in the triangle ACB are given AB, BC, and the angle A; whence $AC = 12·425$.

But the required side AC may be otherwise found independent of trigonometry: For, if BD be made perpendicular to AF, then by similar triangles, $FD = BF^2 \div AF$; and consequently $AC \; (= AF - 2FD) = (AF^2 - 2BF^2) \div AF = (AB^2 - BF^2) \div \sqrt{(AB^2 + BF^2)} = 12·425$.

VIII. QUESTION 469, *by* P. M. *of Durham.*

To describe the circumference of a circle through two given points, and which shall cut off from a given circle an arc equal to an arc given.

Answered by Mr. Lionel Charlton, *of Whitby.*

Having drawn GP to bisect, at right angles, the line AB joining the given points, from any point in it, as P, let the circumference of a circle be described through A and B, to cut the given circle EMND in any two points D and E; through which points draw DEF to meet AB produced in F; then draw FH at an equal distance from the centre C with the chord MN of the arc given; and, perpendicular thereto, draw CQ, meeting GP in Q: So shall Q be the centre of the circle required.

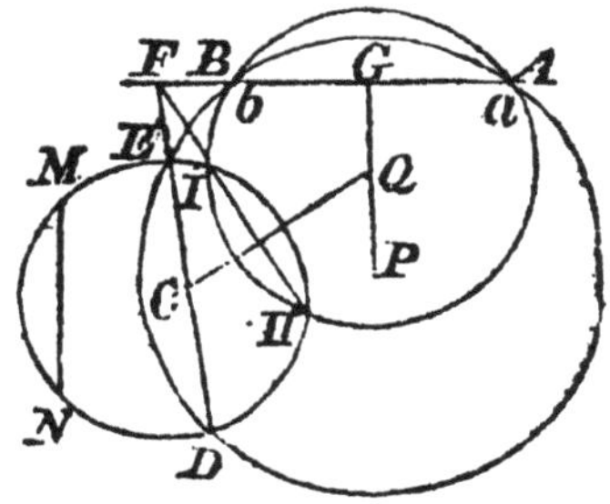

Demonstration. From the centre Q, through H and I, let the circumference of a circle be described, cutting AF in b and a. Then (by Euc. XXXVI. 3.) $Fa \times Fb = FH \times FI = FD \times FE = FA \times FB$: But (by Euc. VI. 2) $Fa \times Fb = FG^2 - bG^2$, and $FA \times FB = FG^2 - BG^2$; therefore $FG^2 - bG^2 = FG^2 - BG^2$; and so, bG being $= BG$, the points b, B coincide, and the circumference of the circle described from the centre Q, through H and I, likewise passes through B and A.

IX. QUESTION 470, *by Mr.* Patrick O'Cavanah, *of Dublin.*

On one of the banks of a certain river stand four windmills, all in the same right line; whereof the distance of the first from the second is known to be 150 yards, of the second from the third 180 yards, and of the third from the fourth 200 yards: Being on the opposite shore, I found by observation, that the two middlemost of them subtended an angle, at my eye, of 15 degrees; and that the angles subtended by the first and second, and by the third and fourth, were equal, the one to the other. Hence I would know (by means of a geometrical construction) not only the breadth of the river, but also my particular distance from each of the four objects, in that station.

Answered by Mr. E. Rollinson.

Let A, B, C, D be the places of the four objects; and upon BC let a segment of a circle BPC be described to contain the given angle of 15°: Make B*a* and DC parallel to each other, taking the former = BA, and the latter = DC; and then draw C*a*O to meet DA, produced, in O; from whence draw OQ to touch the circle in P, which is the point sought.

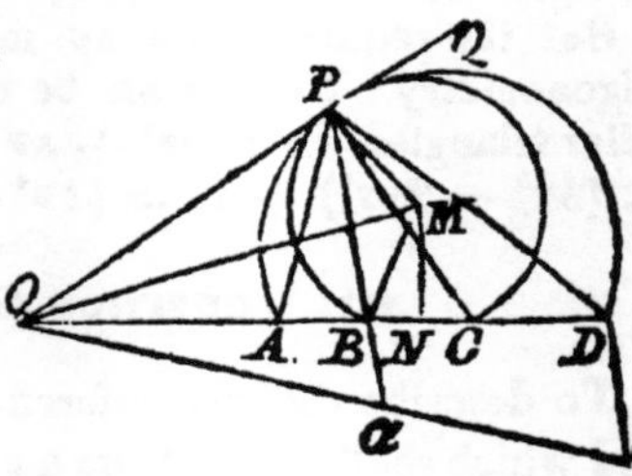

Demonstration. Conceive the circumference of a circle to be described through A, P, and D. By construction and sim. Δs, B*a* (BA) : OB :: DC (DC) : OD; whence by division, OA : OB :: OC : OD, and consequently OA × OD = OB × OC = OP^2, (Euc. III. 36.) whence it is evident, that OQ also touches the circle APD in the point P: But (by Euc. III. 32.) the angle APO = ADP, and BPO = BCP; and consequently APB (BPO — APO = BCP — ADP) = CPD.

Calculation. Having, from the centre M, let fall the perpendicular MN, and drawn MB, MP, NO, &c. it will be as DC — B*a* (50) : BD (380) :: B*a* (150) : BO = 1140; whence NO = 1230. Also as NB (90) : NO (1230) :: tang. NMB (15°) : tan. NMO = 74° 43$\frac{1}{2}$′; whence NOM = 15° 16$\frac{1}{2}$′, and BMO = 59° 43$\frac{1}{2}$′. Again, as sin. MBN (65°) : rad. :: MN : MB (MP) :: sin. NOM : sin. POM = 15°49$\frac{1}{2}$′; whence PMO = 74°10$\frac{1}{2}$′, PMB (= PMO + BMO) = 133° 54′, and BCP (= $\frac{1}{2}$PMB) = 66° 57′: From which the rest are readily found, viz. PB = 639·93, PC = 688·31, APB (or CPB) = 13° 29$\frac{5}{8}$′, PA = 636·49, PD = 788·66, and the perpendicular distance of P from AD (= breadth of the river) = 633·62.

X. QUESTION 471, *by Mr.* E. Rollinson.

A gentleman bought an estate in houses for 1500l. which, being let, brought him in 120l. per annum, clear of all expences and deductions: At the end of ten years (most of the houses being out of repair, and he not choosing to be at the expence of fitting them up) he sold the whole estate again for 800l. The question is, to find what interest he made of his money.

Answered by Mr. J. Honey, *of Redruth, Cornwall.*

Let $a = 1500$l. $b = 800$l. $c = 120$l. and x the required rate of interest: Then will ax = principal and interest, and $ax - c$ = amount after the first payment is deducted. And in the same manner we have $ax^{10} - cx^9 - cx^8 - cx^7 - cx^6 - cx^5 - cx^4 - cx^3 - cx^2 - cx - c = b$ (per question) or $ax^{10} - c\,(x^{10} - 1) \div (x - 1) = b$; therefore $ax^{11} - (a - c)\,x^{10} - bx + b + c = 0$. Solved, $x = 1{\cdot}04142$, &c. and the rate of interest required 4l. 2s. 10d. per cent.

The same answered by Mr. W. Kingston, *of Bath.*

Let x be the rate: Then the amount of 1500l. capital, at the end of 10 years, will be $1500x^{10}$; and annuity of 120l. forborn the same time, will amount to $120 \times \frac{x^{10}-1}{x-1}$. Therefore, we have $1500x^{10} = 120 \times \frac{x^{10}-1}{x-1} + 800$ (by the question), or $x^{10} - \cdot 08 \times \frac{x^{10}-1}{x-1} = \frac{8}{15}$. Put $x = 1 + v$; then x^{10} $(= (1+v)^{10})$ being $= 1 + 10v + 45v^2 + 120v^3 + 210v^4$, &c. our equation, by proper substitution, will become $6{\cdot}4v + 35{\cdot}4v^2 + 113{\cdot}2v^3$, &c. $= \frac{1}{3}$; whence, by reverting the series, v is found $= \cdot 04142$: And the rate of interest sought is 4l. 2s. 10d. per cent.

XI. QUESTION 472, *by Mr.* W. Bevil.

Out of a semi-circular piece of ground, the radius of which is twelve chains, I am to take a garden with a fence of 10 chains (terminating in the semi-circle) so as to contain the most land possible; how many acres will fall to my lot.

Answered by Mr. G. Witchell.

It is demonstrable, even by common geometry (see Simpson's Elem. p. 207, 2d. edit.) that the given length, or boundary EBF (whatever length the right line EF is to have) must form the arch of a circle. In order therefore to the resolution of this, and other problems of the like nature, it will be very convenient to have ready at hand, a proper series, or near approximation for the area of a circular segment (FBF) expressed in terms of the arch BE (a), and its sine DE (s), independent of the radius. Such an approximation is the following one, viz. $2s\sqrt{(\frac{2}{3}a(a-s))} = $ EBF; From whence, by making s variable (and $a = 5$, as given by the quest.) we have $\frac{\frac{4}{3}aa\dot{s} - 2as\dot{s}}{\sqrt{(\frac{2}{3}a(a-s))}}$ = flux. of the segment EBF. But the fluxion of the other segment EAF, if the radius CA ($= 12$) be denoted by r, will, it is well known, be $= \frac{2s^2\dot{s}}{\sqrt{(rr-ss)}}$. Therefore, when the sum of both segments is a maximum, we shall have $\frac{\frac{4}{3}aa\dot{s} - 2as\dot{s}}{\sqrt{(\frac{2}{3}a(a-s))}} + \frac{2s^2\dot{s}}{\sqrt{(rr-ss)}} = 0$; and consequently $\frac{3a}{2} \times (s - \frac{4}{3}a)^2 \times (rr-ss) - s^4 \times (a-s) = 0$: Whence

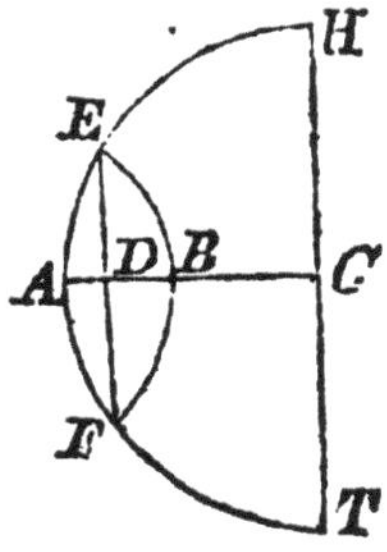

s is found $= 3\cdot844$; from which the area of the segment EBF comes out $= 15\cdot09$, and that of the segment EAF $= 3\cdot26$: And consequently the whole of both $= 18\cdot35$ square chains.*

* Although approximating theorems ought not generally to be used when they are to be put into fluxions, yet they will sometimes answer pretty well, as is the case here. H.

XII. QUESTION 473, *by Mr.* Cha. Wildbore.

'Tis proposed to divide a given number into so many parts, and so proportioned among themselves, that the continual product of the first, the square of the second, the cube of the third, the biquadrate of the fourth, &c. shall be a maximum.

Answered by Lieut. Henry Watson.

Let a represent the quantity given, and x the required number of parts into which it is to be divided; and let the first (or least) part be denoted by y. Then it is very easy to demonstrate, that the 2d part will be $= 2y$, the 3d part $= 3y$, &c. Therefore we have $y + 2y + 3y + 4y \ldots + xy$ (or $x \times \frac{x-1}{2} \times y) = a$: And consequently $y = \frac{2a}{x \times (x+1)}$. Now in order to find such an integer for the value of x, we shall make $y \times (2y)^2 \times (3y)^3 \ldots \times (xy)^x$, or its equal, $2^2 \times 3^3 \times 4^4 \ldots\ldots \times x^x \times \left(\frac{2a}{x.(x+1)}\right)^{\frac{x.x+1}{2}}$, a maximum, let $z-1$, and z be wrote successively therein, instead of x, and let the two quantities thus arising, be made equal to each other: So shall $2^2 \times 3^3 \times 4^4 \ldots (z-1)^{z-1} \times \left(\frac{2a}{(z-1).z}\right)^{(z-1).\frac{1}{2}z} = 2^2 \times 3^3 \times 4^4 \ldots z^z \times \left(\frac{2a}{z.(z+1)}\right)^{\frac{1}{2}z.(z+1)}$; which, if the whole be divided by $2^2 \times 3^3 \times 4^4 \ldots \times (z-1)^{z-1}$, will give $\left(\frac{2a}{(z-1).z}\right)^{(z-1).\frac{1}{2}z} = z^z \times \left(\frac{2a}{z.(z+1)}\right)^{\frac{1}{2}z.(z+1)}$: And, if the indices be now divided by $\frac{1}{2}z$, we shall have $\left(\frac{2a}{(z-1).z}\right)^{z-1} = z^2 \times \left(\frac{2a}{z \times (z+1)}\right)^{z+1}$; whence, by reduction, $\frac{(z+1)^{z+1}}{(z-1)^{z-1}} = 4aa$, or $2 \log. z + (z+1) \log. (1 +$

$\frac{1}{z}) - (z-1) \log. (1 - \frac{1}{z}) = 2 \log. 2a$; from either of which equations the value of z may be found: And the next inferior integer thereunto will be the number of parts required.

Corollary. When a is a large number, z being also large, $(z+1) \times \log. (1 + \frac{1}{z}) - (z-1) \times \log. (1 - \frac{1}{z})$ will then be $= 2$, very near: And consequently $2 \log. z + 2 = 2 \log. 2a$; or, $\log. z + 1 = \log. 2a$; or, by putting $c = 2{\cdot}71828$, &c. (the number whose h. log. is 1), we have $\log. z + \log. c = \log. 2a$: Therefore $cz = 2a$, and $z = \frac{2a}{c}$.

XIII. QUESTION 474; *by Mr.* G. Witchell.

Supposing the moon's declination, as well as the latitude of the place to be given, I desire to know, at what elevation her parallax in altitude increases the quickest.

Answered by Mr. T. Allen, *of Spalding.*

It is evident that the parallax increases the quickest, when its fluxion bears the greatest ratio possible to the fluxion of the angle at the pole P.

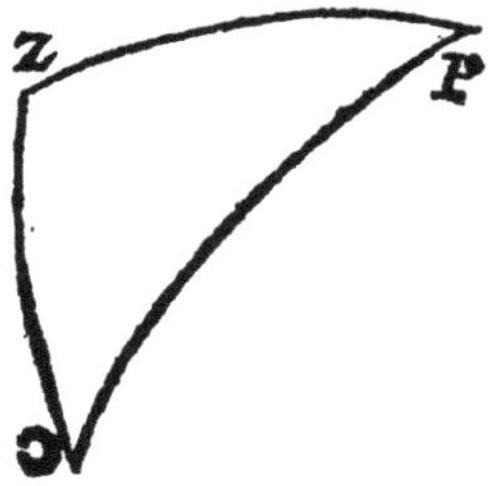

Put a and b for the sine and cosine of P☽ the moon's polar distance, c and d the sine and cosine of PZ (the comp. of lat.) $n =$ horizontal parallax, and $x =$ cosine of P; then will $\sqrt{(1 - (bd + acx)^2)} =$ sine Z☽ the moon's zenith distance, and $n\sqrt{(1 - (bd + acx)^2)} =$ her parallax in altitude; whose fluxion $\frac{-nac\dot{x} \times (bd + acx)}{\sqrt{(1 - (bd + acx)^2)}}$ divided by $\frac{-\dot{x}}{\sqrt{(1-xx)}}$ (the fluxion of the angle P) gives $\frac{nac\sqrt{(1-xx)} \times (bd + acx)}{\sqrt{(1 - (bd + acx)^2)}}$ a max. The fluxion of the log. of which being made $= 0$ (making at the same time $r = bd$, and $s = ac$) we have $- \frac{x}{1 - xx} + \frac{s}{r + sx} + s \times \frac{r + sx}{1 - (r + sx)^2} = 0$; From whence the value of x (and consequently the moon's altitude) may be found.

The same answered by Mr. Witchell, *the Proposer.*

This gentleman substitutes x for the cosine of the moon's zenith distance; and then (the notation of the known quantities being here made to agree with that in the solution above given, by Mr. *Allen*) he finds (by a known theorem) the cosine of the angle at the pole to be $= (x - r) \div s$: and, consequently its sine $= \sqrt{(ss - rr + 2rx - xx)} \div s$; whence (by dividing the fluxion of the one by the other) the fluxion of the angle itself is found $= -\dot{x} \div \sqrt{(ss - rr + 2rx - xx)}$; which (by the quest.) being to $-n\dot{x}\dot{x} \div \sqrt{(1 - xx)}$ (the flux. of the parallax $n\sqrt{(1 - xx)}$) in the least ratio possible, $nnxx \times (ss - rr + 2rx - xx) \div (1 - xx)$ must therefore be a maximum; whence, by taking the fluxion, &c. we have $x^4 - rx^3 - 2x^2 + 3rx + ss - rr = 0$; from which x will be known.

PRIZE QUESTION, *by Mr.* O'Cavanah.

Two persons, A and B, whose chances for winning a single game, are as 3 to 2, the former having seven guineas, and the latter four, enter into play together, on condition that A every game shall stake two guineas to B's one; and that the play between them shall continue till one of them, either by having lost his whole stock, or for want of a complete stake, is obliged to give out. The question now is, to find the gain of B and the disadvantage of A in this agreement; and to point out a general method for the resolution of all questions of this nature.

Note. M. *De Moivre*'s method will not give a true solution; nor has there been any rule or method yet laid down by which such question can be resolved.

Answered by the Proposer.

Let the number of pieces staked at each game by A and B, be denoted by r and s; and let their chances for winning each game, be in the ratio of a to b: Moreover, let Δ denote the expectation of A, when (in the course of the play he has any number (q) of pieces in his possession; and let Δ', or $\overline{\Delta}$ be his expectation, after one more game, when he will have either $q+s$, or $q-r$ pieces, according as he wins or loses. If he wins the game (whereof the probability is $\frac{b}{a+b}$) he will have $q + s$ pieces, and his expectation will then be Δ', which therefore, multiplied by $\frac{a}{a+b}$, gives $\frac{a}{a+b} \times \Delta'$ for his present expectation on that event. But if he loses the game (whereof the probability is $\frac{b}{a+b}$) he will only have $q - r$ pieces, and his expectation being then denoted by $\overline{\Delta}$, his present expectation on this event will therefore be $= \frac{b}{a+b} \times \overline{\Delta}$: And consequently $\frac{a}{a+b} \times \Delta' + \frac{b}{a+b}$

$\times \overline{\Delta}$ will be $= \Delta$ his whole (present) expectation. Whence, putting $n = \frac{b}{a}$ and $m = n + 1$, we have $\Delta' = m\Delta - n\overline{\Delta}$.

Let now, 1, C, D, E, F, G, &c. denote the respective expectations of A, in the case proposed, when he has 1, 2, 3, 4, 5, 6, &c. pieces in possession: Then, r being in this case $= 2$, and $s = 1$, it is manifest, from the equation $\Delta' = m\Delta - n\overline{\Delta}$, that the value of each new term of the series 1, C, D, E, F, G, &c. will be equal to the last (or preceding) term multiplied by m, minus the last but two drawn into n. And so these values are derived, one from another, as in the annexed scheme.

1	1 =	1
2	C =	C
3	D =	mC
4	E =	m^2C $- n$
5	F =	$(m^3 - n) \times$ C $- mn$
6	G =	$(m^4 - 2mn) \times$ C $- m^2n$
7	H =	$(m^5 - 3m^2n) \times$ C $- m^3n + nn$
8	I =	$(m^6 - 4m^3n + n^2) \times$ C $- m^4n + 2mnn$
&c.	&c.	

And, universally, if r be taken to denote the number of any term of this series, reckoned from the beginning, the term itself will be truly expressed by C $\times \left(m^{r-2} - (r-4) \cdot m^{r-5} + n\frac{r-6}{1} \cdot \frac{r-7}{2} m^{r-8}nn - \frac{r-8}{1} \cdot \frac{r-9}{2} \cdot \frac{r-10}{3} m^{r-11}n^3, \&c.\right) - m^{r-4}n + (r-6) \cdot m^{r-7}nn - \frac{r-8}{1} \cdot \frac{r-9}{2} m^{r-10}n^3 + \frac{r-10}{1} \cdot \frac{r-11}{2} \cdot \frac{r-12}{3} m^{r-13}n^4$, &c. Which, when r is taken $= 11$ (the whole number of pieces of both A and B) will become C $\times (m^9 - 7m^6n + 10m^3n^2 - n^3) - m^7n + 5m^4n^2 - 3mn^3 = 11$; because, if A should have all the 11 pieces in possession, the play will then be at an end, and he will then have that sum, *certain*. Now, if in this equation there be wrote $\frac{5}{3}$ and $\frac{2}{3}$ instead of their equals m and n, the value of C will be found $= \frac{376923}{883548}$.

But the required expectation of A, when he has 7 pieces in possession, is found above to be $(m^5 - 3m^2n) \times$ C $- m^3n + nn = \frac{1775\text{C} - 642}{243} = \frac{1780083}{883548} = 4{\cdot}5364$, &c. Therefore his required loss, or disadvantage, will be $= 7 - 4{\cdot}5364 = 2{\cdot}4636$, &c. $=$ 2l. 11s. $3\frac{1}{4}$d.

After the same manner the loss of A and the gain of B may be determined in any other case: But it must be observed, that when B, as well as A, stakes more than one piece at a time, the values of the several terms in the proposed series of expectations will then be expressed by means of as many of the leading terms, as there are units in (s) the stake of B. And if the s last terms of the series so expressed, be made respectively equal to the numbers which they answer to (or stand against), then as many (simple) equations will from thence be obtained, as there are unknown quantities C, D, E, &c. to be determined.

To exemplify the process here pointed out, would, I apprehend, take up too much room: For which reason I am also obliged to omit the invention of a general approximation for the resolution of all problems of this kind; which comes exceeding near the truth in those cases where it is most wanted, that is, where the number of stakes is great.

Questions proposed in 1761, *and answered in* 1762.

I. QUESTION 475, *by Miss* Ann Nicholls,

Old John, who had in credit liv'd,
Tho' now reduc'd, a sum receiv'd:
This lucky hit's no sooner found,
Than clam'rous duns come swarming round;
To th' landlord,—baker,—many more,
John paid in all pounds ninety-four.
Half what remain'd—a friend he lent—
On Joan and self, one-fifth he spent;
And when of all these sums bereft,
One-tenth o'th' sum receiv'd had left.
Now shew your skill, ye learned fair,
And in your next that sum declare.

Answered by Mr. Tho. Sadler.

If x be supposed $=$ the whole sum John received, then will $\frac{x-94}{2}$ $+\frac{x-94}{5}+\frac{x}{10}+94$ or $\frac{8x+282}{10}=x$, per the conditions of the question; whence $x=141$l. the sum required.

Mr. *George Salmon*, late of Mr. *B. Donn*'s school, and several others, observe, that this question is ambiguous, it being doubtful whether $\frac{1}{5}$ of the whole sum received, or only $\frac{1}{5}$ of what remained after the 94l. was paid, was the part thereof spent on Joan and himself, and accordingly find the whole sum received to be either 141 or 235 pounds.

II. QUESTION 476, *by Mr.* J. Hampson.

To find that number, which being any how divided into two unequal parts, the greater part added to the square of the lesser, shall be equal to the lesser part added to the square of the greater.

Answered by Birchoverensis.

Let a and b represent any two given unequal numbers, and suppose ax denotes the greater, and bx the lesser part of the number required; then will be had $aaxx + bx = bbxx + ax$, by the question; whence $x = \frac{a-b}{aa-bb} = \frac{1}{a+b}$, and consequently $\frac{a}{a+b}$ and $\frac{b}{a+b}$ will be the two unequal parts, whose sum will always be $= 1$, the number required.

Mr. *Tho. Barker* puts $x + y$ for the greater, and $x - y$ for the lesser part of the required number, and thence forms the equation $(x+y)^2 + x - y = (x-y)^2 + x + y$, according to the nature of the question; whence he finds $x = \frac{1}{2}$, and consequently the sum of the two parts $x + y$ and $x - y$ equal to 1, the number required; the same as before.

III. QUESTION 477, *by Mr.* Tho. Barker.

Near me two lovely maids reside,
Blest with each grace—their sex's pride;
But should such charms fail to engage,
Without the gold, in this wise age,
The two equations plac'd below, *
Will, when resolved, their fortunes shew.

* Viz. $\left\{\begin{array}{l} xy = 125x + 300y \\ yy - xx = 90000 \end{array}\right\}$ to be solved by a quadratic.

Answered by Mr. Hugh Brown, *of Woolwich.*

Put $300 = a$, and then $xy = 125x + 300y = \frac{5}{12}ax + ay$, and $yy - xx = 90000 = aa$. Whence, substituting $\frac{5ax}{12(x-a)}$ for y, we shall have $144x^4 - 288ax^3 + 263aaxx - 288a^3x + 144a^4 = 0$; or $x^4 - 2ax^3 + \frac{263}{144}aaxx - 2a^3x + a^4 = 0$; and, dividing every term by $aaxx$, and putting $z = \frac{2x}{a} + \frac{2a}{x}$ (according to directions given on p. 156, of Simpson's Algebra, 2d edition) we shall have $\frac{z^2}{4}$

$-2 - z + \frac{283}{144} = 0$, or $z - 4z = \frac{100}{144}$; whence z will be found $=$ $4\frac{1}{6} = \frac{2x}{a} + \frac{2a}{x} = \frac{2x}{300} + \frac{2 \times 300}{x}$; and consequently $x^2 - 2\frac{1}{12} \times 300x = -90000$; from the resolution of which quadratic equation x comes out $= 400$; and from thence y will be found $= 500$.

The same answered by P. M. of Durham.

By transposition $x(y - 125) = 300y$, and $x^2(y - 125)^2 = 300y^2$; but $x^2 = y^2 - 300^2$; therefore, by substitution, $(y^2 - 300^2)(y - 125)^2 = 300y^2$; which, expanded, gives $y^4 - 250y^3 - 2 . 300^2 y^2 + 125^2 y^2 + 2 . 125 . 300^2 y - 125^2 . 300^2 = 0 = (y^2 - 125y - 300^2)^2 - 300^4 - 125^2 . 300^2$; whence $y^2 - 125y - 300^2 = 300 \times \sqrt{(300^2 + 125^2)} = 300 \times 325$, and $y^2 - 125y = 300 \times 625$; which quadratic equation solved, gives $y = 500$, and consequently $x = 400$.

The same answered by Mr. Wm. Embleton.

Put $300 = a$, and $125 = b$; then will $xy = bx + ay$, and $yy - xx = aa$; whence, $y = \frac{bx}{x - a} = \sqrt{(xx + aa)}$, and $bbxx = (x - a)^2 (xx + aa) = x^4 - 2ax^3 + 2aaxx - 2a^3x + aa$: Whence, adding $aaxx$ to each side of this equation, and extracting the square root, there will result $x\sqrt{(aa + bb)} = xx - ax + aa$; from which quadratic equation x and y will readily be found equal 400 and 500 respectively.

IV. QUESTION 478, *by Mr.* W. Chapman.

Having at a certain (unknown) distance, taken the angle of elevation of a steeple, I advanced 60 yards nearer (upon level ground) and then observed the elevation to be the complement of the former to a right angle: Advancing 20 yards still nearer, the elevation now appeared to be just the double of the first. Hence the steeple's height is required?

Answered by P. M. *of Durham.*

Imagine the thing done, and BA to be the steeple whose height is sought, and C, D, E, the three given stations; and suppose the points C, A; D, A; and E, A, to be joined.

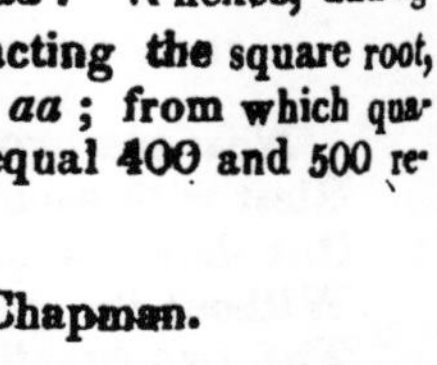

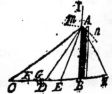

Then, since the ∠AEB is double the ∠ACB (by hypothenuse) and equal to the ∠s ACB, CAE (Euc. 32. I.), they must therefore be equal to each other, and consequently the sides AE, CE, subtending them equal. Moreover, since the right-angled triangles ABC, ABD, have the acute $\angle$s BAC ADB, equal (by hypoth.) they will be similar, and so CB : AB :: AB : BD; whence CB $\times$ BD ($=$ AB^2 $=$ AE^2 $-$ EB^2 $=$ CE^2 $-$ EB^2 $=$) CB $\times$ CE $-$ CB $\times$ BE; whence BD $=$ CE $-$ BE, take away DB, which is common, and there will remain BE $=$ CD $-$ BE, or the double of BE $=$ CD; whence is derived the following

Construction. Bisect CD in F, and FE in G, and produce CE till GB $=$ GC; and about the centre E, with the radius EC, let the arc of a circle be described, and at the point B erect a perpendicular cutting the said arc in A, and BA will represent the height of the steeple required.—— From this construction BA is readily found $=$ 74·162, &c. yards.

The same constructed otherwise by Mr. Da. Hastings.

The three given stations being supposed to be at C, D, and E, as before, produce CE till EH becomes equal thereto, and bisect DH with the perpendicular BI, and from the centre E, with the radius EC (EH), let the arc *mn* be described cutting BI in A, and BA will be the height of the steeple required.

Demonstration. Let the points C, A; D, A; E, A; and H, A, be joined. Then, CE being $=$ EA $=$ EH, by construction, a semi-circle, described about the diameter CH, will pass through the point A, and consequently CAH being a right angle, the $\angle$ADH will be the complement of the $\angle$C to a right angle; and the $\angle$AEB being $=$ the $\angle$C $+$ the $\angle$CAE, will be $=$ twice the angle C, the triangle CEA being isosceles.

The same answered (algebraically) by Mr. R. Mallock, *Writing-master in Lyme Regis, Dorset.*

Let CE (EA) $= a$, EB $= z$, and x and y $=$ the sine and cosine of the $\angle$ACB (rad. 1.); then will $2xy$ and $yy - xx$ be the sine and cosine of the $\angle$AEB; whence by trigonometry, will be readily found

BA $= \frac{x}{y}(z + a) = \frac{y}{x}(z + \frac{a}{4}) = \frac{2xyz}{yy - xx}$; and consequently $z = \frac{3a}{8} = 30$, and from thence BA (the steeple's height) is directly found $=$ 74·16198 yards.

V. QUESTION 479, *by Mr.* S. Kemp.

Having given the vertical angle of a triangle, (104°) and also the length (24) of a line dividing it in the given ratio of 5 to 4, and terminating in the opposite side; to determine the triangle so that the area thereof shall be a minimum?

Answered by Mr. Rich. Gibbons.

Construction. Constitute the angle ABC $=$ the given vertical angle

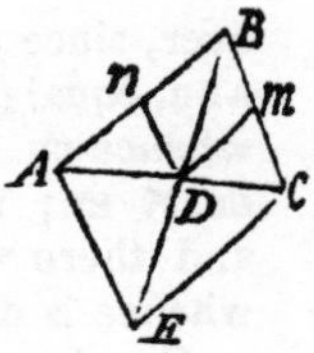

(104°), and divide it in the given ratio of 5 to 4 by the line BE = double the given dividing line (24), and draw EA, EC, parallel to BC and BA respectively, and meeting them in the points A and C, and then ABC will be the triangle required.

Demonstration. ABCE being a parallelogram, by construction, the diagonals BE, AC, will bisect each other (by theor. 12. II. of Simpson's Geom. 2d edition), and consequently BD will be the given dividing line; and DA being = DC, the triangle ABC will be a minimum (by theor. 8. p. 199 and 200 of Simpson's Elem. of Geom. aforesaid).

The same constructed otherwise by Mr. W. Embleton.

The angle ABC being made = the given vertical angle, and BD the given line dividing it in the given ratio of 5 to 4, as before, draw D*m* parallel to BA meeting BC in *m*, and take BA = twice D*m*; then, through the points A and D, let the line AC be drawn meeting BC in C, and ABC will be the triangle required.——For D*m* being $\frac{1}{2}$AB (by construction) CD will be = $\frac{1}{2}$CA, by the sim. triangles CD*m* and CAB, and consequently the triangle ABC a minimum (by theor. 8. p. 199 and 200 of Simpson's Elem. of Geom. 2d edit.).*

An algebraic Solution to the same by Mr. T. Bromhall, *at Mr.* Allen's *School, at Spalding, in Lincolnshire.*

This young gentleman puts D*m* = a, D*n* (parallel to BC) = b, C*m* = x, and s = sine of vertical angle ABC; then he readily finds AB = $a + \frac{ba}{x}$, and $\left(2b + x + \frac{bb}{x}\right)\frac{1}{2}as$ = area of the required triangle, which is to be a minimum: In fluxions, $x^2\dot{x} - b^2\dot{x} = 0$; whence $x = b$, or B*n* = A*n*, and consequently AD = DC, the same as demonstrated above.

* The above cannot be esteemed geometrical constructions, as the method of dividing an angle in the ratio of 4 to 5, is not known. H.

VI. QUESTION 480, *by Mr.* W. Spicer.

The perimeter of a triangle being given (120 feet) and the vertical angle (70°); to determine all the sides thereof, so that the triangle itself shall be the greatest possible.

Constructed by Mr. Da. Hastings.

It evidently appears, from theor. 6. p. 198, of Simpson's Elem. of Geom. 2d edition, that the triangle required will be isosceles; therefore upon AB, equal to the given perimeter, or sum of the three sides, constitute an isosceles triangle with the given vertical angle ACB

(= 70°), and bisect the angles CAB, CBA, with the right lines AD, BD, meeting each other in D, and let the points A, D and B, D be joined; then draw DG and DH parallel to CA and CB respectively; and CDH will be the triangle required; which is too evident to need any further demonstration.

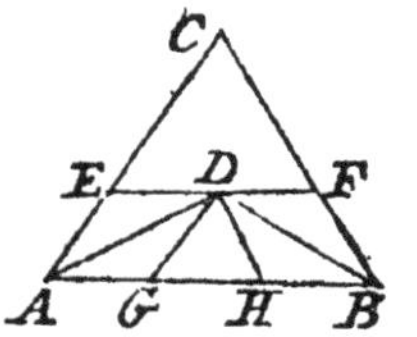

P. M. of Durham, observes, that it is very easy to demonstrate, geometrically, that the two sides comprehending the given vertical angle will be equal; and therefore, premising that, he determines the triangle by the following

Construction. Make ACB = the given vertical angle, and CA and CB each = half the given perimeter; join A, B; and bisect the angles CAB, CBA, by right lines meeting in D; through the point D draw EF parallel to AB, meeting the sides CA, CB in the points E, F; and CEF will be the triangle required.—For the ∠EDA (= DAB, Euc. 29. I.) = ∠EAD (by hyp.) therefore AE = DE (Euc. 6. I.) and consequently CA = CE + ED; and in the same manner CB will be proved = CF + FD; therefore CE + CF + EF = CA + CB, *i. e.* equal to the given perimeter; whence, &c.

The same solved algebraically, by Birchoverensis.

Let s = sine of the given vertical angle ECF (70°), m = 120, the given perimeter, $a + x$ and $a - x$ the two sides comprehending the vertical angle; then will $m - 2a$ = the third side EF, and $(aa - xx) \times \frac{1}{2}s$ = the area of the required triangle, which it is evident by inspection, will be the greatest when $x = 0$, or the triangle ECF is isosceles; and hence the remaining part of the question is easily solved by plain trigonometry, it being, as the sum of the natural sines of the three angles of any plain triangle is to its perimeter, so is the natural sine of any one of those angles to the side corresponding or opposite thereto; whence the sides EC (CF) and EF are readily found = 38·1297, &c. and 43·74059, respectively.

VII. QUESTION 481, *by* P. M. *of Durham.*

Two right lines, and also a third of any order, being given in position, draw another right line intercepted between the line of the indeterminate order and one of the given right lines, also cutting the other, so as to make given angles at the intersection, and have its segments made thereby in a given ratio.

Constructed by P. M. *of Durham, the Proposer.*

Let XY be the line of the indeterminate order; AB, AC, the two right lines given by position and meeting in A; in one of them AB, with which the right line, required, is to make given angles, assume a point B, through which draw a right line, making the given angles with AB, and meeting AC, in D; produce DB to E, so that DB may be to BE in the given ratio; through E, A, draw a right line (produced if necessary) to cut XY in F; and through F draw a line parallel to ED, cutting AB and AC in H and G; and the thing is done.

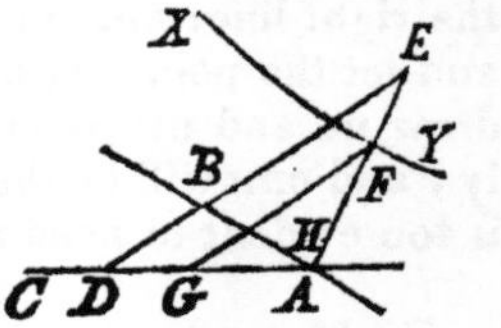

For the right line AB falling upon the parallels DE, GF, intercepted by the same right lines, AD, AE will cut them similarly in B and H (Euc. 2. VI.) and make equal angles with each (Euc. 29. I.); but it makes the given angles with DE, and cuts it in the given ratio in B; therefore also GF, intercepted by XY, and one of the given right lines AC, is cut by the other AB, in H, in the given ratio, and makes given angles therewith at the intersection.

Num. Calc. will vary according to the property of the indeterminate line XY.

VIII. QUESTION 482, *by Mr.* Rich. Mallock.

The area of a triangle being given = 126; the sum of its three sides = 54, and the sum of their squares = 1010; to determine the triangle.

Answered by Mr. R. Butler.

Let $2p = 54$; $s = 1010$; $a = 126$; and x, y, and z denote the three sides of the triangle required. Then, by a known theor. $p(p-x)(p-y)(p-z) = aa$; whence, putting $2p - x - y$ for its equal z, we shall have $(x-p)yy + (2pp + xx - 3px)y + 2ppx - pxx - p^3 = aa \div p$. Moreover, $(2p - x - y)^2 + xx + yy = 4pp - 4px - 4py + 2xy + 2xx + 2yy$ being $= s$, per the question, we get $yy = \frac{1}{2}s - 2pp + 2px - xx + (2p - x)y$; which, substituted for yy in the preceding equation, gives $\frac{1}{2}s - 2pp + 2px - xx + (2p - x)$
$$y = \frac{2ppx - pxx - p^3 - aa \div p}{p - x} + \frac{2pp - 3px + xx}{p - x}y; \text{ but}$$
$\frac{2pp - 3px + xx}{p - x}y$ is $= (2p - x)y$; and therefore our equation becomes
$$\tfrac{1}{2}s - 2pp + 2px - xx = \frac{2ppx - pxx - p^3 - aa \div p}{p - x}, \text{ or } x^3 - 54x^2$$
$+ 953x = 5460$; the three roots of which cubic equation, viz. 13, 20, and 21, will express the three sides of the triangle required, as x stands indifferently for e'er a one of them.

The same answered by Mr. Paul Sharp.

This gentleman puts $a = 126$, $2b = 1010$, $2c = 54$, $x = \frac{1}{2}$ sum of any two sides of the required triangle, and $y = \frac{1}{2}$ their difference;

then will $x+y$, $x-y$, and $2c-2x$, represent the three sides themselves, and $6xx-8cx+2yy+4cc=2b$ (per quest.): Whence $yy=b-2cc+4cx-3xx$. Moreover, $(c-x+y)(c-x-y)(2x-c)c$ is $=aa$; whence $yy=(aa\div c+c^3-2x^3+5cx^2-4c^2x)\div(c-2x)=b-2cc+4cx-3xx$; which, reduced and converted into numbers, becomes $x^3-54x^2+967\cdot25x=5750\cdot25$; whence x will be found $=17$, and $y=4$, and the sides of the required triangle equal to 13, 20, and 21.

The same answered by Nosnihctuh.

The area of any plane triangle being $=$ a rectangle under half its perimeter and the radius of its inscribed circle, it follows, that $126\div27=4\cdot6$ will be the radius of the inscribed circle, in the case of this question, the square of which put $=b$; $1010=n$; $27=c$; and let x denote half the sum, and y half the difference of the two segments of any one of the sides made by the radius of the inscribed circle, drawn to the point of contact thereof, and then will the three sides of the required triangle be represented by $c+y-x$, $2x$, and $c-x-y$; whence (by a theorem on page 116 of Simpson's Select Exercises)

$$\frac{(x+y)(x-y)(c-2x)}{x+y+x-y+c-2x}=(cx^2-cy^2-2x^3+2xy^2)\div c=b,$$

and consequently $y^2=(cb-cx^2+2x^3)\div(2x-c)$. But $(c+y-x)^2+(2x)^2+(c-x-y)^2=n$ (per quest.); whence $y^2=\frac{1}{2}n-c^2+2cx-3x^2=(cb-cx^2+2x^3)\div(2x-c)$; hence, in numbers, we derive the equation $x^3-27x^2+235\cdot25x-682\cdot5=0$; from which (by either of the two first methods explained in Sect. 12 of Simpson's Algebra) x will be found $=10$, and $y=4$, and the three sides of the triangle required 13, 20, and 21.

IX. QUESTION 483, *by Mr.* J. Brampton.

There is a pond in the form of a right-angled triangle, which is intended by the owner for a decoy. Going to survey it I found it so surrounded by bushes to a considerable distance, that the following measures were all that I could take: On the base produced, I measured from the acute angle, 4 chains: Here I could see a tall poplar, which grew on the bank of the hypothenuse; I took its bearing from the chain-line 20°: When I got to the tree I could not see my former station, but found that the perpendicular of the triangle subtended a right angle there: Then I measured from the tree to the angle opposite the base, 5 chains. It is required from these measures to plot the triangle?

Answered by Mr. John Hampson.

Let FDC represent the triangular pond required; in which DB is

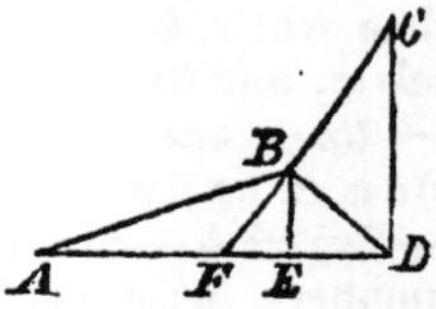

perpendicular upon FC, and AF and BC equal 4 and 5 chains respectively, and the angle BAF $= 20°$.

Put $a =$ AF $= 4$; $b =$ BC $= 5$; m and n equal the sine and co-sine of the angle BAF (20°); $x =$ perpendicular CD, and $y =$ BF.

Then, per similarity of triangles, will be found BE $= by \div x$, FE $= (y \div x) \sqrt{(xx - bb)}$, and consequently $(ax + y\sqrt{(yy - bb)}) \div x =$ AE; whence, by Trigonom. $m : by \div x :: n : (ax + y\sqrt{(xx - bb)}) \div x$; from whence is had $bny = amx + my\sqrt{(xx - bb)}$; but from the similarity of the right-angled triangles FDC, CDB, will be found $by + bb = xx$; from the resolution of which two equations x will be found $= 6{\cdot}0634$; and consequently FD and FC equal to 4·15947 and 7·35296, &c. respectively.

The same answered by Mr. Richard Gibbons.

First, supposing the triangles ABD, AFB, to be similar, will be found the angle ABF ($= \angle$ADB) $= 35°$, and from thence the logarithm of BC too little by 0·0119032; next supposing the same angle ABF $= 36°$, it gives the logarithm of BC too much by 0·0099480: Whence as the sum of the errors is to 60′ the difference of the suppositions, so is the first error to 32′ 42″; which being added to 35° (the first supposition) gives 35° 32′ 42″ = the true measure of the angle ABF; whence AF, BC, and all the angles being given, and now become known, the triangular pond FDC may from thence be easily plotted and determined.

X. QUESTION 484, *by Mr.* Richard Gibbons.

A shell being thrown from a mortar, at an elevation of 30°, the report of its fall was heard at the mortar, just 20 seconds after the explosion: Hence to find the length of the range?

Answered by Mr. T. Allen, *of Spalding.*

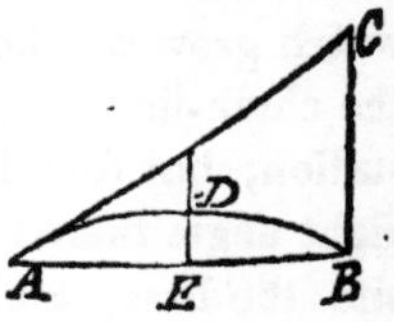

Put $t =$ tang. of the angle of elevation BAC ($= 30°$); $a = 1142$ feet (the velocity of sound per second) $s = 16\frac{1}{12}$ feet (the distance a falling body will descend in a second); and $x =$ AB, the horizontal range required.

Then will $tx =$ BC, and $\sqrt{(tx \div s)} =$ the time of flight; and $x \div a =$ the time of sound's moving from B to A: Therefore $\frac{x}{a} + \sqrt{\frac{tx}{s}} = 20''$ (by the question); whence x is found $= 6033{\cdot}42$ feet, the range required.

If p, (the parameter of the parabolic curve ADB)$=10450{\cdot}7$, and $\frac{1}{2}x$

$=y$; then will $\frac{2y}{p}\sqrt{(\frac{1}{4}pp+yy)}+\frac{p}{2}\times$ hyp. log. $\frac{y+\sqrt{(\frac{1}{4}pp+yy)}}{\frac{1}{2}a}$ $=6353{\cdot}4$ feet, the length of the track ADB.

The same answered by Mr. W. Chapman.

Let $s=16\frac{1}{12}$ feet (the distance a falling body descends in a second); $b=1142$ feet (the space passed through by sound in a second); p and q equal the sine and cosine of 30° (the angle of elevation), and $x=$ the amplitude, or horizontal range required.

Then will $px\div q=$ BC; and $s:1''^2::px\div q:px\div sq=$ the square of the time of flight; whence $c-\sqrt{(px\div sq)}$ will be $=$ the time of the return of the sound (putting $20''=c$); and consequently $1'':b::c-\sqrt{(px\div sq)}:x$; whence $x=bc-b\sqrt{(px\div sq)}$; from which equation x will be found $=6033{\cdot}44$ feet, the amplitude required.

XI. QUESTION 485, *by Mr.* Hugh Brown.

A and B borrow 400l. each, for a certain stated time: At the expiration of which, A, who agreed to allow compound interest, had 463l. 1s. to pay; but the debt of B (who was to pay simple interest only) amounted but to 460l. The time and rate of interest are required.

Answered by Mr. Tho. Barker.

Let $x=$ amount of 1l. in one year, and $y=$ the number of years required; then will $400x^y=463{\cdot}05$l. and $400y\times(x-1)=60$l. (by the quest. and the nature of compound and simple interest), whence

$$y=\frac{3}{20(x-1)};\text{ and consequently } x^{\frac{3}{20(x-1)}}=\frac{463{\cdot}05}{400}=1{\cdot}15762,$$

&c. from which equation x will be found $=1{\cdot}05$; and hence it appears, that 5 per cent. per annum, and three years, are the rate of interest and time required.

The same answered by Mr. Tho. Harris.

Put $P=400$l. $A=463{\cdot}05$l. $a=460$l. $t=$ time, and $r=$ rate of interest required; then will $trP+P=a$ (by the nature of simple interest); whence $t=\frac{a-P}{rP}=\frac{b}{r}$ (putting $b=\frac{a-P}{P}=\cdot15$): Whence, by compound interest, we have $P\times(1+r)^{b\div r}=A$, the debt of A; from which equation, by the help of logarithms, r will be found $=0\ 05$: Whence the time required appears to be three years, and the rate 5 per cent. per annum.

The same answered by Mr. Rich. Holden.

Let P = 400l. A = 463·05l. a = 460l. r = the amount of 1l. in one year, and t = the time required; then will PR^t = A, per compound interest, and $P + Pt \times (R - 1) = a$, per simple interest: Whence will be found $R = \left(\frac{A}{P}\right)^{1 \div t} = \frac{a - P}{Pt} + 1$; and hence, by a table of logarithms and a few trials, t is found = three years, and the rate 5 per cent.

P. M. of Durham, and Mr. *Thomas Allen*, of Spalding, put a = 463·05l. b = 460l. p = 400l. t = number of years or time required, and r = the required rate of interest of 1l. for one year; and then they find $p(1 + r)^t = a$, and $p\,(1 + tr) = b$ (by the nature of compound and simple interest): Whence $\frac{\text{l}.\,a - \text{l}.\,p}{\text{l}.\,(1 + r)} = (t =) \frac{b - p}{pr}$, and consequently $(\text{l}.\,a - \text{l}.\,p)\,\frac{pr}{b - p} = \text{l}.\,(1+r)\ (= \frac{1}{m}\ (r - \frac{r^2}{2} + \frac{r^3}{3}$ — &c.) m being the modulus); and hence, either by a table of logarithms, or reversion, or approximation of series, the value of r will be found = 0·05. Therefore the rate of interest appears to be 5l. per cent. and the time of the loan = three years.

XII. QUESTION 486, *by Mr.* T. Harris.

In sixty-six, * the time declare, [* *Deg. of north lat.*
When day and twilight equal are.

Answered by Mr. Charles Green, *at Greenwich Observatory, and Mr.* Thomas Harris, *of Bugbrook, near Northampton.*

Let x and $-y$ equal the sine and cosine of P☉ (PO) the sun's dist. from the north pole P; s and c equal the sine and cosine of ZP (= 24°) the complement of the given latitude; and $-d$ = the cosine of Z☉ (= 108°); then by a well-known theorem in spherics and the nature of the question, we shall have $cy \div sx$ and $(cy - d) \div sx$ = to the cosines of the angle ZPO and ZP☉ respectively; but the angle ZP☉ is double the angle ZPO (by the quest.), and therefore $2ccyy \div ssxx - 1$ (the cosine of double the angle ZPO) will be $= (cy - d) \div sx$; from the resolution of which equation the value of y will be found = 0·3094, answering to Jan. 28, and Nov. 12, nearly.

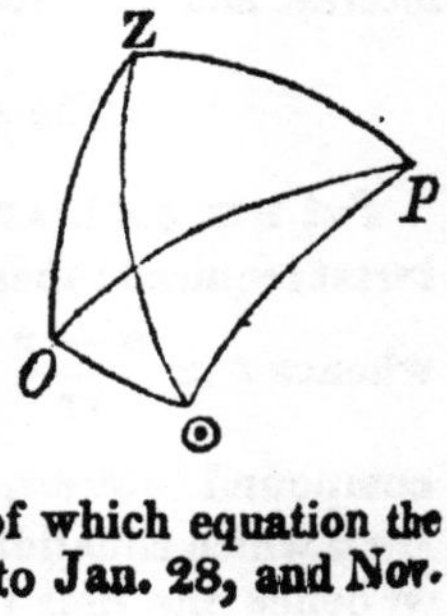

XIII. QUESTION 487, *by Miss* Ann Nicholls, *of Hadham.*

Two places, A and B, are known to lie both under the same meridian: And it is observed, that on Jan. 20, the sun rises 36 minutes sooner at A than at B; and that on May 30, he rises 40 minutes earlier at B than at A: From which data, I demand the latitude of both places?

Answered by Mr. W. Kingston.

Let $d =$ tang. 21° 54′ (the sun's declination May 30.); $a =$ tang. 20° 11′ (the sun's declination Jan. 20.); m and n equal the sine and cosine of 10°; s and c the cosine of 9°; and let the sine and cosine of the ascensional difference in the latitude of the place A, on May 30, and Jan. 20, be denoted by x and y, and z and v respectively; then will $nx + my$, and $cz + sv$, express the sines of the ascensional difference in the latitude of the place B, on the said two days respectively. Then, per spherics, $z \div a = x \div d =$ the tangent of the latitude of the place A, and $(nx + my) \div d = (cz + sv) \div a =$ the tangent of the latitude of the place B, From the first of these equations z is found $= ax \div d$, and consequently v $(\sqrt{(1 - zz)}) = \sqrt{(dd - aaxx)} \div d$; which values substitute for z and v in the second equation, and it becomes $\frac{nx + my}{d} = \frac{acx + s\sqrt{(dd - aaxx)}}{dd}$; whence $a(n - c)x + amy = s\sqrt{(dd - aaxx)}$; and consequently (by involution, &c.) $\frac{nn - cn}{m}xx + \frac{aamm - ddss}{2aam} = (c - n)xy$, or $bxx + p = qxy$ (putting $\frac{nn - cn}{m} = b$, $\frac{aamm - ddss}{2aam} = p$, and $c - n = q$); and this involved again, in order to exterminate y, becomes $bbx^4 + 2bpxx + pp = qqxxyy = qqxx - qqx^4$, or $(bb + qq)x^4 + (2bp - qq)xx = -pp$; whence $x^4 - 2rx^2 = -t$ (putting $\frac{2bp - qq}{bb + qq} = -2r$ and $\frac{pp}{bb + qq} = t$), and consequently $2xx = 2r \mp 2\sqrt{(rr - t)} =$ the versed sine of double the ascens. difference at the place A on May 30; from whence the latitudes of the places A and B are readily found equal to 45° 32′ and 54° 26′, respectively.

XIV. QUESTION 488, *by Mr.* G. Witchell.

To determine the equation of the curve, whose subtangent is every where, equal to the sum of the abscissa and ordinate corresponding; and to find the area thereof, when the two quantities last mentioned are equal, and given?

Answered by Mr. W. Spencer, *of Stannington, near Sheffield.*

Let $x =$ the abscissa, and $y =$ the ordinate of the required curve;

then will $\frac{y\dot{x}}{\dot{y}} = x + y$, (per the quest.) and consequently $\frac{y\dot{x} - x\dot{y}}{yy} = \frac{\dot{y}}{y}$: The fluent of which gives $\frac{x}{y}$ = hyp. log. y; whence x = hyp. log. y^y; the equation of the curve required.

But, to find the area, we have $y\dot{x}$ (the general expression for the fluxion of the area) $= y\dot{y} + x\dot{y}$ (from above); whence fluent of $y\dot{x}$ $= \frac{1}{2}yy + xy$ — fluent of $y\dot{x}$, or twice the fluent of $y\dot{x}$ (= twice the required area) $= \frac{1}{2}yy + xy$, and consequently the area itself $= \frac{1}{4}yy + \frac{1}{2}xy - \frac{1}{4} = \frac{3}{4}yy - \frac{1}{4}$* (when $x = y$).

P. M. of Durham, after having found the equation of the curve and area the very same as above exhibited, adds, moreover, that when $x = y$ ($x \div y$ being then $= 1 =$ h. log. y), y will be $= 1 + 1 + \frac{1}{2} + \frac{1}{2.3}$, &c. $= 2{\cdot}71828$, &c. and consequently the area required (in that case) $= 5{\cdot}36831$.

* The quantity $\frac{1}{4}$ is the proper correction of the fluent, and is thus found: In the equation of the curve (x = hyp. log. y^y or $= y \times$ hyp log y) when $x = 0$, then log. $y = 0$, and $y = 1$; but when $x = 0$, the area ought to be $= 0$; and since, when $x = 0$, and $y = 1$, the area $\frac{1}{4}yy + \frac{1}{2}xy$ is $= \frac{1}{4}$, ∴ $\frac{1}{4}$ is the correction to be subtracted. H.

THE PRIZE QUESTION, *by Mr.* E. Rollinson

To determine the orbit that a planet will describe, when, besides its proper gravitation to the sun, it is urged in the direction of the radius vector by a perturbating force, which is every where in proportion to the sun's attraction, as the cosine of the angle described about his centre, from the commencement of the motion, to any given multiple of the radius.

Answered.

Suppose AP to represent part of the required orbit of the planet commencing motion at A, at the given distance AC from the centre of force C, and put

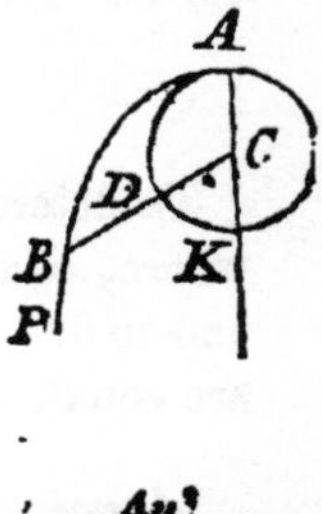

The radius AC (CD) of the circle ADK . . . . $= 1$,
The radius vector CB . . . . $= x$,
The arch AD, measuring the angle ACD, described about the centre of force C $= z$,
The measure of the celerity with which the area ACB increases $= u$,
The measure of the centripetal force tending to the centre C $= Q$;

and then the centrifugal force at B and A will be found to be $\frac{4u^2}{x^3}$ and

$4u^2$ respectively: And $\frac{\ddot{w}}{\dot{z}\dot{z}} = 1 - w - \frac{Q}{4u^4(1-w)^2}$, per p. 139 and 140 of Simpson's Miscellan, Tracts (w being put $= -\frac{1}{x} + 1$, and no force being supposed to act here, besides that tending to the centre of force C, and consequently u a constant quantity).

If now the value of the centripetal force at A be to the centrifugal force *there* in any given ratio of $1 - e$ to 1, then will the centripetal force at A be $= 4uu\,(1-e)$, and consequently that at B $= (1 - e) \times 4uu \times (1-w)^2$ (the centripetal force being, in this case, as the square of the distance inversely); whence we shall have $(1-e) \times 4u^2 \times (1-w)^2 \times \frac{\cos. z}{n}$ = the perturbating force acting on the planet at B (n being the given multiple of the radius) and consequently $(1-e) \times \frac{n - \cos. z}{n} \times 4u^2 \times (1-w)^2$ = the whole force urging it towards the centre of force C in the direction of the radius vector CB, or the whole force whereby it tends to the centre C, in that direction; and this value substituted for Q in the above equation, it becomes $\frac{\ddot{w}}{\dot{z}\dot{z}} = -w + e + \frac{(1-e)\cos. z}{n}$: From the resolution of which equation, w will be found $= e + \frac{(1-e)(B + \frac{1}{2}z)\sin z}{n}$ (B being any constant quantity at pleasure); from whence the orbit AP, and every thing else required, may be readily determined, &c.

All our contributors who have solved this question agree so nearly in the above method of solution, that we cannot attribute it to any one in particular.

Questions proposed in 1762, *and answered in* 1763.

I. QUESTION 489, *by Mr.* T. Baker; *in which, Question* III. *in the last Year's Diary is answered.*

By a quadratic it appears right plain,
Five hundred y——four hundred x explain:
Now, sir, if in return you'll tell to me,
The age and fortune of my charming she;
Which from the giv'n equation * will appear:
I'll strive to do as much for you next year.

* $xy = 59700 + 12y$, $x - y = 2375$; x = her fortune, y = her age.

Answered by Mr. Tho. Sadler.

Multiplying the latter of the two given equations by y, and subtracting it from the former, there results $yy + 2363y = 59700$; from whence y will be found $= \frac{1}{2}(\sqrt{4 \cdot 59700 + 2363^2} - 2363) =$ 25 years, the lady's age; and thence x will come out $= 2400$l her fortune.

Mr. *R. Gibbons* thinks that a great age for such a fortune to lie upon hand; and Mr. *Malachy Hitchins* seems partly of the same opinion, when he says,

Twenty-five years her age I've found,
Her worth just twice twelve hundred pound:
Dear *Baker*, pray don't tarry longer;
I'd rather have one eight years younger.

II. QUESTION 490, *by Mr.* Tho. Atkinson.

Given the sum of the natural sines of the acute angles in a right-angled plane triangle $= 1\frac{2}{5}$ (radius being equal unity) and the base or longer leg of the triangle is $= 40$; required the hypothenuse and perpendicular of the said triangle.

Answered by Mr. Wm. Embleton.

Construction. Through A, the extremity of the right line AD ($= 1\frac{2}{5}$) draw AC, making an angle therewith $= 45°$, and about the other extreme D, as a centre, with the radius (1), describe an arc of a circle cutting AC in C; draw CB perpendicular upon AD, and produce BD till it becomes $=$ BF (40); then draw FE parallel to DC, meeting BC produced in E, and EBF will be the triangle required.

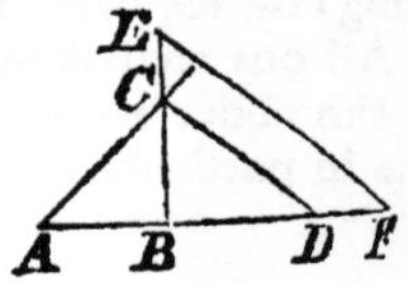

Demonstration. The angle BAC being $= 45°$, and B a right-angle (by construction) BC will be $=$ AB, and consequently the sum of BC and BD (the sines of the angles F and E, to radius DC $= 1$) will be $=$ AD $= 1\frac{2}{5}$, and BF being $= 40$ (per construction) the whole is manifest, &c.

Calculation. As rad. (CD) : $s \angle$A (45°) :: AD (the given sum of the sines) : $s \angle$ACD, the excess of which above ACB (45°) is $=$ BEF, the greater of the two acute angles required.

An algebraic Solution by Mr. John Hudson, *Land Surveyor.*

Let s ($= \frac{7}{10}$) $=$ half the given sum of the sines, and $x =$ half the difference; then will $s + x =$ sine of the greater, and $s - x =$ that of the lesser of the angles E, F; whence per Euc. 47. I. $(s + x)^2 +$

$(1-x)^2$ or $2s^2+2x^2=1$; and consequently $x=\sqrt{\frac{1-2s}{2}}=\frac{1}{10}$; whence the sines of the said two angles appear to be $\frac{6}{10}$ and $\frac{8}{10}$, and the leg EB and hypothenuse EF, 30 and 50 respectively.

III. QUESTION 491, *by Mr.* John Hampson.

Required to find two such non-quadrate numbers, whose product shall be a square, and the said product added to the square of either number shall be a square also.

Answered by Mr. R. Mallock.

Constitute a right angle ACD with two lines AC, CD, respectively measuring the sides of any two known squares at pleasure, and let the points A, D be joined; draw CB perpendicular upon AD, and AB and DB will be the measures of the corresponding numbers required. For BD × AD or BD^2 + BD × AB = DC^2, and AB × AD or AB^2 + AB × BD = AC^2, by a well known property of right-angled triangles, and AB × BD (= BC^2, by Euc. VI. 8.) = (AC^2 × CD^2) ÷ AD^2, which is manifestly a square number, as AC^2 and CD^2 are such, by supposition, &c.

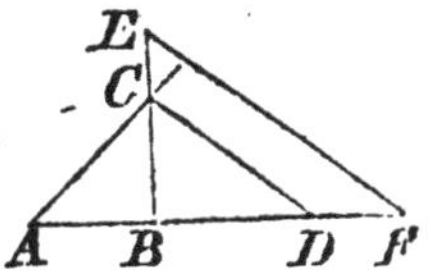

Example. Suppose AC = 3, and DC = 4; then will AD = 5 (per 47 Euc. I.) and thence will be found AB and DB = $\frac{9}{5}$ and $\frac{16}{5}$, respectively; the first equi-multiples of which numbers, in integers, are 9 and 16, but, being square numbers, they are not for the purpose: But the next pair of integers, 18 and 32, (found by multiplying each of the preceding ones by 2) will answer the conditions of the question; and, by proceeding in this manner, innumerable other answers may be discovered.

The same otherwise, by Mr. John Hampson, *the Proposer.*

Let bba and a represent the two numbers required, the product whereof $aabb$ will evidently be a square number; but, per nature of the question, $b^4aa+bbaa$ and $aa+bbaa$ must each be a square number, and consequently $bb+1=$ a square number (as $b^4aa+bbaa$ and $aa+bbaa$ are evidently $=bbaa(bb+1)$ and $aa(bb+1)$ respectively) which suppose $=(b-c)^2$; whence b will be found $=(cc-1)\div 2c$; but as a may be taken at pleasure, suppose it $=4ccd$, and then $bba=d(cc-1)^2$; from whence may be found any number of answers in whole numbers at pleasure, making c not less than 2, and d any non-quadrate number.

Example. If c and d are each $=2$, then will $bba=18$, and $a=32$, the first answers: but if $d=3$, then $bba=27$, and $a=48$, the next answers: And if $d = 5$, then bba will be $= 45$, and $a = 80$, &c. which answers are all in the ratio of 9 to 16. Again, supposing $c = 3$, and expounding d by 2, 3, 5, 6, 7, &c. successively, another set of answers, each in the ratio of 9 to 16, will be obtained; and thus we may proceed as far as we please.

The same answered otherwise by Mr. T. Allen.

Let the required numbers be denoted by x and nx, and then nxx, $(n+1)\,xx$, and $n \times (n+1)\,xx$ being square numbers, by the question, n and $n+1$ must therefore be square numbers likewise; whence, putting z and $z+a = \sqrt{n}$ and $\sqrt{(n+1)}$ respectively, $2az + aa$ $(= (z+a)^2 - zz)$ will be $= 1$, and consequently $z = (1 - aa) \div 2a$, where aa may be any square number less than unity, and x any number at pleasure, &c.

IV. QUESTION 492, *by Mr.* Edw. Kimpton.

Observed three stars A, B, and C, all in the arc of a great circle; the distance of A and B was found to be $= 10°$, and that of B and C $= 20°$: The difference of the azimuths of A and C was 90°; and the middlemost B was the nearest distance possible to the zenith; it is required, from this data, the altitudes of the three stars.

Answered by Mr. Rich. Gibbons.

Let A, B, C represent the three stars, and Z the zenith of the place. Then, the arc ZB being perpendicular upon AC, per the nature of the question, we shall, per spherics, (prop. 35 Emer. Trig.) have sine of AB + B (30°) : sine of AZB + BZC (90°) : : sine of BC — AB (10°) : sine of BZC — AZB (20° 19′ 20″). Therefore AZB $=$ 34° 15′ 20″, and BZC $=$ 55° 9′ 40″; the altitude of the star A $=$ 72° 18′ 14″, of B $=$ 75° 19′ 32″, and that of C $=$ 65° 22′ 23″.

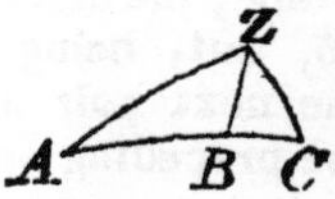

V. QUESTION 493, *by* Nosnihctuh.

To draw a right line through a given point betwixt two right lines given by position, so that the rectangle under the parts thereof, intercepted by that point and those lines, may be a minimum.

Answered by Mr. Da. Kinnebrook, *and Mr.* T. Moss.

Construction. Let P be the given point, and AB and AC the two lines given in position, in which take AD $=$ AE, and let the points D, E be joined; then through P draw FG parallel to DE, and the thing is done.

Demonstration. Join the points A, P, and draw the right line GH

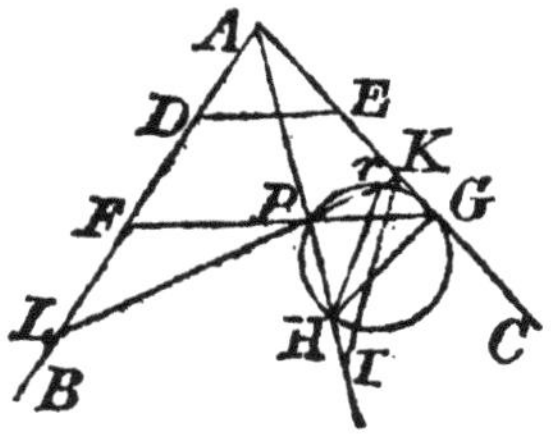

(meeting AP produced in H) so as to make the $\angle$PGH $=$ $\angle$FAP, and through the points H, P, G, conceive the periphery of a circle to be described; then the $\angle$PGH being $=$ $\angle$FAP (by hypothesis), and the $\angle$APF $=$ $\angle$HPG, the $\angle$AFP will be $=$ $\angle$PHG (32 Euc. I.); likewise the $\angle$AFP $=$ $\angle$AGP (by constr.) and consequently the $\angle$AGP $=$ $\angle$ PHG; wherefore the circle touches the right line AC in the point G (32 Euc III): But the triangles APF, HPG, are similar; and therefore AP : PF : : PG : PH; whence PF $\times$ PG $=$ AP $\times$ PH, and consequently, when the rectangle PF $\times$ PG is a minimum, its equal AP $\times$ PH or PH (as AP is a constant quantity) will be a minimum also, which will be in the case above-mentioned, viz. when the line FG is drawn so as to cut off the $\triangle$FAG isosceles; for let LK be any other right line drawn through the said given point P, and from the point r, where that line and the periphery of the circle intersect, draw rH, and parallel thereto draw KI; then, because AC is a tangent to the circle, PK will evidently be greater than Pr: And therefore, as KI is parallel to Hr, PI will be greater than PH, and consequently LP $\times$ PK ($=$ AP $\times$ PI) will every where be greater than FP $\times$ PG or its equal AP $\times$ PH.

The same solved algebraically by Messrs. T. Allen, T. Bromhall, E. Hare, *jun. &c.*

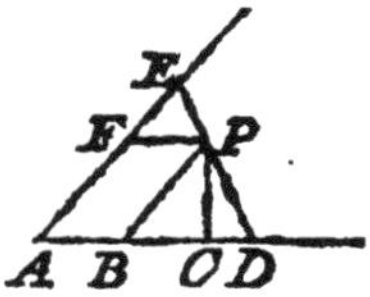

Let P be the given point, and draw BP, FP, parallel to AE and AD respectively, and PC perpendicular upon AD. Put FP $= a$, BC $= b$, PC $= c$, and CD equal x. Then will $\sqrt{(cc+xx)}=$ PD, and per similar triangles, $b + x$ (BD) : $\sqrt{(cc + xx)}$ (PD) : : a(FP) : $a\sqrt{(cc + xx)} \div (b + x) =$ EP; whence PE $\times$ PD is $=$ $a \times (cc+xx) \div (b+x)$, which is to be a minimum, per question; and therefore, in fluxions, $2bx\dot{x} + xx\dot{x} - cc\dot{x} = 0$; from whence x is found equal $\sqrt{(cc + bb)} - b$; and from thence the position of the line ED may be determined.

The same answered otherwise, by Mr. Wm. Embleton.

Put PB $= a$; PF $= b$; BC $= c$; and BD $= x$; and then will PD be found $= \sqrt{(aa + xx - 2cx)}$, by 9. 2. of Simpson's Elem. of Plane Geom. 1st edition: Whence, per similar triangles, BDP, FPE, x (BD) : $\sqrt{(aa + xx - 2cx)}$ (DP) : : b (PF) : $(b \div x)\sqrt{(aa + xx - 2cx)} =$ PE; and consequently PE $\times$ PD will be $= (b \div x) \times (aa + xx - 2cx)$, which, according to the question, is to be a minimum: And consequently, in fluxions, $- aa\dot{x} \div xx + \dot{x} = 0$; whence x is found $= a$, or AE $=$ AD; from which the position of the required line becomes known.

VI. QUESTION 494, *by Mr.* Joseph Fisher.

The gravity of salt water being to that of fresh, as 138 is to 135; in what latitude will a gallon of fresh water be to a gallon of salt water under the equinoctial, as 136 is to 138?

Answered.

The numerical data of this question being misprinted, no solution to it has been received. Some, however, observe, that it is no more than to find x the sine of the latitude of the place where a body, weighing w at the equator, shall equiponderate with another given homogeneous weight W, which is easily found $= \sqrt{\left(\frac{66240}{519} \times \frac{W-w}{w}\right)}$* (the ratio of the diameters of the earth being as 230 to 231, and the gravity and centrifugal force, at the equator, 1 and $\frac{1}{289}$ respectively.

* This expression may be thus found: as 288 : 289 :: w (the weight or grav. at the equator diminished by the centrif. force) : $\frac{289\,w}{288}$ the weight *there* if the centrif. force were taken off, or if the earth did not revolve on its axis; also as 230 : 231 :: $\frac{289\,w}{288} : \frac{231 \times 289\,w}{230 \cdot 288}$ the weight at the pole; hence $\frac{231 \times 289\,w}{230 \cdot 288} - w = \frac{519\,w}{66240}$ is the whole weight gained at the pole: Then, by art. 401 Simpson's Flux. 1 : xx :: $\frac{519\,w}{66240} : \frac{519\,wxx}{66240}$ the weight gained in the lat. whose sine is x, which must be $= W - w$; hence $x = \sqrt{\left(\frac{W-w}{519\,w} \times 66240\right)}$, as above.

Now the limit of the increase being $\frac{519}{66240} = \frac{1}{128}$ nearly, and the increase proposed in the quest. being from 135 to 136, or $\frac{1}{135}$ part only, it is evident that it is proposed within the limit; and, by writing 135 and 136 instead of w and W in the above value of x, it becomes, $x = \cdot 9723207 =$ the sine of 76° 29′ 17″ nearly, the latitude required. H.

VII. QUESTION 495, *by Mr.* W. Spicer.

The legs of a plane triangle being given equal to 30 and 40 respectively; and if a line be drawn from the vertical angle to the middle of the opposite side, the rectangle of the said line and base is a maximum; required the bisecting line and base?

Answered by Mr. W. Embleton.

Let ABC represent the triangle required, and BD the line bisecting its base AC. Now, when BD × 2AD (= BD × AC) is a maximum, $AD^2 - 2BD \times AD + BD^2$ or $(AD \backsim BD)^2$ (= $AD^2 + BD^2$, a constant quantity by theo. 11. book 2. Simpson's Elem. of Plane Geom. 2d edition, diminished by the greatest value possible of the variable quantity 2BD × AD) will manifestly be a

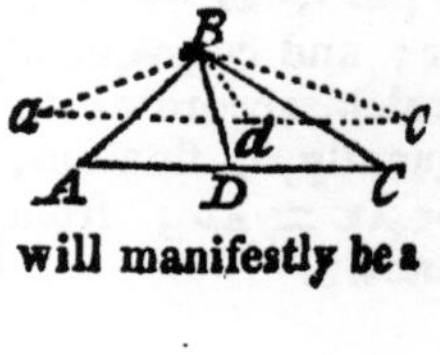

minimum, and consequently AD ∽ BD a minimum, which, it is evident, can only be when it is $=0$; whence it clearly follows that the required triangle will be right-angled at B, because (since AD is now proved $=$ BD $=$ DC) it may be circumscribed by the periphery of a semicircle of that radius described about the centre D;——hence by 47 Euc. I. AC will be found $= 50$; and consequently BD $= 25$.

Corollary. If B*a*, BC, be supposed any other position of the two given sides BA, BC, different from that here determined, then will AD + BD be always greater than *ad* + B*d*; for $AD^2 + BD^2$ being $= ad^2 + dB^2$, and $2AD^2 \times DB \sqsubset 2ad \times dB$ (by what has just now been demonstrated), $AD^2 + 2AD \times DB + DB^2$ will therefore be greater than $ad^2 + 2ad \times db + dB^2$, and consequently AD + DB $\sqsubset$ *ad* + *d*B.

An algebraical Solution to the same, by Mr. T. Bromhall.

Put AB $= 40 = a$, BC $= 30 = b$, AD $= x$, and BD $= y$; then, per 12. 2. Simp. Elm. Plane Geom. 1st edit. $2xx + 2yy = aa + bb$; whence $y = \sqrt{\frac{1}{2}(aa + bb - 2xx)}$, and consequently AC $\times$ BD is $= 2x\sqrt{\frac{1}{2}(aa + bb - 2xx)}$, which being to be a maximum per quest. $4xx \times \frac{1}{2}(aa + bb - 2xx)$ will therefore be a maximum, and consequently, in fluxions, $2aax\dot{x} + 2bbx\dot{x} - 8x^3\dot{x} = 0$; whence $aa + bb - 4xx = 0$, and consequently $2x = \sqrt{(aa + bb)} = 50 =$ AC; and hence it appears that BD $= \frac{1}{2}$AC, and also that ABC is a right angle.

VIII. QUESTION 496, *by* P. M. *of Durham.*

The greatest plane triangle, having a given angle and any other limit, is that whose sides about the given angle are equal; required the demonstration.

Answered by Mr. Rich. Mallock.

The other given limit, here meant, must necessarily be such as the base, the perimeter, the radius of the circumscribing circle, or the like, which would confine the increase of the required triangle within certain bounds or limits, and affect each of the including sides exactly alike, and then their sum may be always denoted by a certain right line determinable by means of the said limit, and consequently (the area of every plane triangle $=$ being the rectangle under any two of its sides drawn into half the sine of their included angle, the triangle sought will be a maximum, when the said right line is so divided, that the rectangle of its two parts may be a maximum, (the sine of the included angle being a constant quantity) *i. e.* when they are equal, which is too well known to need a particular demonstration here.

Mr. *J. Hudson* puts b for the base, a for the sine of the given angle, s and c for the sine and cosine of half the sum, and x and y for the sine and cosine of half the difference of the other two angles, and then the area of the required triangle is readily found $= bb \times (ss - xx) \div$

$2a$, which, it is manifest, will continually increase, as b increases, and x, at the same time decreases, and consequently that it will (under any assigned possible value of b) be greatest, when x is least, that is, when x vanishes, and the required triangle becomes isosceles, whatever the other given limit may be, unless it restrains the including sides from becoming equal at the same time that, per question, they are supposed every way capable of being so, which would be absurd indeed, and incompatible with the true nature of the question!——If either of the including sides, or their difference, or the difference of their squares, or the sum or difference arising from either of them, being added to or subtracted from n times the other (except when n equals unity), or the like, were to be given, then, their affecting the including sides not alike, the question, instead of being limited thereby, would become either absurd or infinite.

The same answered by P. M. *of Durham (the Proposer). Taken from the Diary for the Year* 1764.

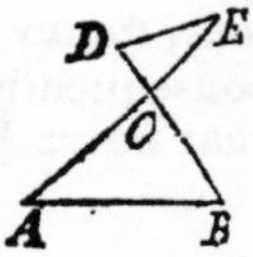

Assume the points D, E, in the sides BC, AC, comprehending the given angle, produced, and let the right line DE be drawn, and then the △ DCE, as also each of its sides, will be given. But the triangles DCE, ACB, are the halves of parallelograms standing about the common angle at C, and parallelograms, standing about a common angle, are directly as the rectangles of their sides (23 Euc. VI), therefore also their halves will be in the same ratio, *i. e.* △DCE : rect. DCE : : △ACB : rect. ACB. Whence the △ACB having a given ∠ and any other limit whatever, is in a constant ratio to the rectangle under the sides about the said given ∠, and consequently when the △ is greatest the rectangle will be the greatest also. But the sides AC, BC, have a constant analogy to the sines of the opposite angles at the base, and may mutually be represented by each other; therefore when the rectangle under the said sides is a maximum, it is manifest that the rectangle under the sines of the angles at the base must be a maximum also. But the sum of these angles being a constant quantity, the rectangle under their sines will be the greatest when the said angles are equal, *i. e.* when the sides subtending them are of consequence equal (6 Euc. I).

IX. QUESTION 497, *by Mr.* Stephen Ogle.

Required to draw a right line through the focus of any given Apollonian parabola, so as to divide the area thereof into two such parts as shall obtain a given ratio.

Answered by Mr. Rob. Butler.

Let DAC represent the given parabola, and EF the given dividing line passing through the focus O, and let the ordinates FR, EL, be drawn

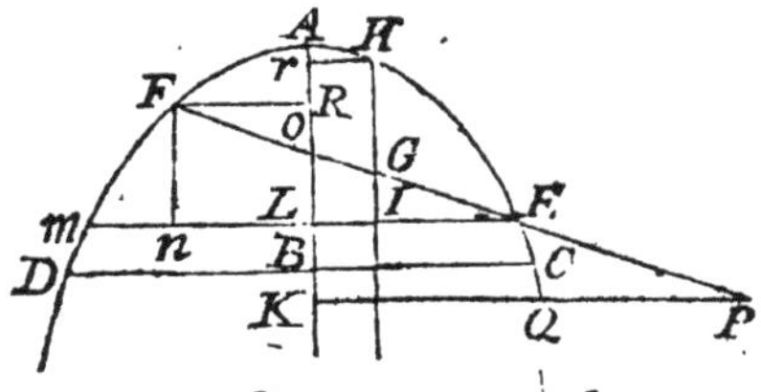

perpendicular upon the axis AL, putting AO $= a$, FR $= x$, and EL $= y$; then, per conics, $xx \div 4a$ = AR, and $yy \div 4a$ = AL ($4a$ being = the parameter); whence RO $= (4aa - xx) \div 4a$ and OL $= (yy - 4aa) \div 4a$; and therefore $\frac{1}{2} x \times \frac{4aa - xx}{4a} + \frac{2}{3} x \times \frac{xx}{4a} + \frac{2}{3} y \times \frac{yy}{4a} - \frac{1}{2} y \times \frac{yy - 4aa}{4a} = \frac{4aax - x^3 + 4aay - y^3}{8a} + \frac{x^3 + y^3}{6a}$ = the area of the figure FAE = a given quantity A, by the nature of the question (because the area of the whole, DAC, and the ratio of the parts DFEC, FAE being given, per the question, the parts themselves will from thence likewise be given); but, by similar triangles, x (FR) : $(4aa - xx) \div 4a$ (RO) : : y (EL) : $(yy - 4aa) \div 4a$ (OL); whence $4aa = (xyy + yxx) \div (x + y) = xy$, which value substituted in the preceding equation of the area, &c. it becomes $(x^3 + y^3) \div 6a + (yxx + xyy - x^3 - y^3) \div 8a = $ A, or $y^3 + 3y^2x + 3yx^2 + x^3 = 24a$A: And hence, by extraction, $y + x$ comes out $= (24a\text{A})^{\frac{1}{3}}$; from the square of which subtracting the equation $4yx = 16aa$, (found from above), and there results $yy - 2xy + xx = (24a\text{A})^{\frac{2}{3}} - 16aa$, and consequently $y - x = \sqrt{((24a\text{A})^{\frac{2}{3}} - 16aa)}$; and hence y is found $= \frac{1}{2}\sqrt{((24a\text{A})^{\frac{2}{3}} - 16aa) + \frac{1}{2} \times 24a\text{A})^{\frac{1}{3}}}$, and $x = \frac{1}{2} \times (24a\text{A})^{\frac{1}{3}} - \frac{1}{2}\sqrt{((24a\text{A})^{\frac{2}{3}} - 16aa)}$; from either of which equations the position of the line EOF is determined.

The same answered otherwise by Mr. Wm. Embleton.

Draw Hr perpendicular upon AK, and put $x =$ Ar; $p =$ the parameter of the principal diameter or axis AK, and suppose the right line FE (passing through the focus O) to be drawn as is required; and let the area of the space FAHE (which is given by the nature of the question) be denoted by a; then, per conics, HG will be found $= \frac{1}{4} \frac{(4x+p)}{4}$, FE $= 4x + p$, H$r = \sqrt{px}$, &c. Whence a (= HG × $\frac{2}{3}$FE × s. ∠HGF) will appear $= \frac{1}{6}\sqrt{p}\ (4x + p)^{\frac{3}{2}}$: And consequently $x = \frac{1}{4} \times \left(\frac{36aa}{p}\right)^{\frac{1}{3}} - \frac{p}{4}$; from whence the position of the line FE may be readily determined.

The same answered more generally by P. M. *of Durham.*

This truly ingenious gentleman supposes the point P, through which the dividing line is required to pass, to be given any where in the plane of the given parabola DAC (either within it or without), and likewise that the right line PEF is drawn as required, cutting the axis AB in O, and dividing the given parabolic area DAC in the given proportion of m to n.

Then bisecting EF in G, and drawing the diameter HG, through G, and the ordinate Em through E, meeting the axis AB in L, and the diameter HG, produced, in I, the area DAC : area EHF :: $m+n : n$ :: BC^3 : EI^3 (cor. 2. 52. Sim. *6*); whence EI is given. Draw PK perpendicular to the axis in K, and meeting the curve in Q, as also Fn perpendicular to the ordinate Em. Then (by a property of the parabola) AO $\times\, p =$ rect. ELn; also (by another property) F$n \times p =$ rect. Enm: But AK $\times\, p = \text{QK}^2$; therefore OK $\times\, p = \text{QK}^2 -$ rect. ELn: Whence $\text{QK}^2 -$ rect. ELn : rect. Enm :: (OK : Fn ::) PK : En, i. e. $\text{QK}^2 -$ rect. ELn $= nm \times$ PK. Call PK $= a$, QK $= b$, EI $= c$, and IL $= x$, then will EL $= c + x$, L$n = c - x$, and $nm = 2x$: Wherefore $bb - cc + xx = 2ax$, and consequently $(a - x)^2 = aa - bb + cc$.

Corollary 1. If the point P be in the curve, then $aa - bb = 0$, and $a - x = c$.

Corollary 2. But if the point P be in the axis, then $a = 0$, and $xx = cc - bb$.

X. QUESTION 498, *by Mr.* Tho. Barker.

Given BE $= 45$, ED $= 21$, and the time of descent of a heavy body through BA $=$ time of descent through BC; required the lengths of the inclined planes AB, BC, and also their position with regard to the vertical line, or plane, BD, so that the rectangle under AE, CD (perpendiculars upon BD) may be the least possible.

Answered.

If BD be supposed to be produced out to F till the time of perpendicular descent along it shall be equal to the time of descent along BA or BC, and the circumference of a circle be described about BF as a diameter, it will, it is very well known, pass through the points A and C (see art 204, p. 230 and 231, of Simpson's Fluxions, 2d edition); whence, DC^2 being $=$ BD $\times$ DF and $\text{AE}^2 =$ BE $\times$ (ED $+$ DF), by the property of the circle, we shall have $\text{AE}^2 \times \text{CD}^2 =$ BE $\times$ BD $\times$ (ED $\times$ DF $+ \text{DF}^2$), which it is evident, by inspection, can admit of neither max. nor minimum; but if it were required to be equal to a given quantity (m^2), then, ED $\times$ DF $+ \text{DF}^2$ ($=$ DF $\times$ (ED $+$ DF)) being $= m^2 \div$ (BE

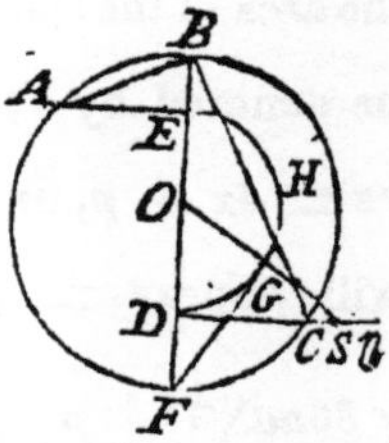

Questions proposed in 1763, *and answered in* 1764.

I. QUESTION 503, *by Mr.* Tho. Sadler.

At Marbury a maid doth dwell,
Whose wit and beauty most excel;
She seems to rival ev'ry fair;
They're few that can with her compare:
Her height, age, fortune, you will know
From th' equations propos'd below.

$$\left.\begin{array}{r} x + 2y + z = 281 \cdot 5294. \\ xyz = 170988 \cdot 22. \\ xx = 4yy + zz. \end{array}\right\} \begin{array}{l} \text{Where } x = \text{her fortune.} \\ y = \text{her height in inches.} \\ z = \text{her age.} \end{array}$$

Answered by Mr. E. Griffiths, *Writing-master, in Ellesmere, Shrops.*

Put $281 \cdot 5294 = a$; $170988 \cdot 22 = b$: and then by transposition, involution, &c. will be found $aa - 2ax + xx = 4yy + zz + 4yz = xx + 4yz$ (per equality, &c.); Whence $aa - 2ax = 4yz$, and consequently $aax - 2axx = 4xyz = 4b$ (per quest.); from the resolution of which equation, x comes out $= 131 \cdot 5294$l. and from thence her age is readily found $= 20$ years, and her height $= 65$ inches.——In the same manner the answer is given by Mr. *J. Hampson*, who thinks it a very great pity that so tall a beauty should have no greater fortune.

II. QUESTION 504, *by the Rev.* Wyvel Blennerhassett.

A ship at a certain port A, observes two islands B and C; B bears from her N.N.E. and C E.S.E. From thence sailing E. by N. 5 miles, finds B and C equally distant from her, and continuing the same course 5 miles farther, she had the said islands in a right line; required the port's distance from the two islands B, C, and also the ship's distance from them at thesecond and third observations.

Answered by Mr. T. Bromhall, *at Mr.* Allen's, *at Spalding.*

Let A represent the port; B, C, the two islands; AD the ship's course; and I, D, her places at the second and third observations: and let the points A, C; A, B; B, C; C, I, and B, I, be joined, and draw DF and IH ∥ BA, and DE and IG ∥ CA respectively; and then, by trigonometry, will be readily found DF (2IH) $= 5 \cdot 55 = b$, and DE $(= 2\text{IG}) = 8 \cdot 316 = c$.

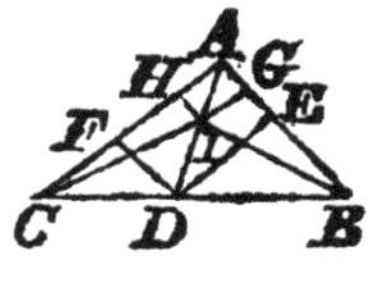

Put AD $(= 10 \text{ miles}) = a$, and CF $= x$; and then, per similar triangles, $x : b :: c : bc \div x =$ EB: Whence CH $= x + \frac{c}{2}$, BG $= \frac{bc}{x} + \frac{b}{2}$, and consequently (47 Euc. I.) $xx + cx +$

$\frac{cc}{4} + \frac{bb}{4} = \frac{bbcc}{xx} + \frac{bbc}{x} + \frac{bb}{4} + \frac{cc}{4}$ ($= \text{CI}^2 = \text{BI}^2$, per quest.) i. e.

$xx + cx = \frac{bbcc}{xx} + \frac{bbc}{x}$: Whence, $(x^3 - bbc)(c + x) = x^4 + cx^3 - bbcx - bbcc = 0$, and consequently $x^3 - bbc = 0$, or $x^3 = bbc$, and $x = \sqrt[3]{bbc} = 6{\cdot}35488$; and hence will be found AC $= 14{\cdot}6695$, AB $= 12{\cdot}8247$, IC ($=$ IB) $= 10{\cdot}8730$, DC $= 8{\cdot}4409$, and DB $= 11{\cdot}0441$ miles respectively.

Mr. *Wm. Embleton* constructs and solves it by means of a semi-cubical parabola only, to which it evidently appertains; and *Sangrado*, jun. constructs and solves it generally, viz. when AI and ID are in any given ratio whatever, by means of the intersection of a circle with a hyperbola described, to the asymptotes AC, AB, through the given point D, &c.

III. QUESTION 505, *by Mr.* J. Hudson, *Land Surveyor.*

There are two places in the same parallel of north latitude, differing in longitude 20°, and their distance on the parallel is known to exceed their distance on the arc of a great circle by 2·335 miles; required the latitude and nearest distance of the said two places, supposing the earth spherical, and 60 miles = a degree.

Answered by Mr. Wm. Spicer.

Put $a = 600$ ($= 10° \times 60 = \frac{1}{2}$ the given difference of longitude in geographical miles), $b =$ nat. sine of 10°, $2c = 2{\cdot}335$ (the given excess of the distance on the parallel above that on the arc of a great circle), and $x =$ the cosine of the latitude required (rad. $= 1.$); and then ax will $= \frac{1}{2}$ the distance of the two places on the parallel, and $bx =$ the sine of half their distance on the arc of a great circle: Whence, putting $v =$ the degrees in the said arc of a great circle, we shall have $60v = ax - c$, per quest. and consequently $x = (60v + c) \div a =$ sine of 32° 43′ 25″; and hence the latitude appears to be about 57° 16′35″, and the nearest distance of the two places 646·3635 geographical miles.

Mr. *J. Hampson* and Mr. *C. Hutton* put $r = 3437{\cdot}64$ ($=$ number of geographical miles in the earth's radius); $y =$ sine of $\frac{1}{2}$ the required distance of the two places on the arc of a great circle (rad. 1.); $a \div b - r = d$; all the rest as above; and then, by a very easy and short process, find $dy - \frac{ry^3}{2.3} - \frac{3,3ry^5}{2.3.4.5} - \frac{3.3.5.5ry^7}{2.3.4.5.6.7} -$ &c. $= c$; from which, by reverting the series y comes out $= {\cdot}09255$: and from thence they find every thing else nearly the same as above.

IV. QUESTION 506, *by* Geometricus.

In two similar right-angled triangles, there is given the base of the one, and the perpendicular of the other, in one sum; and it is proposed to determine the triangles such, that their hypothenuses shall form the legs of another triangle similar to them; and so, that the sum of their three areas may be of a given magnitude.

Answered by Mr. Rich. Gibbons.

Upon ED, the given sum of the base and perpendicular, constitute the rectangle EC = the sum of the three given areas, and also about the same, as a diameter, conceive a semi-circle to be described, cutting AC in B; join the points E, B, and D, B; and EAB, BCD and CDB, will be the three trianles required; which is too evident to need a demonstration.

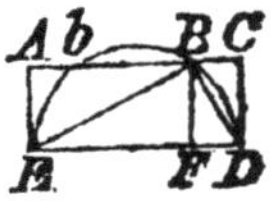

V. QUESTION 507, *by Mr.* W. Spicer.

Given the area of a plane triangle = 235, and its vertical angle = 73° 34′, and the side of its inscribed square = 9·6 to describe the triangle.

Answered by Mr. C. Hutton.

Construction. Produce AB, the side of the given inscribed square, to C, so that AB × BC may be = twice the given area of the triangle, and take BD, ⊥ AC at B, a mean proportionl between AB and BC, and on BC conceive a semicircle BEC to be described, and draw DE ∥ AC meeting the same in E, from whence let EF be drawn ∥ DB meeting AC in F: Then, if on FC a segment FHC of a circle, to contain the given vertical angle, be described, and FG be taken = FB, and GH drawn ∥ FC, meeting the arc FHC in H, and the points H, F, and H, C, be joined, FHC will be the triangle required.

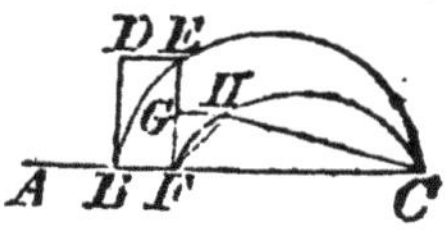

Demonstration. The ∠FHC = the given angle, per construct. and twice the area of the △FHC = FC × FG = (by construct.) FC × FB = (per prop. of the cir.) $FE^2 = BD^2$ = AB × BC = twice the given area, by construct. And (by prob. 3d. part 2d. Simp. Select Exercises) $\frac{AB \times BC}{FC + BF}$ $= \frac{AB \times BC}{BC}$ = AB, the side of the given inscribed square, by construction.

Note. When GH neither cuts nor touches the segment FHC, this quest. is, manifestly impossible.

VI. QUESTION 508, *by Mr.* T. Barker, *of Westhall.*

Supposing A and B to represent two right lines, whereof the sum is given $= s$, and $m \div n$ to denote any given ratio. Required to determine, by a geometrical construction, what A and B themselves must be, so that $A^2 + (m \div n) \times B^2$ may be $= cc$, a given square.

Answered by Mr. Wm. Embleton.

Construction. Upon $AC = s$, the given sum of the two right lines, constitute the rectangle $AE = cc$, and take $AB = AH$, and then AB, BC will be the measures of A and B respectively, and in the given ratio of m to n. For, drawing BF ∥ AH, and producing it and EC till they become each $=$ FE (BC), and joining their extremes G, D, then, supposing $AB : BC :: m : n$, it will be $m : n :: AB \times BC$ ($= BF \times BC = BE$) $: BC^2$: Whence $BE = (m \div n) \times BC^2$; to which equal magnitudes, the equal magnitudes AF, AB^2, being respectively added, it will become $AB^2 + (m \div n) \times BC^2 = AF + BE = AE = cc$, per construction, &c.

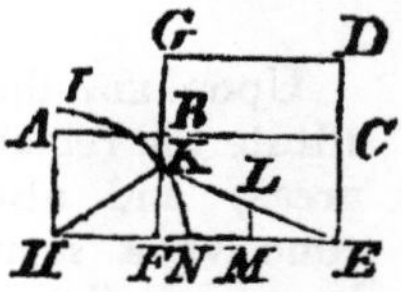

Note. This question will be found useful in the solution of some difficult problems relating to triangles.

The same answered by Mr. Da. Kinnebrook.

Construction. Take $HE = s$, the given sum of the two right lines, and from the centre H, with radius c, suppose the circular arc IN, cutting HE in N, to be described, and draw LM perpendicular any where upon NE, and so as that LM^2 be to $ME^2 :: m : n$, and let the points E, L, be joined by a right line produced, till it meets the said arc in K; from whence draw KF⊥HE, and HF, FE will be the measures of A and B required.——For, joining the points H, K, $n : m :: FE^2 : FK^2 = (m \div n) \times FE^2$ (by construction and sim. triangles, EML, EFK), and (by 47 Euc. I.) $HF^2 + (m \div n) \times FE^2$ ($= FK^2$) $= cc$ ($= HK^2$, per construction).

A more general answer to the same by P. M. *of Durham.*

Suppose in a right line the two points A, B were given, and it was required to determine another point C therein, such that $m \times AC^2 + n \times BC^2$ might be $= p \times DE^2$ (DE being another given right line).

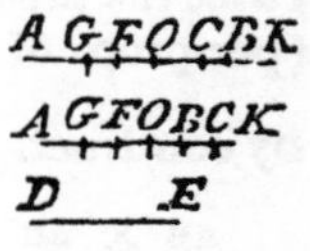

Bisect AB in O, and divide it in F in the given ratio of m to n; take OK, so that $m + n$ may be to p in the duplicate ratio of DE to OK, and, making FG (towards opposite parts of F) $=$ FO, take GC, so that the rectangle under it and CO may be equal to the given rectangle under AK and KB (Simp. Geom. Pr. 6), and C shall be the point required.

Demonstration. For, since AB is bisected in O, rect. $AKB + AO^2$

$=$ OK^2 (6 Euc. 2) and rect. GCO $=$ rect. GOC $+$ OC^2 (3 Euc. 2.) $= 2$ rect. FOC $+$ OC^2, because GO $=$ 2FO (hyp.). But rect. GCO $=$ AKB (hyp); therefore 2 rect. FOC $+$ OC^2 $+$ AO^2 $=$ OK^2, and, multiplying both sides by $(m + n)$, $(m + n) \times 2$ rect. FOC $+$ $(m + n)$ (OC^2 $+$ OA^2) $= (m + n)$ OK^2 $= p \times$ DE^2 (hyp.). Again, because AB is divided in F in the ratio of m to n, and bisected in O, therefore $m + n : m - n$ :: (AO : FO ::) rect. AOC : rect. FOC; and consequently $(m + n) \times$ rect. FOC $= (m - n) \times$ rect. AOC : Whence (($m - n) \times 2$ rect. AOC $+$ $(m + n)$ (AO^2 $+$ OC^2) $= m$ (AO $+$ OC)^2 $+$ n (AO $-$ OC)^2 $=$) $m \times$ AC^2 $+$ $n \times$ BC^2 $= p \times$ DE^2.

Corollary 1. As the point C falls within or without the terms (or points) A, B, the right line AB will be the sum or difference of the right lines sought.

Corollary 2. The numerical solution is had from this equation $(m - n) \times 2$ rect. AOC $+$ $(m + n) \times$ (AO^2 $+$ OC^2) $= p \times$ DE^2, where p is universal, but in the question is limited to the value n.

VII. QUESTION 509, *by Mr.* T. Moss.

A lays B one hundred guineas to one, that in 25 throws with 5 half-pence, he does not throw precisely, 4 heads the first 5 throws, 6 heads the next five throws, 8 the 3d, 10 the 4th, and 12 the last 5 throws : Required the respective values of these two gamesters expectations?

Answered by Mr. W. Spencer.

The throwing of m heads, precisely in n throws with p half-pence, being evidently the same as throwing m heads precisely in $n \times p$ throws with one half-penny only, it follows, from corol. to prob. 5, on page 12, &c. of Simpson's Nature and Laws of Chance, that, if the 5th, 7th, 9th, 11th, and 13th terms of the binomial $1 + 1$ raised to the 25th power, be each divided by $(1 + 1)^{25}$, they will be the respective probabilities required, *i. e.* putting $\frac{25}{1} \times \frac{24}{2} \times \frac{23}{3} \times \frac{22}{4} \times \frac{1}{(1+1)^{25}} = p$; then p, $p \times \frac{21}{5} \times \frac{20}{6}$, $p \times \frac{21}{5} \times \frac{20}{6} \times \frac{19}{7} \times \frac{18}{8}$, $p \times \frac{21}{5} \times \frac{20}{6} \times \frac{19}{7} \times \frac{18}{8} \times \frac{17}{9} \times \frac{16}{10}$, and $p \times \frac{21}{5} \times \frac{20}{6} \times \frac{19}{7} \times \frac{18}{8} \times \frac{17}{9} \times \frac{16}{10} \times \frac{15}{11} \times \frac{14}{12}$ will be the respective probabilities of throwing the number of heads as specified in the question, considering them as independent one of another : and therefore the probability of their all happening together, according to some order or other, will, by corol. to prob. 1st. of Simp. Laws of Chance aforesaid, be expressed by $p^5 \times \frac{21^4}{5^4} \times \frac{20^4}{6^4} \times \frac{19^3}{7^3} \times \frac{18^3}{8^3} \times \frac{17^2}{9^2} \times$

$\frac{16^2}{10^2} \times \frac{15}{11} \times \frac{14}{12}$; but there being $1 \times 2 \times 3 \times 4 \times 5$ $(=120)$ permutations, or different orders or ways in which 5 such events may all happen together, the probability of their happening in the precise order required in the question will therefore be

$$\frac{p^5 \times 21^4 . 20^4 . 19^3 . 18^3 . 17^2 . 16^2 . 15 . 14}{120 \times 5^4 . 6^4 . 7^3 . 8^3 . 9^2 . 10^2 . 11 . 12}$$: From whence the values of A and B's expectations may readily be found, &c.

VIII. QUESTION 510, *by Mr.* Rob. Butler.

Suppose in latitude 52° north, a semi-circular plane, whose radius is 20 feet, to be placed in the plane of the meridian, with its base downwards; at what time in the afternoon, on the 21st of June, will its shadow cover just one acre on an horizontal plane?

Answered by P. M. *of Durham.*

It may be easily shewn, in the projection of any regular figure, whether right-lined or a conic section, that the area of the plane of the figure is to the area of its projection, as the height of the plane above the horizon multiplied by radius, is to the shadow of the said height multiplied by the sine of the angle, which the said shadow makes with the meridian; but this angle is the sun's azimuth; and the area of the plane, its projection, and also its height, are given: And therefore, calling its height $= a$, its shadow $= y$, the given ratio of the plane's surface to its projection as m to n, and likewise, denoting the cosines of the given latitude and declination by c and q, the sine and versed sine of the hour angle sought by x and v, and the cosine of the difference between the lat. and declin. by d, we shall have $m : n :: ar$ (r being rad.) $: y \times$ s. azim. but s. alt. : cos. alt. $:: a : y$, and cos. alt. $: x :: q :$ s. azim. Whence s. alt. $: x :: qa : y \times$ s. azim. and consequently $rr \times$ s. alt. $: qrx :: (ar : y \times$ s. azim. $::) m : n$; but (by spherics) $rr \times$ s. alt. $= rrd - cqv$, and therefore $rrd - cqv : qrx :: m : n$, i. e. $rrd \div cq - v : x :: mr \div c : n$: whence this easy

Geom. Construction. Let AB, AD, represent the complements of the latitude and declination to the rad. AC $(= r)$ and then BE $= c$, DF $= q$, and CG $= d$; produce GB till it cuts the rad. CA produced in H, and take AK a fourth proportional to BH, CH, and AC (12 Euc. 6.). Assume KC, CL in the given ratio of $mr \div c : n$, and draw CL $\perp$ KC at the point C, and through the points K, L draw a right line, cutting the circle in O, and AO will be the measure of the hour angle sought.

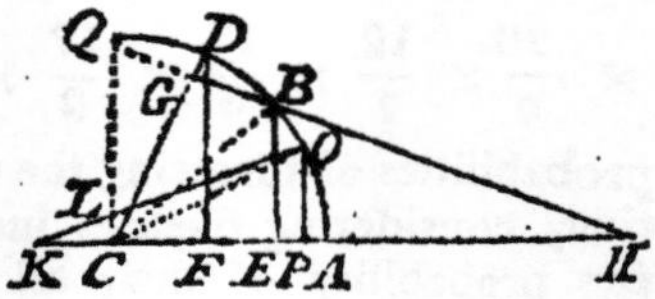

For, producing HQ, CL, to meet in Q, and drawing OP, the sine of the arc AO; then, by similar triangles, DF : DC :: CG : CQ, and DF × BE : rect. DCG :: (BE : CQ :: EH : CH ::) AC : AK $= rrd \div cq$; but KP (AK — AP) : OP :: (KC : CL ::) $mr \div c : n$, i. e. because AK$= rrd \div cq$, and AP is the versed sine of the arc, of which OP is the sine; therefore OP $= x$, and AP $= v$.

The same answered otherwise by Mr. Robert Butler, *the Proposer.*

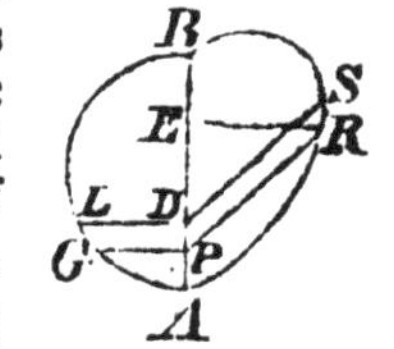

Suppose ARSB represents the required shadow; $s =$ area of the given semi-circle AGB; a and $b =$ sine and cosine of the lat. x and $y =$ those of the sun's alt. z and u $=$ those of his azim. from the north; $d =$ sine of his decl. (rad. $= 1$) and A $= 43560$, the feet in an acre: Then, GP being perpendicular to the horizontal plane ARBA, $(y \div x)$ PG $=$ PR will be the length of its shadow; and drawing DL, DS, indefinitely near, and ∥ PG, PR, and meeting the curves in L and S, respectively, and RE ⊥ AB, then will PG × PD, and PD × ER, be respectively $=$ the areas of the parallelograms PL and PS; but z being the sine of EPR, RE is $= z \times$ PR $=$ $(zy \div x)$ PG, and therefore PD × ER $= (zy \div x) \times$ PG × PD, i. e. the area of the parallelogram PL, drawn into $zy \div x =$ area of the parallelogram PS, and consequently $s \times zy \div x =$ A, let the curve ARB be what it will; but, by spherics, &c. z is found$= \sqrt{(bbyy - (d - ax)^2)} \div by$, which substituted above, gives $(s \div bx)\sqrt{(bbyy - (d - ax)^2)} =$ A: From whence x is found $= \cdot0111641$, and consequently the sun's azim. from the west $= 32° \ 19'$, and the time required $= 9^m \ 16^s$ past 8 o'Clock. No regard is here had to the sun's semi-diameter and refraction; but the answer in such case will be had, by making $d =$ sine of 19′ more than what it here is.

IX. QUESTION 511, *by Mr.* Malachy Hitchins.

In latitude 50° north, there is a cylinder, whose internal diameter is $= 6$ feet, and length $= 12$ feet, lying on the horizontal plane, and making an angle with the meridian $= 20°$ east; at what time of the day, on May 10, 1763, will the greatest part of its internal superficies be enlightened by the sun; and what ratio will that enlightened space bear to the whole internal superficies?

Answered.

Φιλομαθεμαλικος, and some others, (answering this question) make the time sought to be about 12^m past 11 in the morning, and find the illuminated part $= \frac{1}{9}$ the whole internal superficies nearly: but *Philotechnus* finds the illuminated internal surface to be always $=$ a rectangle under the sine and secant of the angle comprehended between the solar ray, issuing from the centre of the sun, and the diameter of the given cylinder, passing through that point of the circumference of its end, whereon the said ray impinges, to radius = the diameter of the said cylinder; from which (as it depends wholly on the sun's altitude and azimuth conjunctly) being fluxed, put = 0, and reduced, the time required, &c. may be determined.

Solution.

As the answer to this question is partly false and partly incomplete, I shall supply the whole solution as below.

I. *To find the Time of the greatest Illumination.*

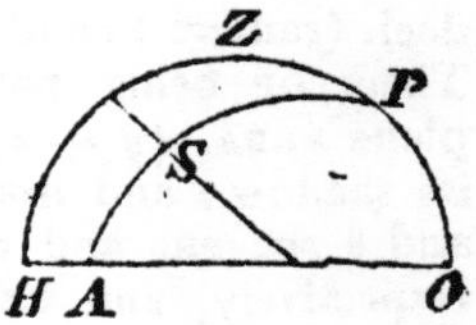

Construction. On the plane of the meridian of the place describe the primitive circle, and therein take Z the zenith and P the pole; draw the horizon HO, on which take HA = 20° the given horizontal angle made by the meridian and the axe of the cylinder; and through A draw the meridian ASP. So shall APZ be the hour angle from noon required when the most surface is illuminated.

Demonstration. It is evident that the curve bounding the illuminated part of the internal surface, is always the shadow of that half or semi-circumference of the end of the cylinder which is turned towards the sun, that is, that semi-circumference whose diameter is perpendicular to the plane drawn through the sun and the axe of the cylinder. It is also evident that the most surface will be illuminated when the solar ray penetrates the farthest distance possible within the cylinder; and this, it is farther evident, will happen when the ray makes the least possible angle with the axe of the cylinder; and this angle being always measured by the arc of a great circle passing through the sun and that point A of the horizon to which the axe of the cylinder is directed, this arc will therefore be a minimum, that is, the arc intercepted by A and the parallel of declination for the given day, is a minimum; in which case it is evident that the arc must be perpendicular to the parallel of declination, and will therefore be a meridian passing through A, as by the construction.

Calculation. In the right-angled triangle AHP, it is as the s. of HP (130°) : radius :: tang. HA (20°) : tang. hour $\angle$P = tang. HA $\div$ s. HP = tang. HA $\div$ sin. lat. = tang. of $25^\circ\ 21'\ 47''$, which answers to $1^h\ 41^m\ 39^s$ and which being taken from 12^h, there remains $10^h\ 18^m\ 21^s$, or nearly 18^m past 10 in the morning is the time of the greatest illumination. And as radius : cos. HA :: cos. HP : cos. PA = $127^\circ\ 50'\ 30''$; from which take PS (= $72^\circ\ 23'$) the sun's polar distance, and the remainder $55^\circ\ 27'\ 30''$ is = AS the measure of the least angle formed by the solar ray and the axe of the cylinder.

Scholium. Hence we may observe that the time is the same whatever the declination is; and that the time given in the original answer is wrong.—It may also be remarked, that if the reasoning in the above, or in what follows, should not be clear to any person, he may assist his imagination by rolling up a rectangular piece of paper into the form of a cylinder, and exposing it to the solar rays, or to the light of a candle.

II. *To find the Quantity of surface illuminated.*

Conceive the cylinder to be cut through on one side in the direction of its axe, and to be spread flat out on a plane; then will the illuminated part appear as in the annexed figure, where AB is the semi-circumference as stretched out in a right line; CB is a quadrant; CD, perpendicular to AB, is the projection of the diameter of the end; or it might be considered as the shadow of the diameter; and in like manner, every other perpendicular EF is the projection or shadow of double the sine of its corresponding arc BE.

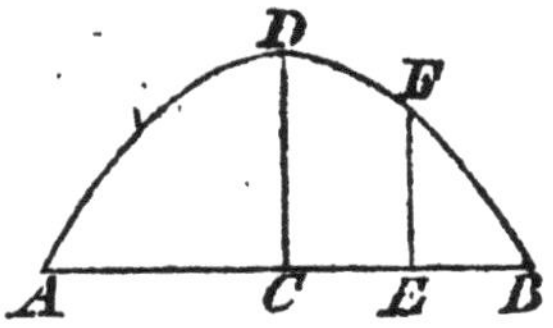

Put now t for the cotang. of the angle formed by CD or by the axe of the cylinder and the solar ray, or the cotang. of the arc AS (55° 27′ 30″) in the first figure to the radius 1; z = the arc CE, and x = its sine to the radius r (3) of the cylinder. Then, by plain trigonometry $2tr$ = CD, $\sqrt{(rr - xx)}$ = S. BE (or cos. CE), and $2t\sqrt{(rr - xx)}$ = EF. But the fluxion of CDFE is $= \dot{z} \times$ EF, and $\dot{z}$ is $= \frac{r\dot{x}}{\sqrt{(rr-xx)}}$; hence $\dot{z} \times$ EF is $= \frac{r\dot{x}}{\sqrt{(rr-xx)}} \times 2t\sqrt{(rr-xx)} = 2tr\dot{x}$; the fluent of which is $2trx$ = the area CDFE. And when E arrives at B, then x is $= r$, and the area CDB is $2trr$, the double of which is ADB $= 4trr = tdd$, putting d for the diameter of the cylinder; which, since d is $= 6$, will be $36t = 24{\cdot}78068$, the illuminated surface required.

Now, the length of the cylinder being double its diameter, its whole internal surface will be $2pdd$, putting $p = 3{\cdot}1416$; and therefore the whole is to the illuminated part, as $2p$ to t, that is, as 6·2832, to ·68835, or as 9·12788 to 1.

Corol. 1. From the above it appears that the projection or shadow (2CDFE $= 4trx$) of a middle zone of the circular end of the cylinder, *upon its internal surface*, is always equal to the shadow or projection, *on the plane of the horizon*, of the rectangle under the diameter and distance between the two rectilinear ends of the zone; and that the projection of the whole circular end, on the internal surface, (that is, the whole illuminated part) is equal to the projection, on the horizontal plane, of a square whose side is equal to the diameter of the cylinder.——Or, it farther appears, that the projection of the middle zone, is to the rectangle under the sine and secant of the complement of the angle made by the solar ray and the axe of the cylinder, to a radius equal to the diameter of the cylinder, as the sine of the arc CE is to the radius. And, also, that the whole illuminated part is equal to a rectangle under the sine and secant of the same angle, to the same radius equal to the diameter of the cylinder: which last is also remarked by *Philotechnus* in the original solution.

Corol. 2. It may be remarked that, as *td* is = DC, therefore the area or illuminated surface *tdd* becomes $d \times$ DC; that is, the diameter of the cylinder drawn into the altitude of the illuminated surface. Which agrees with the rule proved in corol. 1, p. 148 of my Mensuration, from whence the surface might have been at first easily found. H.

X. QUESTION 512, *by Mr.* Joseph Walker.

Two right lines, drawn from the same point, being given both in length and position; it is required so to apply a right line, of a given length, between them, and subtending the given angle, that the part of the one of them, next the said angle, may be to the alternate part of the other, in the given ratio of p to q.

Answered by Mr. T. Moss.

Construction. Suppose DH = one of the given lines comprehending the given angle, and from the centre D with radius DE (= the given line to be applied) conceive the circumference of a circle to be described, and make the ∠HDK = the supplement of the given angle to 180°, taking DK : DH :: $p : q$, viz. the given ratio, and joining the points HK, by a right line cutting the circumference of the said circle in E: From whence drawing EC ∥ DK, meeting DH in C, and producing DH till DA = CH, and CE, till CB becomes = the other given comprehending line, and then, joining the points D, E, the thing is done.—For, by sim. triangles, DK : DH :: CE : CH; and, per construction, DK : DH :: $p : q$; whence, per equality, CE : DA (= CH, per construction) :: $p : q$; and the ∠BCA = ∠ADK (the given angle), and the right lines CA, CB, and DE are = the given lines, by construction.

A Construction to the same by Mr. Da. Kinnebrook, *and* P. M. *of Durham.*

Produce one of the given comprehending lines BC till CG is to the other of them (AC) in the given ratio of p to q, and let the points A, G, be joined, and about the centre C with radius CF (= the given line to be applied) conceive a circular arc to be described cutting AG in F; from whence draw FD ∥ CG meeting AC in D, and, joining the points F, C, draw DE ∥ thereto, and that will be the position of the line required: For by sim. triangles, FD (= CE, per construction) : DA :: (CG : CA ::) p : q, per construct. and DC being a parallelogram (per const.), DE will be = CF = the given line to be applied (per construction).

XI. QUESTION 513, *by Mr.* T. Moss.

Let the ratio of the head and bung diameters be what it will, within certain limits, a spheroidal cask may be so formed, that the diagonal line, such as is now graduated upon gauging rods, will exhibit the true content of the cask: It is proposed to find, by a general method, what those limits are, and how near the head and bung diameters can approach to the ratio of equality, before the above circumstance fails.

Answered by Mr. T. Moss, *the Proposer.*

Let d = the given diagonal of the spheroidal cask required, x = the bung diameter, the ratio of the head and bung as n to 1, and p = ·7854; and then by the well-known theorem, its content will be $\frac{2p}{3} \times \sqrt{(dd - \left(\frac{1+n}{2}\right)^2 \times xx)} \times (2xx + nnxx)$: Which, put into fluxions (supposing n constant) and made = 0, and reduced, gives $x = \frac{2d\sqrt{2}}{(1+n)\sqrt{3}}$ = the bung diameter, in this circumstance, when the cask, under the given diagonal d, is the greatest possible; and this value, substituted for x in the above expression, gives $\frac{16pd^3}{9\sqrt{3}} \times \frac{2+nn}{(1+n)^2}$ = the content of the greatest spheroidal cask having the given diagonal d, and corresponding to any assignable possible value of n: And when it is less than $d^3 \times$ ·6283, the content, as found by the diagonal line, the problem is then manifestly impossible; suppose it therefore equal thereto, in order to determine the limit of n (required), *i. e.* suppose $\frac{16pd^3}{9\sqrt{3}} \times \frac{2+nn}{(1+n)^2} = d^3 \times$ ·6283, or $\frac{2+nn}{(1+n)^2}$ = ·7796; whence n is readily found = ·898, &c.——And hence it appears that the present constructed diagonal line, will not exhibit the true content of a spheroidal cask, when the ratio of the head and bung diameters approaches nearer to an equality than that of ·898 to 1 (or 9 to 10, nearly), in any circumstance whatever.

XII. QUESTION 514, *by Mr.* William Embleton.

Supposing a triangular prism, whereof the dimensions of the sides are 8, 10, and 12 inches, to be laid in still water; I would know in what position, or positions, it will rest, and what part of its surface will be immersed, supposing its specific gravity to be in proportion to that of the water, as 2 to 3.

Answered by Mr. Wm. Embleton.

Suppose ABC represents a section of the given prism parallel to its ends, AED the immersed part thereof, AG, AH, right lines drawn from the angular point A, to the middle points of BC and ED, respectively: Then, if the centres of gravity n, m of the triangles ABC, AED (which are known to be $\frac{1}{3}$ of AG and AH from G and H, respectively) be joined by the right line nm, the prism will, by the laws of hydrostatics (see prop. 87, sect. 9, of Emer. Mechan.) be quiescent, whenever that line becomes perpendicular to the horizontal line ED: But GH will then also be perpendicular to the same (by cor. to theor. 12. B. IV. Simp. Geom. 2d ed.) and so the question be reduced to that of dividing the given triangle ABC into two given parts by the right line ED, so that GH, connecting the middle points of BC and ED, may be perpendicular to the latter of them; which is a problem purely geometrical; and from whence an answer in numbers may be readily derived.——This problem may be of use in determining the structure of a ship, so as to be least subject to roll, &c.

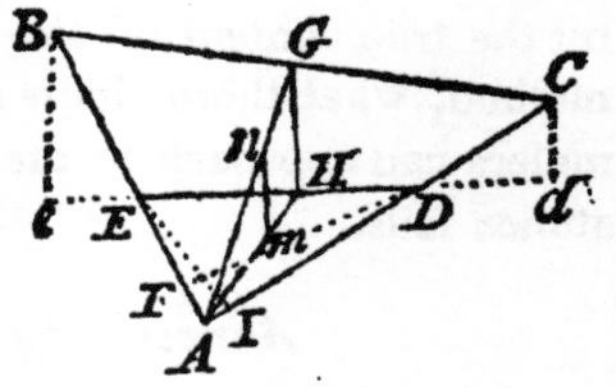

Remark. In order to bring this problem to a final solution; On DE, produced both ways, demit the perpendiculars cd, Be, which will therefore be parallel to GH; also, on AB and AC, the perpendiculars DF and EI.

Since then GB is = GC, and Be, GH, Cd are parallels, He will be = Hd; but HE = HD, ∴ Ee = Dd. Again, since the vertical angles eEB and FED are equal, as also dDC and IDE, the two triangles BeE and DFE are similar, as also the two CdD and EID; hence EB : Be :: ED : EF and DC : Dd :: DE : DI, and as the two means are the same in both, therefore EB : DC :: DI : EF.

Put, now, b = AB, c = AC, m and n = sine and cos. ∠A, x = AE, and y = AD. Then nx = AI, and ny = AF; hence EF = $x - ny$, and DI = $y - nx$; also BE = $b - x$, and CD = $c - y$; therefore the last proportion becomes $b - x : c - y :: y - nx : x - ny$, which produces the equation $x^2 - bx + bny = y^2 - cy + cnx$. But the △AED = $\frac{1}{2}mxy$ is = a, a given magnitude, by the quest. and hydrostatics; therefore y is = $2a \div mx$; which being substituted in the equation, it becomes at last $x^4 - (b + cn)\,x^3 + (c + bn)\,\frac{2ax}{m^2}$

$-\frac{4a^2}{m} = 0$. From whence the value of x may easily be found; and thence that of y. H.

XIII. QUESTION 515, *by Mr.* Abr. Botham.

Suppose a perfectly elastic ball, of 2 inches diameter, to be let fall upon the surface of an hemisphere whose diameter is 30 inches, from the height of 5 feet above the plane of the horizon or base of the hemisphere; to find at what distance from the axis the ball must descend, so that, after being reflected at the surface, it shall impinge upon the horizon at the greatest distance possible from the centre of the hemisphere.

Answered by Mr. Da. Kinnebrook.

Suppose the ball to fall from B, on the point D, of the given hemisphere MDL, and, after reflection, to describe the curve DG, meeting the horizon in the point G; and draw DF a tangent to the same at D: Suppose, also, FG, DE, to be perpendicular to the horizon, and KDI to be a tangent to the hemisphere at D, and put DC ($= 1\frac{1}{4}$ foot) $= r$, $16\frac{1}{12}$ feet $= s$, DE $= x$; and then the direction of the ball DF, after reflection, being such as that the $\angle$BDK $= \angle$FDI (by the nature of reflection), and its velocity therein $= 2\sqrt{(4rs - sx)}$ (as both ball and hemisphere are perfectly elastic), we shall, by the laws of falling bodies, trigonometry, &c. find GC $= \left(\frac{8x \times (4r - x)(2xx - rr)}{r^4} + 1\right) \times \sqrt{(rr - xx)}$, a max. (per quest.) from which, thrown into fluxions, and reduced, &c. x comes out $= \cdot 4646$, &c. and hence every thing else required may be readily found.

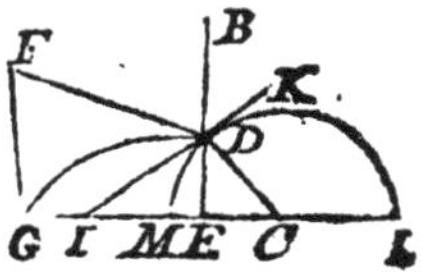

XIV. QUESTION 516, *by Miss* Ann Nicholls.

The latitude pray shew; the hour o'th' night * too tell,
When the three stars below † are in one parallel.‡

* On Jan. 1, 1763. † Sirius, Cor Leonis, and Aliah. ‡ Of Altitude.

Answered by Mr. T. Allen, *of Spalding.*

Put x and y for the sine and cosine of the co-latitude required, v and z for those of the arc of the time that Sirius is then short of the meridian. Also, let a and b, c and d, e and f, be the respective sines and cosines of the given polar distances of the three given stars, Sirius, Cor Leonis, and Aliah; m and n, p and q = the sines and cosines of the difference in right ascension betwixt Sirius and the other two of the said stars, respectively; rad. $= 1$; and then, by spherics, the cosines of their zenith distances will be found equal to $axz - by$, $cx \times (zn - vm) + dy$, and $fy - eqxz + epxv$, respectively; which are all equal, by the quest. From the first and second, y is found $= x \times (az + cmv - cnz) \div (b + d)$, which, substituted in the third, and ordered, gives $\frac{z}{v} = \frac{cm(b+f) + ep(b+d)}{a(d-f) + cn(b+f) + eq(b+d)}$ $=$ tang. of the arc Sirius is short of the meridian at the time required; from whence the hour of the night and latitude of the place, will be easily determined.

XV. QUESTION 517, *by Mr.* Wm. Spencer.

Given $4\dot{x}^2 + 8y^2\dot{x}^2 + x^2\dot{y}^2 + 4y^4\dot{x}^2 - x^3\dot{y}^2 = 0$, to find the value of x when $y = 2$, supposing that $x = 2$, when y vanishes or becomes $= 0$.

Answered by Mr. T. Allen, *and* P. M. *of Durham.*

From the given equation is readily derived $\frac{\dot{x}}{x\sqrt{(x-1)}} = \frac{1}{2} \times \frac{\dot{y}}{1 + yy}$; whence, putting $\left\{ \begin{matrix} a \\ b \end{matrix} \right\} =$ circular arc, whose radius is unity and $\left\{ \begin{matrix} \text{sec. } \sqrt{x} \\ \text{tang. } y \end{matrix} \right\}$, and $c =$ the arc of $45°$, whose sec. $= \sqrt{2}$, the equation exhibiting the relation of the fluents will be found $2a = \frac{1}{2}b$: Which, corrected, according to the condition specified in the problem, becomes $2a = \frac{1}{2}b + 2c$; and hence, by a table of natural or logarithmic sines, &c. the value of a will readily be found $= 1{\cdot}06218$, when $y = 2$; and thence $x = 4{\cdot}21683$, &c. which was required.

THE PRIZE QUESTION, *by Mr.* T. Moss.

Through the angular point C, of a given plane triangle ABC, to draw, geometrically a right line so, that two others AD, BE, being drawn, from the given points A, B to meet the same, and make given angles with the right line AB, whose sum is less than two right angles, their rectangle (AD × BE) may be of a given magnitude.

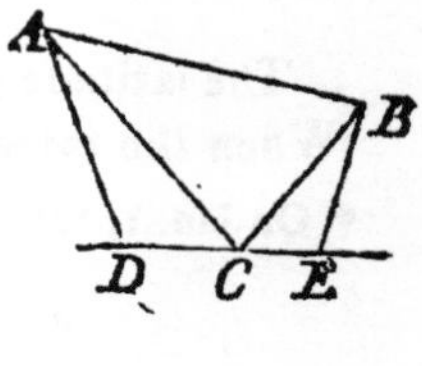

Answered by Mr. Tho. Moss, *the Proposer.*

Imagine DCE to be the position of the line required, and produce BC (see fig. to question) to meet $Ag \parallel$ BE in g, and then, AD × BE being a given quantity (per quest.), it is evident, from sim. triangles BCE, GCF, that AD × FG is a known quantity also, it being to the given rect. AD × BE : : Fg : BE. Moreover, Dk being ⊥ Ag, it is very evident that Fg × AD : Fg × Dk in the known ratio of AD : Dk; whence it follows, that Fg × Dk, or twice the measure of the △ DFg, becomes known. Now, suppose Fm and Dn ⊥ cg, and produce AD to meet BC produced in G, and then it is manifest that cg × (Fm — Dn) = twice the measure of the △ DFg (= Fg × Dk); but cg is known, and consequently Fm — Dn is known also: Hence the problem is reduced to that of drawing a right line from the given point c to cut the right lines Ag, AG, given in position, so that the difference of the perpendiculars Fm, Dn, falling from the points of intersection F, D, upon Bg, may be = a given right line, *i. e.* equal to twice the given measure of the △ DFG divided by the given line cg; of which the following is the

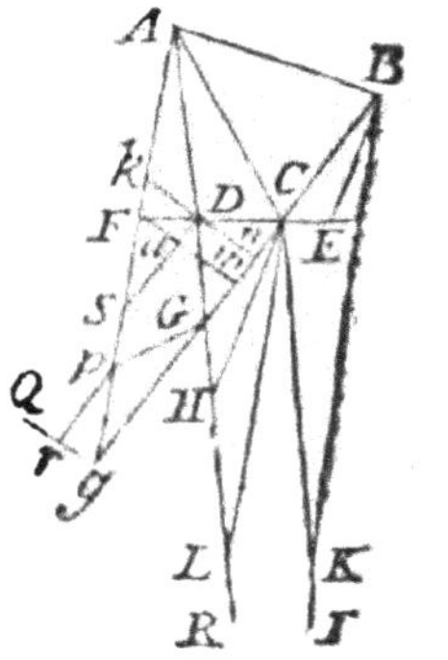

Construction. In gQ ⊥ gB, take gr = the difference of the perpendiculars Fm, Dn, and draw rp ∥ gB meeting Ag in p; join p, G, and draw (by prob. 37, p. 242 of Simp. Geom. 2d edit.), from the given point c, a right line cutting Ag, AG, so that the segments Fp, DG, cut off thereby, may be to each other in the given ratio of Ag to AG, and the thing is done.

Demonstration. Draw Ds ∥ cg cutting Fm in n; then (by const.) Ag : AG : : Fp : DG, and by sim. triangles, and division of ratios, Ag : AG : : sg : DG, whence, by equality, Fp = sg, and, taking sp from each, gp = Fs, and consequently Fu = gr = the given difference, by construct.——This method contains the solution of another problem of equal, if not superior, difficulty.

Mr. *Da. Kinnebrook* draws CH and CI ∥ BE and AD respectively, and produces AD to meet CH in H, and takes CK in CI, so that CH × CK may be = the given rect. AD × BE, and joining B, K, draws CL ∥ thereto, meeting AH produced in L, and then, continuing BC till it meets AL in G, divides AL in D, so that AD × DL may be = AH × GL; from which point D, through C, drawing the right line DCE, and the thing is done.—The demonst. whereof is extremely natural and easy.

The same answered by Mr. Wm. Embleton, *and Mr.* C. Hutton.

Construction. Through the given point C, draw CL ∥ DA meeting BE produced in F; and AL ∥ BE, in L; produce AC to meet BF in G, and take GR so, as that CL × GR may be = the given rect. AD × BE, and upon BR describe the semi-circle BKR: erect RI ⊥ BR, and = to a mean proportional between BF and GR; draw IK ∥ BR, meeting the circumference in K, and KE ∥ RI meeting BF in E, the point through which the required line ECD must pass.

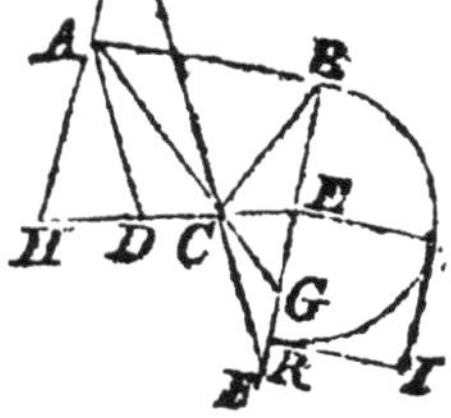

Demonstration. Produce LA to meet ECD, produced, in H; and then, BE (by prop. of the circle) being $= EK^2 \div BE = RI^2 \div BE = BF \times RG \div BE$ (by construc.), and $GE = RE - RG = BF \times RG \div BE - RG \times RG\ (BF - BE) \div BE = FE \times RG \div BE$ (per the fig.), we shall (by theor. 20, B. IV. of Simp. Geom. 2d edit.) have $CL : DA\ (:: HL : HA) :: FE : (GE =)\ FE \times RG \div BE$ (the triangles HCL, ECF, being similar, and the right line ACG making equal angles with the homologous sides thereof, &c. whence, multiplying extremes and means, $AD \times BE = CL \times GR =$ the given rectangle, by construction.

Note. When IK neither cuts or touches the semi-circle BKR, this problem is impossible.

Questions proposed in 1764, *and answered in* 1765.

I. QUESTION 518, *by Mr.* Tho. Sadler.

Old Simon's dead, and Margery is found
A bucksome widow with a thousand pound;
'Cause she hath gold, many a courting go,
Both old and young, the clown and fribbling beau—
She treats them kindly, ladies, you must know.
Her age, and number of sweethearts you'll find,
From the equations which are here subjoin'd.

$$\frac{xy^{\frac{3}{2}}}{\sqrt{(60x + 5y)}} = 71{\cdot}5544. \qquad x = \text{her age.}$$

$$xy - y = 780. \qquad y = \text{her number of sweethearts.}$$

Answered by Mr. Isaac Tarratt.

From the 1st given equation, squared, &c. is derived $x^2y^3 = 60aax + 5aay$ (putting $71{\cdot}5544 = a$, and $780 = b$), and from the second, $x = \frac{b+y}{y}$: Whence $\frac{bb + 2by + yy}{yy} \times y^3 = 5aay + \frac{60aab + 60aay}{y}$; solved, $x = 40 =$ her age; $y = 20 =$ her sweethearts.

II. QUESTION 519, *by Mr.* Rich. Gibbons.

Required the values of x and z, when $x^{z^{\frac{1}{2}}} = z^{x^{\frac{2}{3}}} = 100$?

Answered by Mr. Tho. Walker.

Let X denote the log. of x, Z the log. of z, and L that of 100, and then $z^{\frac{1}{2}} \times x = x^{\frac{2}{3}} \times z = L$ (per the nature of logarithms and the quest.); whence $z = L \div x^{\frac{2}{3}}$. Assume $x = 47 \cdot 5$, and then $x^{\frac{2}{3}} = 13 \cdot 115$, and consequently $\frac{2 \cdot 000000}{13 \cdot 115} = z = 0 \cdot 152421$ the log. of $1 \cdot 4215 = z$, according to this assumption; but $(1 \cdot 4215)^{\frac{1}{2}} \times x$ is $= 1 \cdot 9986 =$ log. of 99·69, which is too little by 0·31: therefore now, suppose x to be $= 47 \cdot 6$, and then z comes out $= 1 \cdot 42$, and the 2d error $= 0 \cdot 16$; whence 0·15 : 0·1 :: 0·16 : 0·106, &c. which added to the last assumed value of x, gives 47·706, &c. for its true value, very near; and thence z is found $= 1 \cdot 42$, &c. And thus, by repeating the operation, we may arrive at any degree of accuracy required.

XII. QUESTION 520, *by Mr.* T. Barker. *addressed to Mr.* M. Hitchins.

SIR,

Your kind advice I wou'd, with joy, obey'd,
But 'tis too late! I've lost the lovely maid,
That charming form, I thought would make me blest,
Is in another's cold embraces prest.
Ingenious algebraist (well known to fame)
Tell all the world my hateful rival's name.*

* From the given equations, viz.
$$\left\{\begin{array}{r} m + w + x + y + z = 37, \\ w^2 - x + y - z = 35, \\ wz + w + z = 53, \\ xy + wz = 59, \\ (my - x) \times wz = 3000. \end{array}\right\}$$

In which the values of m, w, x, y, and z, denote the places, in the alphabet, of the letters composing the required name.

Answered by Mr. Malachy Hitchins.

Could'st thou, dear Baker, sing of Hervey's DEATH,
And not perceive the fleeting state of breath,
Nor know delays to dang'rous issues tend,
Ere thou hast lost thy lover and thy friend?
'Tis now in vain thy mournful tale is told!
Thy charmer's lost!—and her more charming gold!

Mr. *Paul Sharp*, by an easy process, derives the two following equations, viz.

$(43y - xyy - xy - yy - x) \times (59 - xy) = 3000$, and

$$24xy + yxx + xxyy + 6y - 6x - xyy - \frac{(88 + x - y)^2}{xy - 5} = 233;$$

from which, by the method given by the late truly ingenious and profoundly great mathematician Mr. Thomas Simpson, on page 82 of his Essays, he finds $x = 1$ and $y = 19$; and from thence is readily found $m = 4$, $w = 5$, and $z = 8$; which point out the ingenious proposer's rival to be DEATH.

But as solutions this way turn out somewhat troublesome and uncouth, the following method, in point of ease and expedition, may perhaps claim the preference, viz.——The 3d given equation ($wz + w + z = 53$) being evidently no more than to find two whole positive values of w and z such, that their sum added to their product may be $= 53$, it appears, from very little consideration, that they must be 5 and 8, and consequently that their product (wz) is $= 40$; whence it, with certainty, follows (from the 4th given equation) that xy is $= 19$, and consequently that the number 1 must appertain to either x or y, and 19 to the other of them; and thence m is, certainly, found $=$ 4 (from the 1st given equation); whence it also, certainly, follows that $x = 1$, because if it be supposed $= 19$ (and either 1 or 19 it must be) then the last given equation will become impossible. In like manner it appears that w must be $= 5$, because if it be taken $= 8$ (and one of them it must be) then the 2d given equation becomes impossible. Hence the values of m, w, x, y, and z are found to be 4, 5, 1, 19, and 8, answering to the letters (of the alphabet) D, E, A, T, H, respectively.

IV. QUESTION 521, *by Mr.* Joseph Walker.

From the equation $(x + 1) \times (x^2 + 1) \times (x^3 + 1) = 30x^3$, to find the value of x, by means of a quadratic only.

Answered by Mr. Tho. Barker.

From the given equation, the following one viz. $x^6 + x^5 + x^4 - 28x^3 + x^2 + x + 1 = 0$, is readily derived; which, divided by x^3 becomes $x^3 + x^2 + x - 28 + \frac{1}{x} + \frac{1}{x^2} + \frac{1}{x^3}$, or, putting $x + \frac{1}{x} = z$, $(zz + 4z + 10) \times (z - 3) (= z^3 + z^2 - 2z - 30) = 0$: Whence $z - 3 = 0$, and $z = 3$; which is the only possible value of z, in the equation, and which, substituted for z in the assumed equation $x + \frac{1}{x} = z$, gives $x + \frac{1}{x} = 3$; whence $xx - 3x = -1$, and $x = \frac{3 \pm \sqrt{5}}{2} = 2{\cdot}618$, or $0{\cdot}382$; both which numbers answer the conditions of the question.

V. QUESTION 522, *by Mr.* Wm. Embleton.

To divide the given trapezium ADCF into two parts AE, DE (geometrically) by the right line BE, so that AB : EC :: AE : DE :: m : n, i. e. in the given ratio of m to n.

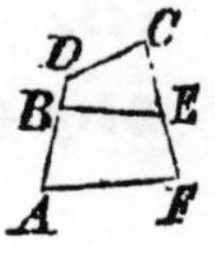

Answered by Mr. C. Hutton.

If the sides FC, AD, of the given trapezium ADCF, be produced to meet in P, the △PDC will be of a given magnitude; and since the said trapezium is to be divided in a given ratio, by the right line BE, each of the parts DE and AE will also be of a given magnitude and consequently the △PBE will be of a given magnitude, which call aa. Then the question is reduced to this, viz. to draw the right line BE so, that AB may be to EC : : $m : n$, and the △BPE = the given square aa.

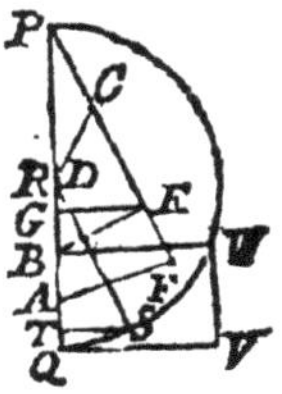

Construction. In PA produced, take AQ to PC in the given ratio of m to n, and on the diameter PQ describe the semicircle PWQ; and from the centre R thereof draw the rad. RS ∥ PF and ST ⊥ PQ. Then if QV be made ⊥ PQ, and = a mean proportional between ($m \div n$) PQ and $aa \div$ ST, and VW and WB be drawn ∥ PQ and QV respectively, we shall have B for one point through which the required line must pass: and if CE be taken to AB in the given ratio of n to m, E will be the other.

Demonstration. Draw EG ⊥ PA: then, by the sim. triangles TSR, GEP, SR : ST : : EP : EP × ST ÷ SR = EG; whence (triangles being the halves of their circumscribing parallelograms, per Elements of Geom.) $\frac{1}{2}$PB × EP × ST ÷ SR or PB × EP × ST ÷ PQ (2RS being = the diameter PQ) will be the measure of the △BPE. But (per construc.) PB × (mEP ÷ n) (= PB × BQ) is = BW^2 = QV^2 = (mPQ ÷ n) × aa ÷ ST (per construction); and consequently each of these equal magnitudes, or spaces, being multiplied by (n × ST) ÷ m × PQ (or repeated that number of times), aa will be found = PB × EP × ST ÷ PQ = the measure of the triangle BPE (the same as found above).

VI. QUESTION 523, *by Mr.* Wm. Spicer.

In a quadrangular piece of ground, whereof the sides are 39, 47, 68, and 57 in a successive order, and the diagonal joining the two shortest of them, 64, 'tis proposed to form a square fish-pond, having its angular points situated in the sides thereof; required a method how to effect the same.

Answered by Mr. C. Hutton.

This quest. may be constructed from prob. 41, on page 246 of Simp. Geom. 2d edit. And if the sides AC, MN, of the given trapezium AMNC, be produced to meet in B, the construction here will be the same as in that prob. (only observing that here C and consequently E fall between A and B); for which reason it seems needless to repeat the construction.

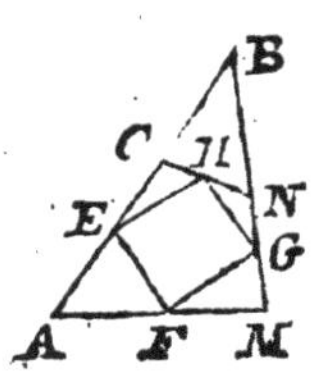

VII. QUESTION 524, *by Mr.* Tho. Barker.

Four spires standing directly north and south, and at the respective distances of 2, 3, and 4 miles from each other, were observed by a traveller, on a road tending to the north-east, the 1st and 2d, and also the 3d and 4th, appearing under equal angles, which they also did a 2d time, after travelling two miles farther on the same road; required his distance from them at each observation?

Answered by Mr. C. Hutton.

From the construc. on page 238 of Simp. Geom. 2d edit. it appears that the locus of concourse of lines drawn from the given points (or places of the spires) A, B, C, D, so that the angles subtended by AB and CD may be equal, will be the circumf. of a circle; which circle IPQ, let be described by that prob.

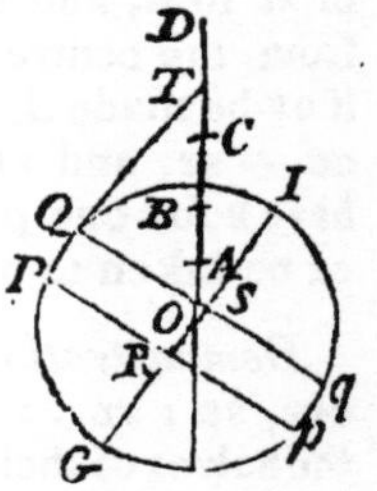

Then to find in which 2 points a right line, drawn in the given direction (north east), shall cut the circumf. of this circle so, that the part of it intercepted by those points may be of the given length (2 miles), through the centre O, draw GOI in the given direction, and on it, from O, each way, set off OR, OS, each = (1 mile) half the given distance of the stations: Then through R and S draw PR and QS ⊥ GI, and the points P, Q, of their intersection with the circumf. of the said circle, will be the two stations (in the road PT) required; which is too evident to need any further demonstration.

VIII. QUESTION 525, *by Mr.* Paul Sharp.

Given the area of a curve $= 198{\cdot}9333 \times aa$ whose equation is $aaxx + y^4 - aaxy = 2axyy$; required the content of the solid generated by the rotation of the said curve about its axis?

Answered by Mr. Rob. Butler.

By transposing the given equation, and extracting the square root, it becomes $yy - ax = \pm a\sqrt{xy}$, or, putting $\frac{b^2x}{z^2} = y$, $\frac{b^4x^2}{z^4} - ax = \pm \frac{abx}{z}$; whence $x = \frac{az^4}{b^4} \pm \frac{az^3}{b^3}$, and $y = \frac{az^2}{b^2} \pm \frac{az}{b}$ (where b may be any quantity at pleasure, and so the curve may be constructed). Suppose $b = 1$; and then $\dot{x} = 4az^3\dot{z} \pm 3az^2\dot{z}$, and $y\dot{x}$ (the flux. of the area) $= 4a^2z^5\dot{z} \pm 7a^2z^4\dot{z} + 3a^2z^3\dot{z}$, and the correct fluent thereof $(\frac{2}{3}z^6 \pm \frac{7}{5}z^5 + \frac{3}{4}z^4 - d) \times aa = 198{\cdot}9333\, aa$ per quest. (d being $=$

$\frac{2}{3} \pm \frac{7}{5} + \frac{3}{4})$; from whence, when a is expressed in numbers, z, and consequently x and y, may be found in numbers also.

In like manner (putting $p = 3{\cdot}1416$) the flux. of the solidity ($pyy\dot{x}$) is found $= p \times (azz \pm az)^2 \times (4az^3\dot{z} \pm 3az^2\dot{z})$, and consequently $pa^3 \times \left(\frac{1}{2} z^8 \pm \frac{11}{7} z^7 \pm \frac{5}{3} z^6 \pm \frac{3}{5} z^5 - n\right) =$ the solidity itself (n being $=$ the sum of the coefficients of the powers of z, &c.).

IX. QUESTION 526, *by* Geometricus.

Required the nature, area and description, &c. of the curve, whose abscissa being the same with that of a given circle, its ordinate shall be every where equal to the difference between the corresponding ordinate and abscissa of the said circle.

Answered by Mr. Da. Kinnebrook.

Let ADB represent the given semicircle, and ACE the required curve, which, it is easy to perceive, will pass through the centre C, of the circle, and meet a tangent to it at B, in the point E, whose distance from B is $=$ the given diameter AB.

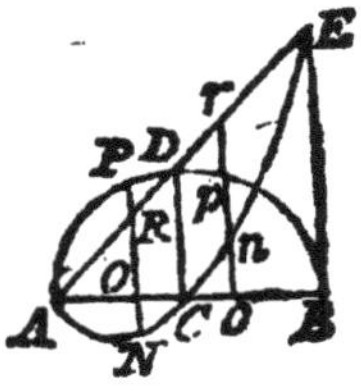

Join the points A, E, and draw the ordinates PO, po, &c. cutting AE in R, r, &c. and AB in O, o, &c. respectively, and take ON, on, &c. thereon every where $=$ PR, pr, &c. and N, n, &c. will be the points in the required curve.——For AB $=$ BE (per nature of the quest. and construct.) and consequently AO $=$ OR, &c. (per sim. triangles) whence PO $\backsim$ AO $=$ PR $=$ NO, &c. (per const.).

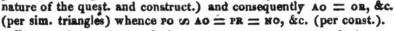

Put, now, AO (or Ao, &c.) $= x$, AB $= 2a$, NO (or no, &c.) $= y$, and then (per prop. of circle and question) $\sqrt{(2ax - xx)} =$ PO, and y (NO, &c.) $= \sqrt{(2ax - xx)} \backsim x$, and $(x \pm y)^2 = 2ax - xx = (2a\sqrt{2} - x\sqrt{2}) \times aax\sqrt{2} \div 2aa$: Whence $(x \pm y)^2 : x\sqrt{2} \times (2a\sqrt{2} - x\sqrt{2}) :: aa : 2aa$, i. e. NR^2 : AR $\times$ RE :: CD^2 : AD^2; and hence it appears, that the required curve ACE is an ellipse, whose centre is D (the rad. DC being perpendicular upon the diameter AB), and AE a diameter thereof, the conjugate to that diameter being DC.——And if an infinite number of parallels to PN be conceived to be drawn, the circular and elliptic areas APDR, ANCO will be divided into an infinite number of parallelograms, whose bases PR, NO, &c. and altitudes are equal; and therefore the circular area APDR $=$ the elliptic area ANCO.——In like manner the area DEB is proved to be $=$ the area CEB; and thence the semi-ellipse ANCED appears to be $=$ the semicircle AGBD.

Mr. *T. Allen*, after proving the nature and description, &c. of the required curve to be the same as above, puts AO, or A*o*, &c. $= x$, the corresponding ordinate ON, or *on*, &c. $= y$, the rad. AC$= a$, and finds the fluxions of the spaces ANO, $cno = \dot{x}\sqrt{(2ax - xx)} - x\dot{x}$ and $x\dot{x} - \dot{x}\sqrt{(2ax - xx)}$ respectively; and the fluent of the former of them (when $x = a$) is $=$ the measure of the circ. quad. ADC $- \frac{1}{2}$AC², and the correct fluent of the latter $= \frac{1}{2}(xx - aa) -$ the measure of the circ. seg. CD*po* $=$ (when $x = 2a$) $\frac{3}{2}$AC² $-$ the measure of the cir. quad. CDB $=$ the measure of the whole space CBE; to which if that of the space ANC ($=$ the measure of the circ. quad. ADC $- \frac{1}{2}$AC²), be added, the area of the whole curve space ANCEBC will be found $=$ AC²; *a thing somewhat remarkable!*

X. QUESTION 527, *by Miss* Ann Nicholls.

On October the 28th, in the latitude 32° north, I observed the brightest of the Pleiades and Aldebaran, both in the same vertical circle; from whence is required the hour of the night, and greatest latitude possible, where such a phænomenon can be observed?

Answered by Mr. T. Allen.

Let A and B be the places of Aldebaran and the brightest of the Pleiades respectively, P the pole, PC the meridian, and CA the vertical circle passing through the two stars. In the spherical triangle ABP, are given AP the polar distance of Aldebaran $= 73°\ 58'\ 54''$, BP that of the brightest of the Pleiades $= 66°\ 58'\ 54''$, and the included angle, the diff. in right ascension of the two stars $= 12°\ 13'\ 30''$, whence the ∠PBC is found $= 60°\ 7'\ 15''$; then in the spherical triangle PBC are given PB and the ∠PBC as above, and PC the comp. of the given lat. $= 58°$, from whence the ∠CPB is found $= 28°\ 55'\ 40''$, which is the arc that the brightest of the Pleiades is short of the meridian at the time required.

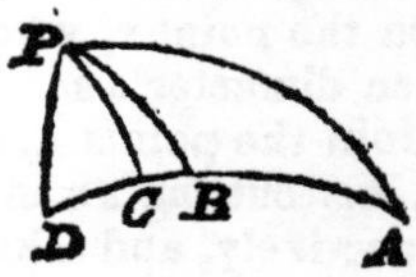

Lastly, the greatest lat. in which this phænomenon can be observed, is, when the ∠PCB becomes $=$ PDB a right angle. Therefore, rad. : sine of PB : : sine of the angle PBD : cosine of 37° 14′ 53″, the greatest lat. required.

XI. QUESTION 528, *by Mr.* Wm. Toft.

At what time on the 10th of June, in the latitude of 12° 30′ north, will the sun's azimuth be the greatest?

Answered by Mr. Wm. Toft, *the Proposer.*

In the orthographic projection of the sphere, on the plane of the meridian, the horizon HO, the equator ÆQ, the prime vertical ZN, the parallel of declination D*d*, &c. will, it is well known, be projected into right lines, and the azimuth circles into ellipses, each intersecting

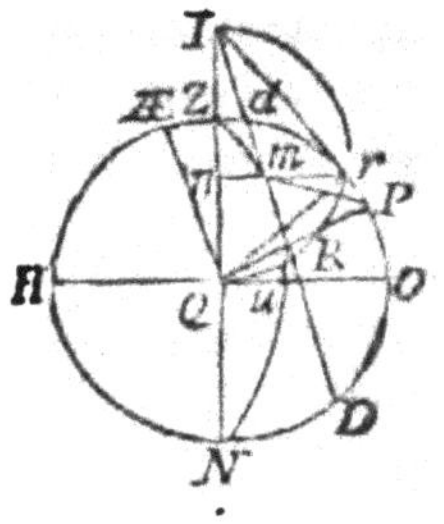

Dd in two points, except that (ZUN) which touches it; and which, therefore, is the azimuth required.

For the sun coming afterwards to the other points of intersection with Dd, will apparently recede back again, and be seen successively on the same points of the compass as before, and so the azimuth decrease again by the same steps it before increased; an appearance surprising enough to those who are ignorant of the true cause thereof.

Therefore to determine the position (or projection) of this azimuth circle, (and consequently to solve the prob.) produce NZ and Dd till they meet in I, and conceive a semicircle to be described upon the diameter IQ, cutting the primitive in r, and let rn be drawn ⊥ NZ, and its intersection m with Dd will be the place of the sun at the time required, or the point through which the projected azimuth circle required must pass.——For, supposing the points m, I; r, I; and Q, r to be joined and the semi-periphery of an ellipse, to the transverse diameter ZN, to be described through the given point m; then will mI be a tangent to it in m, by the property of the ellipse and its circumscribing circle (see theor. 49th of the ellipse in Steel's Conic Sections), because rI is a tangent to the primitive circle in r (QrI being a right angle, by construction).

The method of calculation is from hence also extremely natural and easy; for QR (⊥ Dd) being the nat. sine of the given declin. to rad. of the primitive; QI, and consequently Qn($=$ Qr^2 $\div$ QI, per prop. of the circle) are from thence readily found, the latter of them being the nat. sine of $33° 27'$, the sun's alt. at the time required; and thence nI and nm become known: Whence by another prop. of the ellipse and its circumscribing circle, $nr : nm :: $ QO $: (nm \times$ QO$) \div nr =$ Qu, the nat. cos. $70° 24'$, the sun's azimuth required. And hence the time from noon, when this phænomenon happens, is easily found $= 3^h 55^m$, nearly.

Messrs. *Robert Butler*, *Da. Kinnebrook*, and *Lycidas*, by spherics (from the spherical triangle mZP, where ZP is the complement of the given lat. Pm the complement of the sun's declin. and mZ that of his alt. at the time required), find (sine of ∠ZmP × sine Pm) $\div$ sine PZ $=$ the sine of the ∠PZm, the azimuth required; which, it is manifest, will be greatest when ZmP is a right angle, because the sines of Pm and PZ are constant quantities; from whence the time required is directly found to be either 5^m past 8 in the forenoon, or 55^m past 3 in the afternoon, very near.

XII. QUESTION 529, *by* Plus Minus.

In a certain piece of ground stand two trees, an oak and an ash, distant from each other 10 chains; the piece is surrounded by a ditch of so odd a cut, that if you draw lines from any point of it to the trees, and another from tree to tree, a triangle will be formed, such that the half complement of the angle at the ash to 90°, shall be = the angle at the oak; what is the area of the piece?

Answered by Mr. R. Butler.

Let A represent the ash, O the oak, and with any rad. oq (not exceeding $\frac{1}{2}$OA) and centre q, describe the circumference of a circle cutting OA in n, and on the diam. qA (the complement of oq) conceive the circumference of another circle to be described cutting the former in M, a point in the curve (or figure of the ditch) required; and after this manner any number of other points may be found. For MP being drawn ⊥ OA and the points M, q; M, A; M, n; and M, O joined, the ∠AMP = MqP = double the ∠AOM, as required per the question.

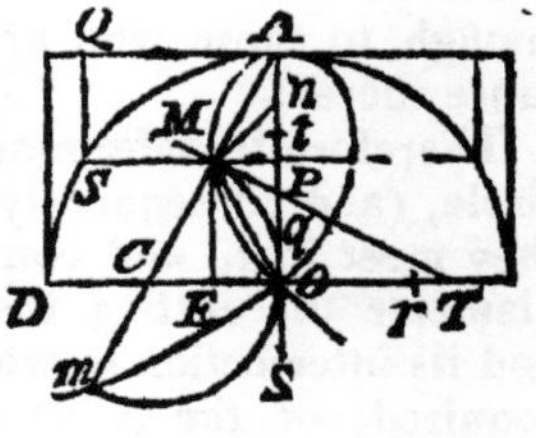

The ditch OMAO being thus constructed, to find the area thereof, put a = OA (= 10 chains), x = abscissa OP, y = ordinate PM; then, per sim. triangles, $a - x$ (AP) : y (PM) :: y (PM) : $\frac{yy}{a-x}$ = Pq;

and $\frac{yy}{x}$ = Pn; whence $nq = \frac{yy}{a-x} + \frac{yy}{x} = oq = x - \frac{yy}{a-x}$, and

consequently $y = x\sqrt{\frac{a-x}{a+x}} = \frac{ax - xx}{\sqrt{(aa - xx)}}$, and $y\dot{x}$ (the fluxion

of the area OMP) $= \frac{ax\dot{x} - x^2\dot{x}}{\sqrt{(aa-xx)}} = \frac{ax\dot{x}}{\sqrt{(aa-xx)}} + \frac{1}{2} \times \frac{aax\dot{x} - 2x^3\dot{x}}{\sqrt{(aaxx - x^4)}}$

$- \frac{\frac{1}{2}aa}{\sqrt{(aa-xx)}}$, and the correct fluent thereof (or area OMP) $aa - a\sqrt{(aa-xx)} + \frac{1}{2}x\sqrt{(aa-xx)} - \frac{1}{2}a \times$ circ. arc whose rad. = a and right sine = x; which, when $x = a$, becomes * $aa - \frac{1}{2}a \times \frac{1}{2} \times 3{\cdot}1416a = 21{\cdot}46$ square chains = the area of the space OMAO; and doubled is 42·92 square chains = the area of the whole space required.

* This expression reducing to $aa \times (1 - {\cdot}7854)$, it may be remarked that OMAO, the semi-area, is equal to the difference between a square whose side is a, and its inscribed circle. Which also agrees with the other solution by *Plus Minus*. H.

The same answered otherwise by Plus Minus, *the Proposer.*

Let A and O be the places of the ash and oak, and M a point in the required curve. 'Tis plain from the data, that the excess of the angle at the fence above 90° = the angle at the oak. Draw MT ⊥ AM, and then TMO = MOA; whence we have the method of describing the required curve, viz draw OD ⊥ AO, and AMC at pleasure: With the centre C, and radius CO describe the circumference of a circle which will cut AC in two points M, m of the required curve. Call AO, a; AP, x; and PM (⊥ AO) y; then (per sim. triangles APM, AOC) CO, or CM = $ay \div x$, and CE $= ay \div x - y$; but $EC^2 + EM^2$ (or PO^2) $= CM^2$, i. e. $2ayy - xyy = x^3 - 2axx + aax$, which is an equation for the hyp^a. defectiva nodata diametrum habens, (see Newton's Enumeratio Lin. 3tii Ord. Species 41^a.) and $y = \frac{\sqrt{x} \times (a-x)^2}{\sqrt{(2a-x)}} = \frac{a}{x^0 \sqrt{(-1+2ax^{-1})}} - \frac{1}{x^{-1} \cdot \sqrt{(-1+2ax^{-1})}}$, which belongs to tab. 2, form 4' cas. 2 and 3, of Newton's Quad. Curv. or $y = \frac{ax - xx}{\sqrt{(2ax - xx)}}$, which belongs to cas. 2 and 3 of form 8; and by either of them, it appears, that if you describe a quad. AD on the centre O, with AO for a radius, and produce PM meeting AD in S, and draw SQ, QA parallel to OA and OD, the area APM shall be equal to the area ASQ; and the whole area of the field $= 2AO^2$ — a semicircle whose rad. is the same AO. That this is the whole of the area is plain, if we look at the equation $2ayy - xyy = x^3 - 2axx + aax$, for when $y = 0$, $x = 0$ or a; so that the trees stand in the fence: And that no other part of the curve beyond O is to be considered as belonging to the field, is also plain; because, by the data, the angle at the fence must never be less than a right one. When y is infinite, $x = 2a$, which gives the asymptote. If you take or, os each $= \frac{1}{2}a$, and set Ar from s to t, you will have the foot of the greatest ordinate.*

* This will easily appear by putting the fluxion of ($y =$) $\frac{ax - xx}{\sqrt{(2ax - xx)}} = 0$. H.

XIII. QUESTION 530, *by* Philotechnus.

A person, aged 35, is desirous of paying 20l. per annum into the insurance office, as long as he lives, on condition that his wife, whose age is 30, shall, after his decease, receive an annuity of 60l. per annum for the remainder of her life after his; required whether the insurer or the insured would have the advantage, interest of money being at 3 per cent?

Answered by Mr. E. Nott.

This gentleman considers the insurer as certain of receiving 20l. per annum during the husband's life, though the wife was to die immediately after the agreement, consequently that (by tab. 5th on page 260 of Simpson's Select Exercises) the present value of his expectation is worth 14·1 years purchase, 282l.

In like manner, the remainder of the wife's life after that of her husband's being (by prob. 15 on page 285 of Simpson's Select Exercises aforesaid) worth 4·7 years purchase, the present value of her expectation will be found to be $60 \times 4{\cdot}7 = 282$l. which, being the same with the other, shews that neither side will have the advantage in this case.

But the proposer, supposing the insurer to be no longer intitled to receive than during the joint existence of the man and his wife, makes the advantage about 22l. on the side of the insured: And if the interest the insurer might make of the money he has a chance of receiving, from time to time, before he is liable to pay any, were to be taken into consideration, the balance would then be found to turn somewhat in favor of him, &c.——His solution is from first principles, but our narrow limits oblige us to omit it.

XIV. QUESTION 531, *by Mr.* Wm. Spencer, *of Stannington.*

To determine the equation, the area, and the length of the curve, whose tangent and subtangent have always the same given difference (d).

Answered by Mr. Rob. Butler.

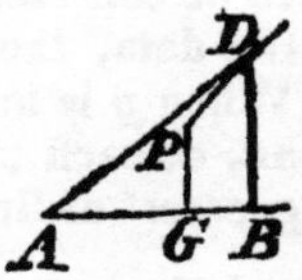

Let PD represent the required curve, whose abscissa GB $= x$, and ordinate BD $= y$, and then $\frac{y\dot{x}}{\dot{y}}$ being $=$ its subtangent AB and $\frac{y\sqrt{(\dot{x}^2 + \dot{y}^2)}}{\dot{y}} =$ its tang. AD, $\frac{\sqrt{y(\dot{x}^2+\dot{y}^2)}-\dot{y}\dot{x}}{\dot{y}} = d$, per question. Whence $\dot{x} = \frac{y\dot{y}}{2d} - \frac{d}{2} - \frac{\dot{y}}{y}$; and y being $= d$ ($=$ GP, the tangent at the vertex) when $x = 0$, per the nature of the question, the correct equation of the fluents will be $x = \frac{yy - dd}{4d} +$ hyp. log. of $\left(\frac{d}{y}\right)^{\frac{1}{2}d}$; the equation of the curve required.

Again, $y\dot{x} = \frac{y^2\dot{y}}{2d} - \frac{d\dot{y}}{2}$, and the correct fluent, or area required, $= \frac{y^3}{6d} - \frac{dy}{2} + \frac{dd}{3}$.

Lastly, $\dot{x}^2 + \dot{y}^2 = \dot{y}^2 \times \frac{(yy - dd)^2}{4ddyy} + \dot{y}\dot{y}$, and $\sqrt{(\dot{x}^2 + \dot{y}^2)} = \frac{y\dot{y}}{2d} + \frac{d}{2} \times \frac{\dot{y}}{y}$; and the correct fluent $\frac{yy - dd}{4d} +$ h. log. of $\left(\frac{y}{d}\right)^{\frac{1}{2}d}$ is $=$ the length of the required curve PD.

Corollary. It hence appears that the curve's length always exceeds the axis by the quantity $d \times$ hyp. log. $(y \div d)$.

XV. QUESTION 532, *by Mr.* Rob. Butler.

To find the sun's longitude, when his declination alters the fastest possible.

Answered by Mr. T. Allen.

Let AEPB (fig. 1.) be one-half of the ellipsis in which the earth revolves about the sun, in the focus S, P the place of the perihelion, and ESm an elliptic sector described in an indefinitely small particle of time t'. Put CP $= a$, AC $=$ CB $= b$, CS $= c$, and let the cosine of ESP, the earth's distance from its perihelion $= x$, rad. $= 1$. Then, by the properties of the ellipsis, ES $= \frac{bb}{a+cx}$, and (by trigon.) E$p = \frac{bb\sqrt{(1-xx)}}{a+cx}$, and S$p = \frac{bbx}{a+cx}$. Therefore, $\frac{bb\sqrt{(1-xx)}}{a+cx} : \frac{bb}{a+cx}$, or $\sqrt{(1-xx)} : 1 :: \frac{abb\dot{x}}{(a+cx)^2}$ ($=$ fluxion Sp) : $\frac{abb\dot{x}}{\sqrt{(1-xx)}\times(a+cx)^2}$, the fluxion or increase (En) of the true longitude in the circular arc EP, whose rad. is ES, in the time t'.

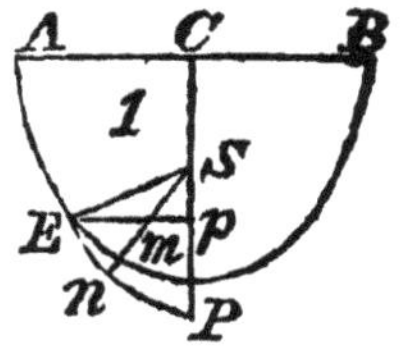

Now, let ♎♑♈ (fig. 2.) represent the southern semicircle of the ecliptic, ♑ the Tropic of Capricorn, P the place of the perihelion, ⊖ the place of the earth in the ecliptic at the time proposed; then will ⊖Q $=$ the sun's declination. Put m and n $=$ the sine and cosine of ♑P, the distance of the perihelion from ♑, p and q $=$ the sine and cosine of ⊖♎Q, and the cosine of P⊖ $= x$ (as above), then will $pnx + pm\sqrt{(1-xx)}$ $=$ the sine of Q⊖ (see art. 478 of Simpson's Flux.) and

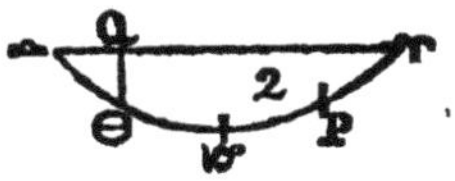

$$\frac{pn\dot{x}\sqrt{(1-xx)} - pmx\dot{x}}{\sqrt{(1-(pnx+pm\sqrt{(1-xx)})^2)}\times\sqrt{(1-xx)}} = \text{the fluxion or}$$

alteration of the declin. in the time t'. Consequently

$$\frac{\sqrt{(1-(pnx+pm\sqrt{(1-xx)})^2)}}{(n\sqrt{(1-xx)}-mx)\times(a+cx)^2}$$ by the quest. must be a min. By making the fluxion of which $= 0$, the value of x may be found.

XVI. QUESTION 533, *by Mr.* T. Allen, *of Spalding.*

Let one end A of a straight inflexible rod AB, 60 feet long, be laid on an horizontal plane, and the other end B on a hemisphere (whose diameter $=$ 30 feet) standing with its base on the said plane, and suppose the end A to advance, in a right line, along the plain, until it reaches and touches the said hemisphere: required the greatest distance the end B will have receded from this position of the rod, and the area of the curve space described thereby, and bounded by the said perpendicular position of the rod, and another right line drawn from the point B, at the commencement of the motion, perpendicular to the same?

Answered by Plus Minus.

Let AB ($=$ 60 feet $= a$) represent the first position of the given rod, and $\alpha\beta$ any other position of it, touching the given hemisphere DBK in t, and draw the rad. ct, and OP, Pβ (perpendicular and parallel to CD), DG ($=$ AB) a tangent to the given circle at D, αE ($\parallel$ DG) meeting βP (produced) in E, and or, through B, perpendicular to CP and DG; and then putting ct ($=$ CD $= 15$ feet) $= r$, CP $= x$, and P$\beta = y$, x (Eα) will be to a ($\alpha\beta$) :: r (ct) : $ar \div x =$ Cα,

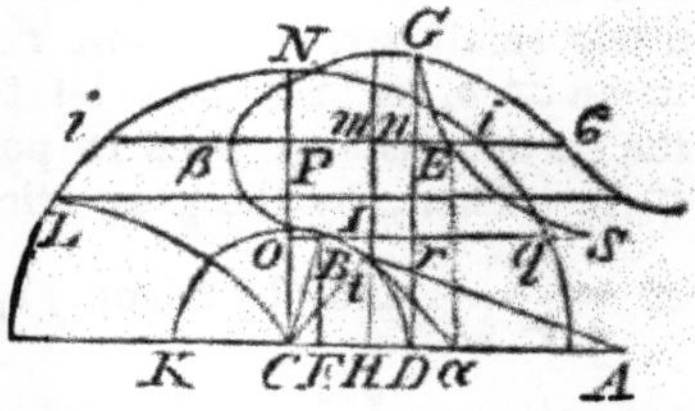

and βE $= y + ar \div x$: but $\beta E^2 + \alpha E^2 = \alpha\beta^2$, and therefore $y = - ar \div x \pm \surd(aa - xx)$; consequently, if about C, with the rad. AB, the circumference of a circle be described cutting CP, βE and Br, produced, in N, i, and q, and also an hyperbola through the point G to the asymptotes CP, CA, meeting Bq, produced, in S, the area GβBr will be $=$ the circular area Nqo — the hyperbolic area Grs $= 1369{\cdot}58$; and the area of any portion βG6 thereof, cut off by a right line $\beta 6 \parallel$ CA will be $=$ the circ. seg. iNi. ——The fluxion of y or $- ar \div x \pm \surd(aa - xx)$, being equated to nothing, gives $x^3 = \pm ra\surd(aa - xx)$; therefore if the cubic parabola, whose equation is $x^3 = ray$, be described through C, it will cut the aforesaid circle in L, the point through which the ordinate passes, which is a maximum ($= 38{\cdot}02$) on one hand, and gives the point of contrary flexure on the other.

XVII. QUESTION 534, *by* Curiosus.

Suppose one end of a thread AB, 100 feet long, is fastened to the circumference of a circle (or cylinder) whose diameter $=$ 4 feet, and, being drawn tight, in the direction of the radius thereof, a ball, weighing 5 pounds, fastened to its other end, and impelled, with a velocity of 20 feet per second, in a direction perpendicular to AB, by a force acting in the plain of the said circle; required the ball's position when it has been 45 seconds in motion, the tension of the string, &c.

Answered by Mr. Da. Kinnebrook.

The ball, in the first part of its motion, moves through the quadrant of a circle, and in the latter part through the involute of the given circle or cylinder; in both which cases the thread is perpendicular to the curve, and therefore cannot affect the ball's motion, and consequently it moves uniformly through both curves. Now the length of the quadrant, to rad. 100 feet, is $= 157{\cdot}08$; whence $157{\cdot}08 \div 20 = 7{\cdot}854$ seconds, the time of its description, and $45'' - 7{\cdot}854'' = 37{\cdot}146''$ the time of the ball's moving in the involute; whence $37{\cdot}146 \times 20 = 742{\cdot}92$ the length of the involute. Put now $a = 100$, $b = 2$, $x =$ the rad. of involution, and $l =$ the length of the involute; then (per page 163 of Simpson's Flux. 2d. edit.) $\frac{1}{2}(aa - xx) \div b = l$; whence $x = \surd(aa - 2bl) =$ (when $l = 742{\cdot}92$) $83{\cdot}836$ $=$ the radius of involution, or the part of the thread or string remaining unwound at the end of 45″: from whence the position of the ball may be found, &c.——The tension of the string is $= (5 \times 20') \div (32\frac{1}{6} \times 83{\cdot}836) = \frac{3}{4}$lb. nearly.

THE PRIZE QUESTION, *by* Philalethes Londinensis.

Supposing the sphere to be projected on the concave surface of an infinitely long cylindric tube of paper, &c. touching it every where on the equator, by the eye at the centre, and afterwards opened, through any of the projected meridians, and spread or stretched upon a plane; required the figure of the rumbs in that projection?

Note. This is proposed with an intent to rectify an error in a mathematical tract wrote by the late Rev. Mr. West, of Exeter, and published anno 1762, which, if unnoticed, might prove of fatal consequence to the navigator, it being pretended to demonstrate there, that the rumbs, in this projection, will be right lines; but the contrary will be proved hereafter.

Answered by Philalethes Londinensis, *the Proposer.*

Let AGPDB be part of the sphere, its centre C, pole P, part of the equator AB, and PA, PB, Pb meridians, of which Pb is infinitely near PB. Let ADe be a loxodromic line that makes equal angles with all the meridians through which it passes, of which let t be the tangent to the rad. of the sphere AC $= r$, and put AB (= AB′) $= y$, AG $= l$, its tangent AG′ $= x$, and then $\sqrt{(rr + xx)}$ (CG′) : r (CA) :: $\dot{y}$ (Bb) : $r\dot{y} \div \sqrt{(rr + xx)} =$ Da, and $r : t :: \dot{l}$ (ea) : $r\dot{y} \div \sqrt{(rr + xx)}$ (Da); whence $\dot{y} = t\dot{l}\sqrt{(rr + xx)} \div rr$: but $\dot{l} = rr\dot{x} \div (rr + xx)$, and therefore $\dot{y} = t\dot{x} \div \sqrt{(rr + xx)}$, and (putting $2{\cdot}30258 = $ M) $y = t$M $\times$ log. $(x + \sqrt{(rr + xx)}) \div r$, the equation expressing the relation betwixt x and y on the globe, or which is the same thing, between the abscissa and ordinate in the proposed chart or projection. But $x + \sqrt{(rr + xx)}$ is = tangent of $\frac{1}{2}(90° + l)$; therefore, lastly, $y = t$M $\times$ (log. cotan. $\frac{1}{2}(90° - d)$ — log. r).——The curve may be very easily constructed by the table of logarithmic tangents; and the preceding figure is exact, when the angle of the loxodromic is 45°.

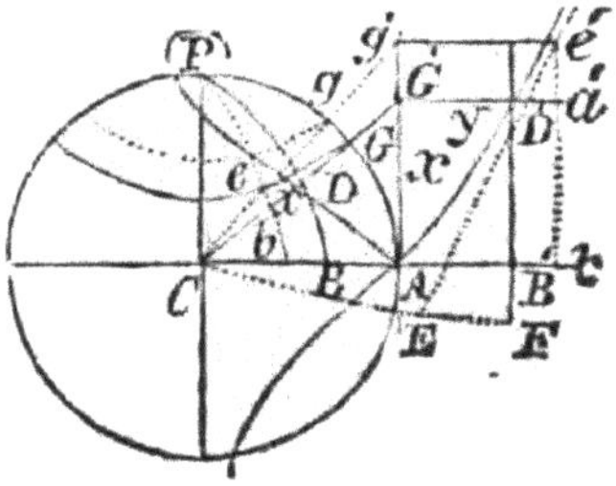

Cor. 1. Because $\acute{e}\acute{a}$: $\acute{a}$D′ :: $\dot{x}$: $\dot{y}$ :: $\sqrt{(rr + xx)}$: t, therefore if through any point D′ of the curve a line be drawn perpendicular to the equator AB′ and the point F be taken therein, so that D′F = G′C, and a line be drawn through F ∥ AB′, on which let the point E be taken such, that FE = the tang. of the constant loxodromic angle, then a line drawn from E to D′ will touch the curve in D′; from whence it follows that all loxodromics in this projection (except those whose directions are north and south or east and west) are mechanical or transcendent curves, having a point of contrary flexure when they cut the equator.

Corol. 2. Since $\dot{y} = t\dot{x} \div \sqrt{(rr + xx)}$, the fluxion of the space AB′ D′ ($= x\dot{y}$) will be $= tx\dot{x} \div \sqrt{(rr + xx)}$, and that space itself $= t\sqrt{(rr + xx)} - tr = t \times$ GG′.

Corol. 3. Since $\dot{y}$ is found $= t\dot{i}\sqrt{(rr + xx)}. \div rr$, if $r\dot{m}$ be put $= \dot{i}\sqrt{(rr + xx)}$ or rm for the sum of all the secants from 0 to l degrees, then will $y = t\dot{m} \div r$, or $ry = mt$, and therefore $r : t :: m : y$. But here m will be the meridional parts of the lat. l for the rad. r; whence the diff. of longitude found by Mercator's sailing corresponds exactly with what is derived from the globe itself.

Questions proposed in 1765, *and answered in* 1766.

I. QUESTION 535, *by Mr.* Tho. Sadler.

Declare ye wits, well known to fame,
A celebrated lady's name;
Whose equal's scarcely to be found
On this terraqueous globe around.
Mount-science she can scale with ease—
Ascend its summit, if she please.
She is the muses' fond delight;
Who're always pleas'd when she's in sight,
Artist's, with ease, her name you'll find
From th' equations here subjoin'd.

viz. $w+x+y+z = 31$, $wz - x - y = 58$, $xy - w + z = 22$, $wy + xz = 157$; where w denotes the 1st letter in the alphabet, x the 2d, y the 3d, and z the 4th, that compose this lady's name; and the 5th letter is the same with the 2d, and the 6th the same with the 3d.

Answered by Mr. Rich. Gibbons.

From the addition of the two first equations is had the sum of the two quantities w, z added to their product $= 89$; which (as they must be whole positive numbers, each not greater than 24, per quest.), it is easy to perceive, will be 4 and 17, and consequently $x + y = 10$ and $xy = 9$, from the 1st and 3d given equations: whence x evidently appears to be 9 and $y = 1$: and these are also equal to the 5th and 6th values required, per quest. Consequently the required numbers are 4, 9, 1, 17, 9, 1, and answer to D, I, A, R, I, A; the lady's name.

The same answered by Mr. J. Ashmore, *and some others.*

If the four given equations are added together, &c. the following one will be derived, viz. $(z + y) \times (w + x) = 268 - 2z$: From which, and the first given equation $(z + y) + (w + x) = 31$, will be found $(z + y)^2 - 2 \times (z + y) \times (w + x) + (w + x)^2 = 8z - 111$; and consequently $z + y = 15\frac{1}{2} + \frac{1}{2}\sqrt{(8z - 111)}$.

Then, as w, x, y, z are to be whole positive numbers, each not exceeding 24 (per question) z must be such a one as will bring out $\sqrt{(8z-111)}$ not only a whole, but also an odd number; which, with very little trouble, will be found to be 17: Whence $y = 1$, and the 1st and 3d given equations become $x + w = 13$ and $x - w = 5$; consequently $x = 9$ and $w = 4$, and the celebrated lady's name required is DIARIA.

II. QUESTION 536, *by Mr.* W. Spicer.

To find two such square numbers, that their sum may be a square, and their difference a cube number, and the sides of the said square and cube equal.

Answered by Mr. W. Spicer, *the Proposer.*

Assume $\frac{1}{2}x^2 + \frac{1}{2}x^3$ and $\frac{1}{2}x^2 - \frac{1}{2}x^3$ for the two required square numbers, whose sum and difference are manifestly a square and cube number, having each the same root x; then, finding two other square numbers $4nn \times (1 + nn)^{-2}$ and $(1 - nn)^2 \times (1 + nn)^{-2}$, such that their sum may be $= 1$, $\frac{1}{2}x^2 + \frac{1}{2}x^3$ will be $= 4nnxx \times (1 + nn)^{-2}$ and $\frac{1}{2}x^2 - \frac{1}{2}x^3 = (1 - nn)^2 \times x^2 \times (1 + nn)^{-2}$, and consequently, the equal root, $x = (6n^2 - n^4 - 1) \times (1 + nn)^{-2} =$ (when n is $\frac{1}{2}$). $\frac{7}{25}$; which value substituted for x in the above assumed expressions, they become $\frac{784}{15625}$ and $\frac{441}{15625}$, two numbers answering the conditions of the question.——Mr. *Wales* denotes them by $9xx$ and $16xx$ (it being evident, from 47 Euc. 1, that they will be in the ratio of 9 to 16); whence (their sum and diff. being $= 25xx$ and $7xx$ respectively) $5x$ will be $= \sqrt[3]{7xx}$ (per quest.) and consequently $x = \frac{7}{125}$, and the required numbers the same as before, &c.

III. QUESTION 537, *by Mr.* W. Toft.

Given ($a =$) the sum of 3 numbers in harmonical proportion, and also ($b =$) their continual product; to find the numbers themselves.

Answered by Mr. S. Woodbey.

Let x, y, and z denote the required numbers, and then, $x + y + z$ being $= a$, $xyz = b$ (per quest.), and $x : z :: x - y : y - z$ (per the nat. of harmonic proportion), $y \times (x + z)$ will be $= 2xz$, or $yy \times (a - y) = 2b$; from the resolution of which, the value of y, &c. may in any case be found.——If $a = 191$ and $b = 254016$, then will y, x, and z come out $= 72$, 63, and 56 respectively.

But Mr. *W. Toft* (the proposer) conceives the three numbers sought to be represented by $z \times (z + v)$, $(z - v) \times (z + v)$ and $z \times (z - v)$ (viz. the numerators of three fractions) $\frac{1}{z-v}$, $\frac{1}{z}$, and $\frac{1}{z+v}$, having their denominators in arithmetic progression, reduced to a common denominator, (it being easy to prove them to be in harmonic proportion), and then, $zz \times (zz - vv)^2$ being $= b$, and $3zz - vv = a$ (per quest.), $2z^3 - az + \sqrt{b}$ will be $= 0$; from whence the value of z, when b is a square number, may be easily found, by the method of divisors, and then v $(= \pm \sqrt{(3zz - a)})$, and the required numbers, will become known.——But when $\sqrt{b}$ is irrational, let it be divided by the greatest square number m^2 it will admit of, and the irrational quotient call n, and then the foregoing equation (supposing $x\sqrt{n} = z$) will become $2nx^3 - ax + m = 0$; from which x may be found, and consequently the values of z and v, and the numbers required.

Example 1. Let $a = 26$ and $b = 576$; then the 1st of the preceeding equations becomes $z^3 - 13z + 12 = 0$; whence $z = 3$ and $v = 1$, and consequently 12, 8, and 6 are the required numbers.

Example 2. Where $a = 13$ and $b = 72$. Then $\sqrt{b} = 6\sqrt{2}$, $m = 6$, and $n = 2$, and the equation in the 2d case becomes $4x^3 - 13x + 6 = 0$; From which $x = 3 \div 2$, $z = 3 \div \sqrt{2}$, $v = 1 \div \sqrt{2}$, and 6, 4, 3 = the three numbers required.

Example 3. Where $a = 39$ and $b = 1944$. Here $\sqrt{b} = 18\sqrt{6}$, $m = 18$, and $n = 6$, and the aforesaid equation becomes $4x^3 - 13x + 6 = 0$, as in the last example, and, of consequence, $x = 3 \div 2$ also: But $z = 3\sqrt{(3 \div 2)}$ whence $v = \sqrt{(3 \div 2)}$, and the required numbers are 18, 12, and 9.

Scholium. The two foregoing general equations being of three dimensions, z and x will each have three values therein, and v as many corresponding ones; as for instance, in example 1, where the values of z are 1, 3 and -4, and those of v corresponding, $\sqrt{-23}$, 1 and $\sqrt{22}$; from whence three sets of terms, answering the conditions of the problem may be found, if those arising from the first and last (viz. from 1 and $\sqrt{-23}$, or from -4 and $\sqrt{22}$) of these corresponding values, can be called real quantities; and, what's somewhat odd, they, as well as the others, answer the conditions of the quest. notwithstanding they are absolutely imaginary: But perhaps neither of them can so properly be called answers, seeing there is one in whole positive numbers, which the problem in some sort tacitly requires.

IV. QUESTION 538, *by Mr.* Tho. Barker.

To determine a point within an equilateral plane triangle, whose perimeter is $= 195$, from which if right lines be drawn to the three angular points thereof, their sum and the sum of their squares shall be $= 113\frac{1}{2}$ and $4337\frac{1}{4}$ respectively.

Answered by Mr. Isaac Tarratt.

Let P represent the required point within the given equilateral plane $\triangle$ ABC, and let the perpendiculars PE, PH, BG, upon AB and AC, be drawn; also let PF be $\perp$ BG, and put $a = 32\cdot5 =$ AG, $b = 56\cdot29 =$ BG, $c = 113\cdot5$, $d = 4337\cdot25$, $h = -4112\frac{1}{4}$, $x =$ BF, and $y =$ PF $=$ HG.

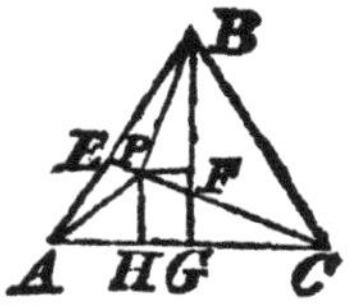

Then (by 47 Euc. 1) $\sqrt{(xx+yy)} + \sqrt{((a-y)^2 + (b-x)^2)} + \sqrt{((a+y)^2+(b-x)^2)} = c$, and $3xx + 3yy - 4bx = h$; whence $y = \sqrt{((4bx - 3xx + h) \div 3)}$; and this value substituted for y, an equation, containing one unknown quantity only, will emerge, from which x comes out $= 40$; and thence y is found $= 5\cdot6$ (fere), and the distances AP, BP, CP $= 31\cdot44$, $40\cdot16$, and $41\cdot43$, very near.

V. QUESTION 539, *by Mr.* Paul Sharp.

In what latitude on the 21st of March at 9 o'clock, will the shadow of an object, standing perpendicular to the horizon, be = to the shadow of the same when standing in a position to the horizon, so as to cast the longest shadow possible at 10 o'clock in the same forenoon.

Answered by Mr. Tho. Bosworth.

Put m, $n =$ sine and cosine of the declin. p, $q =$ cosines of 45° and 30° (the sun's distance from the merid. at 9 and 10 o'clock), x, $y =$ sine and cosine of the required latitude; then, per spherics, $(npy + mx) \times (1 - (npy + mx)^2)^{-\frac{1}{2}}$ and $nqy + mx$ will be the tangent and sine of the sun's alt. at 9 and 10 o'clock respectively; which, whenever an appearance happens, at those times, like what is mentioned in this question, will (as is easy to demonstrate) be equal to each other, or (when the sun is in the line, and $m = 0$, which is the case, very near, at the time specified in the quest.) $py \times (1 - ppyy)^{-\frac{1}{2}} = qy$: From whence $y = \sqrt{\left(\frac{1}{pp} - \frac{1}{qq}\right)} = 0\cdot816496$, the cosine of 35° 15′ 52″ the lat. of the place required; which is the same as found by Mr. *Paul Sharp* (the proposer) by a different method.

VI. QUESTION 540, *by Mr.* Robert Butler.

A gentleman has a triangular garden, whose sides are 50, 60, and 70 poles respectively, in which respectively grow three trees A, B and C, so posited, that if right lines be drawn from tree to tree, a triangle will be so formed that the angles at A, B, and C will be $= 50$, 60, and

70 degrees respectively : Moreover the tree A (in the garden's lesser side) is just 30 poles from the garden's greatest corner. Hence is required the distance of the trees from each other (geometrically), and their situation in the sides of the garden?

Answered by Mr. Rob. Butler, *the Proposer.*

On a right line A*b* (of any assumed length, at pleasure) construct a plane triangle having its ∠s A, *b*, *c* = 50°, 60°, and 70° respectively, and upon A*b* and A*c* describe two segments of circles to contain 78°28′ and 57° 7′ (the garden's greatest angles), and let the diameter AD be drawn, producing it till AD : AL in the given ratio of 2 to 3, and upon AL describe a semicircle cutting the segment A*b* in *e*; through which point and A a right line being drawn, meeting the circumf. AD in *f*, produce it both ways till AF = 20 and AE = 30 poles; join the points *e*, *b* and *f*, *c*, and parallel to those lines let EG and FG (meeting each other in G) be respectively drawn, producing A*b*, A*c*, till they meet the same in B and C; then will EGF be the given triangle, and A, B, and C the places of the trees in the sides thereof.

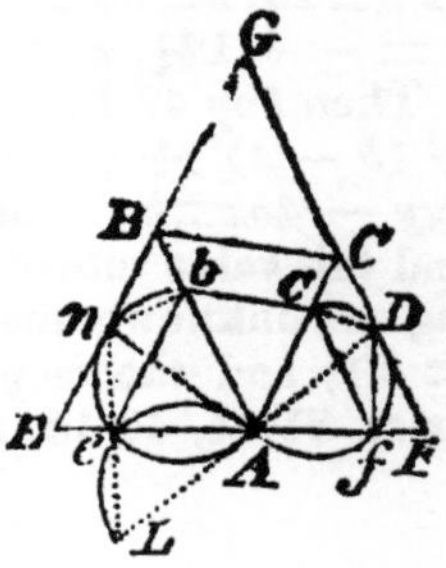

Demonstration. Join the points *e*, L; D, *f*; and B, C: Then per sim. triangles, &c.) A*f* : A*e* (:: AD : AL :: 2 : 3) :: AF : AE (per construction); also (A*b* × AE ÷ A*e* = AB, and (A*c* × AE) ÷ A*e* = AC: Whence AB : AC (:: (A*b* × AE) ÷ A*c* : (A*c* × AE) ÷ A*e*) :: A*b* : A*c*, and consequently BC is ∥ *bc*,* and the △ ABC sim. A*bc*; and the ∠s E, F being respectively = the ∠s *e*, *f* (the given angles per construction) and the side EF = the given side (per construction), EGF will be the given triangle, and A, B, C the true places of the trees in the sides thereof.

Calculation. Draw the diam. A*n*, and let the points *e*, *n*; *b*, *n*; and *c*, D be joined; then (by the common cases of plane trigonometry) AC will be found = 29·92, AB = 32·46, BC = 26·48, BE = 29·89, and CF = 32·46 poles.

* The parallelism of BC, *bc*, may perhaps be proved more geometrically thus: Since, as is proved above, A*f* : A*e* :: AF : AE, or A*f* : AF :: A*e* : AE; but, by sim. triangles, A*f* : AF :: A*c* : AC, and A*e* : AE :: A*b* : AB, ∴ A*c* : AC :: A*b* : AB, and therefore BC ∥ *bc*. H.

Answer to the same by Mr. W. Wales.

Construction. Conceive the given △ DEF to be circumscribed by a circle DKEF, and upon its least side DF constitute a △ DGF similar to the required one. From the vertex E, to the given point A draw the right line EA, and make the ∠ADI = AEF: Then, the intersection I, of the right line DI with the circumference and the point G being joined, make the ∠s EAC, EAB respectively = the ∠s IGF, IGD, and, joining C, B (the point, where the right lines AC, AB meet the

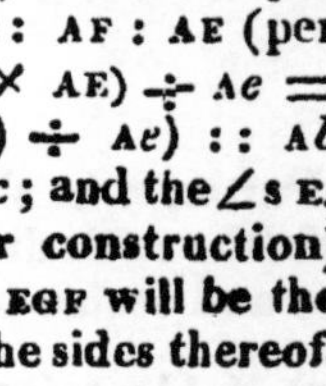

given lines FE, DE), CAB will be the triangle, and A, B, C the places of the trees required.

Demonstration. Produce GI to meet the circumference in K, and let the points D, K and F, K be joined: then (per cor. to 9. 3. of Simp. Geom. 1st. edit.) the ∠s DKF, DEF are equal; and, since the ∠s FDI, AEF are equal (per construction), and FDI, FKI stand on the same arc FI, the ∠s FKI, AEF are equal: Whence the ∠s DKI, DEA are also equal; and, as EAC and IGF, EAB and IGD are equal (per construction), the △s ABE and GDK, ACE and GFK are similar, and consequently AC : FG : : EA : KG : : AB : GD; whence the △s ABC, GDF, having their respective sides proportional, must have their angles equal and be similar.——The given point A being the point of contact of the side DF, with a circle inscribed in the given △ DEF, a very simple construction, in that particular case, may from thence be derived.

VII. QUESTION 541, *by* Jack Hazard.

Three persons A, B, and C toss up, in their turns, and upon equality of skill, a die with two equal faces only, whereof the one is white and the other black, and he that brings up the white face first is to be entitled to a deposit of 30 guineas; Required the value of each person's expectation?

Answered by Mr. Ja. Hazard.

Put the given deposit (30 guineas) $= d$; then, the probability of bringing up a white face at any one throw being constantly $\frac{1}{2}$, the expectation of A on the 1st throw will be $\frac{1}{2} \times d$, which would also be the expectation of B on the 2d throw, were it certain that A would miss the 1st; but the probability of that being also $\frac{1}{2}$, his expectation can only be $\frac{1}{2}$ of $\frac{1}{2}d$. or $\frac{1}{4}d$: In like manner, the value of the expectation of C, on the 3d throw, will be found to be $\frac{1}{8}d$; for, it would be $\frac{1}{2}d$, if it was certain that A would miss the 1st and B the 2d throw, but the probability of that being $\frac{1}{2} \times \frac{1}{2}$, his expectation thereon can only be $\frac{1}{2} \times \frac{1}{2}$ of $\frac{1}{2}d$, or $\frac{1}{8}d$: And by reasoning in this manner, their expectations on the succeeding throws, and consequently the values of their whole expectations (forming three infinite progressions) may be found,

that of A's being $\frac{d}{2} \times : 1 + \frac{1}{8} + \frac{1}{8^2} + \text{\&c.} = \frac{d}{2} \times \frac{1}{1 - \frac{1}{8}} = 18l.$

that of B's $= \frac{d}{4} \times : 1 + \frac{1}{8} + \frac{1}{8^2} + \text{\&c.} = \frac{d}{4} \times \frac{1}{1 - \frac{1}{8}} = 9l.$

and that of C's $= \frac{d}{8} \times : 1 + \frac{1}{8} + \frac{1}{8^2} + \text{\&c.} = \frac{d}{8} \times \frac{1}{1 - \frac{1}{8}} = 4l. 10s.$

Or, the same conclusion may be derived without summation, by considering that, as their corresponding terms are every where in the constant ratio of 1, $\frac{1}{2}$, and $\frac{1}{4}$, their sums will also be in the same invariable ratio.

VIII. QUESTION 542, *by* Curiosus.

An horizontal dial, brought from the coast of France in the late expedition (but without the gnomon), was put into my hands, and being desirous of knowing for what latitude it was made, I found, by means of a pair of compasses, that the angle included between the hour lines of 12 and 3 was exactly = the angle comprehended between those of 4 and 6; from which the latitude may be found, and is here required.

Answered by Mr. W. Wales.

In the spherical triangles PN*a*, PN*b*, and PNC, all right-angled at N, are given their respective angles at the pole P = 45°, 60°, and 90°, and the arc *cb* = *a*N, to find PN, the lat. of the place, the sine of which suppose = x: Then (per spherics) 1 rad.) : x :: 1 (tang. ∠NP*a* = 45°) : x = tang. arc N*a*, and, putting t = tang. ∠NP*b* = 60°, 1 : x :: t : tx = tang. of arc N*b*; whence its cotang. (or the tang. of the arc *bc*) is $1 \div tx$, which, per quest. is = x, and consequently $2xx = 2 \div t = 1{\cdot}15475$, &c. the versed sine of 98° 52′, double the lat. required.

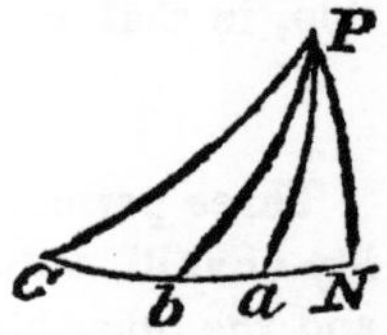

Mr. T. Allen, and several others **by easy short processes, find $x = t^{-\frac{1}{2}} = {\cdot}7598357$ the sine of 49° 26′ 59″, the lat. required; and Mr. *John Potter*, by analogy, without algebra, very ingeniously derives the same conclusion.**

It may also be answered another way independent of algebra, whereby the sine of the arc *ab* comes out = sine of 15° ÷ sine of 105°; from whence the lat. the dial was made for may be readily found.

IX. QUESTION 543, *by Miss* Ann Nicholls.

One night I made observation of an eclipse of the moon, but clouds interposing, I could not see the beginning; however, some time after it cleared up, and at 9h 0′ that evening, I observed her to be exactly 6 digits eclipsed on her lower limb: At 9h 36m 22s the digits eclipsed were 10° 18′, and then clouds appearing again, prevented all further observation. By calculation I found the semi-diameter of the moon = 16m her latitude at the middle = 30m north and her horary motion from the sun = 33m.—From this curious observation, the beginning, middle, and end of this eclipse are required.

Answered by Mr. T. Allen.

Let ABED and CFÆ represent the portions of the lunar orb and the ecliptic, E, B the places of the ☽'s cent. at the 1st and 2d observations,

and AF (= 30′) her lat. at the middle of the eclipse: Then, the ∠CAF being found = 5° 15′ (from Astron. Tables), CA will be = 30·1517′, which put = a, AB = x, BE (= 20′) = c, and CE — CB (= 11·46′) = n; then will $\sqrt{(aa + xx)}$ = CB, $\sqrt{(aa + (c + x)^2)}$ = CE = $n + \sqrt{(aa + xx)}$, and (putting $cc - nn = 2m$), $x = \sqrt{\left(\frac{cc}{4} + \frac{nnaa}{2m} - \frac{m}{2}\right)} - \frac{c}{2} = 11{\cdot}84'$: Whence AE = 31·84′, CE (the semi-diameter of the ⊖'s shad.) = 43·85′, and the measure of the semi-duration AD = 51·7′; and thence are easily deduced $8^h\ 23^m\ 54^s$, $9^h\ 57^m\ 54^s$, and $11^h\ 31^m\ 54^s$ for the beginning, middle, and end of the eclipse, and 11° 8′ 12″ = the greatest quantity eclipsed.

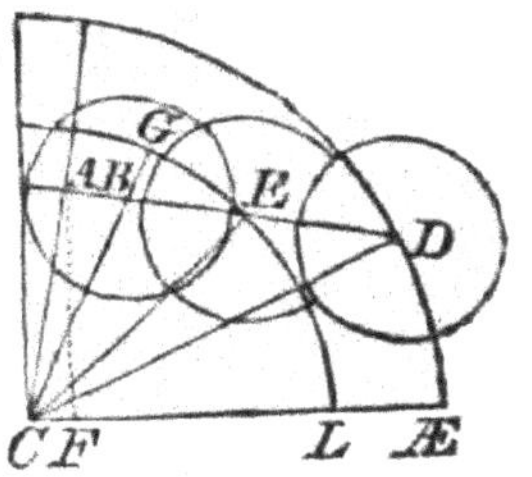

The same answered by Mr. Wm. Wales.

As 60′ : 33′ (the given hor. mot.) :: 36′ 22″ (the time between the observ.) : 20′ = EB = a; also 6 dig. : 16′ (the ☽'s semi-diameter) :: 4 dig. 18′ : 11·466′, &c. = BG = b; whence, putting x = CE (the semi-diam. of the earth's shad.), $a : 2x - b :: b : (2bx - bb) \div a$, half of which = the dist. of CA (⊥ DA and = 30′, the given lat. of the ☽ at A, the middle of the eclipse) from the middle of the base BE of the triangle BCE; whence (per Euc. 47, 1.) $xx - (aa + 2bx - bb)^2 \div 4aa$, = 900 (CA²), and consequently x = 43·6965′, and EA = 31·7708′: Then 33′ (horary motion) : 60′ :: 31·7708′ : 57′ 46″, which added to the time of the first observation, gives $9^h\ 57^m\ 46^s$ for the middle. Also (per 47 Euc. 1.) DA = 51·6107′ = (in time) $1^h\ 33^m\ 50^s$, which subtracted from and added to the middle, gives $8^h\ 23^m\ 56^s$ and $11^h\ 31^m\ 36^s$ for the beginning and end respectively.

X. QUESTION 544, *by Mr.* W. Toft.

A young man, at sea, in working his day's work, forgot to allow for the variation of the compass, and by that means made his difference of latitude too much by 22, and the departure too little by 34 miles; and the distance sailed that day was 100 miles in a direct course, in the north-west quarter: From whence is required the true difference of latitude and departure, and the nature and quantity of the variation.

Answered by Mr. Cha. Hutton.

Construction. About the centre C with rad. CA (= CB ⊥ CA) = 100 (the dist sailed) describe the quad. AB, and take CD, CE thereon, respectively = 22 and 34 (the given err. in lat. and depart.) and let the points D, E be joined; through the centre C and the middle of DE draw the right line Cs, and parallel thereto, and at the dist. of ½DE therefrom, conceive two other right lines to be drawn intersecting the arc AB in T and M; join the points C, T and C, M, and draw TL ∥ AC, meeting CB in L; then is the ∠TCB the measure of the *true* and MCB the *erroneous* course, and the ∠TCM the variation, which is westerly; CL is the *true* diff. of lat. and TL the departure.——For, joining the points T, M, and drawing MS ∥ BC, cutting TL in *r*, and completing the rect. DCE*n*, the △s TM*r* and DCE or DC*n* are equal and similar, because of the equal and perpendicular hypothenuses TM, C*n*, and the parallelism of the legs, *i. e.* T*r* = CE = 34, and M*r* = DC = 22, as by the quest. they ought. The calculation from this construction is extremely easy; whereby the ∠TCM (= variat.) and CL, TL (= *true* diff. of lat. and depart.) come out = 23° 22′, 71·217898 and 70·199816 miles, respectively.

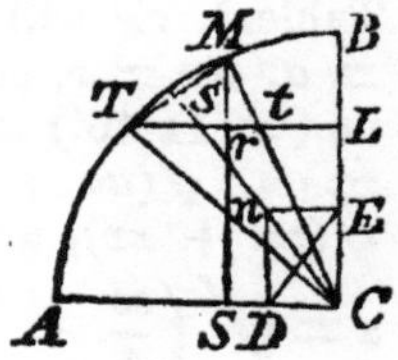

XI. QUESTION 545, *by Mr.* T. Moss.

To investigate a general theorem (without finding the whole content) for determining the true ullage, in ale and wine gallons, of any standing spheroidical cask, whose bung, head diameters, length, and wet inches are given; and that by a rule fit for the practical gauger.

Answered by Mr. Joseph Walker, *Supervisor in the Excise.*

Let h = the head diameter, b = the bung diameter, l = the length, and w = the wet inches of the proposed cask, (all which are given); then will $lb \div \sqrt{(bb - hh)}$ = the transverse of the generating semi-ellipse, and $\frac{1}{2}lb \div \sqrt{(bb - hh)} - \frac{1}{2}l$ = the part thereof intercepted between the vertex of the ellipse and the middle of the head diameter of the cask; and consequently, putting a = ·7854 and x = any variable depth, estimated from the centre of the bottom up the axis, the fluxion of the content corresponding will be $= a\dot{x} \times (bb + \frac{bb - hh}{\frac{1}{4}ll} \times ((l - x) \times x - \frac{1}{4}ll))$; and the fluent, viz. $aw \times (bb - \frac{bb - hh}{\frac{1}{4}ll} \times (\frac{1}{4}ll - (\frac{1}{2}l - \frac{1}{3}w) \times w))$, when $x = w$, divided by 282 for ale, or 231 for wine, gives the ullage, in ale or wine gallons, required; and expressed in words, furnishes the following

General Rule.

Multiply the difference between the semi-length and $\frac{1}{3}$ the wet inches by the wet inches, and subtract the product from the square of the semi-length, and multiply the remainder by the rectangle under the sum and difference of the bung and head diameters divided by the square of the semi-length; then, subtracting this last product from the square of the bung diameter, and multiplying the remainder by the wet inches divided by 359·65 for ale, or 294·11 for wine, the result will be the ullage, in ale or wine gallons, required.

XII. QUESTION 546, *by* Lycidas.

In what point of the ecliptic between ♈ and ♋ does the sun's longitude exceed his right ascension by the greatest diff. possible?

Answered by Mr. T. Allen.

Put $s =$ cosine of 23° 29′, the obliquity of the ecliptic, $x =$ tang. of the ☉'s right ascension to rad. 1; then, per spherics, $x \div s =$ tang. of the longitude, $(x - sx) \div (s + xx)$ (per page 55 of Simp. Trigo.) $=$ tang. of their diff. which, per quest. is to be a max. and therefore its flux. made $= 0$, and reduced, gives $x = \sqrt{s} = \cdot 957693$ the tang. of 43° 45′ 43″ : Whence their greatest diff. = 2° 28′ 35″, and the ☉'s place will be ♉ 16° 14′ 16″.

Astronomicus, by trigonometric reasoning only, finds the sine of the diff. of the ☉'s long. and his right ascension = the rectangle under the sine of their sum and the versed sine of the obliquity of the ecliptic directly, and the versed sine of the supplement of the obliquity of the ecliptic inversely, which, when a max. will evidently become = the rectangle under rad. and the versed sine of the obliquity of the ecliptic directly, and the versed sine of the supplement thereof inversely, = 0·043201, the sine of 2° 28½′, nearly; from which the same conclusion as above exhibited may be found, very near.

XIII. QUESTION 547, *by Mr.* Paul Sharp.

There is a conical tub whose bottom diameter, depth, and top diameter are in arithmetical proportion, whose common difference is 12 inches, and which, when full of water, will empty itself through a circular hole, of 1 inch diameter, in its bottom, in 15·261584 minutes: Quere its content in ale gallons?

Answered by Mr. Paul Sharp, *the Proposer.*

Put $a = 16\frac{1}{12}$ feet, $b = 0{\cdot}005454$ (the area of the circular hole), $c = 3{\cdot}8197$, and x the depth of the tub, in feet; then will $x + 1 =$ top, and $x - 1 =$ bottom diameter, $(3x^3 + x) \div c =$ its content, and $(3x^3 + x) \div bc\sqrt{2ax} =$ the time of evacuation with the 1st velocity (per pa. 173 of Emer. Mechan. 1st edit.): Whence $\frac{3x^3 + x}{b\sqrt{2}}$ $\times \frac{30xx - 20x + 14}{15xx + 30x + 15} = 15{\cdot}261584 \times 60 = 915{\cdot}69504''$ (per page 111 of Emer. Flux. 1st edit.); from which x is found = 4 feet, and the top and bottom diameters = 5 and 3, respectively, and the content = 51·31293 solid feet = 314·42767 ale gallons.

Remark. This solution is false, by the wrong application of the theor. in Emers. Flux. above quoted. The conclusion will be very different when that theor. is properly applied, as any one may easily find. But the same will be more easily and directly deduced from the 1st Art. of our Miscel. cor. 1, pa. 7, thus: Using the same notation as above, we have $\frac{2\sqrt{x}}{b\sqrt{2a}} \times \cdot 7854 \times$

$$\frac{8 \, . \, (x-1)^2 + 4 \, . \, (x-1) \, . \, (x+1) + 3 \, . \, (x+1)^2}{15} = \frac{283\sqrt{x}}{\sqrt{32\frac{1}{6}}}$$

$$\times \; \frac{15xx - 10x + 7}{15} = 915{\cdot}69504; \text{ or } \sqrt{x} \times \left(xx - \tfrac{2}{3}x + \frac{7}{15}\right)$$

$= 18{\cdot}033$: Hence the length x is $= 3{\cdot}4026$, nearly; and therefore the top and bottom diameters $= 4{\cdot}4026$ and $2{\cdot}4026$. H.

XIV. QUESTION 548, *by Mr.* T. Moss.

Supposing there are two given concentric ellipses, whose given difference of transverse diameters is equal to that of their conjugates; 'tis required to demonstrate, in a general manner, whether the part of any common diameter, intercepted between the peripheries of those ellipses, is greater or less than half the said given difference of their transverse or conjugate diameters?

Answered by Mr. T. Moss, *the Proposer.*

Let AaD, EBF be two concentric semi-ellipses, in which EA = Ba = FD. Take ob to OE in the given ratio of oa to OA, and ov : OA :: OB : OE, and conceive the elliptic arcs Eb, Av, respectively similar to Aa, EB, to be described. Draw the semi-diameter om, cutting the similar elliptic arcs in s, n, and r, and let the points r, b; n, v; and s, a be joined; then (alternately) ob : oa :: OE : OA and OB : OV :: OE : OA; whence, by equality and division of ratios, &c. OB — ob : ob :: OV — oa : oa, or alternately, Bb : va (:: OB : oa) :: OB : ov, by equality, from above. Also, by sim. Δs, &c. arising from the similarity of the ellipses,

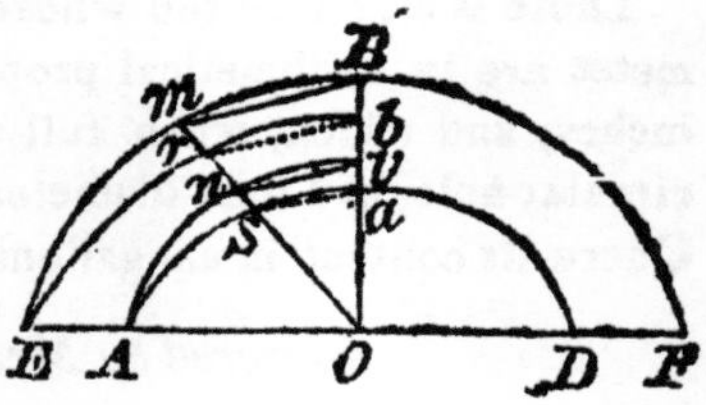

om : on :: OB : ov (:: OE : OA, by supposition),
or : os :: ob : oa (:: OE : OA, by supposition likewise),

and ov : BV :: on : mn; whence, by equality, division of ratios, and alternation, mr : ns (:: or : os :: ob : oa) and so is Bb : va (by equality, from above): Consequently the flux. (or decrease, supposing om to be increasing and indefinitely near to OB) of mr is to the flux. of ns in the constant ratio of Bb to va, and the flux. of om to the flux. of on as OB : ov (because, if the fluents of two variable quantities are always in a constant ratio, their fluxions must necessarily be always in the same ratio): that is, alternately,

BB : flux. of mr :: va : flux. of ns, } and, alternately and
OB : flux. of om :: ov : flux. of on, }

by equality, from above, &c. ov : va :: flux. of on : $(va \div ov) \times$ flux. of on = flux. of ns; also, for the same reason, ov : BV (on : mn, from above) :: flux of on : (BV $\div$ ov) $\times$ flux. of on = flux. of mn; which, as BV is always $\sqsubset$ va, demonstratively proves the increase of mn to be always $\sqsubset$ the decrease of ns, and consequently, that ms is every where $\sqsubset$ Ba, or EA (because, if the said increase and decrease were always equal, ms would be a constant quantity = Ba, and if the decrease was greatest, ms would be $\sqsupset$ Ba; as is very evident).

The certainty of this may also in some measure be evinced by considering the ray om as revolving about the centre O; it being easy to perceive that its intercepted part sm will continually vary, as the inclination to the axis EF varies, and consequently that it must pass through a max. or min. before it can become of the same magnitude again, in E, as it was of, at first, in B.——But it is very easy to demonstrate that it will begin to increase immediately on its departure from the position OB, and be greater than aB the moment before its coincidence, with AE (= aB).——It therefore must pass through a max. between the positions OB, OE, and be every where greater than its two extreme equal values aB, AE.

XV. QUESTION 549, *by Mr.* R. Butler.

A chain's length is 140 feet, and its variable thickness (or law of density) such, that if its ends are fastened to two pins situated in the same horizontal line at the distance of 100 feet from each other, it will dispose itself into a circular form; from whence is required the law of density and the chain's weight, supposing one foot of either end to weigh one pound?

Answered by Mr. Cha. Hutton.

Call the abscissa DE, x; the ordinate EF, y; the arc DF, z; and its tangent t: also call the tangent of the whole arc DB, T. Now, in any curve formed by a perfectly flexible line, the tension or force drawing in direction of the tang. at D (the lowest point) is a constant quantity, which let be denoted by a. Then, the general prop. of any such curve being as 1 (rad.) : tang. $\angle$EFD :: a : weight of the arc DF, (or because the angle at F is, in the circle, measured by the arc DF) 1 : t :: a : at = weight of the arc DF, and 1 : T :: a : aT = weight of the arc DB, aT — at will be = 1lb. (per quest.), supposing DF = 69 feet; whence $a = (\text{T} - t)^{-1}$, and consequently $2a\text{T} = 2\text{T} \div (\text{T} - t)$ = weight of the whole chain ADB.

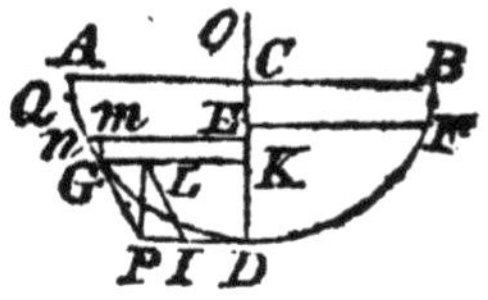

Now, since the arc DB is given = 70, and the sine CB = 50, the rad. OD will be found $= 50{\cdot}99397$, and the length of the semi-circumference $= 160{\cdot}20266$; whence the arcs DB (70) and DF (69) appear to contain 78° 39′ and 77° 31′ 36″, and their tangents T, t to be $= 4{\cdot}9818813$ and $4{\cdot}5210821$, respectively, and $2T \div (T - t) = 21{\cdot}6227$lb. = weight of the whole chain required.——The law of density, or of the section of the chain, is directly as the flux. of the tang. and inversely as the flux. of the arc; for $a\dot{t} =$ flux. of the weight = weight of the length $\dot{z}$; which divided by the length $\dot{z}$, gives $a\dot{t}\dot{z}^{-1} =$ the density or section at any point.

Cor. 1. The law of density is as $a\dot{t}\dot{z}^{-1}$ in all curves; for the manner of finding it is general, supposing t to be the tang. of the angle in which the curve cuts the ordinate.——In the catenary, which is formed by a line of a constant thickness, or whose law of density is a constant quantity, in order to have it as such, in the general expression, $\dot{t}$ must be as $\dot{z}$, and thence t as z; which is a known property of that curve.

Cor. 2. The arc DB can never become a quadrant; because then the density at B would be infinitely great.

Scholium. Hence may be easily deduced the requisites for building the arches of bridges of any known curvature desirable, either for elegance or to suit the situation of the place, &c. and yet be equally strong in all their parts, as if they were catenarian arcs.

Answer to the same by Mr. Rob. Butler, *the Proposer.*

Draw the ordinate GK, and, perpendicular thereto, PL, from P, the point of intersection of the tangents DP, GP; also draw nm, mG, and LI ∥ GK, KD, and GP respectively, and put the abscissa DK $= x$, ordinate GK $= y$, $w =$ weight of the portion DG, and $a =$ the invariable tension at D (the lowest point): Then n being indefinitely near G, Gm $= \dot{x}$, and $nm = \dot{y}$, and, by mechanics, GP, GL, and PL will be as the tension at G and D, and weight of the part DG respectively: Whence $w : a$ (:: PL : GL) :: $\dot{x} : \dot{y}$, and consequently $w \div a = \dot{x} \div \dot{y}$; which is general, let the curve be what it will: But, in the present case, the curve being a circle (whose diam. suppose $= 2r$), $\dot{x} \div \dot{y}$ will be $= y \div (r - x) = w \div a$; that is, the weight ($w$) will be always as the tang. of its corresponding arc DG: Therefore, putting T, t = the tang. of 78° 39′ and 77° 32′, the degrees, &c. contained in the arcs AD (70) and QD (69 feet) respectively, T : t :: w : $w - 1$ (per quest.); whence $w = T \div (T - t) = 10{\cdot}86$lb. which doubled, gives $21{\cdot}72$lb. = the whole weight of the chain required.

THE PRIZE QUESTION, *by Mr.* Da. Kinnebrook.

Suppose a given right-angled plane triangle, whose hypothenuse is an uniform slender rod, and its base parallel to the horizon, to revolve about its perpendicular as an axis, whilst a ring slides freely along its hypothenuse: It is required to determine the time of the ring's descent along the said hypothenuse, its length being = 50 feet, the perpendicular = 40, and the time of one revolution round the axis = 3 seconds?

Answered by Mr. W. Spencer, *of Stannington, in Yorkshire.*

Let BA represent the given rod, revolving about the axis BC perpendicular to the horizontal line AC, D the place of the ring at the end of any time t, and v its velocity, there, in the direction DA.

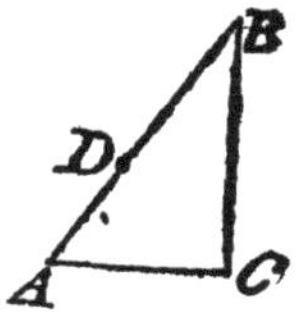

Put AB (= 50 feet) = r, BC (= 40) = p, AC (= 30) = b, $32\frac{1}{6} = g$, 3″ (the time of one revolution) = s, 3·1416, &c. = $\frac{1}{2}c$, and BD = x; then will $bccx \div rss$ = the centrifugal force in a direction parallel to the horizon, and its effect in the direction DA (by the resolution of forces) = $bbccx \div rrss$; which added to $pg \div r$, the true measure of the force of gravity in the same direction, gives $\frac{bbccx}{rrss} + \frac{pg}{r}$ = the whole force accelerating the ring along the rod, at D: Whence, by the principles of motion, $\frac{bbccx\dot{x}}{rrss} + \frac{pg\dot{x}}{r} = v\dot{v}$, and (putting $\frac{bbcc}{rrss} = m$, and $\frac{pg}{r} = n$) $v = \sqrt{(mxx + 2nx)}$; consequently $\dot{t} = \frac{1}{\sqrt{m}} \times \frac{\dot{x}}{\sqrt{(2qx+xx)}}$, putting $q = \frac{n}{m}$; and, taking the correct fluent $t = \frac{1}{\sqrt{m}} \times$ hyp. log. of $\frac{q + x + \sqrt{(2qx+xx)}}{q}$ = (when x becomes = BA = 50) 1·6559″ &c. = 1″ 39‴ 21⁗ &c. the time required.

Cor. 1. The time of descent along the rod in motion, is less than the time of descent along it when at rest, by 18‴ 51⁗ &c.

Cor. 2. When x becomes $= \frac{{}^{*}grss}{pcc}$, the ring, if at free liberty, would no longer slide along the rod, but fly off.

---

* Found by making $bg \div p$ (the gravity of the ring on the rod) = $bccx \div rss$ the horizontal centrifugal force above found. H.

Questions proposed in 1766, *and answered in* 1767.

I. QUESTION 550, *by Mr.* Tho. Sadler.

Ye fair, who can with ease unfold,
What puzzled Oedipus of old,
And can, from algebraic art,
Th' abstrusest of all things impart;
From what you see appear below,*
John's age and fortune you will know;
Who courts young Susan of the Mill,
(But she is more in love with Will.)
Will's young and spruce, but hath no store—
No mouldy sterling to count o'er—
John boasts of gold, and more than that,—
Which makes Sue's heart go pit-a-pat:
She begs the ladies' kind advice,
Were they to chuse, and take their choice,
Whether 'tis best, to marry John,
With all his gold,—or Will, with none.

* $x+y=152$, and $(x-y)^{\frac{2}{3}} \times (x-y)^{\frac{3}{2}} = 8192$; where $y=$ John's age, and x his fortune.

Answered by Mr. Edward Bayley.

Put $x-y=z^6$, and then the 2d given equation becomes z^{13} $(=z^4 \times z^9) = 8192$; whence $z=2$, and $x-y=64$; from which, and the first given equation $(x+y=152)$, is readily found $y=44$, John's age, and $x=108$l. his fortune.

II. QUESTION 551, *by Miss* Ann Nicholls.

I observed a cloud bearing N. E. by E. and took its altitude 17° 4′; some time after I observed the same cloud S. E by S. at which time its altitude was 45° 23′; from whence the point the wind blew from is required?

Answered by Mr. Tho. Vessey, *at Mr.* Allen's *School, Spalding.*

Let P be the place of observation. Draw $Pb \perp PB$, and let BD, bd, be perpendicular to the horizon in the points B and b. Suppose D the place of the cloud at the 1st, and d that at the 2d observation. Put $t =$ tang. of $\angle BPD = 17° 4'$, $T =$ that of $\angle dPb = 45° 23'$, rad. $= 1$; then (per trigonom.) $BD = t \times PB$, and $bd = T \times Pb$; whence, the height of the cloud above the horizon being supposed to continue the same, $t \times PB = T \times Pb$, or $PB : Pb :: T : t$; therefore (the points B, b being joined) the $\angle PBb = 16° 52'$, and consequently the wind blew from N. E. ½ N.

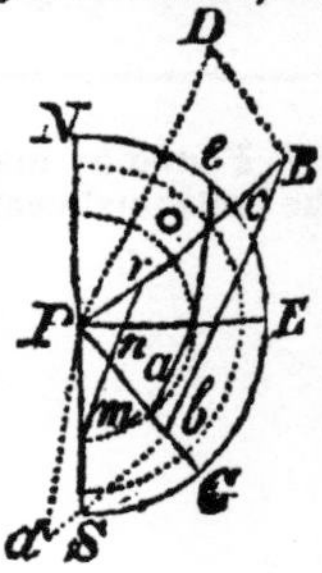

The same answered by Mr. T. Adams, *and others.*

If right lines PC, PG be conceived to be drawn, making ∠s NPC, NPG, with the meridional line NPS, equal to $56\frac{1}{4}°$ and $146\frac{1}{4}°$, the given bearings of the cloud from the north, and the distances *Pr*, *Pn* be taken thereon equal to 3·25729 and ·98671 &c. the cotangents of 17° 4′ and 45° 23′ the cloud's altitude at the two observations, to rad. 1, and the points *r*, *n* be joined, and the right line *rn* produced till it meets NS in *m*, the ∠*rm*N will measure the point of the compass required, which is too evident to need a demonstration, the height of the cloud above the plane of the horizon being the same at each observation; and hence, by trigonom. 4·244 (*Pr* + *Pn*) : 2·27058 (*Pr* — *Pn*) : : 1 (tang. of 45°) : ·53500 = tang. of $\frac{1}{2}$ (∠*Pnr* — ∠*Prn*); ∴ the angle *rm*N = 39° 24′; the same as before.

But Mr. *R. Gibbons*, supposing the direction of the cloud to form the arc of a great circle, gives the following solution. Lay down NC = 5 and SG = 3 points, on the horizontal projection NES, whereof N, E, and S represent the north, east, and south points, and draw the two right circles PC, PG, cutting the parallels of the given altitudes (17° 4′ and 45° 23′) in *o* and *a*; through which points the arc of a great circle *aoe* being described, cutting the horizon in *e*, it will shew the direction of the cloud, and the arc NE, the point which the wind blew from, which is found, * at one operation in spherics, to be = 39° 24′, answering to N.E. by N. $\frac{1}{4}$E.

III. QUESTION 552, *by Mr.* Isaac Tarratt.

A pleasant village, ladies, is defin'd
From the equations underneath subjoin'd,*
Whose pleasing bowers and delightful shades
Are far superior to the woodland glades.
To these retreats the merry nymphs will hye,
With sudden transports of unbounded joy!
And meet their swains, the shepherds of the plain,
Who tune their pipes with a melodius strain:
The choristers will listen there to hear
What Damon says to his beloved fair,
My residence, dear ladies, pray explore,
And you'll oblige your servant evermore.

* $$\left.\begin{array}{l} v + w + x + y + z = 64, \\ v^2 + w^2 = 5 \times (x + y + z) + 30, \\ vwx = 8yz + 6, \\ x^2 + yz = 492, \\ v + w + z = x + y, \end{array}\right\}$$ Where v, w, x, y, and z represent the places of the letters in the alphabet, that compose the town's name.

* Viz. Sine *Po* : tang. *Pa* :: sine *Co* : tang. *Ce* = 16° 51′, which subtracted from CN (= 56° 15′) leaves *Ne* = 39° 24′.

Answered by Mr. B. Sikes.

From the 1st and 5th given equations, $x + y = 32$; and, since v and w are concerned exactly alike in every equation where they enter, let $s = \frac{1}{2}$ their sum, and $r = \frac{1}{2}$ their difference; then, w being $= s + r$ and $v = s - r$, those values substituted in the given equations, by reduction, &c. will be found $ss = \frac{32^2 - 64z + zz}{4} = \frac{5zx + 190x + 16yz + 12}{4x}$: Whence, z and y being exterminated, &c. $x^5 - 53x^4 + 1034x^3 - 10928x^2 + 514320x - 8073216 = 0$; from which x comes out $= 18$, and consequently $y = 14$, and $z = 12$. Hence $w = 15$ and $v = 5$, and the place of Mr. *Tarratt's* residence is Epsom.

Amicus, from the nature of the given equations, very readily determines $x = 18$, and consequently $z = 12$, and $y = 14$; therefore the 1st, 2d, and 3d given equations become $v + w = 20$, $v^2 + w^2 = 250$, and $vw = 75$; from which $v \backsim w = 10$, &c.

IV. QUESTION 553, *by Mr.* Wm. Spicer.

There is a conical spire steeple, whose slant side is = 4 times its base; from the vertex of which, two heavy bodies were let fall at the same time, the one down the perpendicular, and the other along the slant side. Now it was observed that the sound of the body which fell down the perpendicular upon the centre of the base, arrived at the extremity of the slant side just at the same moment of time with the other body; required the steeple's altitude?

Answered by Mr. W. Cole.

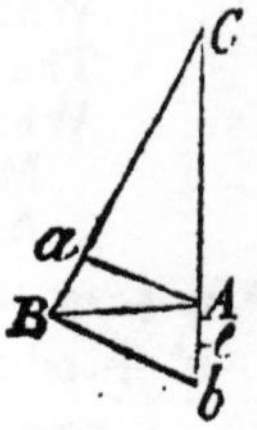

Let AC and BC represent the perpendicular and slant heights of the steeple, respectively. Draw $Aa \perp BC$, and $Bb \parallel Aa$, meeting CA produced in b; then will A, a, and also B, b, be cotemporary positions of the two falling bodies supposing the body at A to continue in motion till the other body arrives at B (as is well known by the laws of descending bodies); and, BC being to BA as 8 to 1, per quest. AB is easily found to be to Ab as $\sqrt{63}$ to 1. Therefore, as $\sqrt{63} : 1 :: 1142$ (the velocity of sound per second) : $1142 \div \sqrt{63}$ = the velocity of the body at e; whence the distance eC will be $= 321\frac{3151}{4053}$, and (per quest.) Ab is found $= \frac{1}{64}$ Cb, and eA $= \frac{1}{127} e$C; therefore, as $127 : 126 :: 321\frac{3151}{4053}$ (eC) : $319\frac{125473}{514731}$ (AC) = the perpendicular height of the steeple required.

Additional Solution.

An algebraic solution to this question may be thus: Put $s = 16\frac{1}{12}$,

$a = 1142$, and $x =$ AB: Then $8x =$ BC, and $x\sqrt{63} =$ AC. By the laws of gravity, $\sqrt{s} : \sqrt{\text{AC}} :: 1'' : \sqrt{(x\sqrt{63})} \div s =$ time of falling through CA, and AC : BC :: $\sqrt{(x\sqrt{63})} \div s : \sqrt{(64x \div s\sqrt{63})}$ = time in CB, also $x \div a$ = time of sound's moving through AB; therefore we have $\frac{x}{a} + \sqrt{\frac{x\sqrt{63}}{s}} = \sqrt{\frac{64x}{s\sqrt{63}}}$: Hence $x\sqrt{63} = a^2 \times (127 - 16\sqrt{63}) \div s = 319{\cdot}249 =$ AC required. H.

V. QUESTION, 554, *by Mr.* Stephen Hodges.

Near the renown'd Lord Viscount Spencer's seat,
Where shady groves project a cool retreat,
Two children at one birth appear'd in view;
From those equations * pray their ages shew.

* $xy^2 + z^3v^3 = 290304$, $x^2u + x^3y^4 = 12521472$, $y = u$, $z = v$, and $v = 2y$; Where x represents the year, y the month, z the day, v the hour, and u the minute A. M. when the eldest was born; and the other was born 4^m after.

Answered by Mr. Isaac Tarratt.

By substituting the values of z, v, and u, in the terms of x and y, we get $xy^2 + 64y^6 = 290304 = a$, and $yx^4 + 8y^7 = 12521472 = b$; from the 1st of which equations x is found $= (a - 64y^6) \div y^2$, and this value substituted in the latter of them, it becomes $(a - 64y^6)^4 \div y^5 + 8y^7 = b$: Solved, $y = 4$. Consequently the eldest was born anno 1760, April the 8th, at 4^m past 8 in the morning, and the other 4^m after that.

If it be considered that x and y must be whole positive numbers, and the latter of them under 12, by the quest. the answer may be easily obtained from the given equations as they stand at first, without any further reduction, &c.

VI. QUESTION 555, *by Mr.* Richard Gibbons.

A tradesman owed his dealer 150l. who agreed to take a certain sum yearly on being allowed lawful interest upon balancing each year's accompt, as the same became due, until the whole was paid, when the debt and interest amounted to 179l. 9s. 10¼d. required the yearly payment, and the time the whole took paying off?

Answered by Mr. Richard Gibbons, *the Proposer.*

The present worth 150l. ($= p$) and the rate of interest 1·05l. ($= r$) being given, put $a =$ the annuity, or yearly payment required, and $t =$ time it was paying off; then there is further given $ta = 179{\cdot}4925$l. to find t and a separately. By the doctrine of annuities and the question, $atr^t - at = ptr^{t+1} - ptr^t$, or, in numbers, &c. $179{\cdot}4925r^t$

$-7{\cdot}5tr^t = 179{\cdot}4925$: Whence $23{\cdot}93233r^t - tr^t = 23{\cdot}93233$: Solved, $t = 6{\cdot}527$ years, and $a = 27l.$ 10s. the annual payments.

VII. QUESTION 556, *by Miss* Ann Nicholls.

There is in lat. 54° north, a town, and in lat, 64° north, a mountain, both under the same meridian, and on Dec. 21st the sun was observed to rise at the same moment of time at the town and on the mountain; from which data I demand the height of the mountain?

Answered by Mr. Felix M'Carthy, *of* Aberdeen.

In the annexed projection, let P represent the north pole, Z the zenith of the place in lat. 54°, z that of the place in lat. 64°, and ⊙ the sun in the horizon in the former of them, on the day proposed; then, in the spherical triangle Z⊙P, is given all the sides to find the angle ZP⊙ = 53° 17′ 27″. Through ⊙ and z describe the arc of a great circle ⊙z; also describe the arc AP, making ⊙A = ⊙Z, and then a spectator at A will it is evident, see the sun rising at the same time with the other two spectators at Z and T, the point where a tang. to the arc Az, at the point A, meets Cz (the semi-diameter of the earth) produced, *i. e.* at the top of the mountain zT required; to determine the height of which, in the spherical triangle ⊙Pz, are given the sides ⊙P, Pz, together with the included angle, to find ⊙z = 96° 45′ 30″, from which subtracting A⊙ = 90°, there remains Az (the measure of the ∠ACz) = 6° 45′ 36″: And thence CT is found = 4009·887 miles (per plane trig.), from which deducting 3982 (the miles contained in Cz, the earth's rad.) leaves zT = 27·887 miles, the required height of the mountain.

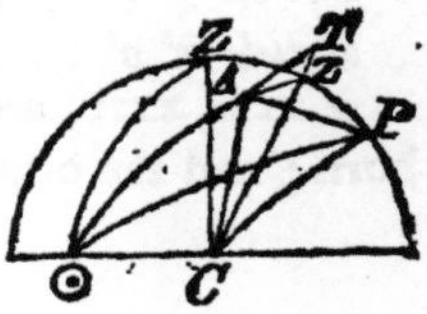

Mr. *Joshua Adams*, allowing 33′ for the ⊙'s horizontal refraction, finds the height = 30·2 miles.

VIII. QUESTION 557, *by Mr.* Tho. Harris.

Say when, * ye astronomic spies,
Bright Sol in the least time will rise?

* Lat. 51° 32′ N.

Answered by Juvenis.

Put a and b = sine and cosine of 51° 32′ (the given lat.), $-c$ = cosine of 90° 32′ 12″ (the dist. of the sun's lower limb from the zenith the moment his upper limb touches the horizon), and x = cosine of the declin. required, rad. = 1; then, per spherics,

$((ax+c)\sqrt{(bb-xx)} - ax\sqrt{(bb(1-xx)-(ax+c)^2)}) \div bb(1-xx)$

= sine of the angle measuring the time the sun is in rising, which (per quest.) is to be a min. In fluxions, &c. and reduced, x comes out = ·0036441 &c. answering to $12\frac{1}{2}'$ nearly, the declination required.

Mathematicus easily reduces this prob. to that of describing the circumference of a circle to touch two circles, given in magnitude and position, and cut the circumference of another circle given in magnitude and position likewise, in a given angle; and from thence readily infers, nearly, the same conclusion as above.

Mr. *Tho. Harris* (the proposer) and Mr. *Isaac Tarratt* observe that "the sun will rise quickest when his azimuth increases the slowest, and that will be when he ascends the horizon in a vertical direction:" and upon this principle they solve it independent of fluxions, and find the declination sought to be 12′ 40″ south; from whence they conclude that the sun rises the quickest the morning before the vernal, and the morning after the autumnal equinoxes.

IX. QUESTION 558, *by* Curiosus.

So to draw a right line, cutting the two sides of a given plane triangle, that the rectangle of the two lower segments shall be equal to a given quantity, and the rectangle of the two upper ones a maximum.

Answered by Mr. T. Moss.

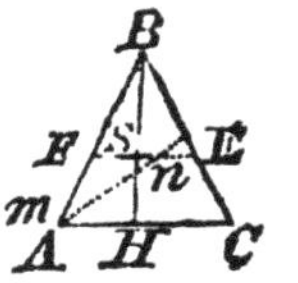

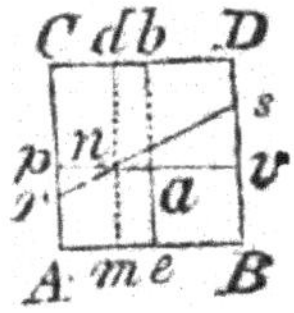

Let ABH (fig. 1st) represent the given plane triangle; produce AH till the right line BC, joining its extreme and the point B, becomes = BA, and upon AB (= one of those equal sides) let a square AD (fig. 2d) be constituted, and conceive a right line rs to be drawn through the same, so that Ar × Bs may be to the given rectangle of the lower segments as BH : BA (fig. 1st) and the area of the part ArsB a min. which, having bisected AB in m and drawn md ∥ AC, it is manifest will be when Ar + Bs is a min. because mn (= $\frac{1}{2}$(Ar+Bs)) × AB is the measure of the space ArsB, and AB is a constant quantity. But it is well known, and may be easily proved geometrically, that the sum of two right lines, under a given rectangle, will be the least possible when they are equal to each other: Whence, taking Ap = the side of a square, whose magnitude is a fourth proportional to BH, BA and the given rectangle of the lower segments, and drawing pv ∥ AB, upon the side AB, of the given △ ABH, set off AF = Ap (Bv), and draw FE ∥ the base AC, cutting BH in s, and the thing is done.

Demonstration. Because by construction, the rect. ApvB (fig. 2) is a min. when the rect. of the lower segments is a given magnitude, thence any rectangular part (Aa) thereof will also be a min. and the rect. of the said segments remain the same. Let, therefore, pa = cp, and draw $eab \parallel$ AC; then because Ape is a min. the square pb, which is a multiple of the $\triangle$ FBE, will be a max. and consequently FBs, being a part of FBE, will also be a max. when AF $\times$ EC ($=$ Ap $\times$ Bv) is a given magnitude. Now, seeing that EF must be $\parallel$ AC, in order that a multiple of the $\triangle m$Bn may be a max. and the rect. under the lower segments of the $\triangle$ ABC a given magnitude (*i. e.* to the given rectangle as BA : BH or FA : sH), it thence evidently follows, that the $\triangle$ FBs (being as BF $\times$ Bs) will be the greatest possible (and the rect. of the lower segments of the $\triangle$ ABH $=$ the given rectangle) when Fs is $\parallel$ AH.

A Fluxionary Answer to the same by Mr. Samuel Vince.

Suppose AB (fig. 1st) to be given $= a$, HB $= b$, the rect. of the lower segments, AF, Hs, $= rr$ (supposing Fs to be the position of the line required), and put AF $= x$, Hs $= y$; then will $xy = rr$, and $(a - x) \times (b - y)$ be a max. per quest. In fluxions $x\dot{y} + y\dot{x} = 0$, and $x\dot{y} + y\dot{x} - a\dot{y} - b\dot{x} = 0$, i. e. $- a\dot{y} - b\dot{x} = 0$, and consequently $ay = bx$; whence x (AF) : y (Hs) :: a (AB) : b (HB), and F$s \parallel$ AH, by the Elem. of Geometry.

X. QUESTION 559, *by Mr.* John Chipchase.

A ball being shot, at an elevation of 76°, from a cannon, whose greatest horizontal range is three miles, struck an object standing at such a distance that, if it had been projected against it in a direction perpendicular to its surface, its force would have been twice as great as it then was; required the object's distance from the cannon, supposing it to stand perpendicular to the horizon?

Answered by Mr. T. Todd.

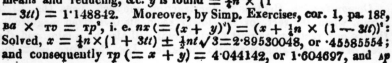

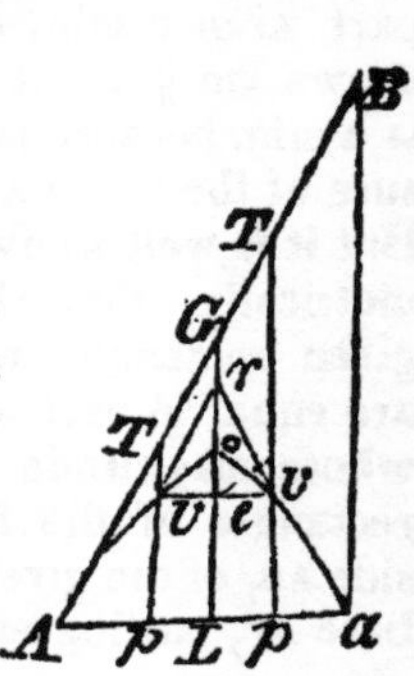

By the nature of projectiles in a non-resisting med. 1. (rad.) : ·4694714, (s. 2 $\angle$ BAa, the given elevation) :: 3 miles (the given greatest rand.) : 1·4084142 miles $=$ Aa, and, by trig. Ba $=$ 5·64884077 miles $= n$. Now, if $t = \cdot 249328$ (the nat. tang. of $\angle$ B $= 14$), $x =$ Tv, and $y = vp$; then by trig. $tn =$ Aa, $\frac{1}{2}tn =$ AL $=$ La, and therefore, by conics, $\frac{1}{4}n$ (GO $=$ OL) : $\frac{1}{4}ttnn$ (La^2) :: $\frac{1}{4}n - y$ (oe) : $\frac{1}{4}ttn \times (n - 4y) = ev^2$; $\therefore$ by 47 Euc. 1, the tang. rv at v, is $= \sqrt{(\frac{1}{4} \times (n - 4y)^2 + \frac{1}{4}t^2n \times (n - 4y))}$; which (by the resolution of forces and the quest.) is to $\frac{1}{2}t\sqrt{(nn - 4ny)}$ ($= ve$) :: 2 : 1; from whence, by multiplying extremes and means and reducing, &c. y is found $= \frac{1}{4}n \times (1 - 3tt) = 1{\cdot}148842$. Moreover, by Simp. Exercises, cor. 1, pa. 188, Ba $\times$ Tv $=$ Tp^2, i. e. nx ($= (x + y)^2$) $= (x + \frac{1}{4}n \times (1 - 3tt))^2$: Solved, $x = \frac{1}{2}n \times (1 + 3tt) \pm \frac{1}{2}nt\sqrt{3} = 2{\cdot}89530048$, or ·45585554; and consequently Tp ($= x + y$) $= 4{\cdot}044142$, or 1·604697, and Ap

$(= t \times (x + y)) = 1{\cdot}008318$, or $\cdot 4000962$ miles, either of those distances answering the question.

XI. QUESTION 560, *by Mr.* T. Moss.

To determine (à priori) the form of a spheroidical standing cask, whose ullage may be truly found by means of the line of segments on the common sliding rule; its whole content being 100 gallons.

Answered by Mr. B. Sikes.

Let b = the bung diameter, l = the semi-length, x = the wet inches, y = the diameter at x dist. from the head on which the cask stands, c = the circular divisor for gallons, and the ratio of the bung to the head diam. as $1 : n$. Then, by the nature of the ellipsis, $yy = nnbb + 2(1 - nn)bbx \div l - (1 - nn)bbxx \div ll$, which multiplied by $\dot{x} \div c$ and the fluent taken, &c. the ullage (or quantity of liquor in the cask, putting $2lv = x$) will be found to be $2lbb \times \frac{2+nn}{3c} \times \frac{3nnv + 6(1-nn)vv - 4(1-nn)v^3}{2+nn}$ where the factor $2lbb \times \frac{2+nn}{3c}$ is known to be = the whole content of the cask; therefore $\frac{3nnv + 6(1-nn)vv - 4(1-nn)v^3}{2+nn}$ will express the fractional part of the ullage, the whole content being the integral quantity, or 1.—— If now the whole length of the cask be supposed = 1, and to be divided into 100 equal parts, v will truly represent the number of parts contained in the wet inches; and, to make the foregoing expression agree with the multipliers, as found, on the sliding-rule, set 100 on the slider to 100 on the line of segments S. S, on the rule, then against any given number v of parts on the slider, is the proper multiplier, or fractional part of the ullage on the rule, or line S. S; which, being put $= m$, and made = the last expression, the value of n will be determined $= \sqrt{\frac{2m - 2vv(3-2v)}{3v - m - 2vv(3-2v)}}$, which will give the form of the cask required.

Let the wet inches divided by the length, or v, be $= \cdot 28$; then m will be found $= \cdot 26$, nearly on the segments S.S. and consequently $n = \cdot 8343$: But if v be taken $= \cdot 72$, m will come out $= \cdot 743$ on the line S. S. and $n = \cdot 8097$; which shews that the line of segments will not give the ullage truly, in any part of the spheroid except the middle; But if n be taken $= \cdot 822$ (= the mean between these two ratios), the line of segments will give the ullage of a spheroid, formed by this ratio, nearer the truth than any other.

Mr. *T. Moss* (the proposer) founds his solution on p. 207, 208, &c. of his Treatise of Gauging, where it is proved, in a general manner, that the ullages of two upright spheroidical casks, having their bung and head diameters in the same ratio, and their wet inches in the ratio of their lengths, will be to each other in the ratio of their whole contents, let their lengths be what they will: Whence it follows that the ullages of every standing spheroidical cask, having some one certain proportion of bung and head diameters, may be truly determined by the sliding-rule, let the lengths of such casks be what they will. Thus, if the ratio of the head and bung diameters of the required spheroidical cask be supposed to be as x to 1 (that being all that, in the present case is required), $1. =$ its content, $y =$ its length, $\frac{1}{3}y =$ the wet inches, and $p = \cdot 7854$; then, by pa. 206 and 207 of the Theory of Gauging, will be found $(1 - (1 - xx) \times \cdot 48) \div (2 + xx) = \cdot 316$ (the ullage, in this case, by the sliding-rule; whence $x = \cdot 82$, very nearly.——If the content of the cask be denoted by a, and every thing else as before, then will $(a - (a - axx) \times \cdot 48) \div (2 + xx) = 316, 31\cdot 6, 3\cdot 16, \cdot 316$, according as a (or the content of the cask on the segment line) is supposed to be divided into 1000, 100, 10, or 1 equal parts; from each of which x comes out $= \cdot 82$, the same as before.

XII. QUESTION 561, *by Mr.* Joseph Fisher.

Admit the earth and moon two perfectly spherical bodies, and the sun's parallax to be 10″; and granting also, that the sun and moon, when conjoined, and both at their mean distances from the earth, elevate the waters of the ocean 12 feet perpendicular, thereby causing the tides: Quere, whether a pendulum clock would gain or lose, and how much per day, by transporting from the earth to the moon; and what length a pendulum should be to make 60 vibrations on the moon's surface, supposing one 39·2 inches to make 60 in a minute on the earth, in lat. 55°?

Answered by Mr. Jos. Fisher, *the Proposer.*

The whole solar force to move the waters of the ocean may be found by a proper calculus of the earth's centripetal force towards the sun, to be to the force of gravity with us as 1 is to 12868200, and Sir Isaac Newton, in his Principia, has shewn that the earth's centrifugal force, arising from its vertiginous motion is to the force of gravity at the equator as 1 is to 289; which force, being calculated, should make the waters under the equator exceed those under the poles in alt. 90640 feet; And hence the elevation of the waters by the sun appears to be $2\frac{1}{38}$ feet; consequently the moon raises them the other $9\frac{37}{38}$ feet, and the solar force to elevate the waters is to the lunar force as 1 : 4·88. The moon's force is to the force of gravity on the earth's surface as 1 is to 2634837, and the moon's density is to the sun's as 1 : 0·191, and the sun's is to the earth's as ·191 is to ·75, both computed from the triplicate proportion of their true diameters, and the simp. propor. of gravity towards each at equal distances, conjunctly; whence the earth's density is to the moon's as ·756 is to 1. Now, the moon's diameter being to the earth's as 1 : 3·65, and the density as 1 is to ·756, the accelerating gravity or force with which a pendulum would be actuated on the moon, is to the same on the earth, under the poles, as 1 is to 2·8; whence the gravity in lat. 55° appears to be 2·7968136, and consequently 34736″, or $9^h\ 38^m\ 56^s$, the time a clock would lose per diem, on the moon's surface; and hence the length of the pendulum required may be easily known.

Note. The moon's centrifugal force, arising from its rotation, is here omitted, being so small as could cause little variation.

XIII. QUESTION 562, *by Mr.* Rob. Butler.

A current of water is discharged by three equal arches or sluices: The first (in shape) a rectangular parallelogram, the second a semicircle, and the third a parabola, having their altitudes equal, and their bases (which are downwards) situated in the same horizontal line, and the water level with the tops of the arches; on this supposition let be shewed the proportion of the quantities discharged by these sluices.

Answered by Mr. Cha. Hutton.

Let VB represent half the parallelogram, AVC half the semicircle, and AVD half the parabola, *i. e.* the halves of the respective sluices or arches, $a =$ AV (the common alt.) $p = \cdot 7854$, &c. then paa is $=$ the area of each of them, $pa =$ AB, $a =$ AC, and AD$= \frac{3}{2}pa$. Put $x =$ VP, and $\dot{x} =$ Pp; then, the water discharged, at any depth x, being as the velocity and aperture, and the velocity as $\sqrt{x}$, we shall have $\dot{x}\sqrt{x} \times$ PQ, $\dot{x}\sqrt{x} \times$ PR, and $\dot{x}\sqrt{x} \times$ PS,

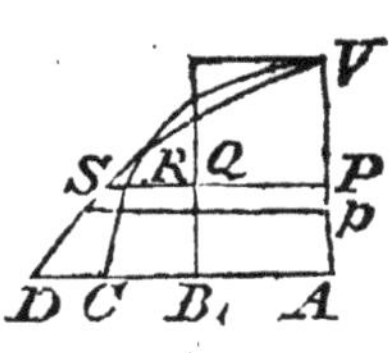

or $pax^{\frac{1}{2}}\dot{x}$, $x\dot{x}\sqrt{(2a - x)}$ and $\frac{3}{2}p\sqrt{a} \times x\dot{x}$ as the fluxion of the quantity of water discharged by the said rectangle, semicircle, and parabola, respectively; the correct fluents of which (when $x = a$) are $\frac{2}{3}paa\sqrt{a}$, $\frac{2}{15}aa\sqrt{a} \times 8\sqrt{2} - 7$ and $\frac{3}{4}paa\sqrt{a}$, and the quantity of water discharged by the rectangle, the semicircle, and parabola, as 1, 1·09847, &c. and $1\frac{1}{8}$ respectively.

XIV. QUESTION 563, *by Mr.* Paul Sharp.

If a vessel, in form of a cone whose base diameter and depth are equal, stand on its base and be filled with water, it will empty itself, through a hole in its base, in the same time that a sphere, whose internal diameter is equal to that of the cone, will empty itself through an equal orifice in its bottom: Quere the demonstration?

Mr. *T. Moss* (the proposer) founds his solution of his Treatise of Gauging, where it is proved, that the ullages of two upright spheroidical c and head diameters in the same ratio, and ratio of their lengths, will be to each other contents, let their lengths be what they the ullages of every standing spheroid tain proportion of bung and head dia by the sliding-rule, let the length Thus, if the ratio of the head spheroidical cask be supposed to present case is required), l = wet inches, and $p = \cdot 7854$; of Gauging, will be found $\cdot 316$ (the ullage, in this very nearly.——If the thing else as before, $= 316$, $31 \cdot 6$, $3 \cdot 16$. on the segment line 1 equal parts; f before.

... Proposer.

... about the centres P, *m*, and *n*, the ... parallel ... below it, ... from the centres P,

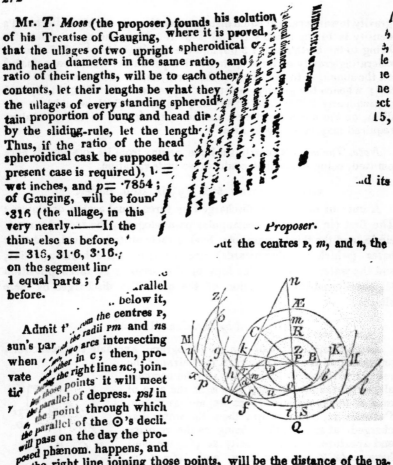

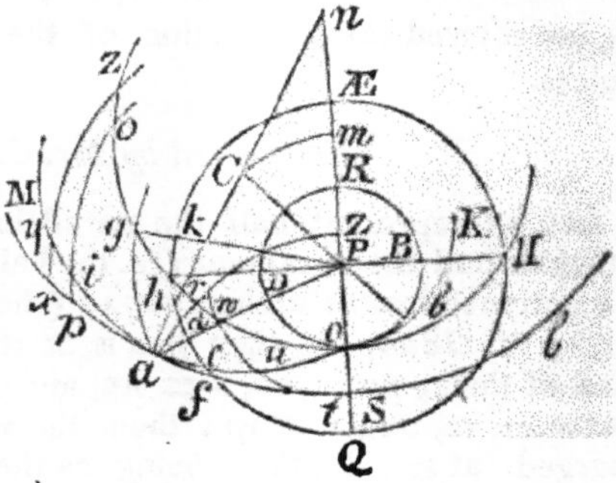

Admit t' ... with the radii PM and *ns* sun's par ... two arcs intersecting when ... each other in C; then, provate ... the right line *nc*, join-tid ... those points it will meet ... the parallel of depress. *psl* in ... the point through which the parallel of the ☉'s decli. will pass on the day the proposed phænom. happens, and *ap*, the right line joining those points, will be the distance of the parallel from the visible pole.

Demonstration. Describe the circle of perpetual apparition, RBOD, and the arcs *tw*, *ag*, *po*, *xz*, &c. of the parallels of declin. intercepted between *psl* (the parallel of depress.) and ZOH (the horiz.); also about the centre C, with the rad. *ca* (= *mo*), let the circ. KM be described, intersecting the parallel of declin. &c. in *u*, *e*, *a*, *i*, *y*, &c. and it will touch the parallels circ. *pl* in *a* (by Euc. 11. 3.): It will also touch the circ. RBOD (by the same), because CP = *m*P (by construct.), and make an angle (*ach*) with the equator = *ohe*, the angle which the horizon makes with the same; and hence the parallel arcs *uw*, *eh*, *ag*, *io*, *yz*, &c. intercepted by them, (by prop. 6th on pa. 121 of Emerson's Trigonom. 1st edit.) will be similar, and consequently passed over by the ☉ in the same time, which will, manifestly, be less than the time in which the corresponding twilight arcs *tw*, *fh*, *po*, *xz*, &c.

will be passed over in, except the twilight arc *ag*, which will be the same, and therefore will be the shortest, &c.

Calculation. Describe the vertical circle *za*, cutting the horizon and equator in *r* and *d* respectively, and the meridional arc PG, intersecting the equator in *k*; then the right-angled spherical triangles *ead*, *hrd*, being equiangular, will be mutually equilateral, and consequently $ad = dr = 9°$; whence (hr being $= ea = hg$, by prop. 6. on pa. 121 of Emer. Trigonom. aforesaid) 1 (rad.) : tang. dr ($\frac{1}{2}$ the depress.) :: cos. $\angle ghk$ = sine of lat. of the place : sine of gk (the required declin.) ——Also, in the spheric $\triangle aed$, sine of $\angle aed$ (or cos. of lat.) : sine of ad ($= 9°$, $\frac{1}{2}$ the depress.) :: 1 (rad.) : sine of ed, the arch of the equat. measuring $\frac{1}{2}$ the duration of the shortest twilight required.*

* Other Solutions may be seen at Question 271.

XVI. QUESTION 565, *by Mr.* Rob. Butler.

Required the content, in ale gallons, of a cask formed by the rotation of a cycloid about its base or longest diameter, and having its head diameter 32, bung 40, and length = 48 inches?

Answered by Mr. Cha. Hutton.

Let VKF be the generat. semicircle, and suppose half the cask to be generated by the revolution of the cycloidal arc VB about DC (= the ord. EB.). Put VF = D, VD = 20 = m, DF ($= d - m$) $= n$, $p = 3{\cdot}14159$ &c. DC $= x$, and BC $= y$; then, by the prop. of the cycloid, &c. $\dot{x}$ will be found $= -\dot{y}\sqrt{((n+y) \div (m-y))}$ and $pyy\dot{x}$ (the flux. of solid.) $= -py^2\dot{y}\sqrt{((n+y) \div (m-y))}$, and the correct fluent (when $y = 16 = a$) is $(15m^2 + 4mn - 3n^2 + 2a \times (5m + n) + 8a^2) \times \frac{1}{24}p \times (m-a)\sqrt{((n+a) \div (m-a))} + (5m^2 - 2mn + n^2) \times \frac{1}{8}p \times (m+n) \times$ arc, whose rad. is 1 and cotangent $\sqrt{((n+a) \div (m-a))}$, = the solidity of $\frac{1}{2}$ the cask required.

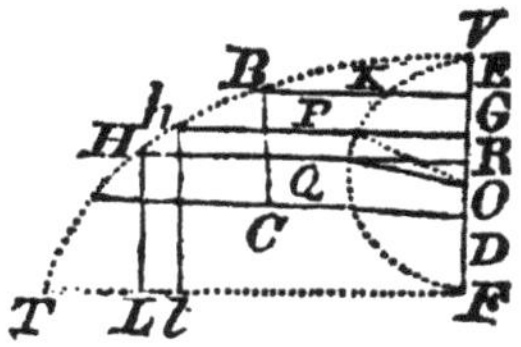

But to find VF, let it be now called z, and VE ($= m - a = 4$) $= c$; then, by the prop. of the cycloid, &c. is found 24 (= BE) $= 2\sqrt{cz} \times : 1 - \frac{c}{3z} - \frac{c^2}{2.4.5z^2} - \frac{3c^3}{2.4.6.7z^3} - \frac{3.5c^4}{2.4.6.8.9z^4} -$ &c. and hence $z = 37{\cdot}34394 = d$, and $n = 17{\cdot}34394$. Hence the above fluent comes out 26434·3196 inches = $\frac{1}{2}$ the content of the cask, and consequently the whole content is 187·4774 ale gallons.

Answer to the same by Mr. Rob. Butler, *the Proposer.*

Put $2b$ = VF ($\frac{1}{2}$ the bung diameter) = 20, VR = 4 ($\frac{1}{2}$ the given difference of the bung and head); then, by prop. of the circle, &c.

RQ $= 8$, and QH $= 16 = d$ (Gh and RH being $\perp$ VF, and cutting the semicircle VKF in P and Q). Also the $\angle$QOV (at the centre of the generating circle) will be found $= 53{\cdot}13°$, and the length of its corresponding arc QV $= 9{\cdot}273$ inches, which put $= n$, and let $p = 3{\cdot}1416$, c $=$ the content, in inches, of the solid generated by the revolution of the space lhVF, about LF, as an axis (hl and HL being $\perp$ TF), the variable quantities x, y, and $z =$ OG, GP and PV, respectively: Then, per the nature of cycloids, $n : d :: z : dz \div n =$ P$h = mz$ (putting $d \div n = 1{\cdot}72544 = m$), and $mz + y =$ the ord. Gh; the flux. of which drawn into $(b + x)^2 = hl^2 =$ GF^2), viz. $(m\dot{z} + \dot{y}) \times (b + x)^2$, will be $=$ c $\div p$. But, by the prop. of the circle, $b\dot{y} = x\dot{z}$, $bb - yy = xx$ and $- \dot{x}y = \dot{s}$ (s being the area of the circular seg. PVG); from which, and the preceding equation, (when the fluents are taken, &c.) will be had $mbbz + (2bb \times (m + 1))\,y + (b \times (m + 2))\,xy - \frac{1}{3}y^3 + b\,(m + 2)\,s =$ c $\div p$: Whence the whole content required appears to be $187\frac{1}{2}$ ale gallons.——It is observable, that though the cask in the first of these solutions is supposed to be generated from the revolution of part of a common cycloid about an ord. $\parallel$ TF, and in the last from the revolution of part of an inflected cycloid around the longest semi-diam. TF itself, yet they exactly agree in their contents.

But if the cask be supposed to be generated from the revolution of part of a common cycloid round the greatest semi-diam. TF, then its length need not be given, being determinable from the nature and other given dimensions of the generating cycloid, as Messrs. *T. Allen*, *Edward Smith*, and *T. Todd justly* observe, and find, by elegant processes, the content in this case, to be 135·2 ale gallons.

THE PRIZE QUESTION, *by Mr.* T. Allen, *of Spalding.*

Supposing a heavy body to descend freely by the force of gravity upon a common cycloidal curve, the radius of whose generating semicircle is 18 feet; and let its distance from the vertex of the cycloid be 12 feet at the commencement of motion. It is required to determine where the body will quit the curve, its distance from the axis when it impinges upon the horizon, and time of its whole descent?

Answered by Mr. Tho. Allen, *the Proposer.*

It is evident that the body will quit the curve when its velocity in direction of the ordinate becomes the greatest possible.

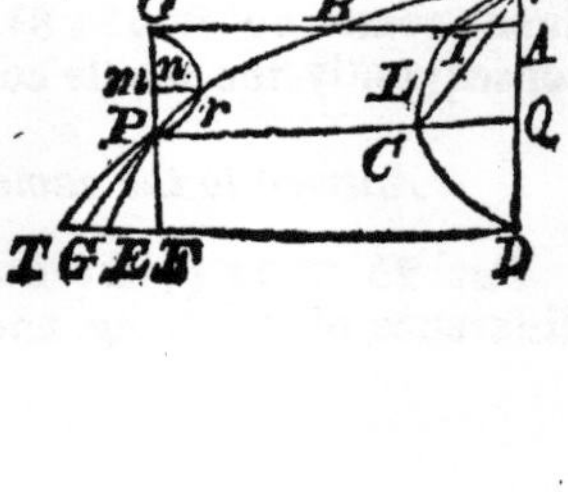

Let B be the place of the body at the commencement of motion. Put VD the diam. of the generating semicircle ($= 36$ feet) $= 2a$, and VQ$= x$; then, since BV is given$= 12$ feet, we have, by the properties of the cycloid, VA $= 1$, and therefore AQ $= x - 1$. Now, since the velocity of the body along the curve, is to the velocity in direction of the ord. as the fluxion of the former is to that of the latter, $\sqrt{((2a - x)(x - 1) \div 2a)}$ will be $=$ the velocity in direction of the ordinate; by making the fluxion of which $= 0$, will be had $x = a + \frac{1}{2} = 18{\cdot}5$ eet.

Put $s = 16\frac{1}{12}$ feet, and $t =$ the time of describing BP; then, per the laws of falling bodies, $\dot{t} = \frac{1}{2}\sqrt{(2as^{-1})} \times x^{-\frac{1}{2}}\dot{x} \times (x - 1)^{-\frac{1}{2}}$, whose fluent gives $t = \sqrt{(a \div 2s)} \times$ hyp. log. of $(\sqrt{x} + \sqrt{(x-1)})^2$; which, when $x = 18{\cdot}5$, will be $3{\cdot}1989''$.

Lastly, when the body ceases to touch the cycloid, as at P, it will describe a parabolic curve, suppose PG. Now, having given the velocity at P $= 34{\cdot}5$ feet per second, very nearly (from what is found above), and likewise the position of the tang. TP, or the $\angle$PTF $= 45°\ 47'$, the dist. GF (supposing FP $\perp$ TD) will be determined $= 12{\cdot}22$ feet, and the time of the body's describing PG $= {\cdot}508''$. Whence GD (the distance the body will impinge upon the horizontal line TD from the axis VD) will be $= 58{\cdot}62$ feet, and the time of the whole descent $= 3{\cdot}7069''$.

The same answered by Mr. Cha. Hutton.

Let VBPE be the cycloid, and VLD the generating semicir. Suppose the body to commence motion at B, and quit the cycloid at P, going off, there, in the curve of a parab. PG. Draw ED $\perp$ VD, and PQ and BA $\parallel$ thereto, meeting the semi-circumf. VLD in C and I, producing the latter of them till it meets FP, $\perp$ DE, produced in O, and upon OP describe the semi-cir. orP. Draw nT a tang. to both the given cycloid VBE, and parab. PG, at P, meeting the semi-circumf. orP in n, and DE, produced, in T, and let nm be $\perp$ OP, and the points V, I and V, C be joined.——Put VA $= a$, VD $(= 36) = d$, $16\frac{1}{12} = s$, and AQ $= x$. Then, by the laws of descending bodies, $2\sqrt{sx} =$ the velocity per second, at Q or P, in the direction PT or VC, and consequently $2\sqrt{(d^{-1}sx \times (d - a - x))} =$ the velocity at P in the horizontal direction. Now, since the body goes off at P in a parab. and the horizontal velocity in a parab. is a constant quantity, the last expression will be constant, and consequently its fluxion $= 0$; whence x comes out $= \frac{1}{2} \times (d - a)$ ($\frac{1}{2}$AD). Also, $a =$ VI$^2 \div d = \frac{1}{4}$BV$^2 \div d = 1$, and VP $(= 2$VC$) = \sqrt{(2d \times (d + a))} (= 2\sqrt{(d \times (a + x))}) = 51{\cdot}613951$. Again, per sim. triangles (putting $\frac{1}{2} \times (d + a) = \frac{1}{2}c$, and $\frac{1}{2} \times (d - a) = \frac{1}{2}g$) will be found FT $= \frac{1}{2}g\sqrt{(c^{-1}g)}$, and $mn = \frac{1}{4}d^{-1}g\sqrt{cg}$: Then, per conics, $d^{-1}gg$ $(= 4mo = 4$QD $\times$ PO $\div$VD$)$ $=$ the parameter of the parabola's axis, and $\frac{1}{2}d^{-1}g\sqrt{(g \times (2d + c))}$ $(= \sqrt{(\text{F}m \times d^{-1}gg)})$ $=$ an ordinate of the axe terminated at the point G of the curve; also, FG $= g\sqrt{g} \div (\sqrt{c} + \sqrt{(2d + c)})$ $(= (2mn +$ ord. to point G$)^{-1} \times 4mn \times$ FT$)$: Whence, (putting $p = 3{\cdot}14159$, &c.) DG $(=$ FG $+$ QC $+$ PC $=$ FG $+$ QC $+$ the circ. arc VC$)$ $= g\sqrt{g} \div (\sqrt{c} + \sqrt{(2d + c)}) + \frac{1}{2}\sqrt{cg} + \frac{1}{180}dp \times$ deg.

in $\angle$VCQ $= 59{\cdot}299337$, the distance from the axis when the body impinges on the horizon.——Lastly, to find the whole time of descent to the horizon. Suppose $t =$ the time of descending through the cycloidal arc BP, then will $\dot{t} = \frac{1}{2}\dot{x}\sqrt{(ds^{-1}x^{-1} \times (a + x)^{-1})}$, and, taking the correct fluent when $x = \frac{1}{2}g$, $t = \frac{1}{2}\sqrt{(ds^{-1})} \times$ hyp. log. of $(d + \sqrt{cg}) = 3{\cdot}199035''$: But, for the time in the parab. the horizontal velocity being constant, and $= g\sqrt{(sd^{-1})}$ feet per second (as found above), $\sqrt{(gds^{-1})} \div (\sqrt{c} + \sqrt{(2d + c)})$ $(= \text{FG} \div g\sqrt{(sd^{-1})}) = \sqrt{(\frac{1}{2}dgs^{-1} \times (d + c + \sqrt{(c \times (2d + c))})^{-1})}$ $= {\cdot}535681'' =$ the time in the said parab. and the sum of these two, or $3{\cdot}734716''$, is $=$ the whole time of descent required.

Questions proposed in 1767, *and answered in* 1768.

I. QUESTION 566, *by Mr.* Tho. Atkinson.

Near Lincoln's city a worthy knight doth dwell,
For wit and valor few can him excel:
In the late war his honor there did shew,
How he was to his king and country true.
The poor to him do cry in their distress—
He rights the indigent and fatherless;
A friend to arts and sciences is he;
An ornament to all society.
Ladies, from hence * declare his worthy name,
And it record in Diary of fame.

* viz. $\left\{\begin{array}{r} w + x + yz = 61, \\ x + y + zw = 501, \\ y + z + wx = 381, \\ w - 4 = x; \end{array}\right\}$ Where the values of w, x, y, and z shew the places, in the alphabet, of the letters composing this gentleman's name.

Answered by Mr. Paul Sharp.

By substituting for w its equal $x + 4$ in the first given equation, the value of x will, from thence, be found $= \frac{1}{2} \times (57 - yz)$, and consequently $w = \frac{1}{2} \times (65 - yz)$; and these values substituted in the 2d and 3d given equations, they become $57 - yz + 65z - yz^2 + 2y = 2 \times 501$, and $y + z + \frac{1}{4} \times (57 - yz) \times (65 - yz) = 381$; from which, by pa. 82 of Simp. Essays, $y = 1$ and $z = 23$; and thence $x = 17$ and $w = 21$, and the honorable gentleman's name is *Wray*.*

Pamphagus observes that this gent. * is a descendant of Sir *Cecil Wray*, who was created a baronet Nov. 25th, 1612.

II. QUESTION 567, *by Miss* Anna Nicholls.

Walking on shore, I was surprized by the flash of a gun, at sea, bearing S.E. by E.: seven seconds after the flash I heard the report, and 4 seconds after that I heard the echo from a castle bearing from me S.W. by W.: the distance of the gun and castle are required?

Answered by Mr. W. Spicer.

Construction. From the △GPD, having its ∠P = 112° 30′ (the sum of the given bearings of the gun and castle), and the containing sides PG, PD equal to 7 × 1142 and (7 + 4) × 1142 respectively (1142 being the number of feet sound passes over in 1″), and draw the right line GC, making the ∠CGD = ∠CDG; then, P being the place of observation, and G the gun, C will be the place of the castle required: For, CG being = CD (per const.), GC + CP will be = PD (the given distance, by const.).

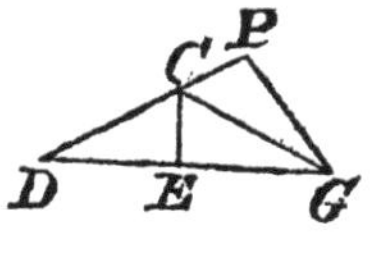

Calculation. In the △GPD, the two sides PG, PD, and included angle being given, the ∠D is found = 25° 18′ 15″, and its double, viz the ∠PCG, is = 50° 36′ 30″; then, in the △PCG, all the angles, and the side PG, are given, to find GC = 9556·4844 and PC = 3005·5156 feet.——And in this manner, nearly, it is also constructed by *Nugo Dargnas*; but Messrs. *J. Chipchase*, *W. Cole*, *J. Osborne*, *Pamphagus*, *J. Paty*, *E. Reed*, and *W. Wales*, after describing the △DPG nearly as before, bisect DG with the perpendicular EC meeting PD in the point C, the place of the castle required.

Mr. *W. Sewell* observes that the point C aforesaid, will be in the periphery of an ellipse, having its foci in P, G, and transverse = PC + GC, as from the well-known property of that curve is manifest!

III. QUESTION 568, *by Mr.* John Chipchase.

There is a town in lat. 50°, and another in lat. 30°, north; and when it was noon, on the 4th of May, at the first of them, the sun was just rising at the second; 'tis required to find the lat. of a port that is equally distant from them both, and 40° W. of the first.

Answered by Mr. John Roper.

Let P represent the north pole, Z the zenith of the place in lat. 30°, *a* that of the place in lat. 50°, ☉ the sun in the horiz. HR of the former, and the merid. P*b* of the latter of them, on the day proposed, *d* the port required, ÆQ the equator intersecting HR in the centre of projec. C, Z*a*, P*d* two arcs of great circles intersecting each other in *n*, and *dm* another arc ⊥ Z*a*, which will bisect it, as Z*d* is = *ad*, per quest. Then, in the right-angled spherical △C*b*☉, are given *b*☉ (the ☉'s declin.) = 15° 58′, and the

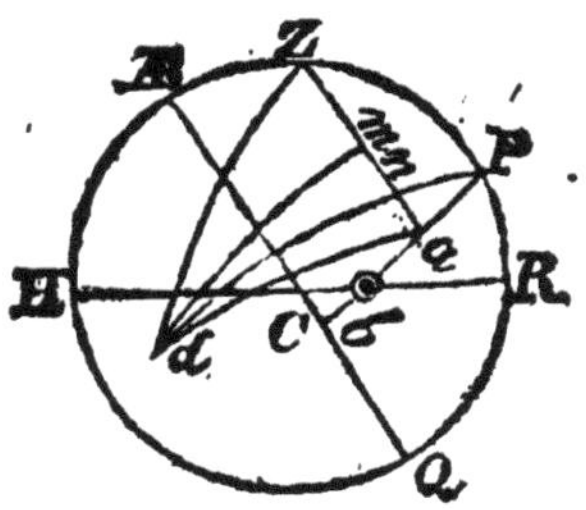

$\angle bc\odot$ (comp. of lat. of the place z) 60°, to find $cb = 9° 30\frac{1}{2}'$; whence Æb (or the $\angle$ ÆPb or ZPa) is $= 99° 30\frac{1}{4}'$: And then, in the spherical $\triangle$ ZPa, the sides ZP, aP with their included angle being given, za will be found $= 73° 4' 39''$, and the $\angle$ Paz $= 63° 13\frac{1}{2}'$. Now, the $\angle$ aPn (the westward bearing of the merid. of the required port d from that of the place a) being given $= 40°$ (per quest.), in the $\triangle$ aPn are given two angles with the interjacent side aP; whence the sides nP, na, and the $\angle$ Pna ($= \angle dnm$) are found $= 35° 12' 4''$, $24° 31' 18''$, and $84° 34' 25''$ respectively; and consequently in the right-angled spherical $\triangle dmn$, $mn\ (= \frac{1}{2}za - na)$, and the $\angle mnd$ being now known, dn is, from thence, found $= 66° 2' 49''$; to which adding nP, and subtracting 90° from their sum, the remainder 11° 14′ 53″ south, will be the lat. of the port required.

IV. QUESTION 569, *by Mr.* Rich. Gibbons.

An annuity being forborn 5 years, at compound interest, amounted to 500l. and now the same hath been forborn 10 years it amounts to 1100l. Required the annuity, and the rate of interest per cent. per annum?

Answered by Masters J. Paty *and* J. Osborne, *Youths of about* 13 *Years of Age, at the Mathematical Academy, Bristol.*

Let $x =$ the annuity, $y =$ the amount of 1l. in one year, $a = 500$, $m = 1100$; then, by the doctrine of compound interest and annuities (see Note, p. 350 of Donn's Arithmetic) $\frac{ay - a}{y^5 - 1} = \frac{my - m}{y^{10} - 1} (= x)$

and consequently (putting $m \div a - 1 = c$, and $m \div a = 2n$) $y^{10} - 2ny^5 = -c$; whence (compleating the square, &c.)* $y = \sqrt[5]{(n \pm \sqrt{(nn - c)})} = 1{\cdot}037137$, &c. and the rate of interest $=$ 3l. 14s. $3\frac{3}{4}$d. per cent. per annum, and the annuity required $=$ 92l. 16s. $10\frac{1}{2}$d.

Pamphagus, observes that 'since the annuity forborn 5 years amounts to 500l. and when forborn 10 years to 1100l. it is manifest that the interest of 500l. for 5 years must be = 100l, Therefore the rate per cent. will be easily found = 3l. 14s. 3d. $\frac{1}{4}{\cdot}19$; and thence, by a well-known theorem, the annuity = 92l. 16s. 10d. $\frac{1}{4}{\cdot}76$.'

---

* The value of y is here brought out by means of a quadratic equa. without any necessity, for it naturally becomes a simple one, thus: Divide the numerators of the terms of the given equation $\left(\frac{ay - a}{y^5 - 1} = \frac{my - m}{y^{10} - 1}\right)$ both by $y - 1$, and the denominators by $y^5 - 1$, &c. and you have $y^5 + 1 = m \div a$; hence $y = \sqrt[5]{m \div a - 1} = \sqrt[5]{(m - a) \div a} = \sqrt[5]{6 \div 5} = 1{\cdot}037137$. H.

V. QUESTION 570, *by Mr.* Rob. Butler.

Given the radii of two circles = 6 and 8, and the distance of their centers = 10 inches; 'tis required to draw a right line 19 inches long, through their point of intersection, so as to terminate in their peripheries, and to determine the length of each segment thereof, intercepted between the said point and peripheries?

Answered by Mr. Wm. Cole.

Construction. Upon AB (the right line joining the centres of the given circles intersecting each other in I) conceive a semicircle to be described, and apply therein, from either A or B, the chord AC = 9·5 (half the given line), and parallel thereto let FG be drawn through I, terminating in their peripheries, and it will be the line required.

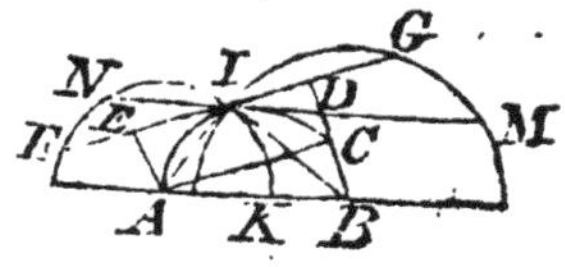

Demonstration. Through C draw BD, and parallel to it, AE, meeting FG at right angles (by 31. 3, and 29. 1. Euc.) and bisecting the chords IG, IF (by 3. 3. Euc.), and then FE + DG will be = ED = AC (= 9·5, per construction), and consequently FG = 2AC = 19.

Calculation. Join the points A, I and B, I: Then, in the △ AIB the sides being given, the ∠ABI is found = 36° 52′ 12″; which taken from the ∠ABC (found = 71° 48′ 18″, from the right-angled △ ACB) leaves 34° 56′ 6″ = ∠IBD: Whence from the right-angled △ IDB, is found ID = 4·5811; and therefore the segment in the greater circle is = 9·1622, &c. and that in the lesser = 9·8377, &c. inches. If the chord AC be inscribed the other way from B, the construction and method of calculation will be the same, and the segment in the greater circle will, in that case, be found = 15·1681, &c. and that in the lesser = 3·8318, &c.

Algebraic Solution to the same by Mr. John Addison.

Put FG (= 19) = a, AI (= 6) = $\frac{1}{2}b$, BI (= 8) = $\frac{1}{2}c$, and x = FI; then will IG = $a - x$, and (per 3 Euc. 3.) EI = $\frac{1}{2}x$, and ID = $\frac{1}{2} \times (a - x)$. Then, the sides of the triangle AIB being in the ratio of 3, 4, and 5, (per quest.), the angle AIB, opposite the longest of them, will therefore be a right-angle, and consequently the right-angled △s AEI, BDI similar; whence $\frac{1}{2}b$ (AI) : $\frac{1}{2}x$ (EI) :: $\frac{1}{2}c$ (BI) : $cx \div 2b$ = BD, and (per 47 Euc. 1.) $(cx \div 2b)^2 + \frac{1}{4} \times (a - x)^2, = (\frac{1}{2}c)^2$, or (in numbers), $x^2 - 13{\cdot}68x = -37{\cdot}8$; solved, x (FI) = 9·8375, or 3·8423; and consequently IG = 9·1625, or 15·1577, &c.

VI. QUESTION 571, *by the late Mr.* W. Toft.

Admit a ship can make her way good, when close hauled, within 6½ points of the wind, being then at N. N. W. and, when she had sailed on the larboard tack 62, and on the starboard tack 75 miles, she is found, by observation, to have altered her latitude 25 miles northing; required the variation of the compass?

Answered by Mr. J. Chipchase.

Construction. From P, the port sailed from, let the right line PN be drawn to represent the magnetical or erroneous meridian. Make the ∠NPT = $4\frac{1}{2}$ and the ∠PTA = 3 points, taking PT = 62 miles (the dist. run on the larboard tack), and TA = 75 miles, (the distance on the starboard tack,) and on the right line joining the points A, P describe a semicirc. and inscribe therein the chord PC = 25 miles (the given difference of lat.), and PC will be the direction of the true meridian, AC the departure, and the ∠NPC the variation required; the demonstration of which is evident from the nature of turning to windward, &c.

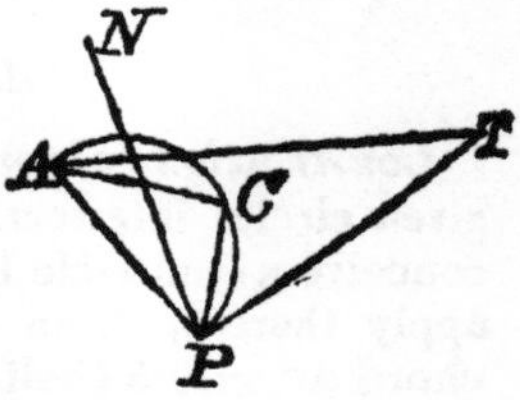

Calculat. In the △APT are given PT, TA together with their included angle, whence AP is found = 41·6694, and the ∠APT = 90° $29\frac{1}{4}'$, nearly; from which take the ∠APC (found = 53° 8′, from the right-angled △ACP) and the remainder 37° $21\frac{1}{4}'$ (= ∠CPT) taken from the ∠NPT = 50°$37\frac{1}{2}'$, per construc.) leaves the ∠NPC = 13° $15\frac{1}{4}'$ the variation required, which is westerly.

VII. QUESTION 572, *by Mr.* Paul Sharp.

There are two places, A and B in north latitude, whose least distance from each other is = $3405\frac{1}{10}$ miles; A the westermost's latitude is more than B's by 10°, and the sum of both their latitudes is = their difference of longitude; required the latitude of each place with the investigation?

Answered by Mr. W. Wales.

Put c = cosine of 10° (the given diff. of lat. of the two places), a= cosine of 56° 45′ 2″ (their given nearest dist.) and x = cosine of their difference of longitude required (rad. = 1.); then, the sine and cosine of the sum of two arcs being every where equal to the sine or cosine of the sum of their complements, the cosine of the sum of the co-latitudes of the said places will be = x (per quest.); and (per Simp. Trig. prop. 27, p. 74) $(2c-2a) \div (c+x) = 1-x$ (the versed sine of their diff. of long.): Whence, by reduction and putting $1-c = 2s$, there results $x = \sqrt{(ss+2a-c)}+s = \cdot 34199$, the nat. cos. of 70°; and so the latitudes are 40° and 30°.——But x may be either affirmative or negative, and consequently the sum of the lat. of the two places 70°, or its supplement to 180°, &c.

VIII. QUESTION 573, *by Mr.* Wm. Spicer.

Given $(0{\cdot}8 \div x^{\frac{1}{3}})^{x^{\frac{1}{4}} \div 0{\cdot}5} = y$, where x multiplied by 100000l. is the fortune of an agreeable young lady; required her fortune when y is a maximum?

Answered by Mr. N. Brownell.

The given equation being $(0{\cdot}8 \div x^{\frac{1}{3}})^{2x^{\frac{1}{4}}} = y$, put $n = \frac{1}{4}$, $z =$ hyp. log. of $(0{\cdot}8 \div x^{\frac{1}{3}})$, and $m = 0{\cdot}43429448$, and then $2zx^n$ (= log. of y) is a max. (per quest.). In fluxions, $2\dot{z}x^n + 2nzx^{n-1}\dot{x} = 0$, or $-\frac{2}{3} \times x^{n-1}\dot{x} + 2nzx^{n-1}\dot{x} = 0$ (because $\dot{z} = -\dot{x} \div 3x$, by the nature of logs.); whence z (= hyp. log. of $0{\cdot}8 \div x^{\frac{1}{3}}) = 4 \div 3$, and $0{\cdot}5790593$ $(4m \div 3)$ is = tabular log. of $(0{\cdot}8 \div x^{\frac{1}{3}})$ or the log. of x = (log. of $0{\cdot}8 - {\cdot}5790593) \times 3 = 7{\cdot}9720921$, and the natural number answering thereto is $0{\cdot}009377608$; which multiplied by 100000 gives 937l. 15s. $2\frac{1}{2}$d. = the lady's fortune.

IX. QUESTION 574, *by Mr.* Samuel Vince.

Given $ax^7y^3 = bx^8y^2 + cx^4y^6 - x^{11} + dy^{10}$, the equation of a curve; to find the area and subtangent.

Answered by Mr. J. Leader.

Substitute vx for y in the given equation, and it becomes, $av^3 \times x^{10} = bv^2x^{10} + cv^6x^{10} - x^{11} + dv^{10}x^{10}$; and thence $x = dv^{10} + cv^6 - av^3 + bv^2$, y (= vx) $= dv^{11} + cv^7 - av^4 + bv^3$, and consequently $y\dot{x}$ the fluxion of the area $= (av^{11} + cv^7 - av^4 + bv^3) \times (10dv^9\dot{v} + 6cv^5\dot{v} - 3av^2\dot{v} + 2bv\dot{v})$; and the fluent $\frac{10}{21}d^2v^{21} + \frac{16}{17}cdv^{17} - \frac{13}{14}adv^{14} + \frac{1}{13} \times (12bd + 6c^2) \times v^{13} - \frac{9}{10}acv^{10} + \frac{8}{9}bcv^9 + \frac{3}{7}a^2v^7 - \frac{5}{6}abv^6 + \frac{2}{5}b^2v^5 - f$ is = the area required (f being the sum of the coefficients of the powers of v).——Then, from the given equation in fluxions, (by transposition, division, and multiplying by y) is had $\frac{y\dot{x}}{\dot{y}} = \frac{2bx^8y^2 - 3ax^7y^3 + 6cx^4y^6 + 10dy^{10}}{11x^{10} - 8bx^7y^2 + 7ax^6y^3 - 4cxy^6}$, the subtangent required.

X. QUESTION 575, *by Mr.* Tho. Barker, *Teacher of Mathematics, and Land Surveyor, at Wissett, in Suffolk.*

Given $ax^n = y^9$, the equation of a curve; it is required to find the exponent n, and the length of the curve, when the distance of the centre of gravity of a solid formed thereby, from the vertex, is $= \frac{25}{34}x$?

Answered by Mr. Pamphagus.

Since $y^9 = ax^n$, y^2 will be $= a^{\frac{2}{9}} x^{\frac{2n}{9}}$, and (by page 204 of Simpson's Flux.) the

$$\frac{\text{flu. of } x^{\frac{2n}{9}+1}\dot{x}}{\text{flu. of } x^{\frac{2n}{9}}\dot{x}} = \frac{2n+9}{2n+18}x =$$

the distance of the centre of gravity of the said solid from its vertex, which (per quest.) is $= \frac{25}{34}x$; and consequently $n = 8$.——Having thus obtained the numerical value of n, we shall, in the next place, have (putting $81 \div 64a^{\frac{1}{4}} = c$) $\dot{y}\sqrt{(1+cy^{\frac{1}{4}})} =$ the flux. of the curve's length; and the correct fluent * thereof

$$\frac{8\times(1+cy^{\frac{1}{4}})^{\frac{3}{2}}}{9c} \times \left(y^{\frac{3}{4}} - \frac{6y^{\frac{1}{2}}}{7c} + \frac{24y^{\frac{1}{4}}}{35c^2} - \frac{16}{35c^3} + \frac{128}{315c^4}\right)$$

will be the length of the curve required?

* See Simpson's Flux. Art. 84.

XI. QUESTION 576, *by Mr.* Wm. Sewell.

Required the dimensions of a cone, which, being suspended by its vertex, shall vibrate as many times in a minute, as it is inches in altitude, its content being 1728 solid inches.

Answered by Mr. T. Todd.

Put $c = 3{\cdot}1416$ (the circumfernce of the circle whose diameter is 1), $s = 1728$ (the given solidity of the required cone), $x =$ its altitude, $y =$ the rad. of its base, and $p = 39{\cdot}2$ inches (the length of a pendulum vibrating seconds); then (per p. 225 of Simp. Flux.) $(4xx + yy) \div 5x =$ the dist. of the centre of oscillation from its vertex, or point of suspension, and (by the doctrine of pendulums, their lengths being inversely as the squares of the number of vibrations made by them in the same time) $(60^2 \times 5px) \div (4xx + yy)$ (the square of the number of vibrations performed by the cone in 1^m) $= xx$ (per quest.); from which and the equation $\frac{1}{6}cxyy = s$, is found $x = (4500p - 3s \div 4c)^{\frac{1}{3}} = 56{\cdot}039$, &c. and $y = \sqrt{(3s \div cx)} = 5{\cdot}426$, &c. inches.

Mr. *Ja. Young* puts $a = 39{\cdot}2$ inches, $s = 1728$, $p = 3{\cdot}1416$, $x =$ the height, and $y =$ rad. of the base of the required cone, in inches; then $\frac{1}{3}pxyy = s$, and (per pa. 239 of Emer. Flux. 1st edit.) $(4xx+yy) \div 5x =$ dist. of the cent. of oscillation from its vertex, or the length of a pendulum isochronal to the cone: Whence, $\sqrt{(4xx + yy) \div 5ax}$ being $=$ the time of one vibration, $\sqrt{(4xx + yy) \div 5ax} : 1 :: 60'' : x$, or $x\sqrt{((4xx + yy) \div 5ax)} = 60$ (per quest.); from whence, and the preceding equation, x comes out $= 56{\cdot}0394$, &c. and $y = 5{\cdot}4264$, &c. inches.

XII. QUESTION 577, *by Mr.* T. Moss.

To investigate a general expression that will exhibit all the different factors, whereby any quantity of spirits, of any strength above hydrometer proof (* as 1 to 1, 1 to 2, 1 to 3, &c. to 1 to 20) being multiplied, the respective products will shew what quantity of water will be necessary to reduce such spirits to any required strength under that proof († as 1 in 2, 1 in 3, &c. to 1 in 20).

* One gallon of water and one of spirits make 2 gallons of hydrometer proof.

† One gallon of water and one of hydrometer proof make two gallons of 1 in 2 under that proof, &c.

Answered by Mr. T. Moss, *the Proposer.*

Let such a quantity of the strong spirits (*i. e.* 1 to 1, 1 to 2, 1 to 3, &c.) as one gallon of water will reduce to hydrometer proof be $= a$, and the quantity when so reduced $= b$ (viz. $b = a + 1$): Moreover, let such a quantity of hydrom. proof as one gall. of water will reduce to the lower strength (*i. e.* 1 in 2, 1 in 3, &c.) $= c$, and the quantity when so reduced $= d$; also, let the required quantity of water necessary to reduce one gallon of the strong to any given strength under hydrom. proof $= x$: Then, $d : c :: 1 + x : (c + cx) \div d =$ the quantity of the hydrom. proof spirits in $1 + x$ quantity (or in one gallon of strong); but $a : b :: 1$ (*i. e.* one gall. of strong spirits) $: b \div a$; whence $(c + cx) \div d = b \div a$, and $x = bd \div ac - 1$.

General Rule.

Let such a quantity of the strong as one gallon of water will reduce to hydrometer proof, be multiplied by such a quantity of the strong as one gallon of water will reduce to the given strength under hydrometer proof: Then, let the product of the two quantities so reduced be divided by the former product; from the quotient subtract unity, and the remainder will be the required factor for multiplying any given quantity of spirits above hydrometer proof, in order to shew the quantity of water necessary to be added to make them of any given strength below that proof.

Suppose, for example, spirits of 1 to 3 were to be reduced to 1 in 7 (*i. e.* to export strength): Here $a = 3$, $b = 4$, $c = 6$, and $d = 7$; therefore, by the above rule, $bd \div ac - 1 = (4 \times 7) \div (3 \times 6) - 1$ $(= 1{\cdot}5555, \&c. - 1) = {\cdot}555$, &c. the required factor.

Note. This prob. will be of great utility to persons concerned in trying the strength of spirits by the hydrometer.

Additional Solution by Mr. Geo. Coughron, *taken from the Miscellanea Mathematica.*

Generally, whether both spirits be above or both below hydrometer proof, or the one above and the other below it.

Let a be the quantity of hydrom. proof. in b gallons of the stronger spirit, c the quantity of the said proof in d gallons of the weaker, m the quantity of the said proof in n gallons of the given standard strength, and x the quantity of the weaker, necessary to mix with b gallons of the stronger, to reduce it to the given strength: Then as $d : c :: x : cx \div d =$ the quantity of hydrom. proof in x gallons of the weaker, and therefore $a + cx \div d = (ad + cx) \div d =$ the quantity of hydrometer proof in $b + x$; whence as $m : n :: (ad + cx) \div d : n(ad + cx) \div dm =$ the quantity of the given strength in $b + x$, viz. $= b + x$; hence $x = d(an - bm) \div (dm - cn)$; therefore the quantity of the stronger is to that of the weaker as b to $d(an - bm) \div (dm - cn)$, universally.

Cor. 1. When $m = n$, or the mixture to be hydrometer, then $x = d(a - b) \div (d - c)$.

Cor. 2. When $c = 0$, or instead of spirit the weaker is only water, then $x = (an - bm) \div m = an \div m - b$, or the ratio of the spirit to the water as 1 to $an \div bm - 1$, which is the same expression (only different letters) with that given by Mr. *Moss*, above.

Scholium. I could easily give the foregoing theorems in words, but I think they are plain enough as they stand: However I shall put down an example in numbers when both spirits are above hydrometer proof; viz. Suppose spirits of 1 to 3, were to be mixed with 1 to 9, so that the mixture may be 1 to 7; here $a = 4$, $b = 3$, $c = 10$. $d = 9$, $m = 8$, and $n = 7$; therefore $x = d(an - bm) \div (dm - cn) = 9(4 \times 7 - 3 \times 8) \div (9 \times 8 - 10 \times 7) = 18$, and consequently $b : x$ as $3 : 18$ or as $1 : 6$; that is, there must be taken 6 times as much of the weak as of the strong.

XIII. QUESTION 578, *by Mr.* Joseph Walker.

To determine the ratio of the axes of a spheroid, to which if the common diagonal rod be applied, in the same manner as in the gauging of a cask, it shall exhibit its true content.

Answered by Mr. Rd. Gibbons.

As this quest. refers to the common diagonal rod, I shall take my example therefrom, where, against 22 inches, is 29 wine gallons or 6699 cubic inches. Then, putting $2x =$ the diameter, and $2z =$ the axis of the required spheroid, $xx + zz$ will be $= 22^2$ (per 47 Euc. 1.), and $4{\cdot}1888xxz$ (the solid content of the spheroid, in inches) $= 6699$ (per question). Whence (xx being exterminated) $484z - z^3 = 1599{\cdot}2647$: Reduced, $z = 20{\cdot}112$; and thence $x = 8{\cdot}917$, and their ratio (or the ratio required) is as $2{\cdot}25$ to 1, or as 9 to 4.

The same answered by Mr. Jos. Walker, *the Proposer.*

Let the content of a spheroid (in inches) be denoted by a, its corresponding diagonal (on the gauging rod) by d, and the semi-axis on which the spheroid is supposed to be generated, by x ($p = 3{\cdot}14159$, &c.). Then, by the known theorem, $(dd - xx) \times 4px \div 3 = a$; which is a general equation. Now, let a, for instance, be interpreted by 60 ale gallons, or 16920 cubic inches, and d the corresponding diagonal (nearly on the rod) by 30 inches; then, the above equation becomes $900x - x^3$ $(= 16920 \div \frac{1}{3} \times 4p) = 4040{\cdot}1165$: Which, having 2 affirmative roots, we therefore have $x = 27{\cdot}425$ and $4{\cdot}61$, nearly; being respectively the semi-lengths of the oblong and prolate spheroid: Whence $(\sqrt{(dd - xx)})$ the semi-diameters thereof will be 12·16 and 29·64. Hence it appears that the axis of every oblong spheroid, whose content may be truly obtained by the common diagonal rod, must be in the ratio of 12·16 to 27·425, but those of the prolate one, as 4·61 to 29·64, very nearly: Which ratios agree with those in Moss's Gauging, pa. 197, and are a confirmation of the truth of the author's principles, with respect to the vast comprehensiveness and utility of the diagonal rod.

XIV. QUESTION 579, *by Mr.* T. Moss.

To determine the number of fifteens that can be made out of a pack of cards, with the investigation.

Answered by Mr. H. Brown, *of the Tower, Mr.* T. Moss, *the Proposer, and Mr.* J. Wore.

Cards.	*Combinations.*		*Cards.*	*Combinations.*	
10,5	16.4	64	8,2,2,2,1	4.4.4	64
10,4,1	16 4.4	256	8,2,2,1,1,1	4 6.4	96
10,3,2	16.4.4	256	7,7,1	6.4	24
10,3,1,1	16.4.6	384	7,6,2	4.4.4	64
10,2,2,1	16.6.4	384	7,6,1,1	4.4.6	96
10,2,1,1,1	16.4.4	256	7,5,3	4.4.4	64
9,6	4.4	16	7,5,2,1	4.4.4.4	256
9,5,1	4.4.4	64	7,5,1,1,1	4.4 4	64
9,4,2	4.4.4	64	7,4,4	4.6	24
9,4,1,1	4 4 6	96	7,4,3,1	4.4.4.4	256
9,3,3	4.6	24	7,4,2,2	4.4.6	96
9,3,2,1	4 4.4.4	256	7,4,2,1,1	4.4.4.6	384
9,3,1,1,1	4.4.4	64	7,4,1,1,1,1	4.4	16
9,2,2,2	4.4	16	7,3,3,2	4.6.4	96
9,2,2,1,1	4.6.6	144	7,3,3,1,1	4.6.6	144
9,2,1,1,1,1	4.4	16	7,3,2,2,1	4.4.6.4	384
8,7	4.4	16	7,3,2,1,1,1	4.4.4.4	256
8,6,1	4.4.4	64	7,2,2,2,2		4
8,5,2	4.4.4	64	7,2,2,2,1,1	4.4.6	96
8,5,1,1	4.4.6	96	7,2,2,1,1,1,1	4 6	24
8,4,3	4.4.4	64	6,6,3	6.4	24
8,4,2,1	4.4.4.4	256	6,6,2,1	6.4 4	96
8,4,1,1,1	4.4.4	64	6,6,1,1,1	6.4	24
8,3,3,1	4.6.4	96	6,5,4	4.4.4	64
8,3,2,2	4.4.6	96	6,5,3,1	4.4.4.4	256
8,3,2,1,1	4.4.4.6	384	6,5,2,2	4.4 6	96
8,3,1,1,1,1	4.4	16	6,5,2,1,1	4.4.4.6	384

Cards.	Combinations.		Cards.	Combinations.	
6,5,1,1,1,1	4.4	16	5,3,3,1,1,1,1	4.6	24
6,4,4,1	4.6.4	96	5,3,2,2,2,1	4.4.4.4	256
6,4,3,2	4.4.4.4	256	5,3,2,2,1,1,1	4.6.4.4	384
6,4,3,1,1	4.4.4.6	384	5,2,2,2,2,1,1	4.6	24
6,4,2,2,1	4.4.6 4	384	5,2,2,2,1,1 1,1	4.4	16
6,4,2,1,1,1	4.4.4.4	256	4,4,4,3	4.4	16
6,3,3,3	4.4	16	4,4,4,2,1	4.4.4	64
6,3,3,2,1	4.6.4 4	384	4,4,4,1,1,1	4.4	16
6,3,3,1,1,1	4.6.4	96	4,4,3,3,1	6 6 4	144
6,3,2,2,2	4.4 4	64	4,4.3,2,2	6.4.6	144
6,3,2,2,1,1	4.4.6.6	576	4,4,3,2,1,1	6.4.4.6	576
6,3,2,1,1,1,1	4.4 4	64	4,4,3,1,1,1,1	6.4	24
6,2,2,2,2,1	4.4	16	4,4,2.2,2,1	6.4.4	96
6,2,2,2,1,1,1	4.4.4	64	4,4,2,2,1,1,1	6.6.4	144
5,5,5		4	4.3,3,3,2	4 4.4	64
5,5,4,1	6.4.4	96	4,3,3,3,1,1	4.4.6	96
5,5,3,2	6.4.4	96	4,3,3,2,2,1	4.6.6.4	576
5,5,3,1,1	6.4.6	144	4,3,3,2,1,1,1	4.6 4.4	384
5,5,2,2,1	6.6.4	144	4,3,2,2,2,2	4.4	16
5,5,2.1,1,1	6.4.4	96	4,3,2,2,2,1,1	4.4.4.6	384
5,4,4,2	4.6.4	96	4,3,2,2,1,1,1,1	4.4.6	96
5,4,4,1,1	4.6.6	144	4,2,2,2,2,1,1,1	4.4	16
5,4,3,3	4.4.6	96	3,3,3,3,2,1	4.4	16
5,4,3,2,1	4 4.4.4.4	1024	3,3,3,3,1,1,1		4
5,4,3,1,1,1	4.4.4 4	256	3,3,3,2,2,2	4.4	16
5,4,2,2,2	4.4.4	64	3,3,3,2,2,1,1	4.6.6	144
5,4,2,2,1,1	4.4.6.6	576	3,3,3,2,1,1,1,1	4.4	16
5,4,2,1,1,1,1	4.4.4	64	3,3,2,2,2,2,1	6.4	24
5,3,3,3,1	4.4.4	64	3,3,2,2,2,1,1,1	6.4.4	96
5,3,3,2,2	4.6.6	144	3,2,2,2,2,1,1,1,1		4
5,3,3,2,1,1	4.6.4.6	576			

The number sought 17264.

Mr *T. Barker* and *Pamphagus* answer it from pages 520, 521, of Birk's Arithmetic, where, they inform us, it is calculated by Major Watson, and the number found to be 17264, or 33528 holes.

Another Solution by Mr. J. Landen, *taken from the Appendix to Dr.* Hutton's *Edition of the Diaries.*

If $(1+x)^4 \times (1+x^2)^4 \times (1+x^3)^4$ (9) $\times (1+x^{10})^{16}$ be actually involved, the co-efficient of that power of x whose exponent is 15 will be the number sought.

To find the coefficient with facility, observe that the above expression is $= (1+x)^4 \times (1+x^2)^4 \times (1+x^4)^4 \times (1+x^8)^4 \times (1+x^3)^4 \times (1+x^6)^4 \times (1+x^5)^4 \times (1+x^7)^4 \times (1+x^9)^4 \times (1+x^{10})^{16} = (1+x+x^2+x^3 \ldots . x^{15})^4 \times (1+x^3+x^6+x^9)^4 \times (1+x^5)^4 \times (1+x^7)^4 \times (1+x^9)^4 \times (1+x^{10})^{16}$.

Now $(1+x+x^2+x^3 \ldots , x^{15})^4$ being $= \left(\frac{1-x^{16}}{1-x}\right)^4 = (1-4x^{16} + \frac{4.3}{1.2}x^{32}$, &c.$) \times (1+4x+\frac{4.5}{1.2}x^2+\frac{4.5.6}{1.2.3}x^3$, &c.) it appears, that, in the value of $(1+x+x^2+x^3 \ldots . x^{15})^4$ the terms wherein the exponents of the powers of x are not greater than 15 are $1+4x+\frac{4.5}{1.2}x^2+$

$\frac{4.5.6}{1.2.3}x^3$, (16) $= \frac{16.17.18}{1.2.3}x^{15} + \frac{15.16.17}{1.2.3}x^{14} + \frac{14.15.16}{1.2.3}x^{13} \ldots\ldots\ldots 4x + 1.$ It appears also (by an easy involution) that, in the value of $(1 + x^3 + x^6 + x^9)^4 \times (1 + x^5)^4 \times (1 + x^7)^4 \times (1 + x^9)^4 \times (1 + x^{10})^{16}$, the terms, wherein the exponents of the powers of x are not greater than 15, are $1 + 4x^3 + 4x^5 + 10x^6 + 4x^7 + 16x^8 + 24x^9 + 38x^{10} + 40x^{11} + 63x^{12} + 128x^{13} + 102x^{14} + 212x^{15}$. It is obvious therefore, that the coefficient sought is $= 1 \times \frac{16.17.18}{1.2.3} + 4 \times \frac{13.14.15}{1.2.3} + 4 \times \frac{11.12.13}{1.2.3} + 10 \times \frac{10.11.12}{1.2.3}$, &c. (13) $= 816 + 1820 + 1144 + 2200 + 660 + 1920 + 2016 + 2128 + 1400 + 1260 + 1280 + 408 + 212 = 17264$.

XV. QUESTION 580, *by Mr.* Cha. Hutton.

Let AIB, RISO be two right lines, any-how intersecting each other within the circle ARBWO, whose diameter is ASW; then, putting AW $= d$, and p, q, r $=$ the cosines of the $\angle$s I, S, A, respectively, the general equation of the circle, defined by AI (x), and RI (y), considered as an abscissa and an ordinate, is $yy + (dq + 2px)y - drx + xx = 0$; Required the demonstration or investigation?

Answered by Mr. W. Wales.

Draw AD, CE $\perp$ OR (C being the centre of the circle), and let the points W, B, be joined. Then, d being $=$ the diameter AW, p, q, r $=$ the cosines of the angles I, S, A (to rad. 1), $x =$ AI, and $y =$ IR (per quest.), $1 : d :: r : dr =$ AB. $\therefore$ BI $= dr - x$. Also, $1 : x :: p : px =$ DI, and $y \pm px =$ DR. Lastly, (because of the parallel lines AD, CE, and by composition of ratios) $1 : \frac{1}{2}d :: q : \frac{1}{2}dq =$ DE, and DR $\mp$ DE $=$ ER $=$ EO $= \frac{1}{2}$RO (cor. 2, 3, Simp. Geom. 2d edit.) $= y \mp \frac{1}{2}dq \pm px$; $\therefore$ IO $= y \mp dq \pm 2px$, and (21·3) $y(y \mp dq \pm 2px) = x \times (dr - x)$; which, properly ordered, gives $yy \mp dqy \pm 2pxy - drx + xx = 0$, a general equation; the upper signs of the 2d and 3d terms obtaining when the $\angle$S is obtuse and the lower ones when the $\angle$I is obtuse, and when they are both acute, the two affirmative signs take place. But, if AB be drawn on the other side of the diameter AW, the contrary must be observed in every case.

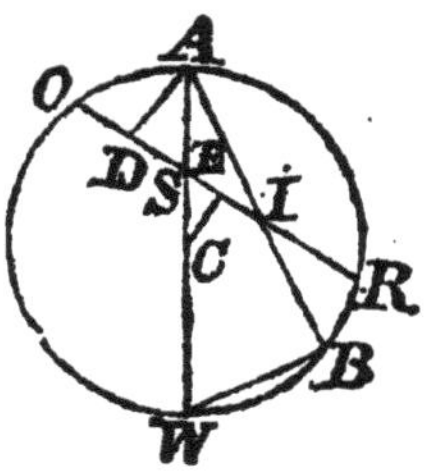

Cor. 1. If AB coincide with AW, $p = q$, and $r =$ radius; $\therefore$ the equation, in this case, is $yy \mp dqy \pm 2pxy - dx + xx = 0$.

Cor. 2. When OR is $\perp$ AW, q vanishes, and the equation becomes $yy \pm 2pxy - drx + xx = 0$.

Cor. 3. Lastly, when both these suppositions take place, the expression becomes $yy - dx + xx = 0$; the common equation.

Answer to the same by Mr. Cha. Hutton, *the Proposer.*

Put $a =$ AB, and s, n, and $c =$ the sines of the $\angle$S I, S and A (the rest the same as above); then (by trigonom.) $cxn^{-1} =$ IS and sxn^{-1} $=$ AS; and (by the prop. of the circle, OS being $=$ AS $\times$ SW $\times$ RS^{-1} $= sxn^{-1} \times (d - sxn^{-1}) \times (y + cxn^{-1})^{-1}$ and OI $\times$ IR $=$ AI $\times$ IB) $(sxn^{-1} \times (d - sxn^{-1}) \times (y + cxn^{-1})^{-1} + cxn^{-1}) \times y$ $= x \times (a - x)$: But $a = dr$, $s = cq + nr$, and $p = cn - qr$; whence, by substitution, $yy + (dq + 2px)y - drx + xx = 0$.

Cor. 1. Hence, $y = -px - \frac{1}{2}dq \pm \sqrt{((px + \frac{1}{2}dq)^2 + drx - xx)}$.

Cor. 2. If $y = 0$, then $x = 0$, or $= dr$ ($=$ AB) as it ought.

Cor. 3. If $x = 0$, then $y = 0$, or $= -dq$.

Cor. 4. If $x =$ AB $= dr$, then $y = 0$, or $= -d \times (2pr + q)$.

Cor. 5. If either S or I be an obtuse angle, the sine of q or p must be changed wherever it is found. A can never be an obtuse angle, and therefore r is always affirmative.

Cor. 6. If AB be on the other side of AW, the sines of p and q will be changed.

Cor. 7. If AB coincides with AW, then $r = 1$, $p = q$, and the general equation is $yy + dqy + 2pxy - dx + xx = 0$; which, when S $=$ a right angle (q being then $= 0$) becomes barely $yy = dx - xx$, the common equation to the circle.

XIV. QUESTION 581, *by Mr.* T. Allen, *of Spalding.*

Let DGB be a semicircle, P a given point in its diameter; to find the nature of the curve DEA, whose ordinate EC shall be always proportional to the area DGP, bounded by the right lines DP, GP, and the circular arc DG.

Answered by Mr. T. Allen, *the Proposer.*

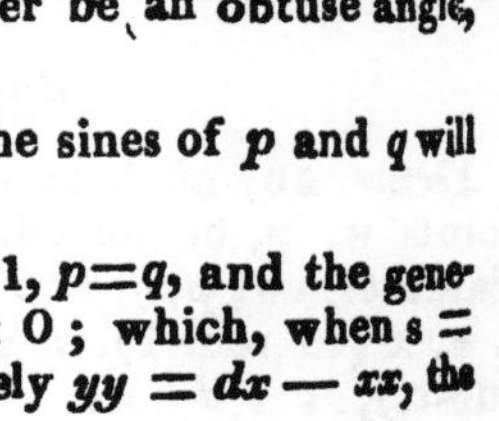

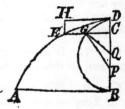

Let DGB be any curve meeting the axis DB in B and let P be a given point in that axis. Then, putting the abscissa DC $= x$, ord. CG $= y$, and DP $= a$, $\frac{1}{2}(y\dot{x} + a\dot{y} - x\dot{y})$ will truly exhibit the fluxion of the space DGP. Wherefore, in the present case, putting DQ (rad.) $= r$, QP $= q$, and the arc DG $= z$, we shall have $\frac{1}{2}(y\dot{x} + a\dot{y} - x\dot{y}) = \frac{1}{2}\dot{x}\sqrt{(2rx -}$

$xx) - \frac{r x\dot{x} - x^2\dot{x}}{2\sqrt{(2rx - xx)}} + \frac{a\dot{y}}{2} = \frac{rx\dot{x}}{2y} + \frac{a\dot{y}}{2}$; whose fluent, $\frac{rz + qy}{2}$,* is the area of the space DGP. If therefore this area be to the ordinate EC in the constant ratio of 1 to m, the ordinate itself will be $= \frac{1}{2}m \times (rz + qy)$.

Corollary. Suppose $m = 2$, and draw DH parallel and EH $\perp$ CE, then will $rx\dot{z} + qx\dot{y} = (arx\dot{x} - qx^2\dot{x}) \div y$ be the flux. of the space EDH; whose fluent is $ar \times (z - y) + \frac{1}{2}(3qry + qyx - 3qrz)$: And therefore the area of EDC will be $= rzx + \frac{1}{2}qyx + ar \times (y - z) + \frac{3}{2}qr \times (z - y)$; which, when $x = 2r$, becomes $\frac{1}{2} \times 3{\cdot}1416rr \times (4r - 2a + 3q)$, the true area of the whole curve DAB.

Answer to the same by Mr. Cha. Hutton.

Let Q be the centre of the semicircle DGB, and draw QG. Put $r =$ DQ ($=$ QG $=$ QB), $d =$ QP, and $n =$ the semicir. DGBD $\div$ AB $=$ the area DGPD $\div$ EC. Then, it is evident that EC $=$ (DGPD $\div n =$) $\frac{1}{2}rn^{-1} \times$ the arc DG $\pm \frac{1}{2}dn^{-1} \times$ GC, according as P is below or above Q.

Cor. 1. If P coincides with Q, $d = 0$, and then EC $= \frac{1}{2}rn^{-1} \times$ DG.

Cor. 2. If P coincides with B, then $d = r$, and EC $= \frac{1}{2}rn^{-1} \times$ (DG $+$ GC).

Cor. 3. If P coincides with D, $d = -r$, and then EC $= \frac{1}{2}rn^{-1} \times$ (DG $-$ GC).

THE PRIZE QUESTION, *by Mr.* Tho. Moss.

The longest side of a trapezium (being the diameter of a circle in which it may be inscribed), the distance intercepted between the extremity thereof and the point of intersection made by that side and that which is opposite, both produced, being given; also the angle formed at the intersection of the diagonal: To construct the trapezium?

Answered by Mr. W. Wales.

Construction. From O, the centre of the circumscribing semicircle, draw the radii OC, OD, making with each other an angle $=$ twice the complement of that formed by the diagonals at the point of intersec-

* This expression is no more than the sum of the two values of the sector DGQ $= \frac{1}{2}rz$, and ΔQPG $= \frac{1}{2}qy$, as given by the common known rules; and therefore all the fluxionary process from the beginning of the solution to this expression, is quite needless, and may be omitted. H.

tion, and through O draw BE, meeting the semi-circumference in B, A and the right line, joining the points C, D, produced in E, so that AE may = the given intercepted distance; then, join the points B, C and A, D, and the thing is done.

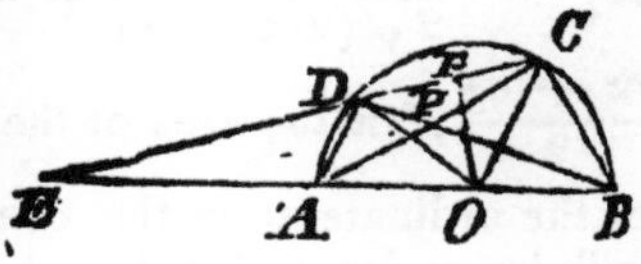

Demonstra. Let the diagonals AC, DB be drawn, intersecting each other in P. The ∠ACB (PCB), being in a semicircle, is a right angle; whence the ∠PBC is the complement of the ∠P, = half the ∠DOC (Simp. Geom. 10. 3. 2d edit.).—The method of calculation, from this construction is extremely obvious and easy.

Construction to the same by Mr. Tho. Bosworth.

Describe the △ECO, having the ∠C = the given angle to be formed at the intersection of the diagonals, the side OC = ½ the given side of the trapezium, and the side OE = OC + the given distance to be intercepted, and with the radius OC describe a semicircle meeting EO, produced in B, and cutting EC in D; join A, D and B, C, and then will be formed the trapezium required.—For, letting fall OF ⊥ DC and drawing the diagonals, intersecting each other in P, it is evident (from the properties of the circle, &c.) that the ∠DBC = ∠FOC, and the ∠ACB = ∠OFC (being both right angles); wherefore the ∠CPB = ∠ECO, the given angle, by construction.

Questions proposed in 1768, *and answered in* 1769.

I. QUESTION 582, *by Mr.* Geo. Lodge.

Diarian sages, you will find,
From the * equations here subjoin'd,
The age and fortune of a fair,
Whose life is blameless, heart sincere.

* viz. $\left\{\begin{array}{l} xxy + \dfrac{\sqrt{x}}{y} = 16386, \\ xyy + \dfrac{x}{\sqrt{y}} = 1056; \end{array}\right\}$ where x represents the number of 10*l.* notes that make up the fortune, and y the square root of the age of this amiable fair one.

Answered by Mr. Rich. Gibbons.

It is manifest (from the data) that x and y (and consequently xxy and xyy) are whole square numbers; whence $xxy + \sqrt{x} \div y$ being $=$ $(16386 =)\ 128^2 + 2$ and $xyy + x \div \sqrt{y} = (1056 =)\ 32^2 + 32$ (per quest.), $\sqrt{x} \div y$ will be $= 2$, and $x \div \sqrt{y} = 32$, and consequently $x = 64$ and $y = 4$. Whence the age of this amiable fair one is 16 years, and her fortune is 640*l*.——All vastly pretty!

II. QUESTION 583, *by Mr.* W. Spicer.

A heavy body descending freely by the force of gravity on an inclined plane, whose length is 400 feet, descends 111 feet of the said length in the last second; required the altitude of the said plane and the time of descent?

Answered by Mr. Paul Sharp.

By the laws of descending bodies and the division of ratios, $\sqrt{400} - \sqrt{289} : \sqrt{400} :: 1'' : \sqrt{400} \div (\sqrt{400} - \sqrt{289}) = 6\frac{2}{3}''$, the whole time of descent; and the distance that would be perpendicularly descended in the same time is $714\frac{22}{27}$ feet: Whence the altitude of the plane required is readily found $= (400^2 \div 714\frac{22}{27} =)\ 223\frac{161}{193}$ feet.

III. QUESTION 584, *by Master* J. Paty, *at the Mathematical Academy, Bristol.*

Supposing a cow to bring forth a she calf at the age of two years, and then to continue yearly to do the same, and every one of her brood to bring forth a she calf at the age of two years, and afterwards yearly likewise; how many may spring from the old cow and her brood in 40 years?

Answered by Mr. W. Spicer.

From the nature of the question, it appears that the increase in the first year will be 0, in the 2d year 1, in the 3d year 1, in the 4th year 2, in the 5th year 3, in the 6th year 5, and so on to 40 years or terms (each term being = the sum of the two next preceding ones); whence, the two last terms are 39088169 and 63245986, and the sum of them all (or the whole series) $= 2 \times 63245986 + 39088169 - 1 = 165580140$, the increase required.

IV. QUESTION 585, *by Mr.* Jer. Ainsworth.

Required to find a fraction such, that being taken from its reciprocal the remainder shall be a square?

Answered by Mr. Tho. Barker, *of Wisset.*

Let $1 \div x$ be the required fraction; then will $x - 1 \div x = aa$, a square number (per quest.), and consequently $x = \frac{1}{2}\sqrt{(a^4 + 4)} + \frac{1}{2}aa$; whence it appears that if such a fraction a be found as that its biquadrate added to 4 may be a square number, the conditions of the question will be answered, &c.

V. QUESTION 586, *by Miss* Ann Nicholls.

I made observation, last spring, at two places under the same meridian, differing in latitude 4°, and found the sun to rise exactly at 5 o'clock in the one and at 4 in the other, and that the difference of his meridian altitude at the said places, at the times of observation, was 2°; the latitude of each place and the days of observation are required?

Answered by Mr. Wm. Gawith.

Since the difference in latitude of two places situate under the same meridian is universally equal to the difference of the sun's meridional altitudes at those places for the same day, it is manifest that during the interval of the two observations, the sun had changed his declination 2°. This being premised, let a and b represent the sines of the two given ascensional differences, c and d the tangents of 2° and 4° respectively, and x the tangent of the sun's declination at the first time of observation: Then (by Crakelt's translation of Mauduit's Trigonometry, p. 31) $(x + c) \div (1 - cx)$ will be the tangent of the declination at the second time of observation; and (by spherics) $a \div x$ and $(b - bcx) \div (c + 1)$ the tangents of the less and greater latitudes respectively: the differences of which being 4° (per quest.), $(a + dx) \div (x - da)$ will be $= (b - bcx) \div (c + x)$, or $xx \times (d + bc) + x \times (a + cd - b - dabc) = - ca - dab$, and $x =$ tang. 7° 54′ 50″, corresponding to April 10th: Whence the other day of observation will be found to be April 15th; and the two latitudes 48° 48′ 19″ and 52° 48′ 19″.

VI. QUESTION 587, *by Mr.* C. Smith.

A designer, having occasion to delineate in perspective the longest half of an ellipse coinciding with the horizontal plane, desires to know the true breadth of its projection, when projected on a perspective plane touching (both it and its vertical) the nearest end of the transverse diameter, and making an angle with the said diameter of 50°; the distance of the eye from the perspective plane being 100, the transverse diameter 32·92, and the semi-conjugate 11 feet.

The direction of the visual line from the eye to the perspective plane not being given in proposing QUESTION 587, has prevented any of our ingenious correspondents answering it: but were it sufficiently limited, nothing more would be required than to find the breadth of the section of a given elliptical cone, made by a plane passing through the extremity of the greater diameter of its base in a given direction; which may be easily affected, as *Pamphagus* justly observes.

VII. QUESTION 588, *by Mr.* Jos. Dymond.

Given the right line bisecting the vertical angle of a plane triangle,

and terminating in its base, and the perpendiculars falling thereon (produced) from the extremities of its base; to determine and construct the triangle.

Answered by Mr. J. Chipchase.

Construc. On the right line DG take DB and BE respectively equal to the greater and the less given perpendiculars, and erect thereon, at B, the perp. BH = the given bisecting line: Produce DG till DF : EF :: DB : BE (*i. e.* in the ratio of the given perpendiculars), and through the points F, H let a right line be drawn, meeting EC, DA (perpendicular to DF) in C and A; join the points A, B and C, B, and ABC will be the triangle required.

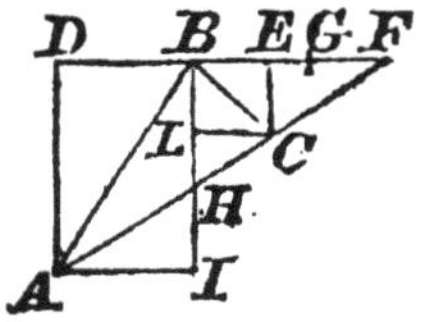

Demonstra. Draw AI and CL ⊥ BH (produced) in I and L; then, DF being to EF :: DA : EC (per sim. △s) :: DB : BE (per construc.) and the ∠s D and E equal (being both right), the △s BDA, BEC will be similar (Euc. 6. 6.), and consequently, taking the equal ∠s DBA, EBC from the right ∠s DBH, EBH, their remainders, viz. the ∠s ABH and CBH will be equal; whence BH manifestly bisects the ∠ABC, and AI, CL are respectively = DB, BE (the given perpendiculars, per construction).

The method of calculation from this construction is exceedingly easy.

VIII. QUESTION 589, *by* Pamphagus.

Given the base, one of its adjacent angles, and the line bisecting the vertical angle of any plane triangle; to determine the said triangle.

Answered by Master Jer. Osborne, *at the Mathematical Academy, Bristol.*

Suppose the base given $= a$, the bisecting line $= b$, the sine and cosine of the given angle at the base $= s$ and c, and the sine of half the vertical angle $= x$ (rad. $= 1$). Then, (per trigonom.) the sine of the angle the line bisecting the vertical angle makes with the base will be $= s\sqrt{(1 - xx)} + cx$, and that of the vertical angle itself $= 2x\sqrt{(1 - xx)}$; also $(bs\sqrt{(1 - xx)} + bcx) \div s$ will be found = the side adjacent to the given angle, $bx \div s =$ the segment of the base (made by the said bisecting line, and) adjacent to the same, $sa \div 2x\sqrt{(1 - xx)}$ = the other side, and $(as - bx) \div s =$ the other segment. Whence (Euc. 3. 6.) $abss\sqrt{(1 - xx)} + abscx - bbsx\sqrt{(1 - xx)} - bbcxx = abss \div 2\sqrt{(1 - xx)}$, or (when properly reduced) $x^6 \times 4b^2 - x^5 \times 8abs + x^4 \times 4aass - 4bbss - 4bb) + x^3 \times (4abs^3 + 8abs + xx \times (4bbss - 4aass) - 4abs^3 x + a^2 s^4 = 0$; from whence x may be determined.

IX. QUESTION 590, *by* Nujo Dargnas.

To determine a point, from which if right lines be drawn to the three angular points of a given plane triangle, the sum of their squares shall be equal a given space (*aa*), and if from the said point lines be drawn to the three sides of the triangle (produced if necessary) making given angles therewith, the rectangle contained under the line drawn to one of the sides and a given line (*b*) shall be equal to the sum of the rectangles under the other drawn lines and given lines (*c*, *d*).

Answered by Nujo Dargnas, *the Proposer.*

Construc. Let ACB be the given triangle: the given lines *b*, *c*, *d* are referred to AB, AC, and BC respectively. From any point L in AB, draw LM, LN making the angles ALM, ANL equal to the given angles to be made with the sides AB, AC respectively; take OL : LN :: *c* : *b* and LP : PM :: OL : LM, and draw AP meeting BC in H. In like manner from any point *a* in AB draw *ad*, *ac* making the angles B*ad*, B*ca* = the given angles to be made with the sides AB, BC respectively: Take W*a* : *ac* :: *d* : *b* and *ab* : *bd* :: W*a* : *ad*; join B, *b*, meeting AC in R, and join R, H. Again, bisect AB in E; join E, C and take EG = $\frac{1}{2}$ GC, and describe a circle with a radius GY such that $3GY^2 = aa - AB \times AE - 2EC \times EG$, meeting RH (produced if necessary) in D, the point required.

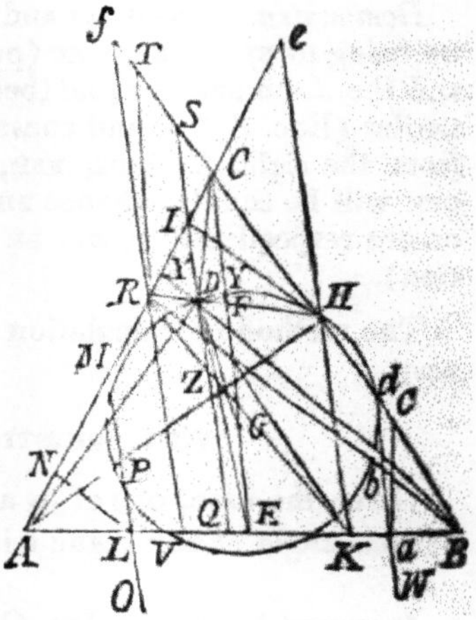

Demons. From R and H draw RV, HK ∥ DQ or ML; and RT, HI ∥ DS or *ac*; and DX or LN respectively: Join D, E; D, G and R, K meeting DQ (produced if necessary) in Z, and draw DF ⊥ CE; produce, if necessary, HK, RV to meet AC, CB produced in *e* and *f*.——Then $AD^2 + BD^2 = AB \times AE + 2ED^2$ (Simp. Geom. 2. 3.) $= AB \times AE + 2EF^2 + 2DF^2$, and adding to each $CD^2 = CF^2 + FD^2$, $CD^2 + AD^2 + BD^2 = AB \times AE + 2EF^2 + CF^2 + 3FD^2$. But $CF^2 + 2EF^2 = CD^2 - 3DF^2 + 2ED^2 = DG^2 + CG^2 - 2CG \times GF$ (CD^2, Euc. 13. 2.) $+ 2DG^2 + 2EG^2 + 2CG\ (4EG) \times GF$ ($2ED^2$ (Euc. 12. 2.) $- 3DF^2 = 3FG^2$ ($3DG^2 - 3DF^2$) $+ 2EC \times EG$ ($CG^2 + 2EG^2$); whence $AD^2 + BD^2 + CD^2 = AB \times AE + 3FG^2 + 2EC \times EG + 3FD^2 = AB \times AE + 2EC \times EG + 3GY^2$ ($3DG^2$) $= aa$, the given space (per construc.). Again, KH : H*e* :: LP : PM :: OL : LM, and IH : H*e* :: LN : LM; whence, by equality, KH : IH :: OL : LN :: *c* : *b* (per construction), and consequently KH × *b* = IH × *c*. Also, VR : R*f* :: *ab* : *bd* :: W*a* : *ad*, and RT : R*f* :: *ac* : *ad*; by equality,

VR : RT :: $wa : ac$ :: $d : b$ (per construction); whence VR $\times\ b =$ RT $\times\ d$. Again, DZ $\times\ b$: KH $\times\ b$ :: ZR : RK :: DR : RH :: DX $\times\ c$: IH $\times\ c$: But KH $\times\ b =$ IH $\times\ c$; whence DZ $\times\ b =$ DX $\times\ c$. Also, QZ $\times\ b$: RV $\times\ b$ :: KZ : RK :: DH : RH :: DS $\times\ d$: RT $\times\ d$; But RV $\times\ b =$ RT $\times\ d$; whence QZ $\times\ b =$ DS $\times\ d$, and therefore DS $\times\ d +$ DX $\times\ c = ($QZ $\times\ b +$ DZ $\times\ b =)$ DQ $\times\ b$.

If the point D (required) falls without the $\triangle$ ACB, the preceding conclusion will become DS $\times\ d -$ DX $\times\ c =$ DQ $\times\ b$, or DX $\times\ c -$ DS $\times\ d =$ DQ $\times\ b$, according as the said point is situate on the left or right hand side of the line EC; the demonstration whereof is exactly the same as above.

X. QUESTION 591, *by Mr.* Paul Sharp.

Required the area of the common parabola, whose abscissa is $= 10$, and radius of curvature, at the bounding ordinate, equal to three times a tangent drawn from thence and terminating in its axis produced?

Answered by Mr. J. Addison.

Put the given abscissa $(= 10) = a$, and the corresponding ordinate $= y$; then will $2a$ be the subtangent, $\sqrt{(4aa + yy)}$ the tangent and $yy \div a$ the latus-rectum of the parabola required, and (per Simp. Flux. pa. 75, 2d edit.), $(4aa + yy)^{\frac{3}{2}} \div 2ay =$ the radius of curvature at its bounding ordinate; whence $(4aa + yy)^{\frac{3}{2}} \div 2ay + y = 3\sqrt{(4aa + yy)}$ (per question); solved, $y = 9{\cdot}5$, and the area required $=$ 126·66, &c.

And according to this method, nearly, the answer is also given by Mess. *T. Barker*, *Pamphagus*, *T. Robinson*, *Alex. Rowe*, and *S. Vince*; but a press error having unluckily happened in proposing this question, viz. *and* instead of *at* before "the bounding ordinate, &c." it is become neither so neat nor simple as was intended; for, according to the proposer's substitution and meaning, a is $=$ the given abscissa, $4x =$ the parameter, $2\sqrt{(aa + ax)}$ the tangent, $(2a + 2x)\sqrt{(ax + xx)} \div x$ the radius of curvature at the ordinate, and $(2a + 2x)\sqrt{(ax + xx)} \div x = 3 \times 2\sqrt{(aa + ax)}$, or $a + x = 3\sqrt{ax}$; whence $x = 1{\cdot}459$, the parameter $= 5{\cdot}836$, the bounding ordinate $= 15{\cdot}27874$, and the area $= 101{\cdot}85826$.

XI. QUESTION 592, *by Mr.* J. Chipchase.

There is a bason, in form of a semi-globe, filled with spring water and placed on horizontal ground, and a person can see a piece of money laid at the bottom of it to the distance of nine feet from its axis, when quite full; but when filled to only half the whole depth, he can see it no further than to the distance of 6 feet therefrom: Required the height of the person's eye and the diameter of the bason?

Answered by the Rev. Mr. C. Wildbore.

Let ABK represent half the bason, and join A, K; then shall the $\angle$GAK $=$ BAK (45°) $=$ the angle of refraction at the 1st station: Make AI $= \frac{1}{4}$AK, so shall GAI $=$ the corresponding angle of incidence, the sines being to each other as 3 : 4; whence BI $=$ $(1 \div 2\sqrt{2})$ AB $=$ ·35355333AB.

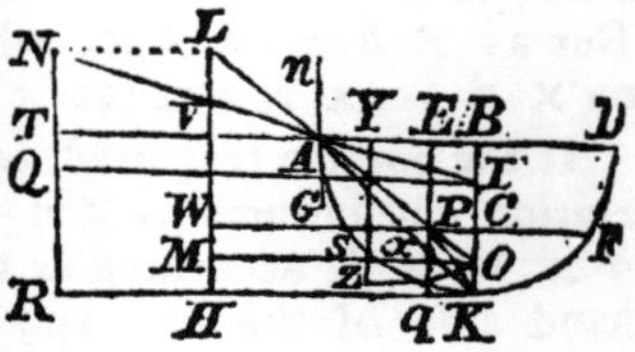

Now let CS be the surface of the water, BY $= \frac{1}{4}\sqrt{7} \times$ AB, $xz = \frac{1}{8}$AB, and from the centre Y, on the line cx, and with the line (or distance) xz, describe the conchoid of Nicomedes zo, and from the point O, where it cuts BK, draw OL through A; join K, P; so shall OPq be the angle of incidence and KPQ that of refraction, at the 2d station. For the common equation of the conchoid is CO $\times$ BY $\div$ BO $(= \frac{1}{4}\sqrt{7}$ CP) $= \sqrt{xz - \text{CO}^2}$), or $28\text{CP}^2 = 9\text{AB}^2 - 64\text{CO}^2$, or 64OP^2 $(= 64\text{CP}^2 + 64\text{CO}^2 = 9\text{AB}^2 + 36\text{CP}^2 = 36\text{CK}^2 + 36\text{CP}^2 = 36\text{KP}^2$, or $8\text{OP} = 6\text{KP}$: Whence, as 8 : 6, or 4 : 3 :: KP : OP, and consequently the sines of the angles OPq, KPq are in the same ratio, and must be the required angles.——The equation of the conchoid gives CO $=$ ·287087953 $\times$ AB. Also, by sim. $\triangle$s, $1 : \frac{1}{2} +$ ·287087953 :: 6 (OM) : LM $=$ 4·7225274, and AB : BI, or $1 : \dfrac{1}{2\sqrt{2}}$ :: QI (9) : NQ $=$ 3·1819800; whence LM $-$ NQ $=$ 1·5405474 $=$ IO $=$ (·287087953 $+ \frac{1}{2} - 1 \div 2\sqrt{2}$) $\times$ AB, and consequently AB $=$ 3 553456 (the radius of the bason), and IO $-$ CO $=$ IC $=$ ·5203914; which give NQ $+$ IC $+$ CK $=$ 5·4790994 feet, the height of the spectator's eye.

XII. QUESTION 593, *by Mr.* J. Bennett.

To determine the nature of the curve, whose tangent terminated every-where by it and an indefinite right line CD, is a constant quantity $= 100$; also, to determine the length of the part thereof intercepted between its highest point and that point whose height, above the said right line CD, is $= 20$?

Answered by Mr. T. Allen.

Let the abscissa CE $= x$, ordinate EF $= y$, curve AF $= z$; and let GF $=$ DB $= 100 = a$, AC $= 20 = c$. Then will the subtangent GE $= y\dot{x} \div \dot{y}$, and $y^2\dot{x}^2 \div \dot{y}^2 + yy = aa$; whence $\sqrt{(\dot{x}^2 + \dot{y}^2)}$ $(= \dot{z}) = a\dot{y} \div y$, whose fluent, corrected, gives $z = a \times$ hyp. log. $(y \div c)$; which when $y = a = 100$, becomes $= 100 \times$ hyp. log. of $5 = 160{\cdot}94379$, the length of the curve AFB. From the equation

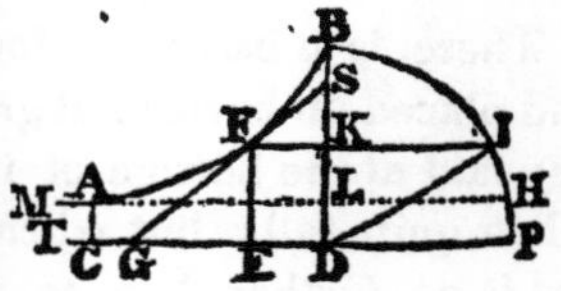

$\frac{y^2\dot{x}^2}{\dot{y}^2} + yy = aa$, we also have $= \sqrt{(aa-yy)} \times \frac{\dot{y}}{y} = \frac{aa\dot{y}}{y\sqrt{(aa-yy)}}$ $- \frac{\dot{y}}{\sqrt{(aa-yy)}}$; where taking the fluents, properly correcting them, and putting $d = \sqrt{(aa-cc)}$, will be had $x = \frac{1}{2}a \times$ hyp. log. of $\left(\frac{a-\sqrt{(aa-yy)}}{a+\sqrt{(aa-yy)}} \times \frac{a+d}{a-d}\right) + \sqrt{(aa-yy)} - d$, the equation of the curve required.

Corollary. Upon the centre D, with the radius DB, describe the circular arc BIH. Make DL = AC, and draw LH ⊥ BD. Then will the area LBIHL = the area of the whole curve AFBDCA. For, putting DK $= y$, KI $= \sqrt{(aa-yy)}$, and the flux. of LHIK will be $\sqrt{(aa-yy)} \times \dot{y} = y\dot{x}$, from what is given above; therefore, &c.

Answer to the same by the Rev. Mr. Cha. Wildbore.

Let TD be the indefinite right line, MFB the required curve, and GF the given tangent at F; then at B the vertex (from the nature of the curve) BD = FG = 100 = a. Produce GF till it meets BD in S, and let fall FK ⊥ BD; then, putting BK $= x$ and FK $= y$, SK will be $= y\dot{x} \div \dot{y}$, and SK : FS : : FE (KD) : GF (a): Whence $\dot{y} = \frac{\dot{x}\sqrt{(2ax-xx)}}{a-x}$ $\left(= \frac{u^2\dot{u}}{aa-uu} = \frac{aa\dot{u}}{aa-uu} - \dot{u}\right.$, by putting GE $= \sqrt{(2ax-xx)} =$ $u)$; and, restoring x, $y = \frac{1}{2}a \times$ hyp. log. of $\frac{a+\sqrt{(2ax-xx)}}{a-\sqrt{(2ax-xx)}}$ $- \sqrt{(2ax-xx)} = 131{\cdot}263155$ when FE = 20. Now, let $z =$ the curve FB; then $\dot{z} = \left(\sqrt{(\dot{x}^2+\dot{y}^2)} =\right) \frac{a\dot{x}}{a-x}$, and z correct $= a \times$ hyp. log. $\frac{a}{a-x} = 160{\cdot}94379$, as required.——Whence these consectaries.

1. Because, when x is taken $= a$, the expressions for y and z become infinite, therefore the curve will continually approach nearer to the line TD, but can never meet with it.

2. A value which this infinite area can never exceed, may be found from the above expression for $\dot{y}$, thus: TE's flux. is $= -\dot{y}$, which multiplied by FB (= KD $= a - x$) gives $-\dot{x}\sqrt{(2ax-xx)}$ for the flux. of the infinite area ETMF. With GF (GB) radius describe the quadrant BIP, and produce FK to meet it in I; then the area of the circular segment DKIP, will be the fluent of this expression, or = the area ETMF; and when FE becomes = BD, the infinite area MFBDT = the whole quadrant BDP; which is also known from other principles, by considering the invariable line GF, as moving from an horizontal position to a vertical one.

3. Because $y\dot{x}$ is the fluxion of the area FEK, twice the area of the quadrant BDP will be $=$ the fluent of $\frac{1}{4}ax \times$ hyp. log. of $\frac{a + \surd(2ax - xx)}{a - \surd(2ax - xx)}$, when $x = a$; a conclusion not very easy to be derived another way.

XIII. QUESTION 594, *by Mr.* S. Vince.

To find the sum of the infinite series $\frac{1}{1.5.v^5} + \frac{1}{1.2.3.7.v^7} + \frac{1.3.3}{1.2.3.4.5.9.v^9} + \frac{1.3.3.5.5}{1.2.3.4.5.6.7.11.v^{11}} +$ &c.

Answered by the Rev. Mr. C. Wildbore.

In the given series substitute x for $\frac{1}{v}$ and take the fluxion; which will give $x^3\dot{x} \times : x + \frac{x^3}{1.2.3} + \frac{1.3.3.x^5}{1.2.3.4.5} + \frac{1.3.3.5.5.x^7}{1.2.3.4.5.6.7} +$ &c. $= x^3\dot{x}z$ (z being the arc of the circle whose sine is x $\left(\frac{1}{v}\right)$ and rad 1.): Whence, taking the fluents and restoring $\frac{1}{v}$, $\frac{1}{1.5.v^5} + \frac{1}{1.2.3.7.v^7} + \frac{1.3.3}{1.2.3.4.5.9.v^9} + \frac{1.3.3.5.5}{1.2.3.4.5.6.7.11.v^{11}} +$ &c. $= \frac{8z + (2 + 3vv) \times \surd(vv - 1) - 3zv^4}{32v^4}$, the sum of the series required.

Corollary. This method of solution may be extended to any series of the form $\frac{Ax^n}{n} + \frac{Bx^{n+m}}{m+n} + \frac{Cx^{n+2m}}{2m+n} +$ &c. by means of artifices similar to those made use of in Simpson's Fluxions, part 2d, sec. 3d and 6th, where A + B + C + &c. expresses any known series; and whereof the summation of the extremely difficult series $x + \frac{x^2}{4} + \frac{x^3}{9} + \frac{x^4}{16} +$ &c. proposed by the very sagacious Mr. *L*—— in the Ladies' Diary for 1760, is given as an example; but, for want of room, we are obliged to omit it.

XIV. QUESTION 595, *by Mr.* Tho. Moss.

To draw a right line from the given point E, in the base FA, produced of the given triangle ACF, intersecting AC in n, and meeting CF in E; so that drawing the right line AE, the triangle AnE may be the greatest possible?

Answered by Mr. Cha. Hutton.

Construction. Parallel to CF draw RB, meeting CA, produced, in B; ⊥ CA draw AD, meeting the circle on CB in D; make $Bn = BD$, and n will be the point through which RnE must pass.

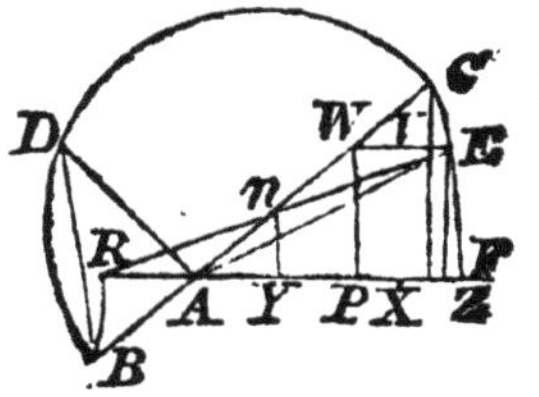

Demonstration. Draw EW ∥, and EZ, WP, nY ⊥ AF ——Then, by construction, $BA \times AC = (DA^2 = DB^2 - BA^2 = Bn^2 - BA^2 = (Bn + BA) \times (Bn - BA) =) (nB + BA) \times An$, or $An : AB :: AC : nB + BA :: nC (AC - An) : nB ((nB + BA) - AB)$ :: (by sim. △s) CE : BR, or $An : CE :: AB : BR ::$ (by sim. △s) $WC : CE$; whence $An = WC$.

Again, by sim. △s, WC (An) : Cn (AW) :: nY : ZE (WP), or $WC : CE :: Cn \times nY : CE \times EZ$, But when RnE moves about the centre R, and cuts the lines AC, CF it is known that the fluxion of EC : the flux. of Cn :: $CE \times ER : Cn \times nR$ :: (by sim. △s) $CE \times EZ : Cn \times nY$; whence the flux. of CE : flux. of Cn :: EC : CW (by equality of ratios), and consequently the flux. of EZ = the flux. of nY; in which case, it is well known, that their diff (EZ — nY) is a max. But this difference drawn into $\frac{1}{2}$ the given line AR is = the area of the △ AnE; and hence the said triangle is a maximum.

Corollary. It appears from the demonstration that $An = WC$; whence $nY = CV$, and $nY + EZ =$ the whole perpendicular CX.

Construction to the same by the Rev. Mr. C. Wildbore.

At the given point R erect the perpendicular DR (p. 13, Euc. 6.) a mean proportional to RA, RF, and take LR and RT = DR: Draw NT ∥ CF, and through N draw RE; join A, E, and the thing is done.

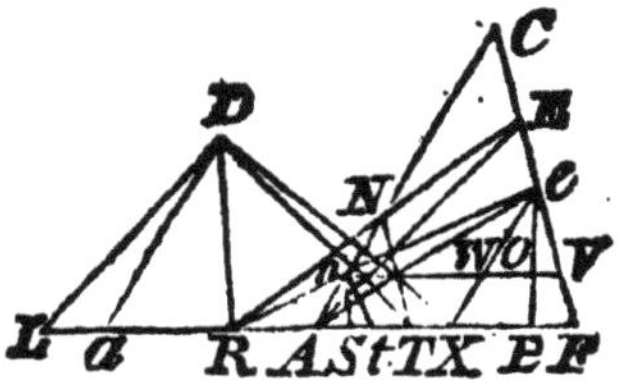

For, let Re be any other position of the required line; intersecting AC in n. Draw nt ∥ CF, ex to CA, nv to RF, and let fall the perpendiculars ns, eP: Then, the △s RnA, ReA, on the common base AR, are in the ratio of their perpendiculars ns, eP; consequently the △ Ane (their diff.) will be as $eP - ns = eo$, which must, therefore, be a maximum. But the △ vew being similar to the given one ACF, the ratio of eo to wv is given; therefore $wv = nv (tF) - nw (AX) = RF - Rt - RX + RA$ (per fig.) must be a maximum: But RF + RA is invariable; therefore $Rt + RX$ must be a minimum. By sim. △s, RA : RT :: RX : RF; whence DR must be likewise a mean proportional to Rt, RX. Take $Ra = Rt$, and join D, a; then the ∠s *a*DL, TDX being equal, DT greater than D*a*, and the ∠DTX greater than D*a*L, TX is evidently greater than *a*L, and consequently *a*x than LT; wherefore R*t*+RX is a minimum when each of them becomes equal to RT, as per construction.

Corollary. If R*t* + RX instead of a minimum, or the △A*ne* instead of a maximum, had been supposed = a given quantity, from what is shewn above the prob. is reduced to this, viz. In any right-angled △ *a*DX, are given the perpendicular DR and the base *a*x, to construct the triangle; the method of doing which is well known: And the construction will be equally easy, when, instead thereof, the ratio of the △R*n*A to the △R*e*A or of *n*s to *e*P or of A*n* to *e*F is given; for, in all these cases, the ratio of *a*R : RX is given.

XV. QUESTION 596, *by Mr.* Cha Hutton.

Let NO be the fixed and MP the semi-revolving axe of any spheroid, AGBQA any section of it by a plane inclined to the axes, and NPO another section through its axes and perpendicular to the former section: It is required to find the figure of the section AGB and the solidity of the part cut off the spheroid by it, both in the case of the oblong and oblate spheroid, supposing the distances DM, ME, of the perpendiculars BD, AE, from the centre M, to be 5 and 12, and the axes 50 and 30.

Answered by Mr. W. Crakelt.

If IK, CH be two ordinates to the fixed axis ON; MR a semi-diameter to the ordinate BA, and OS, RS two tangents at the points O, R. Then (by Deidier's Géométrie, p. 401) will be had BC × CA : IC × CK :: $RS^2 : OS^2$ :: B*b* × *b*A : *cb* × *b*H: But (per prop. of the circle), IC × CK = CQ^2 and *cb* × *b*H = be^2; consequently BC × CA : B*b* × *b*A :: CQ^2 : be^2, the property of the ellipse.

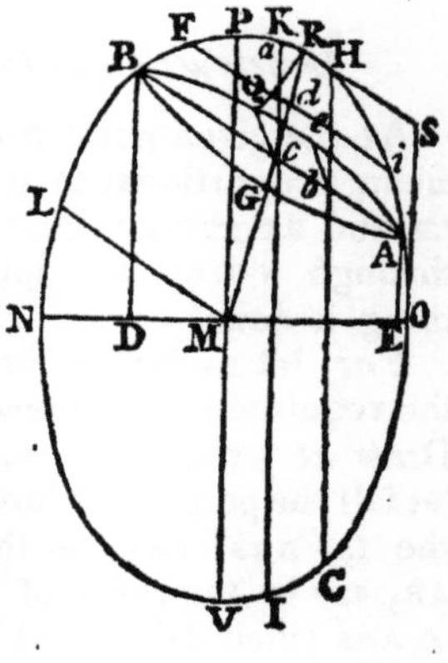

Having thus discovered the figure of the section AGBQ, let 3·14159 &c. be put = p, the semi-fixed axe NM = m, the semi-revolving axe PM = n, the semi-diameter MR = r, its semi-conjugate LM = s, the semi-ordinate BC = t, its corresponding abscissa RC = u (all which may, it is evident, be looked upon as given quantities), and any variable part R*d* of RC = x. Then (per prop. of the ellipse) rr (MR^2) : ss (LM^2) :: $2rx - xx$ (2MR − *d*R) × *d*R) : ($sd \div rr$) × ($2rx - xx$) = dF^2, and thence $dF = (s \div r)\sqrt{(2rx - xx)}$. Also (by trig.) rad. (1) : sin. ∠R*da* (= $mn \div rs$) : x (R*d*) : R*a* = $mnx \div rs$ (R*a* being perpendicular to FB)); whence the fluxion of the solidity (as the semi-conjugate of the section made by the plane

FD is $= (n \div r)\sqrt{(2rx - xx)}$) will be $(pmnn \div r^3) \times (2rx\dot{x} - x^2\dot{x})$: the fluent whereof $((pmnn \div r^3) \times (rxx - \frac{1}{3}x^3))$, when $x = u$, gives $(pmnn \div r^3 \times (ru^2 - \frac{1}{3}u^3))$ for the solidity required, equal in the oblong spheroid (when $m = 25$, $n = 15$, $r = 15{\cdot}0028901$, and $u = {\cdot}6418766$) 31·88508, and equal in the oblate spheroid (when $m = 15$, $n = 25$, $r = 24{\cdot}9938494$, and $u = 5{\cdot}3937069$) 1272·97097, &c.

Cor. When $x = 2r$, the solid BRA becomes $= \frac{4}{3}pmnn =$ the whole spheroid.

THE PRIZE QUESTION, *by Mr.* Tho. Allen.

If a chain 40 feet long, consisting of exceedingly small links of equal density, be put over a pulley (void of friction) with the two ends A and B hanging down, to the respective distances of 21 and 19 feet from the pulley, and in these circumstances, suppose the chain to put itself in motion. It is required to determine in what time the end A will arrive at an horizontal plane 200 feet below the pulley?

Answered by Mr. Tho. Allen, *the Proposer.*

Let $c = 40$ feet, the whole length of the chain, $a = 21$, $x =$ any variable part passed over the pulley in the time t, $v =$ the velocity of the chain at the end of that time, and $s = 32\frac{1}{6}$ feet. Then will $a + x$ be as the gravity of the descending, and $c - a - x$ that of the ascending part; therefore $a + x - (c - a - x) = d + 2x$ (by putting $d = 2a - c$) will be as the force acting on the chain, and $c : s :: d + 2x : s(d + 2x) \div c =$ the velocity that would be generated in 1″, by that force. Therefore 1″, (time) : $s \times \frac{d + 2x}{c}$ (velocity) $:: \frac{\dot{x}}{v}$ $(\dot{t})$: $\dot{v}$, the velocity generated in the time $\dot{t}$; $\therefore s \times \frac{d\dot{x} + 2x\dot{x}}{c} = v\dot{v}$; where taking the fluents, &c. $v = \sqrt{\frac{2s(dx + xx)}{c}}$. Therefore $\dot{t} = \sqrt{\frac{c}{2s}} \times \frac{x^{-\frac{1}{2}}\dot{x}}{\sqrt{(d+x)}}$; whose corrected fluent gives $t = \sqrt{\frac{2c}{s}} \times$ hyp. log. of $\frac{\sqrt{x} + \sqrt{(d + x)}}{\sqrt{d}}$; which, when $x = 19$ (or the chain quits the pulley will be $= 2{\cdot}908137$ seconds.

To find the remaining part of the time, put $p = 25{\cdot}33$ (the value of v $\left(= \sqrt{\frac{2s(dx + xx)}{c}}\right)$ when $x = 19$), $z =$ the space descended, after quitting the pulley, in the time T, and V = the velocity at the end of that time. Then, if the chain descended from rest, V^2 would be $= 2sz$; therefore, in the present case, $V^2 = 2sz + pp$: $\therefore V = \sqrt{(pp}$

$+ 2sz)$, and $\dot{t}\ (= \dot{z} \div v) = \dot{z} \div \sqrt{(pp + 2sz)}$; whose corrected fluent gives $T = (\sqrt{(pp + 2sz)} - p) \div s$; which, when $z = 160$, becomes $2{\cdot}46342''$. And hence the whole time required will be $=$ $5''\ 22{\cdot}3'''$; very nearly.

The same answered by Mr. W. Crakelt.

If 40 feet (the length of the whole chain) $= l$, 2 feet (the diff. of the two ends A and B hanging down from the pulley at first) $= d$, $32\frac{1}{6}$ feet the velocity generated at the earth's surface in $1''$, by gravity) $= s$, $x =$ a small part of the chain ascended or descended in any variable time t, and $v =$ the corresponding velocity of the point A at the end of that time (in seconds). Then, let the weight of the chain be what it may, $d + 2x$ will represent the motive force acting thereon; and therefore (since $l : d + 2x :: s : (s \div l)(d + 2x)$, and $1'' : \frac{\dot{x}}{v}\ (= \dot{t})$ :: $\frac{s}{l}\ (d + 2x) : \frac{s\dot{x}}{l}\left(\frac{d + 2x}{v}\right) = \dot{v})\ v\dot{v} = \frac{s}{l} \times\ (d + 2x)\ \dot{x}$; or, by taking the fluents, &c. $v = \sqrt{\left(\frac{2s}{l} \times (dx + xx)\right)} =$ (when $x = 19$) $25{\cdot}3322916$, &c. and consequently $\dot{t}\ \left(= \frac{\dot{x}}{v}\right) = \left(\frac{2s}{l} \times (dx + xx)\right)^{-\frac{1}{2}} \times \dot{x}$; and, by taking the correct fluents, $t = \sqrt{\frac{l}{2s}} \times$ hyp. log. of $\dfrac{\frac{1}{2}d + x + \sqrt{(dx + xx)}}{\frac{1}{2}d} =$ (when $x = 19$) $2{\cdot}90825683$, &c. seconds, the time elapsed when the chain quits the pulley. Now the height from which a heavy body must fall freely from rest, to acquire the foregoing velocity, will (by Simpson's Select Exer. pa. 184) be $= vv \div 2s = 9{\cdot}975$ feet, and the corresponding time $=$ $v \div s = {\cdot}78753238''$: Then $200 - 40 + 9{\cdot}975 = 169{\cdot}975$ feet, and the time in which a heavy body would descend through that dist. $=$ $3{\cdot}25090661''$; whence $2{\cdot}90825683'' + 3{\cdot}25090661'' - 0{\cdot}78753238''$ $= 5{\cdot}37163106''$, the time in which the end A will reach the horizontal distance given; which was required.

The same answered by the Rev. Mr. Cha. Wildbore.

Let APB, at its commencement of motion, and EPD when the end A has descended to E, be positions of the chain; then it is manifest that the motive force at E will be as $\text{PE} - \text{PD} = d\text{E} = \text{A}b + 2\text{AE}$ which multiplied into $\dot{\text{AE}}$, is known to be as half the fluxion of the square of the velocity at E; wherefore that velocity will be as $\sqrt{2} \times \sqrt{(\text{A}b \times \text{AE} + \text{AE}^2)}$, and $\dot{\text{AE}}$ divided by this will be as the fluxion of the time, and the time itself (bisecting Ab in e, and taking eg $(= \sqrt{(\text{A}b \times \text{AE} + \text{AE}^2)})$ a mean proportional to AD and AE) will be as $(1 \div \sqrt{2}) \times$ hyp. log. of $(gE \div eb)$: But the real motive force at E, measured by the distance that might be uniformly gone over in 1″ with the velocity at B, is to dE, as the force of gravity to AP + PB, or as $32\frac{1}{6} : 40$, or $193 : 240$; therefore the velocity of the end A per second, when B arrives at P, $= \sqrt{(193 \times 399 \div 120)} = 25{\cdot}332287$ feet, and the time $= 3{\cdot}6882542 \sqrt{(120 \div 193)} = 2{\cdot}9082566''$; and hence, by the laws of descending bodies, only, (without the help of fluxions) the remaining part of the time is easily found $= 2{\cdot}463375''$, and consequently the whole time of descent required $= 5{\cdot}371631''$.

Corollary. If B be supposed to ascend along an inclined plane; then, the sine of the plane's inclination to the horizon, to rad. 1, being called s, the motive force at E will be as $PE - s \times PD = PA + AE - s \times PB + s \times AE = (1+s) \times \left(\frac{PA}{1+s} - \frac{s \times PB}{1+s} + AE\right)$; whence it is manifest that the velocity and time may be found with the same ease in this case, as before. Or, if A descends, or A and B descend and ascend along two differently inclined planes, the method of solution will be still the same.

Questions proposed in 1769, *and answered in* 1770.

I. QUESTION 597, *by Mr.* Tho. Sadler.

Dear ladies, you with ease may find *
A matchless hero's name,
Who was beloved by mankind,
And mounted up to fame:
To serve his country boldly dar'd
Hot sulphur, smoke, and fire,
And long campaigns' fatigue he shar'd,
To conquer proud Monsieur.

* viz. From the equations $\begin{cases} w+x+y+z = 52 \\ wx + yz = 360 \\ wz + xy = 280 \\ xy + xz = 315 \end{cases}$: Where w, x, y, and z denote the places of the letters in the alphabet composing the gentleman's name.

Answered by Mr. Wm. Spicer.

From the 1st given equation w is $= a\ (52) - x - y - z$; and this value substituted in the other three, they become $ax - xx - xy - xz + yz\ (= 360) = b$, $az - xz - yz - zz + xy\ (= 280) = c$, and $ay - xy - yy - yz + xz\ (= 315) = d$: Whence, by addition,

$a \times (x+z) - (x+z)^2 = b + c$, $a \times (z+y) - (z+y)^2 = c + d$, and $a \times (x+y) - (x+y)^2 = d + b$, and completing the square, &c. $x+z = \frac{1}{2}a - \frac{1}{2}\sqrt{(aa - 4(b+c))} = 20$, $z+y = \frac{1}{2}a - \frac{1}{2}\sqrt{(aa - 4(c+d))} = 17$, and $x+y = \frac{1}{2}a - \frac{1}{2}\sqrt{(aa - 4(b+d))} = 25$, and consequently $x = 14$, $y = 11$, $z = 6$, and $w = 21$; which shew the matchless hero's name to be WOLF.

II. QUESTION 598, *by Master* J. Paty, *at the Mathematical Academy at Bristol.*

A gentleman erecting a house, whose breadth was 28 feet and back wall 6 feet 9 inches higher than the front, had by him a sufficient quantity of rafters for the front, each 24 feet long, which, to save timber, he was unwilling to cut, and therefore orders his builder to make the back rafters of such a length as will make the declivity of them and front alike, for uniformity; but being at a loss, is desirous of hav- it proposed in the Diary, that he may know how to proceed.

Answered by Master J. Spencer, *at Mr.* Allen's *Boarding School, in Spalding.*

Put BE, the diff. of the heights of the walls ($= 6·75$) $= a$; BA, the breadth of the building ($=28$) $= b$; AD ($=$DC) the length of the front rafters ($= 24$ feet) $= c$; and BC $= x$; then, drawing DF ∥ EB, FC will be $= \frac{1}{2} \times (b + x)$, EC $= \sqrt{(aa + xx)}$, and per sim. Δs, x (BC) : $\sqrt{(aa + xx)}$ (EC) : : $\frac{1}{2} \times (b + x)$ (FC) : c (DC); $\therefore \sqrt{(aa+xx)} = 2cx \div (b + x)$, and consequently $x = 7·345$, very nearly; and hence the length of the back rafters are easily found $= 14·02288$, &c. feet.

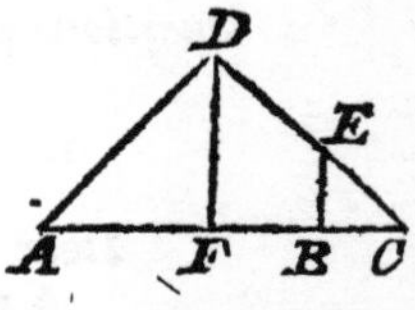

This is nearly the same as the prize question for the year 1744, to which, several constructions were given the year following. H.

III. QUESTION 599, *by Mr.* Wm. Spicer.

Suppose A lends B 2000l. at 5l. per cent. per ann. simple interest, which B is to pay again in the following manner, viz. 1l. immediately down at the end of the first year, 2l. at the end of the second year, 3l. at the end of the third year, and so on, increasing 1l. every year; required at what time B's debt will be the greatest, and also in what time A will be indebted to B the sum of 85l. 15s.

Answered by Mr. Wm. Spicer, *the Proposer.*

Put $p = 2000$, $r = 0·05$, $b = 85·75$l. and $x =$ the time in which B's debt will be greatest; then (by progression) $\frac{1}{2} \times (x+1) \times (x+2) + \frac{1}{6}r \times (x^3 + 3x^2 + 2x) =$ the sum of B's yearly payments and their interest, and $p + prx =$ the amount of the principal p in the same time, and consequently $p + prx - \frac{1}{2} \times (x+1) \times (x + 2) - \frac{1}{6}r \times (x^3 + 3x^2 + 2x) =$ the sum he will then be indebted, a max. per quest.——In fluxions, and reduced, $x = 45{\cdot}184086$, &c. which shews that B's debt will be the greatest just before the 46th payment is made. Again, supposing x, now, to denote the time when A is indebted to B the sum of 85·75l. (the rest remaining the same as before), then will $\frac{1}{2} \times (x+1) \times (x+2) + \frac{1}{6}r \times (x^3 + 3x^2 + 2x) - p - prx = b$ (per question); $\therefore x^3 + 63x^2 - 11818x = 250170$, and $x = 93$ years, the time required.

Remark. This solution seems to be false. For, the sum of the payments being $1 + 2 + 3 + \ldots\ldots x = \frac{1}{2}x \,.\, (x+1)$, and the sum of the interests of these payments $= r \,.\, (x-1) + 2r \,.\, (x-2) + 3r \,.\, (x-3) + \ldots\ldots (x-1) \,.\, r\,(x-(x-1)) = rx \times : 1 + 2 + 3 \ldots\ldots (x-1) - r \times : 1^2 + 2^2 + 3^2 + \ldots\ldots (x-1)^2 = \frac{1}{2}rxx \,.\, (x-1) - \frac{1}{6}rx \,.\, (x-1) \,.\, (2x-1) = \frac{1}{6}r \,.\, (x^3 - x)$, $\therefore$ the annual payments and their interests together are $\frac{1}{2}x \,.\, (x+1) + \frac{1}{6}r \,.\, x^3 - x)$, which taken from $p + prx$, and the flux. of the remainder made $= 0$, we obtain $x = \sqrt{\left(2p + \frac{1}{3} + \frac{1}{rr} - \frac{1}{r}\right)} - \frac{1}{r} = \sqrt{4380\frac{1}{3}} - 20 = 46{\cdot}18408$, the time when the debt is greatest.

Again, the root of the equation $\frac{1}{2}x \,.\, (x+1) + \frac{1}{6}r \,.\, (x^3 - x) - p - prx = b$, or $x^3 + 60x^2 - 11941x = 250290$, is $x = 94\frac{1}{2}$ years, the time when 85·75l. is due to B.

The time when the whole debt is just cleared, will be the root of this equation $\frac{1}{2}x \,.\, (x+1) + \frac{1}{6}r \,.\, (x^3 - x) = p + prx$, or $x^3 + 60x^2 - 11941x = 240900$, where $x = 94{\cdot}066$, the time when the debt is cleared. H.

IV. QUESTION 600, *by Mr.* Paul. Sharp.

A gentleman having a certain number of guineas and moidores, was asked how many he had; to which he replied, that the square root of the guineas multiplied by the cube root of the moidores, when the product is the greatest possible, will be $= 39{\cdot}1918$; required how many he had of each sort?

Answered by Mr. J. Addison.

It is easily proved, that two numbers to have the condition required in this question (viz. that the cube root of the one multiplied into the square root of the other may be a max. &c.) must be in the ratio of 2 to 3, and consequently that if $xx =$ the number of guineas, $\frac{2}{3}xx$ will $=$ that of the moidores; whence $x \sqrt[3]{\frac{2}{3}xx} = 39{\cdot}1918$ (per quest.), and $xx = 96$, and $\frac{2}{3}xx = 64$, the numbers required.

Messrs. *J. Chipchase*, *I. Dalby*, *E. Parnel*, and *Paul Sharp*, the proposer, put x and y for the number of guineas and moidores, and x

= the sum of them both; then will $y = z - x$, and $\sqrt[3]{(z-x)} \times \sqrt{x}$ be a max. (per quest.). In flux. and reduced (z being constant), $x = \frac{3}{5}z$: Whence $y = \frac{2}{5}z$; and, $\sqrt{\frac{3}{5}z} \times \sqrt[3]{\frac{2}{5}z}$ being $= 39{\cdot}1918$ (per quest.), $z = 160$, and consequently $x = 96$, and $y = 64$, the same as before.

V. QUESTION 601, *by Mr.* T. Moss.

If one side AB of any plane triangle ABC, be divided into two parts, and from the point of division (E) two right lines be drawn parallel to, and terminating at the other sides, and the said points of termination be joined by a right line, and there be likewise drawn another right line, as DH or MN, parallel to either of the two aforesaid parallel lines, so as to intersect the other line, and terminate in the sides of the triangle; then the two extreme parts Hm, Dn or Nn, MC of the three parts into which the line, so drawn, is divided, will always be in the ratio of the two parts BE, AE of the line AB first divided; required the demonstration?

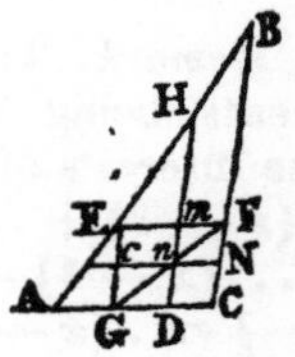

Answered by Mr. E. Williams, *Capt. in the Royal Artillery.*

By the construction of the proposer's figure (see the last year's Diary) the following analogies are evident, viz. BF : Hm :: FE : mE :: EG (FC) : CG (Dn); therefore Hm : Dn :: (BF : FC) :: BE : AE. In like manner AG : MC :: GE : CE :: EF (GC) : mF (Nn), and consequently MC : Nn :: (AG : GC) :: AE : EB.

VI. QUESTION 602, *by Mr.* J. Edwards, *of Magdalen College, Camb.*

A gentleman has a right-angled triangular garden, at the right angle of which grows a tree 40 feet high, whose shadow, I observed, on the 21st of June, at 2^h 30^m P. M. terminated in the hypothenuse of the said triangle in a perpendicular direction, and measuring the the garden, I found that the difference of both its sides from the hypothenuse was 15 and 30 yards: Quere the latitude of the place, and area of the garden?

Answered by the Rev. Mr. Crakelt.

Construction. Let BA (30 yards) represent the difference betwixt the hypothenuse and less leg, and CA (15) that betwixt the hypothenuse and greater; then, at the point A, erect the ⊥ AD, meeting BD, making with AB an angle of 45°, in D, and join the points D, C: This done, from B apply BE (to DC) = BA, and from the point F, where CF drawn ∥ BE meets DB produced, let fall the ⊥ FG upon AE produced; so will CFG be the right-angled triangular garden required.

Demonstration. Produce DA till it meets FH, drawn ∥ AG, in the point H: Then, by similarity of △s, AB : HF :: DB : DF :: EB : CF; but (by construc.) EB = AB and consequently CF = HF = AG = AC + CG = AB + BG = AB + FG.

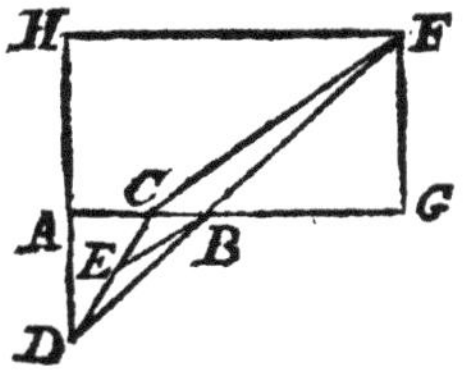

The calculation from this construction is extremely easy, whereby the length of the tree's shadow, at the given time, comes out = 36 yards, and consequently the sun's altitude = 20° 19′ 23″; whence, per spherics, the latitude required is readily found to be 37° 33′ 38″ south, &c.

VII. QUESTION 603, *by Mr.* Wm. Gawith.

In a right-angled plane triangle ABC, suppose the two legs AB and BC = x^5 and y^5, the hypothenuse AC = $(xx+yy)^{\frac{3}{2}} \times axy$, and the perpendicular BD upon the hypothenuse = x^3; to determine the triangle, by quadratics.

Answered by Mr. G. Coughron.

By the data and properties of a right-angled triangle, $x^{10}+y^{10} = (xx+yy)^3 \times aaxxyy$, and $x^5y^5 = (xx+yy)^{\frac{3}{2}} \times axy \times x^3$. Substitute zxx for yy and its powers in the first equation, and it becomes $x^{10} \times (z^5+1) = x^{10} \times (z+1)^3 \times aaz$; ∴ $z^5+1-(z+1)^3 \times aaz = 0$; which, divided by $zz \times (z+1)$ and reduced, gives $zz + \frac{1}{zz} - (1+aa) \times \left(z+\frac{1}{z}\right) + 1 - 2aa = 0$; or, putting $v = z + \frac{1}{z}$; $vv - (1+aa) \times v - (1+2aa) = 0$; ∴ $v = \sqrt{\left(\frac{(1+aa)^2}{4} + 1 + 2aa\right)} + \frac{1+aa}{2}$, which put = $2c$, and then, $z + \frac{1}{z}$ being = $2c$, z will be found = $\pm\sqrt{(cc-1)} + c$. Lastly, from the second equation is had $z^{\frac{5}{2}} x^{10} = x^8 \times (z+1)^{\frac{3}{2}} \times az^{\frac{1}{2}}$; ∴ $x = (z+1)^{\frac{3}{4}} \times \frac{\sqrt{a}}{z}$; which, as z is now known, will likewise be known, and from thence the triangle becomes known.

VIII. QUESTION 604, *by Mr.* J. Dymond.

Given the line bisecting the vertical angle of a plane triangle and terminating in the base, the perpendicular falling thereon from one of

the angles at the base, and the other angle at the base; to construct the triangle.

***Answered by Mr.* T. Moss.**

Construction. **Upon the given bisecting line AB describe a segment of a circle to contain the given angle of the required triangle, and in BQ ⊥ AB, take Bn = the given perpendicular. Bisect AB in *m*, and through the points *m* and *n* draw a right line meeting the arch of the said segment in I: Join I, A, and I, B, and draw D*n*C ∥ AB; in IB, produced, take BH = BC; Draw AH, and AHI will be the triangle required.**

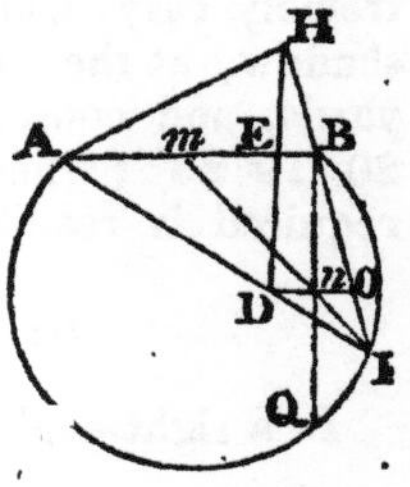

Demonstration. **Draw HD cutting AB in E: Then, because (by construc.) AB is bisected in *m*, and DC is ∥ AB, D*n* is = *n*C; also (by construction) BH is = BC; whence HD is ∥ B*n*, and consequently HE ($= \frac{1}{2}$HD) = B*n*, the given perpendicular, by construction. Moreover, because the given line AB is ⊥ HD and bisects it at E, the ∠EAD = EAH; and AIH is = the given angle, by construction.**

IX. QUESTION 605, *by Mr.* J. Turner.

In any ellipsis the transverse axis is the greatest of all the diameters, and the conjugate axis the least; and, of any other diameters, that which is nearest to the transverse axis is greater than those which are farthest from it. The demonstration hereof is required, geometrically, from the consideration of the solid?

***Answered by the Rev. Mr.* Crakelt.**

Let ABC be the representation of any upright cone; ADEA that of any elliptic section made therein; AC = EC = the semi-transverse of such section; DC = C*e* its semi-conjugate; F*o* any other semi-diameter; GHI, K*o*L, MNE the diameters of three circular sections made by planes passing through the points D, F, E, and EO a parallel to the axis of the cone, BP: Then, by similarity of △s, L*o* : IC :: EC — C*o* (E*o*) : EC, and K*o* : GC :: EC + C*o* (A*o*) : EC (AC), and consequently L*o* × K*o* = (by prop. of the circle) F*o*^{2} : IC × GC or DC^2 :: EC^2 — C*o*^{2} : EC^2, the principal prop. of the ellipse: But DC^2 or IC × GC = (by similarity of triangles and the nature of parallels) IC × MN or EN = IC × HO = (HO + 2OI) × HO = HO^2 + HO × 2OI, and EC^2 = OI^2 (CH^2) + HO^2 + HO × 2OI + EO^2, and consequently EC^2 — DC^2 = OI^2 + EO^2 = EI^2,

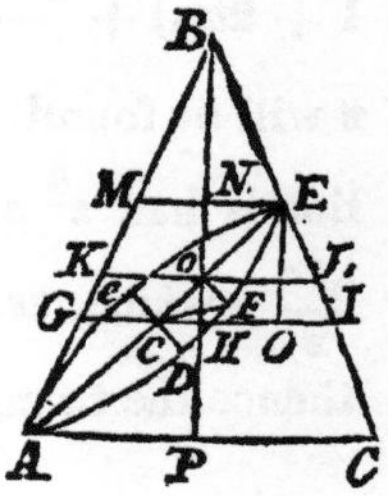

or $DC^2 = EC^2 - EI^2$: Wherefore $EC^2 \times FO^2 = DC^2 \times EC^2 - DC^2 \times CO^2 = DC^2 \times EC^2 - EC^2 \times CO^2 + EI^2 \times CO^2$, or, by adding $EC^2 \times CO^2$ to both $EC^2 \times FC^2 = EC^2 \times DC^2 + EI^2 \times CO^2$: Whence, as a similar equality will obtain for all the other points of the curve, DFE, and the greater CO^2 becomes the greater will $EC^2 \times FC^2$ become likewise, the truth of the prop. is manifest.

X. QUESTION 606, *by Mons.* Fermat.

Let ANB be a semicircle described upon AB as a diameter, and AD, BC two perpendiculars, each equal to AN, the chord, of 90°. Also, let any point E be assumed in the semi-circumference, and two lines ED, EC be drawn from thence to the points D, C, intersecting the diameter in two points O, V: then $BO^2 + AV^2$ will be, universally, equal to AB^2; required the demonstration?

Note. This question was proposed by *Mons. Fermat* to *Dr. Wallis*, as may be seen at page 188 of the Commercium Epistolicum, published at Oxford in the year 1658, but was never yet publicly answered.

Answered by Clericus.

Join the points B, N; D, C; and draw ES ⊥ DC intersecting AB in Q: Then, per sim. △s, DA : AO :: ES : DS = AQ, and BC (= DA) : BV :: ES : SC = BQ, and consequently $DA \times DA : AO \times BV :: ES \times ES : AQ \times BQ = EQ^2$ (by property of the circle): Whence $2AO \times BV$ (= $2DA^2 \times EQ^2 \div ES^2$) = $AB^2 \times EQ^2 \div ES^2$, because $2DA^2$ (= $2AN^2$) = AB^2 (47 Euc. 1.). But $OV^2 : DC^2(AB^2) :: EQ^2 : ES^2$ (per sim. △s OEV, DEC), and consequently OV^2 (= $AB^2 \times EQ^2 \div ES^2$) = $2AO \times BV$ (= $2AO \times (OB - OV)$) = $2AO \times OB - 2AO \times OV$; ∴ $OV^2 + 2AO \times OV = 2AO \times OB$, and $AO^2 + OV^2 + 2AO \times OV + OB^2 = AO^2 + 2AO \times OB + OB^2$; i. e. $(AO + OV)^2 + OB^2 = (AO + OB)^2$, or $AV^2 + BO^2 = AB^2$.

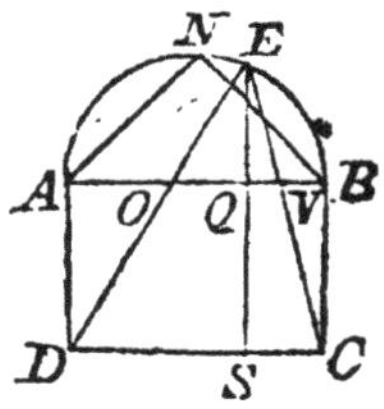

XI. QUESTION 607, *by the Rev. Mr.* Wm. Crakelt.

In a plane triangle there are given one of the angles at the base, the length of a line drawn from the other angle to the middle of its opposite side, and the perpendicular from the given angle upon its opposite side a maximum; to determine the triangle, by construction.

Answered by the Rev. Mr. Lawson.

Construction. On AB the given line, describe a segment to contain the given angle. Bisect the circumference in D, and join A, D; D, B: Continue DB till BE equals BD; then draw AE cutting the circumference continued in G. Bisect again the circumference ADG in C,

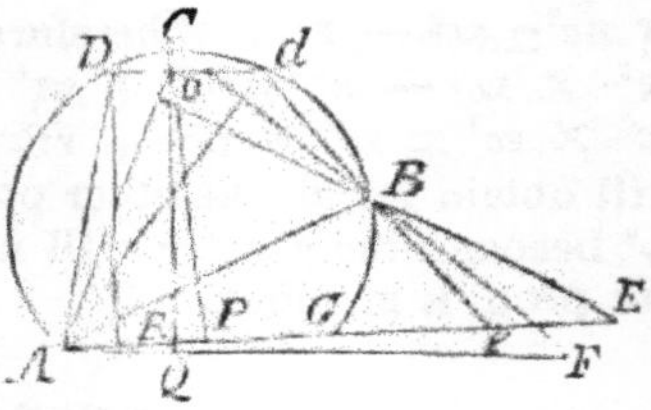

and join A, C; C, B; and continue CB till BF equals CB; then join A, F, and I say ACF is the triangle required, whose perpendicular CQ is a maximum.

Demonstration. From C let fall CP ⊥ AG, and it will be the greatest perpendicular in the segment ACG. But CQ is still greater, because the locus of the remaining angle of the required triangle (viz. EF*e*) is a circumference similar to ADB, to which AF is a tangent.

Note. If the perpendicular be given in magnitude, the same construction may be used; and if P*o* be set off on PC equal thereto, and through *o*, D*od* drawn ∥ AG, the points D and *d* will each give a solution to the problem.

Construction to the same by Mr. Crakelt, *the Proposer.*

Let AB be supposed the given line drawn from the unknown angle at the base to the middle of its opposite side, and having produced it to C so that BC may be = AB, describe upon BC a segment of a circle capable of containing an angle = the given one at the base, and draw the tangent thereto AD: Then, drawing DC, through A draw a parallel thereto to meet DB produced in E, and the △ AED, formed thereby, will be that required.

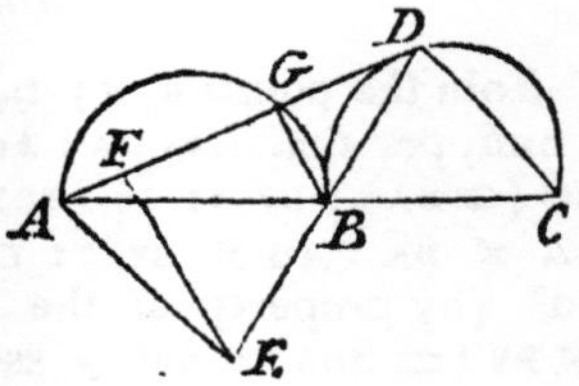

Demonstration. The line AE being ∥ DC, the angle AEB will be = the angle CDB = the given angle at the base (by construction), and the △s ABE, CBD similar; whence AB : BC :: BE : BD; but AB = BC (by construction); wherefore BE = BD. And, that the perpend. EF will in this case be a maximum, is very plain, since in all other (possible) cases the line AD will cut the segment BDC in two points, and fall nearer the point B, and consequently not give the ⊥ BG or its double, EF, a maximum.

Scholium. If it was required that EB should be to BD in a given ratio, or that the perpendicular instead of a maximum should be of a given magnitude, the construction would be the same very nearly.

XII. QUESTION 608, *by Mr.* J. Chipchase.

There is a hollow cylinder with a circular hole in its side $\frac{1}{10}$ of an inch in diameter, and at the distance of 3 feet from its bottom, out of which when the cylinder is full, the water will spout to the distance of 5 feet from its bottom on an horizontal plane, but after it has continued running for the space of 15′, it will only spout to the distance of 3 feet: Required the content of the cylinder?

Answered by the Rev. Mr. Wildbore.

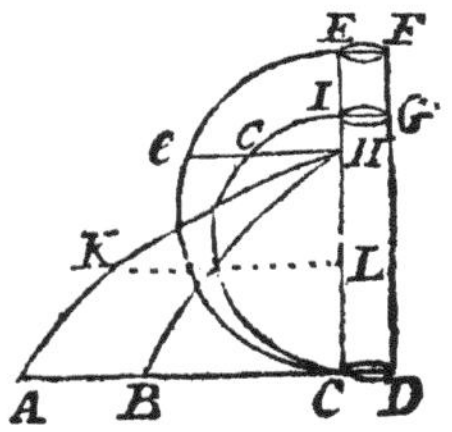

If HL = EH, KL will = 2EH, and (per conics) KL^2 : HL, or $4EH^2$: EH :: AC^2 : HC ; ∴ HE = $AC^2 \div 4HC$ = 25 inches; and in like manner is found IH = 9 inches. Also (per Newton's Princip. edit. opt. p. 329, 330, 331) the times in which EH and IH may be emptied will be = $100EF^2\sqrt{2EH} \div \sqrt{193}$ and $100EF^2\sqrt{2IH} \div \sqrt{193}$ respectively; the difference of which is evidently = 900^s (the time in which EI is run out at H) per question; whence EF = 6·64872, and the content of the cylinder 2117·8266 cubic inches.

Answer to the same by Mr. T. Allen.

Let CEFD represent the cylinder, H the hole, and HA and HB the spouting water, when the surface occupies EF and IG respectively. Upon the diameters EC, IC describe the semicircles E*e*C, IcC, and draw *ec*H ⊥ EC: Then, it is well known that the horizontal distances AC, BC, of the spouting fluid, are respectively = 2 × *e*H and 2 × *c*H. Therefore, since *e*H and *c*H are given = 2·5 and 1·5 feet respectively, EH and IH are easily found = 2·0833, &c. and 0·75 feet: Therefore CE, the height of the cylinder, is = 5·0833, &c. feet. Put now, EH = *h*, F = ·0000545415 the area of the orifice in feet, $s = 16\frac{1}{12}$, t = time of emptying EFGI with the first or greatest velocity, IH = c, EI = 1·33 &c. = d, the given time $900'' = a$, p = ·785397, and EF (the diam. of the cylinder) = x. Then, (per Emers. Mech. 1st edit. pa. 173,) $tF\sqrt{2hs} = pdxx$; ∴ $t = pdxx \div F\sqrt{2hs}$: And, per Emers. Flux. 2. edit p. 139, $\dfrac{2pdxx}{F\sqrt{2hs}} - \dfrac{2pdxx\sqrt{c}}{F\sqrt{2hhs}} = a$; whence, &c.

XIII. QUESTION 609, *by Mr.* Wm. Wales.

To determine the declination of that star whose change in azimuth, in a given latitude, is the greatest possible in a given time, reckoned from that of its rising.

Answered by Mr. W. Wales, *the Proposer.*

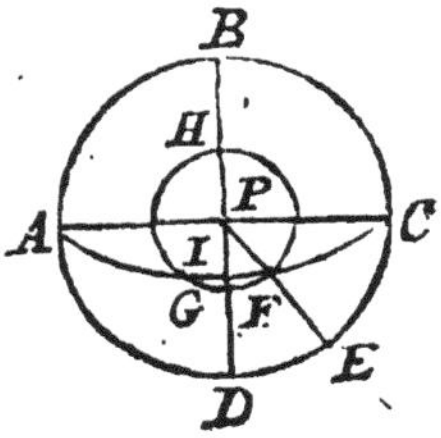

Projection. Describe the primitive ABCD for the equator, and draw BD for the meridian; from D, set off DE = $\frac{1}{2}$ the arc of the equator passed over in the given interval, and join P, E: Describe the horizon AGFC intersecting PE in F; then, from P, the pole, as a centre, through F, describe the parallel of declination FGH, which shall be that of the star required.

Demonstration. It is a thing too well known to need demonstrating here, that every phænomenon whatever changes its azimuth the fastest possible when on the meridian, and with equal velocities at equal distances therefrom; consequently, the change in azimuth will be the greatest possible during the whole of any given interval when the phænomenon is on the meridian precisely at the middle of that interval, or when the semi-diurnal arc DE, turned into time, is $= \frac{1}{2}$ the given interval.

Scholium. It is manifest that when the given interval exceeds 12 hours, the declination will be of the same name with the latitude, but of a contrary name when the interval is less.

Remark. It may at first sight seem to some, as if this method of projection would fail in particular cases; such as, if it was required to find the sun's declination when his change in azimuth was the greatest possible in any given time, reckoned from his rising, and less than the shortest day in the latitude given: But on more mature consideration this will be found a mistake; for in all those cases, the sun's declination must be the greatest possible, and of a different name with the latitude, because then the whole of the given interval is the nearest possible to the meridian, which is the very principle on which the preceding projection is founded.

I scarce need add that cos. FPI : rad. :: tang. AP (the given lat.) : tang. PF the co-declination required.

XIV. QUESTION 610, *by Mr.* C. Hutton.

To find the radius of a circle, whose area shall be equal to the surface of an elliptic spindle, or to that of any frustum or segment of it, having given the ellipse from which it is generated, and the distance of the centres of the ellipse and spindle?

Answered by Mr. T. Allen.

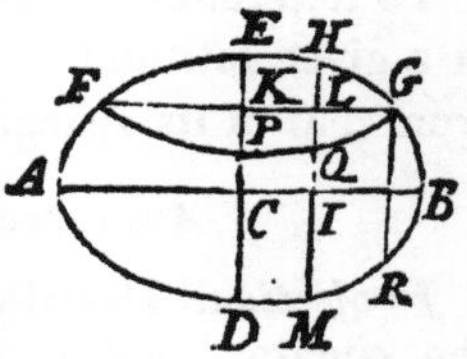

Let AEBD represent the given ellipse, and FEGP the spindle, generated by the revolution of the elliptic arc FEG about its axis FG. Put CB $= a$, CE $= b$, KC $=$ LI $= d$, CI $=$ KL $= x$, IH $= y$, arc EH $= z$, and $p = 3{\cdot}14159$; then, per conics, $y = (b \div a)\sqrt{(aa - xx)}$, and therefore HL $=$ LQ $= (b \div a)\sqrt{(aa - xx)} - d$: Whence $\dot{z}\ (= \sqrt{(\dot{x}^2 + \dot{y}^2)}) = (\dot{x} \div a)\sqrt{((a^4 - ccxx) \div (aa - xx))}$ (by putting $bb - aa = -cc$) and $(2pb\dot{x} \div aa) \times \sqrt{(a^4 - ccxx)} - 2pd\dot{z}$ = the flux. of the spindle; whose fluent is $(pb \div aa)\sqrt{(a^4x^2 - c^2x^4)} + (pbaa \div c) \times$ circ. arc, whose rad. is 1 and sine $cx \div aa$, $- 2pdz$, a general expression for the surface, either of the whole, or of any frustum of the spindle: But the two first terms thereof express the surface of the portion EHMD of the spheroid AEBD; wherefore it is evident that the radius of a circle

$= \text{the surface} \left\{ \begin{matrix} \text{EHQP} \\ \text{EGQ} \\ \text{HGQ} \end{matrix} \right\}$ will be a mean proportional between the surf. $\left\{ \begin{matrix} \text{EHMD} \\ \text{EGRD} \\ \text{HGRM} \end{matrix} \right\} - 2p \times \text{KC} \times \left\{ \begin{matrix} \text{EH} \\ \text{EG} \\ \text{HG} \end{matrix} \right\}$ and $\frac{1}{p}$.

XV. QUESTION 611, *by Mr.* Tho. Allen.

The sum of the infinite series $\frac{1}{2.4.(2+3)} + \frac{1.3}{2.4.6.(4+5)} + \frac{1.3.5}{2.4.6.8.(6+7)} + \frac{1.3.5.7}{2.4.6.8.10.(8+9)} +$ &c. is required, with the investigation.

Answered by the Rev. Mr. Wildbore.

The terms of the proposed series being multiplied separately by x^5, x^9, x^{13}, &c. respectively, and the fluxion taken, it becomes $\frac{x^4\dot{x}}{2.4} + \frac{1.3x^8\dot{x}}{2.4.6} + \frac{1.3.5x^{12}\dot{x}}{2.4.6.8} +$ &c. $= \frac{\dot{x}}{x^4} - \frac{\dot{x}}{2} - \frac{\dot{x}\sqrt{(1-x^4)}}{x^4} = \frac{\dot{x}}{x^4} - \frac{\dot{x}}{2} - \frac{\dot{x}\sqrt{(1-x^4)}}{x^4} - \frac{\frac{2}{3}\dot{x}}{\sqrt{(1-x^4)}} + \frac{\frac{2}{3}\dot{x}}{\sqrt{(1-x^4)}}$; the fluent of which when $x = 1$, is $\frac{e + \sqrt{(ee - 2f)}}{3} - \frac{5}{6} = \cdot 040685$, the sum required; where $e = \frac{1}{4}$ the periphery of the ellipsis whose semi-axes are $\sqrt{2}$ and 1, and $f = \frac{1}{4}$ of that of the circle, whose radius is 1.

THE PRIZE QUESTION, *by* Plus Minus.

Let a thread MA*m* be fixed by its ends to the ends M, *m* of a ruler, whose middle point is C, so as to slide over a tack fixed in the point A, whilst the point C of the ruler keeps constantly in the line AC; then will the ends M, *m* of the said ruler describe a certain curve, which revolving about its axis AC, will generate a solid, the ratio of whose content to that of its circumscribing cylinder is here required; the length of the thread being to that of the ruler in the subduplicate ratio of 4 to 3?

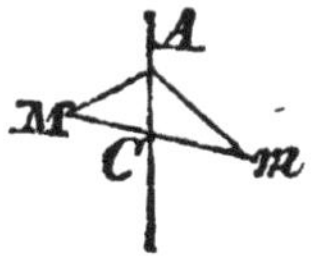

Who'er is unable this truth to discover,
Either is *not at all*, or an *unhappy* lover.
Happy lovers do all (tho' not skilled in the arts)
Know the *solid content* of *two conjugate hearts*.

Answered by Mr. Tho. Allen, *of Spalding.*

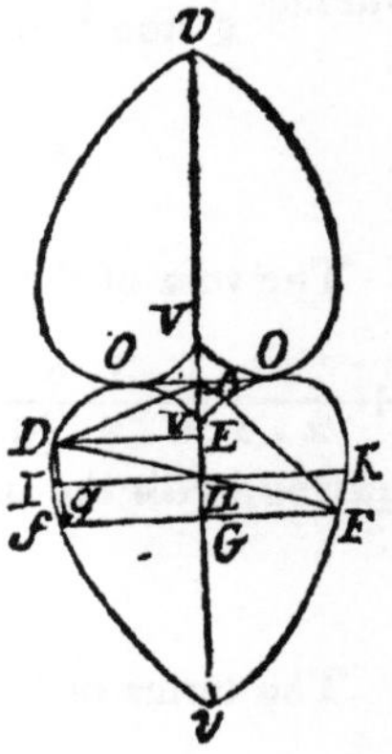

Let DF ($= 2a$) represent any indefinite position of the ruler, and DAF ($= s$) the corresponding position of the thread. Put AH, the distance of the tack A from the middle of the ruler $= z$, HG $=$ HE $= x$, and $p = 3{\cdot}14159$, &c. then will $FG^2 = DE^2 = aa - xx$ (47 Euc. 1.), AE $= z - x$, and AG $= z + x$, and consequently $\sqrt{(aa - xx + (z+x)^2)} + \sqrt{(aa - xx + (z-x)^2)} = s$; $\therefore z = \sqrt{\dfrac{s^4 - 4ssaa}{4ss - 16xx}}$ (generally) $= \dfrac{1}{2\sqrt{(1-xx)}}$ in the present case, where $a = \frac{1}{2}\sqrt{3}$ and $s = 2$ (per quest.). Further, when the extremity of the ruler passes the point O, the fluxion of EA, or $\dot{x} - \dot{z}$ will evidently be $= 0$; therefore $-\frac{1}{2}x\dot{x} \times (1 - xx)^{-\frac{3}{2}} + \dot{x} = 0$, or $4 \times (1 - xx)^3 - xx = 0$; in which equation one root, or value, of x (which is that here required) is evidently $= \sqrt{\frac{1}{2}}$. Now, $p \times (aa\dot{x} - x^2\dot{x} + (\frac{1}{2}x^3\dot{x} - \frac{1}{2}aax\dot{x}) \times (1 - xx)^{-\frac{3}{2}})$, $p \times (x^2\dot{x} - aa\dot{x} + (\frac{1}{2}aax\dot{x} - \frac{1}{2}x^3\dot{x}) \times (1 - xx)^{-\frac{3}{2}})$ and $p \times (aa\dot{x} - x^2\dot{x} + (\frac{1}{2}aax\dot{x} - \frac{1}{2}x^3\dot{x}) \times (1 - xx)^{-\frac{3}{2}})$ being the fluxions of of the solids whereof OVO, IOAOK, and IvK are the respective sections (IK being the greatest ordinate), their fluents, by proper correction and substitution, will be $p \times : \dfrac{2}{3\sqrt{\frac{1}{2}}} - \dfrac{2 + \sqrt{3}}{4}$, $p \times : \dfrac{2}{3\sqrt{\frac{1}{2}}} - \dfrac{5}{8}$, and $p \times : \dfrac{5}{8} + \dfrac{\sqrt{3} - 2}{4}$; the sum of the two last of which minus the first gives $p \times \dfrac{\sqrt{3}}{2} =$ the content of the solid required; the ratio of which to that of its circumscribing cylinder will be as $1 : \dfrac{\sqrt{3}}{2} + \dfrac{3}{4}$, as is evident from what has been determined above.

Ingenious *Plus Minus*, you see how I prove
Your *heart*'s *true content* from what's given above:
And since I'm so happy this truth to discover,
I hope you'll esteem me, a happy true lover.

The same answered by the Rev. Mr. Crakelt, *of Northfleet, in Kent.*

Suppose the lines drawn as in the preceding diagram, and the ruler to be moved from its first position *Vv*, into the position DF, and put FA + AD (the length of the thread) $= 2a$, FA $-$ AD $= 2x$, DF (the length of the ruler) $= 2b$, DE $=$ FG $= y$, and $p = 3{\cdot}14159$, &c.

Then, by theorem 10, 11, book 2d, of Simpson's Geometry, 2d edit. and the nature of the problem, will be found $GV = a + b - a\sqrt{\frac{aa - bb}{aa - bb + yy}} - \sqrt{(bb - yy)}$, $EV = b - a + a\sqrt{\frac{aa - bb}{aa - bb + yy}} - \sqrt{(bb - yy)}$, and supposing the solid to be divided into two parts by a plane coinciding with its greatest ordinate, the sum of the fluxions of those parts $= \frac{2py^3\dot{y}}{\sqrt{(bb - yy)}}$; the correct fluent whereof $\left(-\frac{2p}{3}yy\sqrt{(bb - yy)} - \frac{4pb^2}{3}\sqrt{(bb - yy)} + \frac{4pb^3}{3}\right)$ gives (when $y = b$) the solidity required $= \frac{4pb^3}{3} = \frac{8pb^3}{6}$, the solidity of a sphere, whose diameter is $=$ the length of the ruler. Again, $(aa - bb + xx - ax) \times (aa - bb + xx)^{-\frac{1}{2}}$ representing the value of AE, its flux. made $= 0$, and properly reduced, gives $x^3 + (aa - bb) \times x = (aa - bb) \times a$, or (since $bb = \frac{3aa}{4}$, by the data) $x^3 + \frac{1}{4}a^2x = \frac{1}{4}a^3$; and the root $\frac{1}{2}a$, by substitution, shews that, in the present case, the points E, A coincide, or that the altitude of the circumscribing cylinder is $=$ VA, and consequently that its solidity is to that of the generated solid as $1 + \frac{2}{\sqrt{3}}$ to $\frac{4}{3}$.——The several curious corollaries deduced from this solution, our narrow limits oblige us to omit.

Mr. *James Yeret* produces FG to meet the curve again in f; then, retaining the last substitution, and joining D, f and drawing Hg ⊥ thereto, $\sqrt{(bb - yy)} =$ EH $=$ HG (per 47 Euc. 1.) and $4pyy\sqrt{(bb - yy)} =$ the fluxion of the solid generated by the revolution of the space VDgfV about the axis VV; the correct fluent whereof $\left(\frac{4p}{3} \times (b^3 - (bb - yy)^{\frac{3}{2}})\right)$ when $y = b$, gives $\frac{4p}{3} \times b^3 = \frac{p}{6} \times 8b^3$, the content of the whole solid required, let the length of thread be what it will, &c.

The ingenious proposer, *Plus Minus*, after his curious solution, makes the following remark, viz. 'I have given the length of the thread in proportion to that of the ruler as $2 : \sqrt{3}$, that the conjugate hearts may osculate, as in o and o; had the thread been shorter, they would have intersected; had it been longer, they would not have touched; and had it been much longer, they would not have had this form of an heart, but would have been two ovals. Yet in all cases

the solid generated by the rotation of one of the curves is equal to a sphere, whose diameter is equal to the length of the ruler.'

General Answer to the same, by the Rev. Mr. Wildbore.

The curve being generated by the motion of the point *m* from D to O, let the semicircle DFO, whose rad. is ED ($=\frac{1}{2}$**OD** $= \frac{1}{2}$**M*m*** $=$ c*m*) be supposed to be generated by that of the point F. Through the centre E draw EF ∥ c*m*, and then, the ∠s *m*CD, FED being equal, *m* and F must be cotemporary positions of the two points. Moreover, p . B˙D . Bm^2 $= p$. B˙G . Bm^2 $+ p$. G˙D . Bm^2 is evidently = the flux. of the solid generated by the revolution of B*m*D about BD $=$ p . B˙G . GF^2 $+ p$. G˙D . GF^2 (because B*m* $=$ GF) : But p . G˙D . GF^2 = the fluxion of the spheric segment generated by the revolution

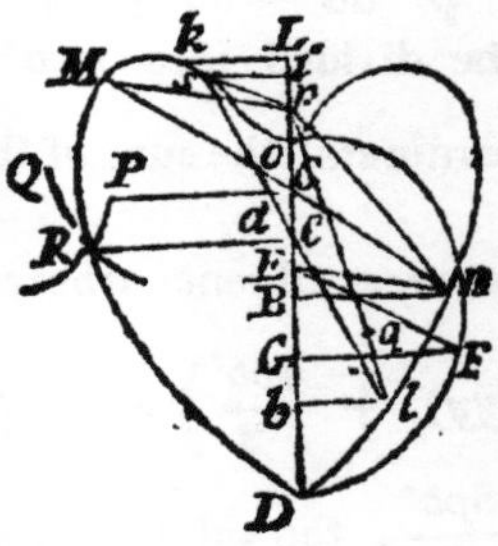

of GFD about GD ; whence the corrected fluent of p . B˙G . GF^2 + the said segment = the said solid ; but let the relation of BG to GF be what it will, when each of them is = to nothing at the same time, the corrected fluent of p . B˙G . GF^2 must necessarily be = 0 : In which case therefore, the segment will be = the solid. Now, *m* and F evidently arrive at O in the same time, making GF and BG each equal to nothing; and, by consequence, the sphere generated by OFD = the solid generated O*m*D.——Which property is not at all affected by AM ∔ A*m*, but it is universal to the whole family of curves, when c*m* is a constant quantity. Further, M*m* is evidently equal to the greatest diameter of the solid, or to that of its circumscribing cylinder, and LD = its height; therefore the ratio of the solidities will be $\frac{1}{3}$OD : $\frac{1}{2}$LD.

Now, to determine the solid answering to any particular data; from the centre *i* with the rad. *i*Q $= \frac{1}{2}$. (AM + A*m*) describe a circle, and from A as a centre, with the power Ad^2 $=$ *i*Q $\sqrt{(i\text{Q}^2 - \text{c}m^2)}$, describe the equilateral hyperbola PR, and where it cuts the circle at R let fall the perpendicular R*e* ; at *i* erect the ⊥ *is*, and through *e* draw *sl*, which will be = M*m* ; for (per trigonom.) IQ : A*e* : : *ie* : $\frac{1}{2}$ A*l* — A*s*) = *ql* (taking A*q* = *i*Q) and $(ql + \text{A}q)^2 = \text{A}l^2 = (\text{A}e + eb)^2 + es^2 - ie^2 = \text{A}e^2 + 2\text{A}e \,.\, ei + es^2$; ∴ $\text{A}e^2 \,.\, \text{A}q^2 - ie^2 \,.\, \text{A}e^2 = \text{A}q^2 \times (\text{A}q^2 - es^2) = \text{A}d^4$: But $\text{A}q^2$ $(= i\text{Q}^2) - ie^2 = \text{R}e^2$ (47 Euc. 1.); therefore R*e* . A*e* = Ad^2, a known property of the hyperbola, whose asymptote is AD and power Ad^2. It is moreover evident that, when A*i* is the greatest possible, or when *i* coincides with L, the hyperbola will not cut but touch the circle, AL being universally the height of the circumscribing cylinder above AD ; and, in the particular case proposed, because $\sqrt{(i\text{Q}^2 - \text{c}m^2)} = \frac{1}{2}i\text{Q}$, Q$d$ = Re ∓ Ad, $i\text{Q}^2 - \text{A}d^2 = \text{AD}^2$, $\frac{1}{2}i\text{Q}^2 = \text{A}d^2 = ic^2$, A*i* = 0, AL = 0, AD is the height of the circumscribing cylinder, and the ratio above will become $\frac{1}{3}$OD : $\frac{1}{2}$AD or 1 : $\frac{3 + 2\sqrt{3}}{4}$: : the solid content of the heart to that of its circumscribing cylinder.

Cor. 1. The fluxion of the area BmD $=$ B'D . Bm $=$ B'G . GF $+$ G'D . GF $=$ B'G . GF $+$ flux. of the circ. segment GFD; whence, when BG and GF each $= 0$, the area BmD $=$ circular segment GFD; wherefore the area of the whole circle $=$ that of the curve.

Cor. 2. From what is shewn above, the solidity of any segment of the heart, cut perpendicular to the axis OD, is easily derived, being $=$ the spheric segment above $+$ the correct fluent of p . B'G . GF^2 $=$ $\frac{1}{6}p \cdot \text{GD} \times (3\text{FG}^2 + \text{GD}^2) + \frac{1}{4}p \times \left(\frac{\text{AE}^4}{\text{AC}} + \text{AE}^2 \cdot \text{AC} - \text{AE}^3\right) =$ the solid generated by the revolution of BmD about BD.

Cor. 3. In the sane manner it will be found that the fluent of B'G : GF is $= \frac{1}{2}\text{AE}\sqrt{(\text{AE}^2 - \text{AC}^2)} - \frac{1}{4}\text{AE}^2 \times$ hyp. log. of $\frac{\text{AE} - \sqrt{(\text{AE}^2 - \text{AC}^2)}}{\text{AE} + \sqrt{(\text{AE}^2 - \text{AC}^2)}}$; which $+$ the circ. segment GFD is $=$ the area of the curve BmD.

Cor. 4. If cm is not $=$ ED, but in a given ratio, $n \cdot cm =$ ED, then will $\frac{1}{nn} \times$ globe $=$ the solid, as is evident from what is demonstrated above; and the same universal property may be still farther extended.

Scholium. The above universal properties may be applied to the finding a number of fluents of very difficult forms; but the subject is much too copious for this place.

Can I mistake the problem, miss the prize!
On whom my Lucy's choicest blessings rise;
For whose content she labours and she lives:
Friendship, truth conjugate she daily gives
Entwin'd with love unspotted;——then may I
E're guide her on to bliss;——to endless joy.

Note. The initials answer the *Prize Enigma.*

Questions proposed in 1770, *and answered in* 1771.

I. QUESTION 612, *by Mr.* T. Sadler.

In Whitchurch now a maid doth dwell,
Her neighbours stile her, bonny Nell;
She likes to live a maiden's life,
And won't consent to be a wife,
Tho' Harry, Richard, Ben, and Joe
To Nelly oft a courting go.

Above a score sue for the maid,
Fop, clown, and leering ganymede,
But to them all 'tis Nelly's song,
To marry she is yet too young.
Her age and fortune you will find
From the equations * here subjoin'd.
Come artists say, from what is said,
Is Nelly old enough to wed?
For should the maiden stay too long,
She may forget her fav'rite song.

* $\begin{cases} x^2 + y^2 = 13900 + x^3, \text{ and} \\ x^8 + y^3 = 654508000000 + y^4 \end{cases}$: To find x her age in years, and y her fortune in pounds.

Answered by Mr. Wm. Wilkin.

Put $a = 13900$ and $b = 654508000000$; then, from the 1st equation, $y = (a + x^3 - x^2)^{\frac{1}{2}}$, which value substituted in the 2d gives $x^8 + (a + x^3 - x^2)^{\frac{3}{2}} = b + (a + x^3 - x^2)^2$: From whence x is found $= 30$, her age in years, and thence $y = 200$, her fortune in pounds.

II. QUESTION 613, *by Mr.* Paul Sharp.

Given the difference of the transverse and conjugate diameters of an ellipse $= 8$, and the difference of the length, of the greatest inscribed parallelogram and radius of curvature at the end of the conjugate diameter $= 3{\cdot}28426$, to find the said diameters and radius of curvature at the said point.

Answered by Mr. Wm. Spicer.

Let $x =$ the semi-transverse; then will $x\sqrt{2} =$ the length of the greatest inscribed parallelogram, and (by art. 71 of Simpson's Flux.) $xx \div (x - 4) =$ rad. of curvature at the extremity of the conjugate diam. whence (putting $a = 3{\cdot}28426$) $xx \div (x - 4) = x\sqrt{2} - a$, per quest, and $x = 20$: Consequently the transverse diameter $= 40$, the conjugate $= 32$, and the rad. of curvature $= 25$.

III. QUESTION 614, *by Mr.* Wm. Gawith.

Given $(x^{15} + y^{15} + z^{15})^3 + (x^3 + y^3)^2 = a$, $(x^{15} + y^{15} + z^{15})^3 \times (x^3 + y^3 + z^3)^3 = b$, and $(x^3 + y^3)^4 + (x^3 + y^3 + z^3)^3 = c$; to find x, y and z by means of quadratics.

Answered by Mr. Geo. Coughron.

Put $(x^{15} + y^{15} + z^{15})^3 = u$, $(x^3 + y^3)^2 = v$, and $(x^3 + y^3 + z^3)^3 = w$; then the given equations will become $u + v = a$, $uw = b$, and $v + w = c$: Take the last equation from the first, and $u - w$

$= a - c$, from which and the 2d equation u is found $= \frac{1}{2} \times (\sqrt{((a - c)^2 + 4b)} + a - c)$, $w = \frac{1}{2} \times (\sqrt{((a - c)^2 + 4b)} - a + c)$, and from thence $v\ (= a - u) = \frac{1}{2} \times (a + c - \sqrt{((a - c)^2 + 4b)})$; whence $x^{15} + y^{15} + z^{15}\ (= u^{\frac{1}{3}})$, $x^3 + y^3\ (= v^{\frac{1}{2}})$ and $x^5 + y^5 + z^5\ (= w^{\frac{1}{3}})$ all become known, and from the 2d and 3d of these equations $z = (w^{\frac{1}{3}} - v^{\frac{1}{2}})^{\frac{1}{3}}$; whence $x^{15} + y^{15}\ (= u^{\frac{1}{3}} - z^{15})$ also becomes known, which put $= r$ and $x^3 + y^3\ (= v^{\frac{1}{2}}) = 2s$, and then (by quest. 102, p. 63 of Simp. Select. Exerc.) $x^3 - y^3$ will easily be found $= 2 \times ((\frac{1}{10} rs^{-1} + \frac{4}{5}s^4)^{\frac{1}{2}} - s^2)^{\frac{1}{2}}$ (supposing x to be greater than y) which put $= 2d$, and then x and $y = (s + d)^{\frac{1}{3}}$ and $(s - d)^{\frac{1}{3}}$ respectively.

IV. QUESTION 615, *by Mr.* John Chipchase.

In the midst of the Market-place at Stockton, is erected a fine column of the Doric order, on the top of which stands an urn 5 feet in height, that appears the greatest possible to a person (the height of whose eye is 5½ feet) standing at the distance of 32 feet from the base of the column, on level ground: From whence the column's height is required?

Answered by the Rev. Mr. Crakelt.

Construction. Take AB = 32 feet (the dist. of the observer from the column) and in an indefinite perpendicular thereto, at the point A, the dist. AC = 5 feet (the length of the urn): Bisect AC in D, and take DE = DB; also take EF = AC and AG = 5½ feet, the height of the observer; then GF will be the required height of the column.

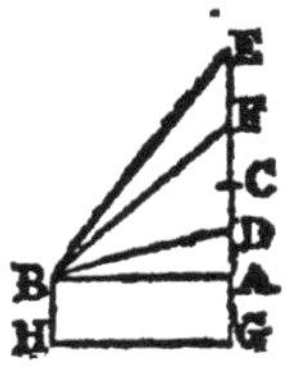

Demonstration. Since $BA^2 = BD^2 - AD^2 = (BD + AD) \times (BD - AD) = (ED + DA) \times (ED - DC)$ (by construction) $= EA \times EC = EA \times AF$, a circle described through the three points F, E, B will have the line AB for a tangent, and by Simp. Algebra, p. 358, 2d edit. the line FE will appear under the greatest angle to an eye at B, and consequently FA + AG or FG will be the perpendicular height of the column.*

* See Mr. Crakelt's remark on this question, at the end of the solution to question 660.

V. QUESTION 616, *by Mr.* Wm. Spicer.

Required the area of the exponential curve whose equation is

$$(x)^{x^2} = b^{\frac{y^{\frac{1}{2}}}{a}}\ ?$$

Answered by Mr. J. Spencer, *Discip.* T. Allen.

Put z and a for the hyp. logs. of x and b respectively, and then $zs^2 = ay^{\frac{1}{2}} \div x$; and $y = a^{-2} z^2 x^6$; whence $y\dot{x} = a^{-2} z^2 x^6 \dot{x}$, and the fluent (by art. 341 of Simp. Flux.) is $\frac{1}{7}a^{-2} x^7 \times (z^2 - \frac{2}{7}z + \frac{2}{49}) =$ the area of the curve required.

VI. QUESTION 617, *by Mr.* Samuel Vince.

If the subtang. of a curve be expressed by $\frac{(dx - x^4)^2\, y\dot{y}}{(a - bx^2)^2\, \dot{x}}$; required the value of the semi-ordinate y thereof, when the abscissa x is $= s$ given quantity s?

Answered by Mr. Geo. Coughron.

Per quest. $\frac{(dx - x^4)^2 \times y\dot{y}}{(a - bx^2)^2 \times \dot{x}} = \frac{y\dot{x}}{\dot{y}}$; $\therefore \dot{y} = \frac{a\dot{x} - bx^2\dot{x}}{dx - x^4} = \frac{a\dot{x}}{dx} + \frac{a}{d} \times \frac{x^2\dot{x}}{d - x^3} - \frac{bx\dot{x}}{d - x^3}$, and, taking the correct fluent, $y = \frac{a}{d} \times$ hyp. log. of $x + \frac{a}{3d} \times$ hyp. log. of $\frac{d}{d - x^3} + \frac{b}{3r} \times (\text{N}\sqrt{3} + \text{M} - \text{M}')$ (where N = the circular arc whose sine is $\frac{x\sqrt{\frac{3}{4}}}{\sqrt{(rr + rx + xx)}}$ to rad. 1, $r = d^{\frac{1}{3}}$ and M, M′ = the hyp. logs. of $\frac{r - x}{r}$ and $\frac{\sqrt{(rr + rx + xx)}}{r}$ respectively, see arts. 329 and 330 of Simpson's Flux. 2d edit.); whence the value of y may be determined let that of x be what it will.

The Rev. Mr. *Wildbore* solves it thus, viz. $\dot{y} = \dot{x} \times \frac{a - bx^2}{dx - x^4}$ (per question) $= \frac{a\dot{x}}{dx - x^4} - \frac{bx\dot{x}}{d - x^3} = \frac{a\dot{x}}{dx - x^4} - b\dot{x} \times$:

$$\frac{-\frac{1}{3}x + \frac{1}{3}d^{\frac{1}{3}}}{d^{\frac{1}{3}}x^2 + d^{\frac{2}{3}}x + d} + \frac{\frac{1}{3}}{d^{\frac{1}{3}}x - d^{\frac{2}{3}}} = \frac{a}{d} \times \frac{\dot{x} - 4d^{-1}x^3\dot{x}}{x - x^4 d^{-1}} + \frac{a}{d} \times$$

$$\frac{4d^{-1}x^2\dot{x}}{1 - x^3 d^{-1}} + \frac{b\dot{x}}{3d^{\frac{1}{3}} \cdot (d^{\frac{1}{3}} - x)} + \frac{b}{3d^{\frac{1}{3}}} \times \frac{t\dot{t}}{cc + tt} - \frac{1}{2}b \times \frac{\dot{t}}{cc + tt}$$

(putting $cc = \frac{3}{4}d^{\frac{2}{3}}$ and $t = x + \frac{1}{2}d^{\frac{1}{3}}$), the correct fluent of which is

$$y = \frac{a}{d} \times \left\{ \text{h. l.}\left(x - \frac{x^4}{d}\right) - \frac{4}{3} \times \text{h. l.}\left(1 - \frac{x^3}{d}\right)\right\} - \frac{b}{3d^{\frac{1}{3}}} \times \text{h. l.}$$

$\left(d^{\frac{1}{3}}-x\right)+\frac{b}{6d^{\frac{1}{3}}}\times$ h. l. $(cc+tt)-\frac{b}{2c}\times$ circular arc whose tang. is $\frac{t}{c}$ to rad. $1+\frac{b}{2c}\times$ circular arc tang. $\frac{1}{\sqrt{3}}$ and rad. 1; which, placing s for x, gives what is required.

VII. QUESTION 618, *by Mr.* J. Addison.

Suppose a triangular prism, whose length is 12 feet, and base a right-angled isosceles triangle, the longest side of which is also 12 feet, to be placed perpendicularly in a stream of water 12 feet deep, with its right angle directly facing the stream; required the velocity of the stream per sec. when its force against the prism is =387072lb. avoirdupoise.

Answered by Mr. J. Addison, *the Proposer.*

Let ABC represent a section of the prism parallel to its ends, and BD and AD be parallel and perpendicular to AC respectively; then, the number of forcing particles of the fluid acting upon the side AB, being as BD, and their force as the square of the sine of the angle of incidence BAD (45°), by mechanics, the whole force upon the two sides thereof AB, BC, will manifestly be = half that upon its base, AC (supposing AC, directly, to face the stream): But, putting v = the velocity of the water (per second) reckoned in feet, and $s=16\frac{1}{12}$ feet, the force against AC is $=(144vv\div 2s)\times 62\cdot5$lb. Av. (per prop. 107 of Emerson's Mechanics, 1st edit. a cubic foot of water weighing $62\frac{1}{2}$lb. Av.); therefore $(144vv\div 4s)\times 62\cdot5$ = the force against ABC = 387072lb. Av. per quest. Whence $v=52\cdot6008$, &c. feet, the velocity of the stream per second, which was required.

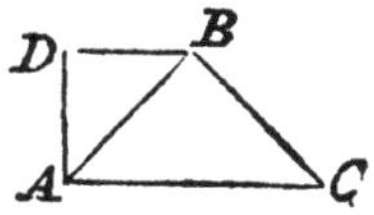

The Rev. Mr. *Wildbore* has also favoured us with a curious solution to this quest. founded on the principle, that the effect which the pressure of the water has to move the prism in the direction of the stream shall be equal to that of a weight of 387072lb. Avoir. from which, and the consideration of the action of the back water, he proves that the effect of the pressure against low-arched bridges, in time of floods, such as those made by Mr. *Brindley* for conveying his aqueducts over rivers, is not so great as is commonly imagined, not varying according to the depth of the water above, but nearly as the square of the difference of the depth above and below the bridge: Also, that the arch ought in this case to be a parabola, that so its width may be every where proportional to the quantity of water which is to be discharged through it; for the water runs with the greatest velocity at the bottom, and when the upper part of the arch is choaked up with hay, ice, &c. if it be thus formed it will discharge the most water when and where there is the greatest necessity for it, &c. but we cannot spare room for the solution at present.

VIII. QUESTION 619, *by Mr.* Steph. Ogle.

Given an angle A in magnitude, and a point P in position; to draw a right line PLS in such a direction, that PLS + M may be = to AS + AL, M being a given right line.

Answered by the Rev. Mr. Wildbore.

If the conchoid ruler revolve about the pole P upon the given line ASW with the given line or dist. SM = M, till PM = AL + AS, it will give the position, or positions, of PM required, with much more ease than any solid construction derived from the following

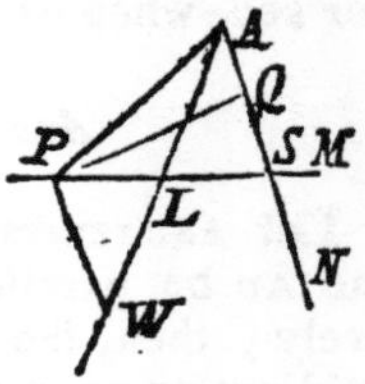

Analysis. Join A, P, and draw PQ perpendicular and PW parallel to AS. Put AP $= a$, AW $= b$, AQ $= g$, PW $= d$, AS $= x$, and AL $= y$; then per sim. triangles, $b - y$ (AW — AL) : d (PW) :: y (AL) : x (AS), whence $x = dy \div (b - y)$. Also, by Euc. 2. 13, and the conditions of the question, $x + y - \text{M} = \sqrt{(aa + xx - 2gx)}$; reduced, $y^3 - (2d + 2\text{M} + b) \cdot y^2 + (2\text{M}d + 2\text{M}b + \text{M}^2 - a^2 - 2gd) \cdot y + (a^2 - \text{M}^2) \cdot b = 0$: Which equation not being of the kind that admits of reduction by any of the methods of divisors, except in particular cases, it should seem as if the problem could not generally be reduced to a plane one.——But the roots of the equation may be found with great facility, after the manner of Newton and Leibnitz, by means of the above mechanism.

IX. QUESTION 620, *by Mr.* Edw. Williams, *Captain in the Royal Artillery.*

The same construction remaining as in Mons. Fermat's theorem (see quest. 606); it is required to demonstrate, geometrically, when OV will be a maximum.

Answered by the Rev. Mr. Crakelt.

Construction. From the vertices, D, C, of the given perpendiculars, AD, BC, to the middle N of the semicircle ANB draw the right lines, DN, CN, and the part, OV, which they intercept on the diameter AB will be the maximum required.

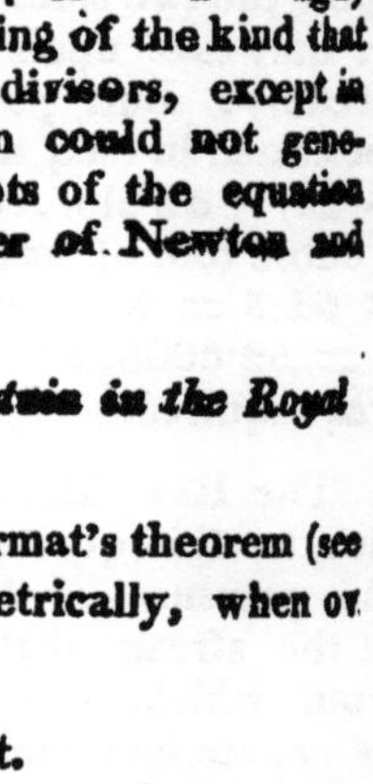

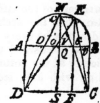

Demonstration. For to any other point, E, of the said semicircle draw the right lines DOE, CVE, and from N and E, let fall the perpendiculars NQS and EGF upon DC, and draw EC ∥ BA; then by similarity of triangles, NS : NQ :: DC : OV, and

EG : EF : : *ov* : DC, or by taking the rectangles of the corresponding terms, NS × EG = (NC + CS) × CQ : NQ × EF = (NC + CQ) × CS : : *ov* : OV : But (NC + CQ) × CS is greater than (NC + CS) × CQ ; wherefore OV will be greater than *ov*.

Scholium. Whatever the segment ANB, and the lengths of the perpendiculars AD, BC are, the construction and demonstration will hold good.

The Rev. Mr. Wildbore's *Demonstration is thus, viz.*

At any given dist. whatever, draw AB ∥ to DC and E*h* ⊥ to the same; then 'tis self-evident that the higher the point *e* or E is taken above *h*, the greater will *ov* or OV be; through *e* draw a right line parallel to DC, in which take any point F; join FD, FC, cutting AB in *r* and *s* : Complete the parallelograms DG, DH ; then, F*e* being = GH, and PC (⊥ to FEH) common (to the Δs CGH, CF*e*, DF*e*) BI will be = *sv* = *ro*, and adding *os* (common), *rs* will evidently be = *ov* ; therefore 'tis manifest that OV will be the greatest at the vertex, and the same at equal heights above DC, let *gh* and the curve AFE be what they will.

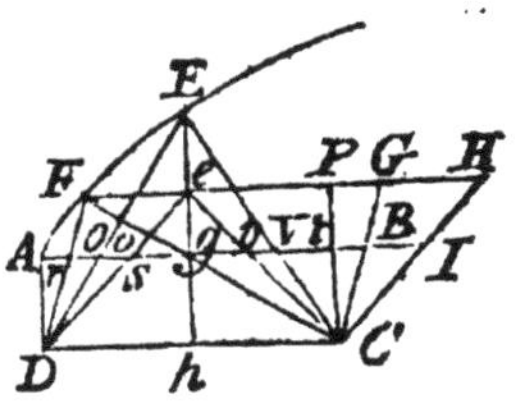

X. QUESTION 621, *by the Rev. Mr.* Wildbore.

Given the right line DC and the angle D, it is proposed geometrically to determine the point B so, that BA, DA being in a given ratio, the sum of BA and BC may be a given quantity.

Answered by Mr. Steph. Ogle.

Construction. Make the ∠CDI = the given ∠D, and draw CE to terminate in DI, so as to be to DC in the given ratio of BA to DA ; produce it till CF = the given quantity, and through F draw KH ∥ to DC meeting DI in H ; join C, H, and draw EG to terminate therein so as to be = EF ; then, drawing CB, BA ∥ to EG, CE respectively, the thing will be done.

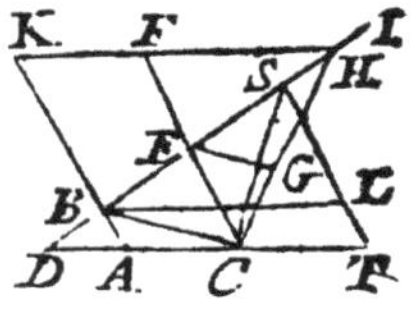

Demonstration. Produce AB to meet KH in K ; then, per sim. triangles, EF : BK (: : HE : HB) : : EG : BC, where the antecedents EF, EG being equal (by construction) the consequents BK, BC must be equal also : Consequently AB + BC (= AB + BK) = CF, the given quantity by construction.

XI. QUESTION 622, *by the Rev. Mr.* Crakelt.

Given the right line bisecting the vertical angle of any plane triangle, and terminating in its base, and the rectangle under each side and its adjacent segment of the base made by the said bisecting line; to construct the triangle.

Answered by the Rev. Mr. Crakelt, *the Proposer.*

Construction. Let P and Q represent the sides of two squares equal in magnitude to the two given rectangles, and find a fourth proportional, DC, to the given bisecting line, P and Q; perpendicular to which erect CE $= \frac{1}{2}$ the said bisecting line, and therewith as rad. describing a circle, draw DGEF: In CD produced take DI $=$ DF, and on CI as diam. describe the semicircle CKI; erect the perpendicular DK, and having taken thereon Dn $=$ Q, draw nl, nm $\parallel$ to KI, KC respectively; then with DC, DG, and Dm constitute the $\triangle$DHC, and in DH produced take HB $=$ GF, and join the points B, C: Circumscribe the $\triangle$BDC with a circle, and produce CH to meet it in A; then, joining the points A, B, ABC will be the required triangle.

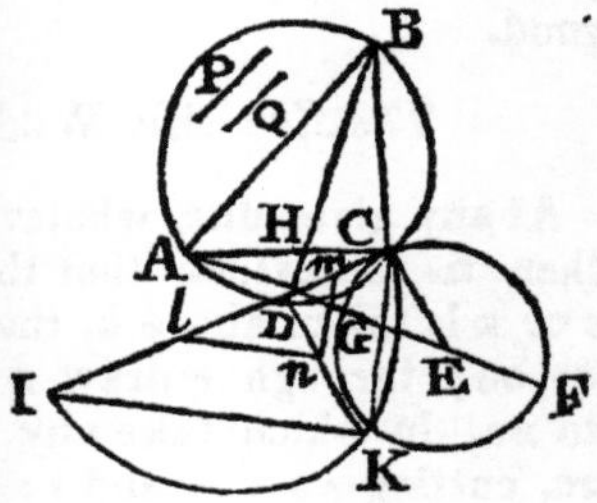

Demonstration. By construction BD $\times$ DH $=$ DF $\times$ DG $=$ DC^2; whence the $\triangle$s BDC and HDC are similar, and consequently BD : DC :: BC : CH: But, by reason of parallels we also have BD (ID) : DC :: lD : CH (Dm); whence BC $=$ Dl, and BC$\times$CH $=$ Dl $\times$ Dm $=$ (since the $\angle lnm$ is right) Dn^2 $=$ Q^2, by construction. The $\angle$ABD $=$ $\angle$ACD $=$ $\angle$DBC; whence BH ($=$ GF $=$ 2CE $=$ the given bisecting line) bisects the $\angle$ABC: Therefore CB : HB :: DC (AD) : AH, and CH : HB :: DC : AB, and consequently CB $\times$ CH : HB^2 :: DC^2 : AH $\times$ AB: But, by construction, HB^2 : P^2 :: Q^2 : CD^2; wherefore, *ex æquo perturbatè*, CB $\times$ CH : P^2 :: Q^2 : AH $\times$ AB; but, from what has been already proved, CB $\times$ CH $=$ Q^2, and of consequence AH $\times$ AB $=$ P^2.

XII. QUESTION 623, *by Mr.* T. Moss.

If from the extremities of the diameter of a circle ever so many chords are drawn, two and two, intersecting each other in an ordinate perpendicular to that diameter, the chords joining the extremities of every corresponding two of them, being produced, will all intersect the said diameter produced in one point: Quere the demonstration?

Answered by Mr. T. Moss, *the Proposer.*

Demonstration. Let any two (unequal) chords BH, DC be drawn, intersecting each other in the point m; thro' which point draw the perpendicular (or semi-chord) FI; draw the radius EH, and also FH and DH: Then it is evident that the $\angle$FHB is equal to $\angle$BHF, each being $=$ $\angle$FDm ($=$ $\angle$BDC). Moreover, the $\angle$EBH $+$ $\angle$FHB

= ∠EFH ; but ∠EBH is = ∠EHB : Therefore ∠EHB + ∠BHP (FHB) = ∠EHP = ∠EFH ; whence the ∠E being common, the △s EHP and EHF are similar, and so HE : EF : : PE : HE, or EB (HE) : EF : : PE : EB (HE), or (dividedly) EB : EB — EF (FB) : : PE : PE — BE (BP), or, alternately, EB : PE : : FB : BP : Again, by division of ratios, EB : PE —EB : : FB : BP—FB, i. e. EB : BP : : FB : BP—FB ; in which circumstance it is demonstrated (see theo. 15, p. 74, Simp. Geom. 1st edit.) that if any two right lines be drawn from the points F, P, meeting any where in the periphery of the circle described with the rad. EB, they will be in the constant ratio of FB to BP, and consequently a right line (BG or BH) drawn from B to the point where the two said lines meet in the periphery will always bisect the angle formed by those lines. Therefore, through any other point *n*, in the perpendicular FI, let now two chords be drawn from B and D to meet the periphery in G and L ; then, having drawn GF and GL, it is evident from the former part of this demonstration, that the ∠FGB (= FD*n*) = ∠BGL, and consequently that (by the above-mentioned theorem) GL produced must necessarily meet DB produced, in the very same point P where HC produced met it.

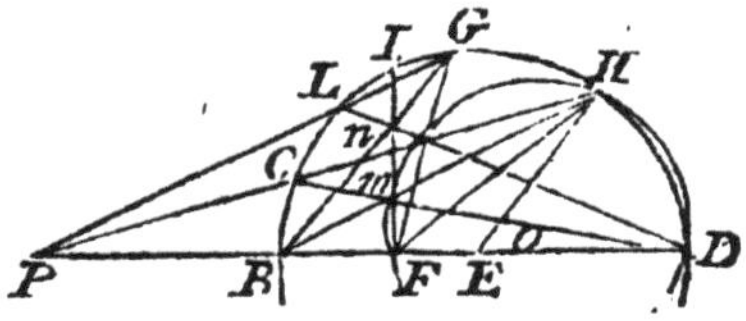

The Rev. Mr. Wildbore's *Demonstration is as follows, viz.*

In the given ordinate FI take any point *m*, and draw D*m*C, B*m*H ; and HCP meeting DB produced in P. Bisect D*m* in O, and with OD rad. describe the circ. F*m*D, which per Euc. 3, 31, will circumscribe the quadrilateral F*m*HD ; then, per 1, 32, the ∠HBE = ∠BPH + ∠BHP, the ∠HED = ∠EHP + ∠EPH, and, per 3, 21, the ∠CHB = ∠CDB (*m*DF) = *m*HF ; therefore the ∠PHE = ∠EBH (BHE) + ∠PHB (CDB), the ∠HPD = ∠HBE — ∠CDB, and the ∠EFH = ∠EBH + ∠*m*HF (*m*DB) : Consequently the ∠EFH = ∠PHE, and the ∠E being common, the △s FEH, HEP (per 1, 32,) are similar, and (per 6, 4,) FE : EH (= BE) : : EH : PE ; where 'tis manifest that FE and EH will continue the same wherever in the given ordinate FI the point *m* is taken, and consequently that their third proportional PE must continue the same likewise.

Cor. 1. If, in like manner, a circle be described about the quadrilateral BC*m*F, it will appear that the ∠FC*m* is = ∠*m*BF = ∠HC*m* ; ∴ CD bisects the ∠FCH : And it is shewn above that BH bisects the ∠CHF ; ∴ IF bisects the ∠CFH, and *m* is the cent. of the circle inscribed in the △CFH. It is evident likewise that PI is a tangent at I.

Cor. 2. Because FE : BE = EH : : FH : PH : : BF : BP ; therefore by conversion and division, BP : BP — BF : : BE : BE — FE = BF. Wherefore if, from any two given points P and F, two right lines in a given ratio are to be drawn, divide PF in that ratio in B, and take

BE : BF :: BP : BP — BF; then two lines meeting any where in the circumference of the circle described with the rad. BE, will be in the required ratio, which is the property laid down at p. 336 of Simpson's Alg. 3d. edit. and applied to the solution of many difficult problems by that ingenious author.——And with the same ease may other curious and useful proportions be derived.

XIII. QUESTION 624, *by Mr.* J. Powle.

Given $1^3 . 1^2 + 2^3 . 3^2 + 3^3 . 5^2 + 4^3 . 7^2$ &c.... to x terms $=$ 845815702, to find x.

> Where x is the age of my partner for life:
> A fond mother, true friend, and good wife.

Answered by Mr. Cha. Hutton.

Put s for the given sum of x terms of the series; then the $x+1$ term or $\dot{s}$ is $= (x+1)^3 . (2x+1)^2 =$ (putting $z = x+1$) $z^3 . (2z - 1)^2 = 4z^5 - 4z^4 + z^3 = 4\dot{z}\underset{1\,2\,3\,4}{zzzz} - (40\dot{z} + 4) . \underset{1\,2\,3}{zzzz} + (100\dot{z}^2 + 24\dot{z} + 1) . \underset{1\,2}{zzz} - (60\dot{z}^3 + 28\dot{z}^2 + 3\dot{z}^3) . \underset{1}{zz} + (4\dot{z}^4 + 4\dot{z}^3 + \dot{z}^2) . z$; and, by taking the integrals, &c. we then obtain $s = \frac{1}{60} \times : 40x^6 + 72x^5 - 5x^4 - 50x^3 - 5x^2 + 8x$, the sum of the series proposed ($\dot{z}$ being $=1$, and $x = z - 1$); from which the remaining part of the question might be easily determined, &c.

And by other different but ingenious methods the sum of the series is likewise determined by Mr. *G. Coughron*, and several others, but the ingenious proposer having, by some mischance, given us a wrong number (845815702), none of them could find the right age, that sent by the proposer along with the quest. in the year 1767 being 48.

Answer to the same by the Rev. Mr. Wildbore.

The proposed series $1^3 . 1^2 + 2^3 . 3^2 + 3^3 . 5^2 + 4^3 . 7^2$.... to x terms is evidently $= 1^3 . (2 . - 1)^2 + 2^3 . (2 . 2 - 1)^2 + 3^3 . (2 . 3 - 1)^2 + 4^3 . (2 . 4 - 1)^2$.... to x terms $= 4 \times (1^5 + 2^5 + 3^5 .. x^5) - 4 \times (1^4 + 2^4 + 3^4 .... x^4) + 1^3 + 2^3 + 3^3 .... x^3 =$ (by case 3d on p. 206 of Simp. Alg. 2d edit. &c.) $\frac{2}{3}x^6 + \frac{6}{5}x^5 - \frac{1}{12}x^4 - \frac{5}{6}x^3 - \frac{1}{12}x^2 + \frac{2}{15}x$, which suppose $=$ 8458957024*, and then $x = 48$, the truly valuable lady's age required by (the proposer) Mr. *Powle*.

* *Note.* This truly ingenious gentleman having discovered this number so as to give the value of x a whole number, it happens luckily to be the very number that gives $x =$ the age meant by the proposer, &c.

XIV. QUESTION 625, *by* Plus Minus.

Just after reading in the book of Genesis of the tower of Babel, whose top was to reach up to heaven, I fell asleep, and dreamed, that

I was got upon the top of this tower. When St. Paul's clock, in London, began to strike 12, I saw my own shadow fall on that church, and some time after saw it quit the earth's surface, just at the spot where a man seemed to be very hard at work. I called out (as loud as I could bawl) "Surely, friend, 'tis time to leave off"; "I think so too," says he, "for our clock wants but a quarter of 7." This adventure happened on the first of May, as I knew by the garlands on the May-poles all about the country, and so the sun's declination was 15° 19′. Hence I would know the height of the tower, the place where it stood, and the habitation of my friend aforesaid.

Answered by Mr. W. Wales.

From the data of this quest. it appears that the tower stood to the south of London, and under the same meridian, and that the sun set at $6^h\ 45^m$ by the time at the labourer's meridian, which let the arc ZPH represent, where Z is the zenith, or geographical situation, P the pole, and ⊙ the sun in the horizon HC. Draw P⊙, the sun's polar distance, the vertical Z⊙ and PLB for the meridian of London,

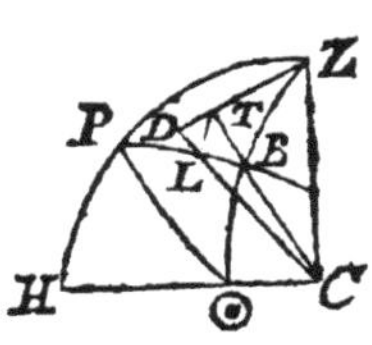

and the point B where these two last intersect will be the situation of the tower, and L that of London. Now, in the right-angled spherical △PH⊙, having given P⊙ and the ∠P, HP, the labourer's latitude, is readily found = 35° 27′ 46″ north; and if ZT be a tangent to the arc ZB in Z, and meeting CB produced in T, TB will be the height of the tower. Join T, L, and from T on CL produced, let fall the perpendicular TD, and the ∠DLT will be = the ⊙'s zenith dist. at London on the given day, whose sine and cosine denote by r and p, the sine and cosine of ZP by s and c, those of LP by a and b, the cosine of ∠PZB (= H⊙) by v, and DL by x. Then $1 + x =$ DC, and $rx \div p =$ TD; also (Euc. 47. 1.) $\sqrt{(\text{DC}^2 + \text{DT}^2)} =$ TC: Moreover TC : 1 :: TD : sine ∠DCT = sine LB, whose cosine also is thence known; and (Simp. Trig. p. 55) cosine PL × cos. LB — sine PL × sine LB = cos. PB. Again, since TC is the secant of ZB; its cosine = 1 ÷ TC, from whence its sine also is determinable, and (per Anderson s theo.) sine ZB × sine ZP × cos. ∠PZB + cos. ZB × cos. ZP = cos. PB. These two values for the cos. PB expressed in symbols, according to the preceding notation, and equated, give $b + (x \div p)\ (pb - ar) = sv\sqrt{(2x + xx \div pp)} + c$; but $pb - ar$ and sv are each = the sine of the ⊙'s declination, for which putting d, and $b - c = m$, the equation will be $m \div d + x \div p = \sqrt{(2x + xx \div pp)}$; whence $x = pm^2 \div 2d\ (dp - m) = 5{\cdot}9433341$. Hence TC and the ∠TCD are readily found = 8·1928755, and 32° 4′; and thence the tower's height = 7·1928755 radii of the earth, and its latitude 19° 27′ north; and lastly, the labourer's longitude = 95° 19′ east.

Answer to the same by the ingenious Proposer, **Plus Minus.**

In fig. 1, P represents the north pole, ☉ that place of the earth to which the sun was vertical when the shadow quitted its surface, Z the place of the tower, and M that of the man. It is evident that at the time aforesaid the sun must be in the horizon of M, and therefore ☉M $= 90°$; ☉P is the comp. of declin. of the sun, and the ∠☉PM $= 101° 15'$: Hence the man's lat. will be found $= 35° 27' 46\frac{1}{2}''$ north. Find the ∠P☉M and call its cos. q, and let the sine and cos. of ☉P be s and d respectively. Then, to find expressions for the sine and cosine ☉Z and the cosine of ZP, consider, fig. 2d, where A represents the top of the tower, Z its base, L London, P the north pole, C the earth's centre, and M the man's place (which is here put in the meridian of L to prevent confusion in the diagram): Draw AL cutting the axis CP in F, and let fall CD, AE ⊥ to AL and CP, and join CL, CM, which call r, and AL, z; moreover, let the sine and cosine of the sun's merid. alt. at London be m and n ($=$ LD and CD). Then AC $= \sqrt{(zz + 2mz + rr)}$, and the sine of the ∠CAM ($=$ ☉Z) $= rr \div \sqrt{(zz + 2mz + rr)}$, and its cosine $= r\sqrt{(zz + 2mz)} \div \sqrt{(zz + 2mz + rr)}$. Now the ∠s FCD, FAE being equal to the sun's declination, $s : d :: n$ (CD) $: dn \div s$ (FD), $s : r :: n : rn \div s =$ CF, $r : d :: z + m - dn \div s$ (AF) $: (dsz + dsm - ddn) \div rs$ (FE), and FE + FC (or CE) $= (dz + dm + sn) \div r$ (putting ss for $rr - dd$): Hence the cosine of ZP $= (dz + dm + sn) \div \sqrt{(zz + 2mz + rr)}$, and (by means of an excellent theorem in the Lady's Diary 1732, pa. 8,) $qrrs \div \sqrt{(zz + 2mz + rr)} + drr\sqrt{(zz + 2mz)} \div \sqrt{(zz + 2mz + rr)} = rr \times (dz + dm + sn) \div \sqrt{(zz + 2mz + rr)}$; whence $z = (dm + sn - sq)^2 \div (2ds \times (q - n))$. —Hence the rest will be found as follows, viz. The height of the tower AZ $= 7{\cdot}370827$, radii of the earth $= 29364{\cdot}296$ miles, its latitude $= 19° 22'$ N. and the man's longitude from London $= 95° 26' 30''$ E. nearly.

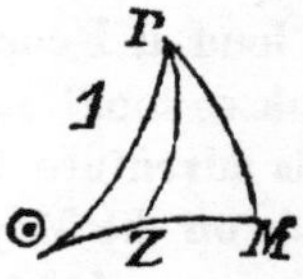

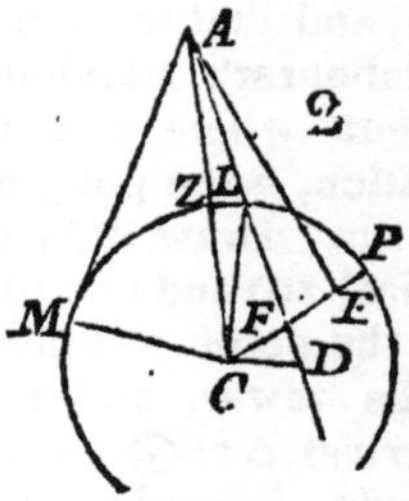

XV. QUESTION 626, *by Mr.* Tho. Allen.

Let ADB be a given semi-ellipse, CnH a circular arc described on the centre A with the radius AC $=$ the semi-conjugate, and suppose the right line AnM cutting the circle in n and the ellipse in M, to revolve about the point A, until it coincides with AB. Let, also, QM, Pn be drawn perpendicular to AB in the points Q and P, and on Pn (produced when necessary) let

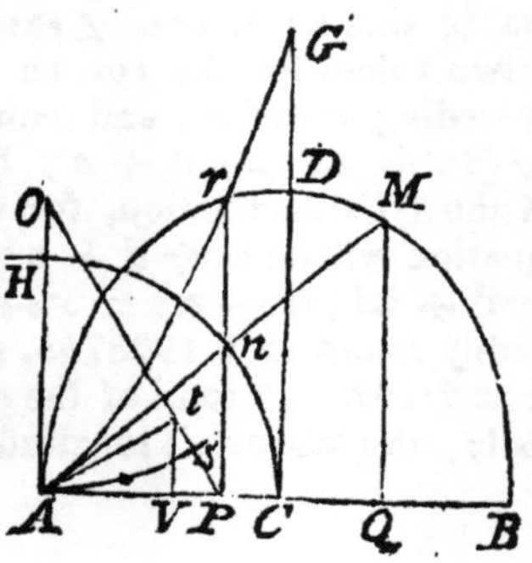

PR be taken always $=$ AQ; then, whilst the point M describes the semi-elliptic arc ADB, the point r will describe the curve line ArG. It is required to determine the content of the solid generated by the revolution of the said curve about its axis AC.

Answered by Mr. Geo. Coughron.

Put AP $= x$, Pr (AQ) $= y$, AC $= a$, CD $= b$, and $3{\cdot}1416 = p$, then will P$n^2 = aa - xx$, and, per prop. of the ellipsis, QM$^2 = b^2a^{-2} \times (2ay - yy)$: Whence, by means of the sim. Δs APn, AQM, will be found $y = 2abbxx \div (a^4 + ccxx)$ the equation of the curve, cc being $= bb - aa$; consequently $pyy\dot{x}$ (the flux. of the required solidity) $= 4paab^4x^4\dot{x} \div (a^4 + ccxx)^2$, and (taking the flu.) the solidity itself, when $x = a$, will be either $2pa^3b^4c^{-5} \times ((3aac + 2c^3) \times (aa + cc)^{-1} - 3a \times$ cir. arc whose tan. is $\frac{c}{a}$ and rad. 1), or $2pa^3b^4 c^{-5} \times ((3aac - 2c^3) \times (aa - cc)^{-1} - \frac{3}{2}a \times$ h. log. of $\frac{a+c}{a-c}$), according as b or a is the semi-transverse, &c.

Answer to the same by the Rev. Mr. Wildbore.

The figure being drawn as directed in the quest. erect the $\perp$ AO $=$ AC$^2 \div \sqrt{}$(DC^2 — AC^2), with which, as rad. and cent. O, describe the arc of a circle, and join O, P cutting it in s; let fall At, tV $\perp$ upon OP and AC respectively, then will AV (by sim. triangles) $=$ AP . AO^2 . OP^{-2}, which, because AO^2 + AP^2 = OP^2, is evidently the fluent of A$\dot{\text{P}}$. AO^2 . (AO^2 + AP^2 — 2AP^2) OP^{-4}. Take any where a right line M = 2AC . DC^2. (DC^2 — AC^2)$^{-1}$ = 2AO^2 . DC^2 . AC^{-3}; then, by conics, &c. AM^2 = 2DC^2. AQ. AC^{-1} — AQ^2 . AC^2 . AO^{-2} and, by means of similar triangles, M : AQ :: AO^2 + AP^2 : AP^2; therefore AQ = Pr = M $\times$ AP^2 $\times$ OP^{-2}, Pr^2 (= M^2 . AP^4 . OP^{-4}) = M^2 $\times$ ((AO^2 + AP^2)^2 — $\frac{3}{2}$ AO^2 . (AO^2 + AP^2) + $\frac{1}{2}$AO^2 . (AO^2 + AP^2 — 2AP^2)) $\times$ (AO^2 + AP^2)$^{-2}$ and the solidity generated round AP, or the fluent of p . rP^2 . A$\dot{\text{P}}$, is p . M^2 $\times$ (AP — $\frac{3}{2}$ the arc As + $\frac{1}{2}$ AV), which was required.

Corollary. Pr (= M . AP^2 . OP^{-2}) being = M $\times$ (1 — AO^2 $\times$ OP^{-2}), the fluent of Pr . A$\dot{\text{P}}$ is M . (AP — the arc As) = the area of the curve ArP.

Plus Minus puts AC $= c$, CD $= t$, AP $= z$, and Pr $= v$, and then finds the equation of the curve ArG to be $v = 2cttzz \times (c^4 \pm mmzz)^{-1}$ where $mm = tt - cc$, and is affirmative or negative according as CD is the longer or shorter semi-axis. In the former case, the curve ARE is the conchoidal hyperbolism of an ellipsis with a diameter without a centre, and that diameter is a line perpendicular to AC in the point A, and its asymptote is parallel to AC; and, in the latter case, the curve is the hyperbolism of an hyperbola with a diameter, its two parallel asymptotes being perpendicular to AC and equidistant from A, and the 3d parallel to AC, which is the tangent at the vertex of the axis, viz. A. If $tt = cc$, or $m = 0$, i. e. if the semi-ellipsis AMD becomes a semicircle, the curve degenerates into a common parabola, and AC is still a tang. at A the vertex of the axis. The fluxion of the solid required is $4pc^2t^4z^4\dot{z} \times (c^4 \pm m^2 z^2)^{-2}$, and the fluent, or solid itself, is easily found from Emerson's forms, &c.

THE PRIZE QUESTION, *by the Rev. Mr.* Wildbore.

A lady sent her servant to a well famous for its excellent water, who filled the bucket, and drew it up very uniformly and steadily; but when the bottom arrived at land, found that the last drop of water was running out of it; for some unlucky boys had bored an hole therein 1·066105 inch diameter. Hence the time it was in drawing up is required, the diameter and depth of the bucket being each of them one, and the depth of the well to the surface of the water 29 feet?

Answered by the Rev. Mr. Wildbore, *the Proposer.*

Suppose SS to be the surface of the water, L the land, O the bottom of the bucket when filled with water, B the bottom of it when the part AS is drawn out of the water, C the surface of the water in the bucket in this position, LS $=$ 348 inches $= n$, AB $=$ SO $=$ AD $= 12 = a$, $1{\cdot}066105 = c$, $193 = d$, $s =$ the velocity per second with which the bucket is drawn up, AS $=$ BO $= r$, and AC $= x$. Then $\sqrt{d} : 2d :: \sqrt{\tfrac{1}{2}}cs : d^{\frac{1}{2}} \sqrt{(2r - 2x)} =$ the velocity per second of descent of a column of water whose height is CB and immersed part SB, owing to its own pressure, which increased by s gives $s + d^{\frac{1}{2}} \sqrt{(2r - 2x)} =$ the real velocity per second with which the water runs out of the orifice: Wherefore $c^2a^{-2} \times (s + d^{\frac{1}{2}} \sqrt{(2r - 2x)}) =$ the relative velocity per second, with which the surface of the water descends in the bucket at C, and $s - c^2a^{-2} \times (s + d^{\frac{1}{2}}\sqrt{(2r - 2x)}) =$ the absolute velocity per second, with which the surface of the water ascends in the well.

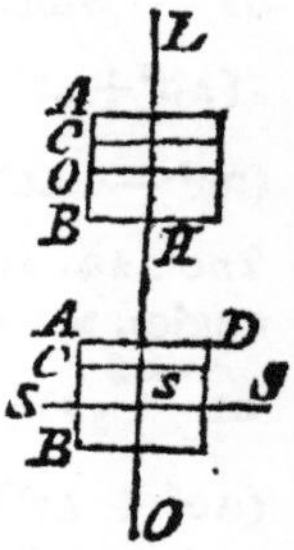

Whence $\dfrac{aa \cdot (\dot{r} - \dot{x})}{(aa - cc) \cdot s - c^2d^{\frac{1}{2}} \sqrt{(2r - 2x)}} =$ the fluxion of the time of ascent from S to C or from O to B; the fluent of which corrected by taking r and x each $= 0$, will, when AS $=$ BO $= r = x$ (putting $\gamma = \frac{(aa - cc) \cdot s}{cc\sqrt{d}}$), give $\frac{aa}{cc\sqrt{d}}$ ($\gamma \times$ hyp. log. of $\frac{\gamma}{\gamma - \sqrt{(2a - 2x)}} \sqrt{(2a - 2x)}) = \frac{a}{s}$, the time in which the bucket ascends out of the water.——When it has ascended to x, suppose O the surface of the water therein, $b = a - x =$ CB, and y $=$ CO; then, in the same manner as before, will $\frac{cc}{aa} \times (s + d^{\frac{1}{2}} \times \sqrt{(2b - 2y)}) =$ the relative velocity per second with which the surface of the water descends in the bucket at O, and the fluxion of time of ascent from S to H $=$ that of descent in the bucket between those two positions $= \frac{aa\dot{y}}{cc \times (s + d^{\frac{1}{2}}\sqrt{(2b - 2y)})}$; the fluent of which, when $y = b = a - x$, or when H arrives at L (putting $\frac{s}{\sqrt{d}}$ $= e$), is $\frac{aa}{cc\sqrt{d}} \times : \sqrt{(2a - 2x)} + e \,.$ hyp. log. of $\frac{e}{e + \sqrt{(2a - 2x)}}$ $= \frac{n}{s} =$ the time of ascent of the bottom of the bucket from the surface of the water till it arrived at land: From which equation and that above x is found $= \cdot 4557$, $s = 12$, and $(n + a) \div s = 30''$ the time required.

In this solution the motion of the bucket is supposed to commence when the water therein, and that in the reservoir, are both at rest, with their surfaces on the same horizontal level.

Corollary. $\frac{\text{AD}^2 \sqrt{2\text{AB}}}{(\cdot 088842)^2 \sqrt{(16 \cdot 0833)}} = 44 \cdot 67768''$ is the time in which the bucket would be emptied at rest.

Scholium. From the same principles may one part of the theory of the curious wheel for raising water upon the river Limat, at Zurich, in Switzerland, be determined.

Additional Solution.

This is the solution in which was proposed that new principle of effluent water, which occasioned that long altercation on the subject, in our Miscellany, by two learned and ingenious gentlemen. This dispute was carried on, by both parties, with that true gentlemen-like candour, openness, and modesty, which has done them much honour. And of their several ingenious arguments, every reader is left to avail himself, and to adopt that principle which to him seems to be the true one.

The above solution is given on the principle of the velocity of the effluent stream being increased by that of the vessel itself. And we shall here subjoin the solution on the contrary principle; that every person may take which he pleases.

Let a, c, d, n and s denote the same quantities as in the above original solution; also, $x =$ any variable alt. of the surface of the water in the bucket above that in the well, while the bucket is ascending out of the water; and t the corresponding time. Then, $\sqrt{2dx} =$ the velocity of the issuing stream, and $\therefore c^2a^{-2}\sqrt{2dx} =$ the velocity of the descending surface in the bucket; consequently $s - c^2a^{-2}\sqrt{2dx}=$ the velocity per sec. with which the surf. of the water in the bucket ascends above that in the well; $\therefore s - c^2a^{-2}\sqrt{2dx} : 1'' :: \dot{x} : \dot{x} \div (s - c^2a^{-2}\sqrt{2dx}) = \dot{t}$; hence t is $= \frac{-a^2}{dc^2}\sqrt{2dx} + \frac{a^4s}{c^4d}$ $\times$ hyp. log. $\frac{s}{s - c^2a^{-2}\sqrt{2dx}}$ which by the question will be $= \frac{a}{s}$ when the bottom of the bucket arrives at the surface of the water in the well, or when the bucket is just totally emerged from the water in the well.

Now the remaining part of the time, or that in which x depth of water will issue from the bucket, is $\frac{a^2}{dc^2}\sqrt{2dx}$, by prob. 1, art. 1, Miscel. and this must be $= \frac{n}{s}$, by the question. Adding now this to the former time, we have $\frac{a^4s}{c^4d} \times$ h. l. $\frac{s}{s - c^2a^{-2}\sqrt{2dx}} = \frac{a+n}{s}$ the whole time of emptying. In this equation write $\frac{c^2dn}{a^2s}$ for $\sqrt{2dx}$ its value by the equation $\frac{a^2}{c^2d}\sqrt{2dx} = \frac{n}{s}$), &c. and we have $\frac{a+n}{a^4} \times c^4d = s^2 \times$ hyp. log. $\frac{s^2}{s^2 - c^4dna^{-4}}$; or, in numbers, $1{\cdot}879796 = s^2 \times$ com. log. $\frac{s^2}{s^2 - 4{\cdot}1841096}$. From hence we find $s = 7{\cdot}9667$. And $\therefore \frac{a+n}{s} = 45{\cdot}18809''$ the whole time required, on this principle. H.

Questions proposed in 1771, *and answered in* 1772.

I. QUESTION 627, *by Mr.* Tho. Sadler.

Young Hodge, a homely country swain,
Long courted Susan of the plain,

But never could a method find
To bring the fair one to be kind.
Her pride is center'd in herself;
She calls him clown and country elf,
And mimics fashion with an air:
For dressing few with her compare.
This charmer's name with ease you'll find
From what is underneath subjoin'd.*
O! tell the way to make her kind.

* $\left\{\begin{array}{ll} x + y^2 + z = 201, \\ x^2 + yz = 37, \\ z^2 + x^2y = 130, \end{array}\right\}$: To find x, y, and z; their values expressing the places of the letters in the alphabet that compose her name.

Answered by Mr. Hen. Clark.

It is evident from the first equation that y cannot be greater than 14 nor less than 12 (because x, y, and z are to be whole positive numbers by the nature of the quest. not exceeding 24); therefore supposing $y = 14$, xx from the second equation becomes $= 37 - 14z$, which value substituted in the third equation it becomes $zz - 196z = -388$; whence, by compleating the square, &c. $z = 2$, a whole number; consequently $x = 3$, $y = 14$, and $z = 2$, answering to the letter's in the alphabet COB, the lady's name required.

II. QUESTION 628, *by Mr.* Wm. Gawith.

Given $(x^5+y^5+z^5)^5 + (u^5 + z^5)^5 = a$, $x^5 + y^5 + u^5 + 2z^5 = b$, $(u + z)^2 \times (x^5 + y^5 + z^5) = c$, and $(u^5 + z^5) \times (x + y)^2 = d$, to find x, u, y, and z by quadratics.

Answered by Mr. G. Coughron.

The two first given equations being the sum and the sum of the 5th powers of $x^5 + y^5 + z^5$ and $u^5 + z^5$, it will be found by quest. 48, pa. 105 of Simpson's Alg. 2d edit. that $x^5 + y^5 + z^5$ and $u^5 + z^5 = \frac{1}{2}b \pm \sqrt{(\sqrt{(\frac{1}{5}ab^{-1} + \frac{1}{20}b^4)} - \frac{1}{4}bb)}$, the upper of the double signs giving the value of the one, and the lower the other of them; these values being put $=m$ and n respectively, from the 3d and 4th equations, by division, &c. will be had $u + z = c^{\frac{1}{2}} m^{-\frac{1}{2}}$, which call $2c$, and $x + y = d^{\frac{1}{2}} n^{-\frac{1}{2}}$ which call $2r$; and from thence, as above, will be found u and $z = e \pm \sqrt{(\sqrt{(\frac{1}{10}ne^{-1} + \frac{4}{5}e^4)} - e^2)}$; consequently $x^5 + y^5 = m - z^5$ becomes known, which call s, and then from it and $x+y=2r$, x and y come out $= r \pm \sqrt{(\sqrt{(\frac{1}{10}sr^{-1} + \frac{4}{5}r^4)} - r^2)}$, the affirmative sign giving the value of the greater, and the negative sign that of the lesser of them.

III. QUESTION 629, *by Mr.* G. Cetii.

If any plane triangle be circumscribed by a circle, and have another circle inscribed therein, and if from any one of the angular points a straight line be drawn to the centre of the inscribed circle, and produced to meet the periphery of the circumscribing one: Then will the sum of the sides which include the angle from whence the straight line is drawn, be to the said straight line, so produced, as the third side of the triangle is to that part of the straight line which is produced beyond the centre of the inscribed circle. It is required to demonstrate this geometrically?

Answered by Mr. Isaac Dalby.

Demonstration. The line bisecting the $\angle$A of the proposed $\triangle$BAC being produced to meet BC in S and the circumf. of its circumscribing circ. in D, let the points B, D and C, D be joined, and right lines drawn from B, C to G the cent. of the inscribed circle: Likewise draw GE, GF $\parallel$ AB, AC meeting BD, CD in E and F; then (E. 3, 21,) the $\angle$DBC being $=$ $\angle$DAC $=$ $\angle$BAD and also $=$ $\angle$EGD (by construction), the $\angle$SBG $=$ $\angle$GBA (Euc. 4, 4,) and the $\angle$DGB $=$ the $\angle$s GAB, GBA (E. 1, 32,) BD will be $=$ DC (Euc. 1, 6,) $=$ DG, and the $\triangle$s BSD, DEG sim. and equal; $\therefore$ GE $=$ BS: And in the same manner GF is proved to be $=$ CS; and consequently, the trapeziums BACD, EGFD being sim. and alike situated, AB $+$ AC : AD :: EG $+$ GF (BS $+$ SC $=$ BC) : DG.

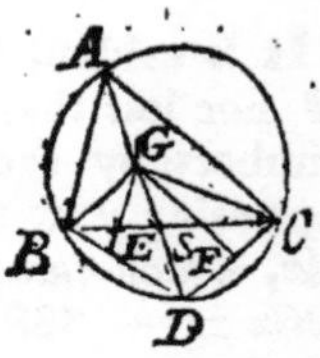

IV. QUESTION 630, *by Mr.* R. Mayo.

To determine the locus of all the places on the earth which have the same altitude of the sun with any given place, at a given instant of time.

Answered by Mr. G. Cetii.

As the ☉'s alt. is const. and given, its complement, or the ☉'s zenith dist. will be const. and given also. Now, nothing can be plainer than that the ☉'s zenith dist. is always equal to the dist. between the given place and that to which the sun is vertical at the given time; whence this dist. is given and constant, and therefore if the point to which the ☉ is vertical at the given time be found, and round it, as a pole, a circle be described through the given place, it will be the locus required.

Corollary. Hence, if it was required to find the locus of all the places where the sun rises or sets at the same time on a given day, it is manifest it will be a great circle, whose pole is the point to which the sun is then vertical.

V. QUESTION 631, *by Mr.* Wm. Spicer.

A shell was discharged from a mortar, which in its flight just touched the top of a steeple, and in four seconds of time after fell at the distance of $3262\frac{6}{7}$ feet from the bottom of the steeple: From whence the report of its fall was heard at the mortar just 12 seconds after the explosion: Required the steeple's height?

Answered by Mr. G. Coughron.

Let AEB represent the path of the shell, AC its direction at the commencement of motion, EF the steeple; and let DG be parallel and BC perpendicular to the horizontal line AB. Put FB ($=$ DG) $= 3262\frac{6}{7}$ feet $= a$, the time of its description $= 4'' = b$, the velocity of sound per second $= 1142 = c$, and the time of the shell's flight, or the time of its describing AEB or AC $= x$; then will $b : a :: x : ax \div b =$ AB, and consequently $x + ax \div bc = 12'' = t$; whence $x = bct \div (a + bc) = 7''$, AB $= ax \div b = 5710$ feet, BC $= 7^2 \times 16\frac{1}{12} = 788\frac{1}{12}$ feet, DE $= 3^2 \times 16\frac{1}{12} = 144\frac{2}{3}$ feet, AF $= \frac{3}{7} \times 5710 = 2447\frac{1}{7}$, FD ($=$ AF $\times$ BC $\div$ AB) $= 337\frac{3}{4}$, and the steeple's height FE ($=$ FD $-$ DE) $= 193$ feet.

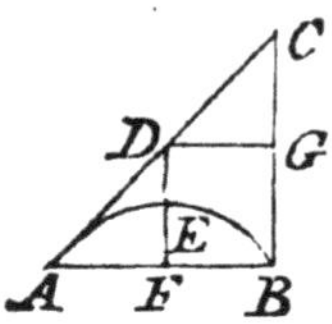

The same, independent of Algebra, by Mr. Cha. Hutton.

The time in which sound moves from B to F is $=$ FB $\div 1142 = 3262\frac{6}{7} \div 1142 = 2\frac{6}{7}''$; consequently the horizontal velocity of the shell is to that of sound as $2\frac{6}{7}$ to 4. But, as $4 + 2\frac{6}{7} : 4 :: 12 - 4 - 2\frac{6}{7} : 5\frac{1}{7} \times 4 \div 6\frac{6}{7} = 3''$, the time of the shell's flight from A to E or F, so that the whole time of its flight is $7''$. Also, AF $= \frac{3}{7}$AB, FD $= \frac{3}{7}$BC; and, by the law of falling bodies, $1^2 : 7^2 :: 16\frac{1}{12} : 16\frac{1}{12} \times 49 =$ BC, and $1^2 : 3^2 :: 16\frac{1}{12} : 16\frac{1}{12} \times 9 =$ ED; whence FE ($=$ FD $-$ ED $= \frac{3}{7}$BC $-$ ED $= 16\frac{1}{12} \times 21 - 16\frac{1}{12} \times 9 = 16\frac{1}{12} \times 12$) $= 193$ feet, the height of the steeple required.

VI. QUESTION 632, *by Mr.* Paul Sharp.

Suppose a tub in the form of the frustum of a common parabolic conoid, whose top diameter $= 4$, bottom diameter $= 2{\cdot}52982$, and depth $= 3$ feet, to be filled with water: Required the time of its emptying through a circular hole of 1 inch diameter in its bottom?

Answered by Mr. Isaac Dalby.

Let AGEB represent the tub, and suppose the parabolic conoid to be compleated. Put the top diameter ($= 4$) $= d$, the bottom diameter

$(= 2{\cdot}52982) = b$, the depth $(= 3 \text{ feet}) = h$, the part of the axis DR $= n$, $\cdot 7854 = a$, $t =$ the time in which the tub's circumscribing cylinder would empty itself through the same orifice with the first or greatest velocity uniformly continued, x $(= r\text{D})$ any variable altitude of the running water's surface above the bottom of the tub, and $y =$ the time sought. Then, by the nature of the curve, $\text{NS}^2 = d^2 \times (n+h)^{-1} \times (n+x)$, and by reasoning as in examples 4, 5, on pa. 139, 140, &c. of Emers. Fluxions, 2d edit. $\dot{y} = -((n+b) \times \sqrt{h})^{-1} \times (tnx^{-\frac{1}{2}}\dot{x} + tx^{\frac{1}{2}}\dot{x})$; the correct fluent whereof, when $x = 0$, is $y = 2t \times (3n+h) \div (3n + 3h) = 14' \; 4{\cdot}3476''$ the time required.

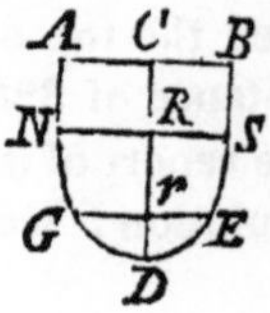

Corollary. The times of emptying the tub and its circumscribing cylinder are as $\frac{2}{3}$ to 1 : and if the tub was inverted, the time through an equal orifice would be $2t \times (3n+2h) \div (3n+3h) = 18' \; 45{\cdot}769''$.

The Rev. Mr. Wildbore *answers it thus:*

'From the data are easily found $p = \frac{4}{3}$ the parameter of the generating parabola, and the depth wanting to compleat it at bottom $= 2$ feet $= a$. Let c $(= 3 \text{ feet}) =$ the depth of the vessel, $bb = 1 \div (4 \times 144) = \square$ of rad. of the orifice, $x =$ any variable height of the water above it, $y =$ the ord. corresponding, $d = 16\frac{1}{12}$ feet, and $t =$ the time required; then, by the laws of hydrostatics, $\frac{bb}{yy}\sqrt{2dx} = \frac{bb\sqrt{2d}}{p} \times \frac{\sqrt{x}}{a+x} =$ the velocity of descent along the axis of the vessel; by which dividing $-\dot{x}$, it gives $\dot{t}$, whose corrected fluent, when $x=0$, is $t = (p\sqrt{2c}) \times (a + \frac{1}{3}c) \div bb\sqrt{d} = 844{\cdot}3476'' = 14{\cdot}07246''$ the time required.'

For the solution of this question, Mr. *Hutton* refers to article 1 of his New Mathematical Miscellany, where the subject of such exhaustions is fully treated on.

VII. QUESTION 633, *by Mr.* Tho. Moss.

Having given the hypothenuse of a right-angled plane triangle, and the part of the leg intercepted between the right angle and a point where a perpendicular on the middle of the hypothenuse meets the said leg, to determine the triangle.

Answered by Mr. T. Moss, *the Proposer.*

Construction. On BE the given intercepted part of the leg produced, take EC (by prob. 18, pa. 107 of Simp. Geom. 2d. edit.) such that BC × EC may be = the rectangle under the given hypothenuse and its half, and about C, as a centre, with the given hypothenuse as rad. describe an arch of a circle cutting BQ ⊥ CB in A; then, join the points A, C and the thing is done.

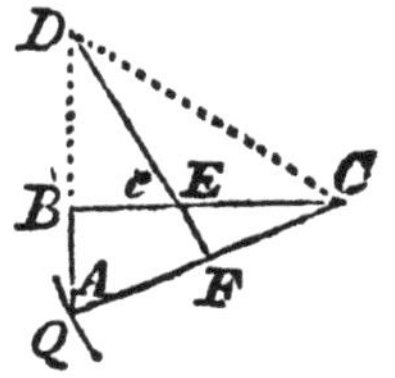

Demonstration. Upon the middle of AC erect the perpendicular F*e* meeting BC in *e*; then, by sim. triangles, AC : BC :: *e*C : FC, and consequently (by Euc. 16, 6.) AC × FC = BC × *e*C = BC × EC (per construction); whence *e*C = EC, and the points *e*, E coinciding, B*e* = BE (the given intercepted part, per construction), and ABC is a right angle, per construction.

Corollary. If AB, FE be produced to meet in D, and CD be drawn, it evidently appears that DE bisects the ∠BDC, and that the bisecting line BD is to the hypothenuse CD in the given ratio of BE to FC ($\frac{1}{2}$AC); whence it manifestly follows that the preceding construction is, in effect, the very same as that of a right-angled triangle having the hypothenuse and a line bisecting one of its acute angles, and terminating in the opposite leg, given.

VIII. QUESTION 634, *by Mr.* S. Clark.

Required to draw a right line CB from a given point C between two right lines DE, DF given in position to terminate in one of them (DE) so, that drawing another right line BA from the point of termination (B), parallel to another right line DG given, likewise, in position to terminate in the other of them (DF), the difference of the lines, CB, BA, so drawn may be equal to a given right line (MN).

Answered by Mr. G. Coughron.

Construction. Draw CH and HI ∥ DF and DG. On HI take IK = the given diff. MN, and through K draw LKP ∥ DF; join C, L and with the rad. HK and cent. H, intersect CL in *o*, O, and ∥ HO and DG draw CB and BA, which will be the lines required.

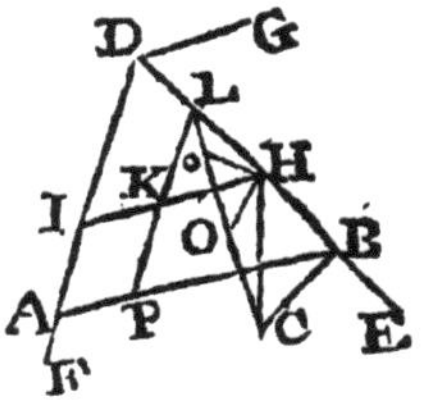

Demonstration. By sim. △s, HK : BP (:: LH : LB) :: HO (= HK) : BC; consequently BC = BP = AB — AP = AB — IK = AB — MN, and therefore AB — BC = MN, as it ought.

Corollary. The rad. HK cutting CL in two points shews that the prob. admits of two answers.——When MN is ⊏ HI the construction is still the same, only HI must be continued beyond H till it is = MN; and when CB is to be greatest, MN must be set from I to the left hand, on HI produced, and then the rest of the construction will be the same as above.

Construction to the same by Mr. Cha. Hutton.

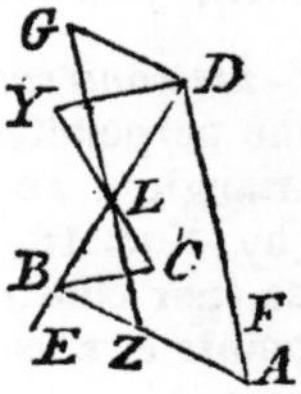

Take DG = MN, and draw GZ ∥ DF cutting DE in L. Draw CLY so that DY = DG: Then, drawing CB, BA ∥ DY and DG, meeting DE and DF in B and A respectively, and the thing will be done.—— For by similar triangles, DL : LB :: DG : BZ :: DY : BC; but DG = DY; therefore BZ = BC, and ZA = DG.

IX. QUESTION 635, *by the Rev. Mr.* Crakelt.

Two of the sides of a plane triangle, together with the line drawn from the angle they include to the centre of the inscribed circle being given, to determine the triangle by construction.

Answered by the Rev. Mr. Wildbore.

Construction. Upon BO the given bisecting line, produced, take BD a fourth proportional to BO and the two given sides of the triangle, and on the diameter DO describe a circle; to which from B set off the two given sides BC, BA (the latter of them intersecting the circumf. suppose in E); join A, C, and ABC is the required triangle.——For (per Euc. 3, 37,) DB is a fourth proportional to OB, AB, and EB, and (by construct.) to OB, AB, and BC; therefore EB = BC, and ∠CBO = ∠EBO, and the sides about the equal angles being (by construction) proportional, the △s AOB, DCB (per Euc. 6, 6,) are similar, and (per 3, 21,) ∠CDO (= BAO) = ∠CAO; ∴ (per 4, 4,) O is the centre of the inscribed circle, and the triangle is that required.——And hence by way of corollary is easily deduced an elegant demonstration to quest. 3d.

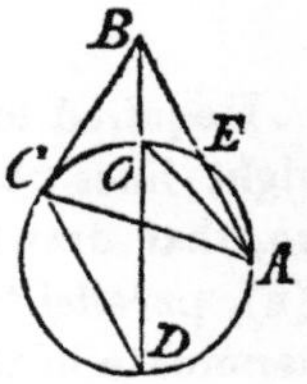

X. QUESTION 636, *by Mr.* T. Moss.

To draw a right line through the given point E in the side BF of the given plane triangle BFC meeting BC in G and CF produced in H so, that the sum of the perpendiculars Gm, Hn drawn from the points of intersection G, H to the said side BF may be equal to a given right line PQ.

Answered by Mr. G. Cetii.

Construction. At a distance from BF = PQ and parallel thereto draw DK, and produce BC and FC to meet it in K and I. Make IT = 2FE, and take TD so, that TD × DK = BE × EF; then draw DE, meeting CF produced, in H, and the thing is done.

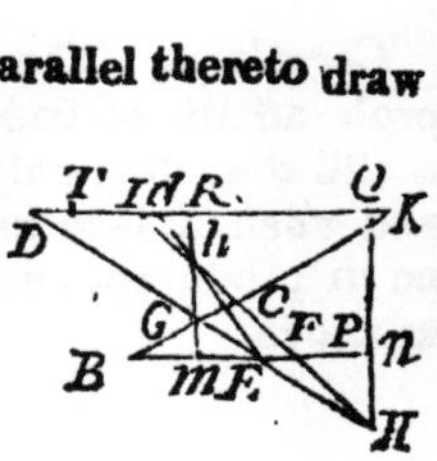

For DT : EF :: BE : DK (by construction), and (Simp. Geom. 20, 4,) Gm : PQ — Gm (GR) :: BE : DK : also PQ + nH : nH :: DT + 2EF

(DI) : EF, or by division, PQ — Hn : Hn :: DT : EF ; therefore, by equality, PQ — Hn : Hn :: Gm : PQ — Gm, or by compos. PQ : Hn :: PQ : PQ — Gm. Now the antecedents being here equal, the consequents will be so also, and therefore PQ = Gm + Hn.

But if FC be produced towards I, take Id so, that Id × Kd = BE × EF, and draw dE for the line required; the demonstration of which being much more simple than that of the above case, needs not pointing out.

XI. QUESTION 637, *by Mr.* J. Chipchase.

A gentleman has a direct south-reclining dial in his garden. Its style is a straight pin erected perpendicular to the plane of the dial, at the distance of 18 inches from its centre, the shadow of which returns at a particular time of the day, for several days in summer. On the 21st of June the arch described by the returning shade contained 4° 24′ 40″, and the height of the sun above the plane of the dial at 3^h 17^m 57^s that afternoon, was = to his altitude above the horizon: Hence is required the latitude of the place and height of the pin that the top of its shade may shew the true hour of the day?

Answered by Mr. W. Wales.

Projection. Let HZPN be the meridian, P the elevated pole, and PO the semi-axis of the sphere. Describe DC, the parallel of the ☉'s declination for the given day, and the great circle QAR to touch it, making the ∠Q, which it forms with the meridian, equal to half the arc the returning shade describes, and Q will be the pole of the dial-plane, OL. Draw the hour-circle PBS, answering to 3^h 17^m 57^s and cutting DC in B, the place of the sun at that time. Round B, as a pole, describe thro' Q, the less circ. TQZ, cutting the merid. again in Z, which will be the zenith of the place required; and therefore ZP is the complement of its latitude. Lastly, make OE = 18 inches, and draw EX ⊥ OL, meeting PO in X, and EX will be the height of the pin, and the △OEX the proper gnomon of the dial.

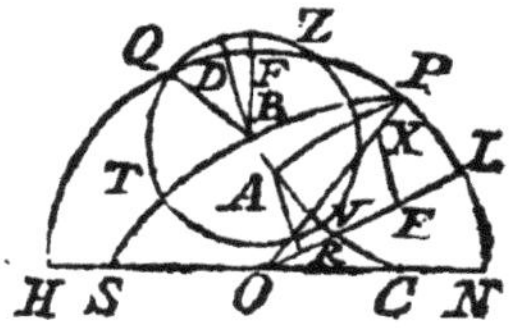

Demonstration. It is plain that the motion of the shade will be direct until the sun comes to A, *i.e.* into the vertical to the dial-plane, QAR, which touches the parallel of declination; after which it can go no farther westward, but will return; and when the sun gets on the meridian at D, will have described an arc = the measure of the ∠AQP, and which is therefore = half the arc described by the returning shade. Again, Q and Z being the poles of the planes OL, HN, and BQ = BZ by construction, it is plain that the altitudes of the point B above those planes are equal. Lastly, OL being the plane of the dial, PO the semi-axis of the earth, and OE = 18 inches, EX must be the height of the pin; for its top must be in the right line drawn through the pole and the centre of the dial.

Answer to the same by Mr. J. Chipchase, *the Proposer.*

It is evident that the plane of this dial is parallel to the horizon of some place within the tropics, and that the time of the shadow's returning will be when the sun's azimuth from the north is the greatest possible at that place. Let LO be the horizon of the place, which is parallel to the plane of the dial (see fig. preceding), Q the zenith thereof, P the north pole, DC the parallel of declination on the 21st of June, A the place of the sun when his azimuth from the north is the greatest possible, and the $\angle$QAP will therefore be a right one. Put the sine of PQ $= x$, sine of AP $= s$, its cosine $= c$, and then the sine of the $\angle$AQP or arch RL, will be $= s \div x$, and the sine of the amplitude OV $= c \div x$, and its cosine, or the sine of the arch VL, $= \sqrt{(1 - (cc \div xx))}$. The arch RV or RL — VL described by the returning shade, being $= 4° 24' 40''$, call its sine b, and then (per Emer. Trig. prop. 6, b. 1.) $\frac{sc}{xx} - \sqrt{\left(\left(1 - \frac{cc}{xx}\right) \times \left(1 - \frac{ss}{xx}\right)\right)} = b$;

which reduced gives $x = \sqrt{\frac{1 - 2bsc}{1 - b}} = \cdot 9743693$ the nat. s. of 77°;

$\therefore$ the height of the pole above the dial-plane is 13°; and radius : 18 inches :: tan. 13° : 4·156 inches, the height of the pin required. Describe a merid. to make an angle of $3^h\ 17^m\ 57^s$ or 49° 29′ 15″ with HQP, cutting the parallel of declin. in B, and describing BF $\perp$ QP, make FZ = FQ, and Z will be the zenith of the place where the dial is fixed: For, by the question at $3^h\ 17^m\ 57^s$ the height of the ☉ above the plane of the dial was equal to his alt. above the horizon of the place where it stands; B is the ☉'s place, and QB, ZB are the complements of his alt. above the plane of the dial and horizon of the place where it is fixed, at that time, which are equal by construction.——Lastly, co-tang. BP : rad. :: cos. $\angle$FPB (49° 29′ 15″) : tang. FP = 56° 13′ 30″; whence ZP = 35° 27′, the comp. of the required latitude.

XII. QUESTION 638, *by the Rev. Mr.* Wildbore.

What right-angled plane triangle is that, the numerical value of whose double area being subtracted from that of each of its sides severally shall leave three square numbers?

Answered by the Rev. Mr. Crakelt.

Let the legs of the required triangle be represented by x and nx; then are $x - nxx$, $nx - nxx$ and $\sqrt{(xx + nnxx)} - nxx$ to be square numbers (per quest.): Suppose $x - nxx = xx$; then will $nx = 1 - x$, by which means two of the conditions of the question will be fulfilled, and therefore what remains is to make $\sqrt{(xx + nnxx)} - nxx = \sqrt{(1 - 2x + 2xx)} + xx - x$ a square number. Now, to effect this, imagine $1 - 2x + 2xx = (mx - 1)^2$, and x will be $= \frac{2m-2}{m^2-2}$, $\sqrt{(1 - 2x + 2xx)} + xx - x = \frac{m^4 - 4m^3 + 6m^2 - 4}{(m^2-2)^2}$, and m greater than 2: but, since $m^4 - 4m^3 + 6m^2 - 4$ is not a square number, feign it $= (r - 1 - m^2 + 2m)^2$, and m will be $=$ $\frac{2r - 2 + \sqrt{(2r^3 + 2r + 4)}}{2r}$, and r greater than 3; but $2r^3 + 2r + 4$ not proving a square number, suppose it $= (2 + 2vr)^2$, and r will be $= vv + \sqrt{(v^4 + 4v - 1)}$, and v greater than 1; let $4v - 1 = z$, and $v^4 + 4v - 1$ will be $= \frac{z^4 + 4z^3 + 6z^2 + 260z + 1}{256}$, which being supposed $= \left(\frac{1 + 130z - zz}{16}\right)^2$ we shall get $z = \frac{4223}{66}$, and thence $\frac{18465217}{20590417}$, $\frac{18325825}{20590417}$, and $\frac{2264592}{20590417}$ for the sides of the triangle required.

Answer to the same, by the Rev. Mr. Wildbore, *the Proposer.*

Let x and $1 - x =$ the legs of the triangle: then will $\sqrt{(1 - 2x + 2xx)} =$ the hypothenuse, and $x - xx =$ the double area, which subtracted from these three values of the sides, leaves xx, $1 - 2x + xx$, and $\sqrt{(1 - 2x + 2xx)} - x + xx$ for the three square numbers required, the two first being necessarily squares; and in order to make the last a square, put $v \cdot (1 - x) + x = \sqrt{(1 - 2x + 2xx)}$, and the said square will become $(1 + 4v^3 - v^4) \times (1 + 2v - vv)^{-2}$; therefore, if $1 + 4v^3 - v^4$ be a square number less than unity the conditions of the question will be answered. Make $v = ba^{-1}$; then must $a^4 + 4ab^3 - b^4 =$ a square, or, putting $a = b + c$, $c^4 + 4bc^3 + 6bbcc + 8b^3c + 4b^4 =$ a square; and in order to take off the two fisrt terms and the last by transposition, make it $=$ a square whose side is $cc + 2bc - 2bb$, and we find by reduction, $3c = -8b$, or in the least integers, $c = 8$ and $b = -3$: but b must not be a negative quantity; therefore let $b = d - 3$ and $a = b + c = d + 5$, and substituting these values for a and b we obtain $4d^4 + 16d^3 + 24d^2 + 1040d + 4 =$ a square suppose its side $= 2dd - 260d - 2$, in order to destroy the first and two last terms by transposition, and we find $d = \frac{4223}{66}$, $b = \frac{4025}{66}$, $a = \frac{4553}{66}$, $v = \frac{4025}{4553}$, and the sides of the triangle required $=$ three fractions whose numerators are 2264592, 18325825, and 18465217, and com. denom. 20590417; which is the answer (though somewhat differently deduced) given by Ozanam at pa. 604 of his Nouveaux Elemens d'Algebre (which has been handed to me

the representation of a great circle which touches the lesser one, and has a given arc intercepted between the point of contact and its intersection with some other given great circle.

Answered by the Rev. Mr. Wildbore.

Projection. Describe ACRP the primitive, ATB the given lesser circ. POC the given great one, and the right circ. BDR touching ATB, suppose in B: From B set of BD = the given intercepted arc, and through D describe OD ‖ to ATB, cutting the given circ. POC in O; then, through O describe the great circ. OT, touching ATB in T, and it will be that required;—Because the two arches OT, BD, touching the lesser circ. ATB, and being bounded by the same parallel OD, are equal.

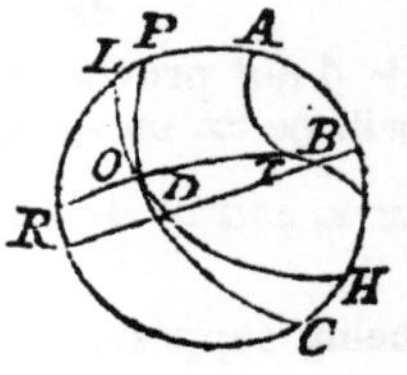

XV. QUESTION 641, *by Mr.* Tho. Allen.

Suppose ABF to be a certain spiral line described by a body in motion, beginning at A, and having the following property, viz. $x = \frac{N^{2z} + 1}{2N^{z}}$, where AC = CD = 1, CB, any distance from the centre C, $= x$, circ. arc AD $= z$, and N = the number whose hyp. log. = 1: Required the length of the said spiral, or the space described by the revolving body, when $x = 1000$?

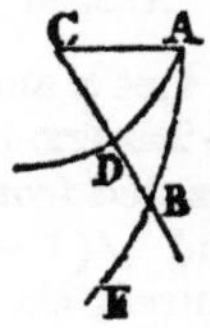

Answered by the Rev. Mr. Crakelt.

From the given equation ($x = \frac{N^{2z} + 1}{2N^{z}}$) by multiplying, compleating the square, and taking the hyp. log. is had $z =$ h. l. of $(x + \sqrt{(xx - 1)})$. But $1 =$ CD : x (CB) :: $\dot{x} \div \sqrt{(xx - 1)}$ (the angular velocity of a body at D) : $x\dot{x} \div \sqrt{(xx - 1)} =$ the angular velocity of the same body at B; whence the fluxion of the curve's length will be $= \dot{x} \times \sqrt{(2xx - 1)} \div \sqrt{(xx - 1)}$, and the length itself = the arch of that equilat. hyperbola, which hath two for each of its diameters and $x - 1$ for its abscissa = (when $x = 1000$) to 1414·2132088, &c.

Answer to the same by the Rev. Mr. Wildbore.

Description. With the centre C and vertex A describe the equilat. hyperbola AGK; draw the asymp. CL and ord. EG. Take BH = EG, AR = the hyp. log. of CH, and the circ. arc AD = AR; then, drawing a right line from C through D, take upon it CB = CE, and the point B will be in the spiral: and thus may points be determined therein *ad libitum*.

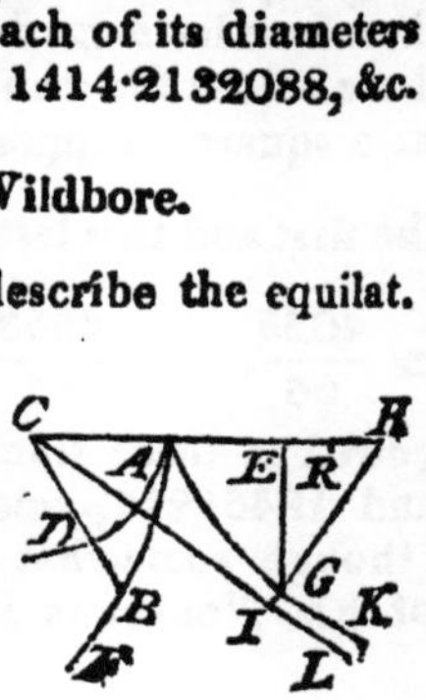

Demonstration. By conics, CE = (CH² + CA²

(1)) $\div 2CH$, $CH = CE + \sqrt{(CE^2 - 1)}$, and by construc. $AR = AD =$ the hyp. log. of CH; ∴ CH = N to the AD power, and consequently $CE = CB = \frac{N^{2z} + 1}{2N^z}$, as per question. Moreover, the flux. of the spiral AB is well known to be $= \sqrt{(\dot{C}E^2 + CE^2 \cdot \dot{A}D^2)}$, and $\dot{A}G = \sqrt{(\dot{C}E^2 + \dot{E}G^2)}$: Likewise, $EG^2 = CE^2 - 1$, $EG \cdot \dot{E}G = CE \cdot \dot{C}E$, $(CE \cdot \dot{C}E) \div EG\ (\dot{E}G) + \dot{C}E = \dot{C}H$ (because EG = EH), $EG \cdot \dot{C}H = CH \cdot \dot{C}E$, $\dot{C}H \div CH = \dot{A}R = \dot{A}D = \dot{C}E \div EG = \dot{E}G \div CE$; ∴ $CE^2 \cdot \dot{A}D^2 = \dot{E}G^2$, and $\dot{A}B = \dot{A}G$, and because they both begin together, AB = AG.

Computation. Because when CE = 1000, both it and EG are very great in comparison of CA, the common series and approximations for AG are useless; but here the series given at p. 511 of Simp. Flux. may be applied to very great advantage: For, the diff. betwixt the part of the asymp. beyond I and that of the curve beyond G being exceedingly small, or quite inconsiderable in comparison of the diff. betwixt CI and AG, CI — AG = the sum of that series very nearly = ·32111; ∴ this being subtracted from $CI = CH \times \frac{1}{\sqrt{2}}$ (HGI being ⊥ CL) gives AG = 1113·8921 = AB, the length of the curve required.

Corollary. The hyp. log. of CH (1999·9995) being = 7·6009024 = AD, this divided by the whole circumference of the circle, shews that the body will have made 1·209725 revolutions round the centre when its distance therefrom is 1000.

THE PRIZE QUESTION, *by the Rev. Mr.* Wildbore.

In the latitude of London, what is the declination of that star, whose azimuth increaseth the least possible in two hours after rising; and on what day of the year does the sun's azimuth increase the most possible in two hours after rising?

Answered by Mr. Wm. Wales.

Let HZPN be the meridian, HQN the horizon, ÆQ the equator, Z the zenith, and P the pole. Make QF = the measure of the given interval, and, having described the circle of perpetual apparition, N*ob*, describe, through F, the great circ. EFD to touch it. Then by prop. 13, b. 2, of Theodosius's Spherics, the arcs of all circles ∥ N*ob*, or ÆQ, intercepted between the circles HQN, EFD are similar and described in the same time with QF.

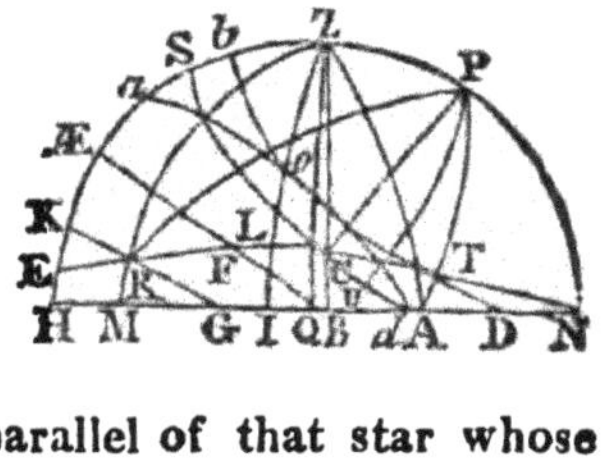

Suppose, now, the circle KRG to be the parallel of that star whose change in azim. GM, is a maximum, and SCA the parallel of that star whose change in azim. AB is a minimum in the given interval of time, ZRM and ZCB being verticals passing through R and C, the places of the two stars when situated in the circle EFD, and ZLI the vertical whose pole is D. By the above prop. of Theodosius, QG = FR ; ∴ MG, the change of azim. in the given time, is the diff. between QM and FR, and so QM—FR is to be a max. ; but when QM — FR is a max. IM —LR is a max. because QI and FL are constant quantities. Now IM is the measure of the ∠RZL ; therefore when the change in azim. is a max. the diff. between the ∠RZL and its opposite side, RL, is a max. and consequently its sine is so too : But the ∠L is right, and the side ZL given ; also, by spherics, rad. : sin. ZL :: tang. Z : tang. LR, and by composition, division, and equality of ratios, rad. + sin. ZL : rad. — sin. ZL :: tang. Z + tang. LR : tang. Z — tang. LR :: sin. (Z + LR) : sin. (Z — LR) (by prop. 4, p. 58, Simpson's Trig.). Now, the two first terms of this analogy being constant, the last will manifestly be a maximum, when the third, or sin. (Z+LR), is so, *i e.* when Z + LR = 90° ; therefore the sine of their difference (Z — LR) when a maximum = (rad. — sin. ZL) ÷ (rad.+sin. ZL) ; and hence, having the sum and difference of Z and LR, those arcs themselves are known, and of course every thing else in the △RZL ; and, as the ∠LZP is known by the construction, the ∠RZP will be known, and we shall have in the triangle RZP, the sides RZ and ZP with the included angle, to find RP, the parallel's distance from the elevated pole, = (in the case given) 190° 27¼′, or its declin. 19° 27¼′ south ; and this parallel the sun occupies Nov. 18th and Jan. 22d.

Again, because FC = QA, the change in azim. BA = FC — QB, which is therefore to be a minimum ; or LC — IB is to be a minimum ; or, which amounts to the same thing, IB — LC must be a maximum, and hence the point C will be determined exactly in the same manner that the point R was, and we shall have in the △CZP, the sides CZ and ZP with the included angle to find CP, = 58° 55⅙′, and so the declin. required is 31° 4⅜′, north.

Scholium 1. The declin. of the parallel SCA will, it is plain, be of the same name with the latitude of the place until the circle EFD comes into the position *aod*, that is, to touch the circle of perpetual apparition in the prime vertical, ZQ : For then, the intersecants N*d*, *do* being equal, *od* + *d*Q = 90°, which has been proved to be the case when the change in azim. is a maximum or min. and consequently the circle N*ob* is the parallel required, and the star rises due north or south ; after which the declination of the parallel SCA becomes of a different name from the latitude of the place till the time given is 18^h when it becomes again of the same name therewith.

2. In like manner the declin. of the parallel KRG will be of a different name until the given time be 6^h, after which it becomes of the same name with the lat. until the circle EFG touches the circle of perpetual apparition in the western prime vertical, and then it will be again of a different name from the latitude. And all these determ. hold equally true of a star moving from the horizon at setting, only the declin. here said to be of the same name with the latitude will

there be of a contrary one, and *vice versa*; for it is only considering them as stars rising to the Antipodes.

3. In all these conclusions I have supposed the given latitude was not less than 45°; for if it be, the given time may be such as to cause the circle EFD to cut the meridian between the zenith and elevated pole; in which case the question may admit of more answers than one; but they cannot always be exhibited by this, or, I believe, any other general method. It may, however, be easily done by others, adapted to the particular cases where this fails; but your limits are much too narrow for so copious a subject.——I must, before I take my leave, return thanks to the ingenious proposer for bringing this question again under consideration, as I had passed it over in too hasty a manner before; the reasons for which you are well acquainted with.

The same answered by the Rev. Mr. Wildbore.

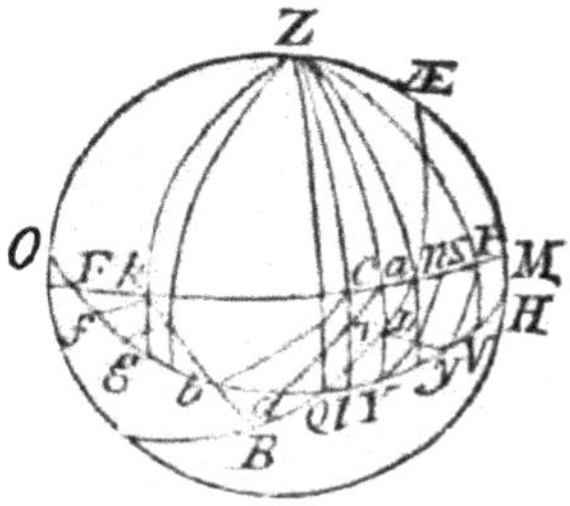

The annexed scheme is an oblique representation of the concavity of the eastern hemisphere, bounded by the meridian, OZÆH of the place, HO the horizon, ÆQ the equinoctial, and the given hour arch is set off from Q to n; at n the $\angle an$T is made = TQY, and thus is the great circle FnM described. Then by Gregory's Astron. book 2, prop. 40, the arches of the equinoctial and its parallels intercepted between this circle and the horizon are similar, and aT = TY, $an \doteq$ QY, QT = Tn, $an = d$Q, $ac = db$, $ak = df$, &c. Now, suppose the circle FanM to be generated by the motion of a point both ways from a; then, when the sun or star rises at d, its alteration of azim. in the time Qn will be dY, and when at b it will be bl, and because the $\angle$s at Y, l, and a are right, Yl will be ⊏ ac; but ac is = bd: Therefore bl is ⊐ dY, or alteration of azim. when the phœnom. rises at b, will be ⊐ when it rises at d, the difference being lY — ca: But in the right-angled isosceles spheric △ FYa, lY is the decrem. of FY, and ca of Fa; and if k be taken for any other position of the moving point, if in the right-angled spherical △ Fgk, the decrem. of Fk is ⊐ that of Fg, the said alteration of azim. may still decrease by taking the point k nearer F, for the same reason as at a: But if the decrem. of Fk be equal to that of Fg, it can decrease no further; for at the next point, the decrem. of Fk being greatest, the said quantity will be upon the increase; therefore, when the required increase of azim. is a min. the decrem. of Fk is = that of Fg.——And in the very same manner, it appears that when the point moves the other way from a (M being placed at the other intersection of the circle FaM with the horizon) the alteration of azim. as the declin. varies, will increase till the decrements of PM and VM are equal, and then it will be a maximum.——And hence, per spherics and pa. 280 of Simpson's Flux. is easily found $\angle k\mathrm{v}g = 18^\circ\ 32'$, $kg = 13^\circ\ 10'$, $\angle \mathrm{v}kg = 76^\circ\ 50'$, $\mathrm{v}g = ka = 44^\circ\ 14'$, $\mathrm{v}k = g\mathrm{v} = 45^\circ 46'$, $an = 11^\circ\ 51'$, $kn = 56^\circ\ 5'$, $k\mathrm{B} = 31^\circ\ 5'$ the star's declin. N. required; $gd = 22^\circ\ 4'$, and thence $fg = 22^\circ\ 10'$, the min. azim. in 2 hours from rising corresponding, $\mathrm{SP} = 20^\circ\ 32'$, $y\mathrm{v} = 25^\circ\ 14'$ the max. increase (or alteration) in 2 hours, and $\mathrm{D}y = 19^\circ\ 27\frac{2}{3}'$, S. the corresponding declination answering to Nov. 18, or Jan. 22.

Answer to the same by Mr. Cha. Hutton.

Let P be the north pole (see fig. 1st. preceding), Z the zenith, HN the horizon, AC a great circle passing through the object at its rising and 2 hours after; describe the perpendicular circle PU, and the circles PA, PC, ZA, and ZCB. Put a = sine of 15° or $\angle$APU, b = its tang. c = sine of lat. or cos. of ZP, d = its cos. or sine of ZP, x = sine declin. or cos. of AP or PC, and z = cos. of declin. or sine of AP. Then, in the right-angled spherical $\triangle$PUA, $1 : a :: z : az$ = s. AU or $\frac{1}{2}$AC, and hence $2az\sqrt{(1 - aazz)} \div (1 - 2aazz)$ = tang. AC; also $1 : b :: x : bx$ = cotang. PAU; and hence $(1 + bbxx)^{-\frac{1}{2}}$ = sin. and $bx \times (1 + bbxx)^{-\frac{1}{2}}$ = cos. PAU. In the quadrantal $\triangle$AZP, $z : 1 :: c : c \div z$ = cosine and $(1 \div z)\sqrt{(zz - cc)}$ = sine PAZ: Whence $(c - bx\sqrt{(zz - cc)}) \div z\sqrt{(1 + bbxx)}$ = s. ZAC or cos. CAB. And in the right-angled spherical $\triangle$ABC, 1 : tang. AC :: cos. CAB : tang. AB $= 2a\sqrt{(1 - aazz)} \times (c - bx\sqrt{(zz - cc)}) \div (1 - 2aazz) \times \sqrt{(1 + bbxx)}$ =, (since $a \div b = \sqrt{(1 - aa)}$, and putting $e = (1 - 2aa) \div 2aa$), $(1 \div b) \times (c - bx\sqrt{(dd - xx)}) \div (e + xx)$ = a maximum or minimum —— This in flux. &c. gives $2cx\sqrt{(dd - xx)} = b \times ((dd + 2e) \cdot xx - dde)$, or $x^4 - \cdot 3780667xx = \cdot 0296788$; and hence $x = \sqrt{(\cdot 2668459 \text{ or } \cdot 1112207)} = \cdot 5165713$ or $\cdot 3334977$, answering to 31° 6′ 15″ and 19° 28′ 52″, the two declinations required.

Additional Solutions.

1st. *By* Astronomicus, *taken from the Gentleman's Diary, for* 1804, *where the question was proposed rather more generally as follows:*

To find the declination of that star, whose change in azimuth is a *maximum* or *minimum* in a given time, reckoned from the time that it transits a given almucantar in a given latitude. Ex. gr. Suppose the latitude of London, the time one hour, and the almucantar 15° above the horizon.

Let Z be the zenith of London, P the north pole, $mz\mathrm{P}n$ the meridian and mn the given almucantar; let S be a star passing the given almucantar, and let σ mark the position of the same star an hour afterward: draw PQ an arch of a great circle = PZ, and making with the meridian an angle ZPQ equal to the horary angle corresponding to the given interval of time; and describe the great circles PS,

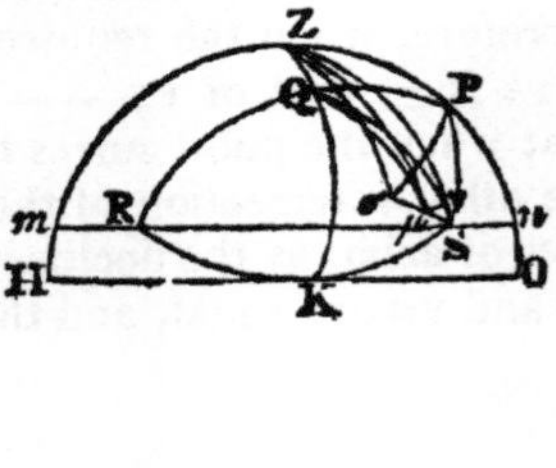

Pσ, zs, zσ. Then because ∠QPZ = SPσ, therefore ∠ZPσ = ∠QPS ; but PQ = PZ, and PS = Pσ; consequently the two spherical triangles ZPσ and QPS are equal in every respect; and ∠PQS = ∠PZσ; In order that szσ the change in azimuth be a maximum or minimum, it is necessary that the fluxion of the angle PZS, corresponding to an indefinitely small variation in the position of the star s, be equal to the fluxion of the angle PZσ, or to the fluxion of the angle PQS ; or which amounts to the same (having drawn a great circle through z and Q) it is requisite that the fluxion of the angle szκ be equal to the fluxion of the angle SQK. Let μ be a star passing the almucantar indefinitely near s, and draw the great circles zμ, Qμ, and let μν be perpendicular to QS ; then according to what has been shewn szμ = SQμ. Put ż = szμ, then μs = ż × sin. zs, and μν = ż × sin. QS ; therefore sin. zs : sin. QS :: μs : μν :: Rad. : (sin. QSμ =) cos. ZSQ = sin. QS ÷ sin. zs. Let arc zs = a, PZ = PQ = λ, horary angle SPσ = QPZ = h, zQ=b, and QS = x; then cos. ZSQ = sin. x ÷ sin. a; but from the triangle zsQ we get cos. ZSQ = (cos. b — cos. a × cos. x) ÷ (sin. a × sin. x); whence sin. x ÷ sin. a = (cos. b — cos. a × cos. x) ÷ (sin. a × sin. x) and 1 — cos. b = cos.² x — cos. a × cos. x. But 1 — cos. b = 2 sin.² ½b = 2 sin.² λ × sin.² ½b ; therefore the arch x will be determined from the quadratic 2 sin.² λ × sin.² ½b = cos.² x — cos. a × cos. x. The arch QS being determined, let a small circle be described round Q as a pole at the given distance QS, and it will cut the given almucantar in two points s and R (on opposite sides of the great circle zQ, and at equal distances from it) which will be the positions of the required stars; the star at s being that with the maximum, and the star at R, that with the minimum change in azimuth. The arches PS and PR being computed the declinations of the stars will be known.

Calcula. To find zQ=h, we have sin. ½b=sin. λ + sin. ½h, whence zQ = 9° 19'. Take tang. φ = (sin. λ × sin. ½h ÷ cos. a) × √8, then cos. x = (cos. a ÷ cos. φ) × cos.² ½φ ; hence x = 72° 24'. The other root of the quadratic is of no use in the present enquiry. Again cos. zsQ = sin. x ÷ sin. a ∴ zsQ = 9° 19' : and sin. szQ = (sin. zsQ ÷ sin. QZ) × sin. QS ; therefore szQ = 72° 24' ; likewise cot. PZQ = cos. λ × tan. ½h, and PZQ = 84° 7'. In the triangle PZS there are now known the two sides PZ, zs, and included angle SZP, = PZQ — szQ = 11° 43', whence PS = 37° 42', and the star's declination 52° 18' north. And in the triangle PZR there are known PZ and zR, and ∠ PZR = PZQ+szQ = 156° 31', whence RP = 110° 25' and the star's declination 20° 25' south.

2nd. By Mr. Skene, *taken from Davis's Mathematical Companion for* 1805.

Let z be the zenith of the place, P the pole of the equator, s' the situation of the star when it transits the given almucantar, and s its situation after the given time. Describe the great circles zP, zs, zs', PS, PS'; let the arch Pq be described equal to PZ, and making with it the spherical angle qPz equal to SPS' ; draw also the arches zq, qs ; and we shall have $qs = ZS'$, $PQS = PZS'$, and $PZS - PZS' = PZS - Pqs$ a maximum or minimum. Then as $PZ = Pq =$ the co-latitude of the place, and $ZPq = SPS'$ are given, Zq and the equal angles PZq, PqZ are given. Also, because $PZS - Pqs$ is $= 360^\circ - PZq - PqZ - SZq - SqZ$ is a maximum or a minimum, and $360^\circ - PZq - PqZ = 360^\circ - 2PZq$ is given, the sum of the angles SZq, SqZ, is a minimum or maximum, and the zenith distance $ZS' = qs$ is given by the conditions of the question. We have, therefore, in the spherical triangle ZSq, the base Zq and one of the sides qs given, to construct it when the sum of the angles at the base is a minimum or a maximum, which is easily done by means of one of Mr. Cotes's theorems, or otherwise by a stereographic projection of the triangle. By either method we obtain this equation: $\sin. qs \times \cos. ZSq = \sin ZS$; but, by spherics, $\sin. ZS \times \sin. qs \times \cos. ZSq = \cos. Zq - \cos. ZS \times \cos. qs$, and $\cos. Zq = 1 - \sin.^2 ZP \times \text{v. sin. } ZPq$. Wherefore, $\sin. ZS \times \sin. qs \times \cos. ZSq = \sin.^2 ZS = 1 - \sin.^2 ZP \times \text{v. sin. } ZPq - \cos. qs \times \cos. ZS$, or $\cos. ZS (\cos. ZS - \cos. qs) = \sin.^2 ZP \times \text{v. sin. } ZPq$. But the $\text{v. sin. } ZPq = 2 \sin.^2 \frac{1}{2} ZPq$; consequently $\cos. ZS (\cos. ZS - \cos. qs) = 2 \sin^2. ZP \times \sin.^2 \frac{1}{2} ZPq$; from which we derive the following method of calculation. Let $B^2 = 2 \sin.^2 ZP \times \sin.^2 \frac{1}{2} ZPq$, $\text{tang. } \phi = 2B \div \cos. qs$, and we shall have $\cos. ZS = B \text{ co-tang. } \frac{1}{2}\phi$, or $= - B \text{ tang. } \frac{1}{2}\phi$, the zenith distance of S having two values, one less and the other greater than 90°. Now the $\cos. ZSq = \sin. ZS \times \sin. \frac{1}{2} Zq \div \sin. qs = \sin. ZP \sin. \frac{1}{2} SPS'$, $\text{co-tang. } PqZ = \text{tang. } \frac{1}{2} SPS' \times \cos. ZP$, and $\sin. SqZ = \sin. ZS \times \sin. ZSq \div \sin. Zq$; wherefore ZSq, Zq, PqZ, and SqZ, are given, and hence $Pqs = PqZ \pm SqZ$. But ZP, qs, are given by the question, and therefore PS, the star's polar distance, is given.

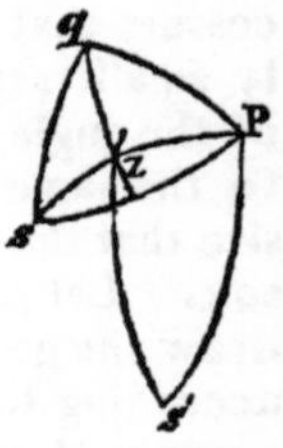

For example; suppose the latitude 51° 31′ N., the time one hour, and the almucantar 15° above the horizon, or $ZP = 38^\circ\ 29'$, $SPS' = ZPq = 15^\circ$, and $ZS' = qs = 75^\circ$.

We find the log. $B = 9{\cdot}060204$, $\phi = 41^\circ\ 35'\ 36''$, $ZS = 72^\circ\ 23'\ 43''$ only, the other value leading to an absurdity; $ZSq = 9^\circ\ 19'\ 23''$, $Zq = 9^\circ\ 18'\ 57''$, $PqZ = 84^\circ\ 6'\ 58''$, $Zqs = 72^\circ\ 32'\ 6''$, $Pqs = 156^\circ\ 39'\ 4''$, or $11^\circ\ 34'\ 52''$, and the star's declination equal to 20° 26′ 30″ S., or 52° 19′ 30″ N., nearly.

Had we solved this question algebraically, and taken PS for the unknown quantity, we should have obtained an equation of the fourth order, as is manifest from our analysis, for ZS has two values as well as Pqs, and therefore PS has four. In this manner the only solution I have met with is conducted, producing a very complex biquadrate equation having all the terms.*

In the particular case of this question, when the given almucantar coincides with the horizon, as originally proposed in the Diary; by putting $qs = 90^\circ$ in our solution, we get $\cos. ZS = \sin. ZSq = \pm \sin. ZP$

---

* The author, we presume, alludes to a solution given to this question in the Gentleman's Diary for 1782.

sin. $\frac{1}{2}$ZPQ $\sqrt{2}$, sin. $\frac{1}{2}zq$ = sin. ZP sin. $\frac{1}{2}$ZPQ, co-tang. PQZ = cos. ZP tang. $\frac{1}{2}$ZPQ, sin. SQZ = sin. ZS × cos. ZS ÷ sin. zq = sin. 2ZS ÷ 2 sin. zq, PQS = PQZ ± SQZ, and sin. declin. = cos. PQS × sin. ZP. Ex. gr.: let ZP = 30° 29′, and SPS′ = ZPQ = 30°. Then ZS = 76° 50′, or 103° 10′, ZSQ = 13° 10′, zq = 18° 32′ 12″, ZQP = 78° 9′ 15″, ZQS = 44° 11′ 30″, PQS = 122° 20′ 45″, or 33° 57′ 45″, and the star's declination equal to 19° 26′ 45″ S. when the change in azimuth is a maximum, and 31° 4′ 30″ N. when a minimum.

Questions proposed in 1772, and answered in 1773.

I. QUESTION 642, *by Mr.* John Shadgett.

In friendship two sisters together reside,
With virtue replete; each a stranger to pride:
Maria for beauty with Venus may vie,
And Cloe for wisdom Minerva defy:
Maria is prudent in ev'ry degree,
Whilst Cloe is court'ous, good natur'd, and free.
From what's under-written,* their ages I ask:
Resolve it, dear ladies, nor think't a hard task.

* Given $\begin{cases} x^2 + xy + y^2 = 1087, \\ x^4 + x^2y^2 + y^4 = 45777295; \end{cases}$ to find the value of x the age of Maria, and that of y the age of Cloe.

Answered by Mr. Jos. Cowley, *and several others.*

Put $xx + yy = v$, $xy = w$, $1087 = a$, and $45777295 = b$, and then the given equations become $v + w = a$, and $vv - 2w^2 + w^2 = b$, from the former of which $v = a - w$, and that value substituted in the latter of them, &c. gives $w^3 - w^2 - 2aw = b - aa$; whence $w = 357$, and consequently from the first given equation, $xx + 2xy + yy = a + 357$ and $xx - 2xy + yy = a - 3 \times 357$; and thence x and $y = \frac{1}{2}\sqrt{(a + 357)} \pm \frac{1}{2}\sqrt{(a - 3 \times 357)} = 21$ and 17, or 17 and 21, the two ages required. Mr. *Joseph Cowley*, and Messrs. *John Aspland* and *William Smart*, put $s = \frac{1}{2}$ the sum, and $d = \frac{1}{2}$ the difference of them, and then find by an easy process, $64d^6 - 52128d^4 + 4074476d^2 - 55467952 = 0$; whence $d = 2$, and the ages required come out the same as above.

Mr. *George Coughron*, by transposing xy in the 1st equation, and squaring and taking it from the 2d, deduces the cubic equation $(xy)^3 - (xy)^2 - 2axy = b - aa$; from which $xy = 357$, and $x + y$ $(= \sqrt{(xx + 2xy + yy)}) = \sqrt{(a + xy)} = 38$; and thence x and y are found the same as before.

of the segments of the base, and the difference of the angles at the base being given to construct the triangle?

Answered by Mr. J. Chipchase.

Construction. On AE the given diff. of the segments of the base, describe the segment of a circle capable of containing an angle = the given diff. of the angles at the base, and take AG = the given diff. of the base and perpendicular, and bisecting GE in F, erect the perpendicular EI = EG, and draw the right line FIB cutting the circumference in B; from whence upon AE produced, let fall the perpendic. BD, and having taken DC = DE, let the points A, B and B, C be joined, and ABC will be the triangle required.

Demonstration. Join the points E, B; then the angles BEC, BCE are equal, and the ∠ABE is the diff. between the angles at the base BAC, BCA (BEC) by 32, Euc. 1, which is of the given quantity by construction. Also, because DC = DE by construction AE is the diff. of the segments of the base, which by hypothesis is of the given magnitude likewise. Lastly, EI being = EG = 2EF (by construc.) DB will be = 2DF=2DE + 2EF = CE + EG = CG; whence the diff. of the base AC and perpendicular DB is = AG, the given difference by construc.——In this solution the base is supposed to be greater than the perpendicular; but if the perpendicular be greatest, the construction will be the same, only AG must be set off the contrary way on the base produced.

The Rev. Mr. Crakelt's *construction is as follows, viz.*

Upon DB, equal to the given diff. of the segments of the base, describe the segment of a circle capable of containing the given diff. of the angles at the base; and having taken thereon and on an indefinite perpendicular bisecting it, EN and EF respectively equal to the half and the whole of the diff. betwixt the perpendicular and the base, through N and F draw a right line to meet the circular arc in A: then draw AGC ∥ to BD, and join the points A, B; C, B, and ABC will be the triangle required.

Demonstration. On AC let fall the perpendiculars H, BI, and draw DA: then since, by 29, Euc. 1, and 29 of 3d, AG=GC, and HG (by 34, Euc. 1.) = DE = EB (by construction) = GI, therefore will AH = IC, and AI — IC = HI = DB. Moreover, by sim. triangles, FE : EN :: FG : GA; but by construction FE is = twice EN; wherefore FG will be = twice AG = AC; and consequently BI — AC = EG — AC = EF; and the diff. of the angles BCA, BAC is manifestly = the ∠BAD.

VI. QUESTION 647, *by Mr.* T. Moss.

From the given point E in the side DR produced, of the given rectangle BDRF, to draw a right line EA cutting the sides RF, DB, thereof in a and C, and the side FB produced in A so, that the trapezium CDRa may be to the $\triangle$ ABC so formed, in the given ratio of m to n.

Answered by Mr. T. Moss, *the Proposer.*

Construction. Bisect DR in r, and take rG $=$ Er; also, in GE produced, take EN a fourth proportional to m, n and DR, and upon GN describe a semicircle, and erect the perpendicular EM meeting its circumference in M: then in FB produced, take BA $=$ EM; join E, A and the thing is done.

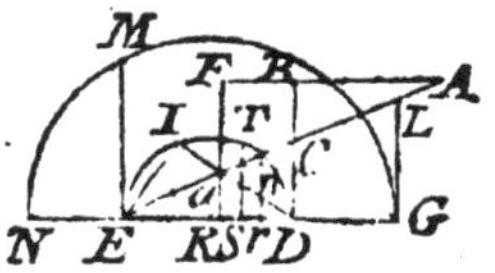

Demonstration. Draw GL $\perp$ GE meeting AE in L: then, by the prop. of the circle and construc. the rectangle NEG $=$ EM^2 $=$ AB^2, and therefore EG : AB :: AB : EN; but by sim. $\triangle$s, EG : AB :: GL : BC; whence, by equality, AB : GL :: EN : BC; but by construction, Er $=$ rG, Rr $=$ rD, and GL $\parallel$ rn; wherefore GL $=$ 2rn, and consequently the rectangles contained by AB, BC, and EN, 2rn, and also their halves, are equal; whence m being to n (:: DR : EN by construc.) :: DR $\times$ rn ($=$ trapez. CDRa) : EN $\times$ rn ($=$ $\frac{1}{2}$AB $\times$ BC $=$ $\triangle$ ABC).

Mr. *G. Coughron* constructs it by taking BA so, that m may be to n :: ED^2 — ER^2 : BA^2; then drawing EA, the thing required will be done. For, by similarity of $\triangle$s, the $\triangle$ EDC : $\triangle$ ERa :: ED^2 : ER^2, and dividedly, the trapez. CDRa : ED^2 — ER^2 :: $\triangle$ EDC : ED^2 :: $\triangle$ ABC : AB^2, and alternately, the trapez. CDRa : $\triangle$ ABC (:: ED^2 — ER^2 : AB^2) :: m : n, per construction.

Answer to the same by the Rev. Mr. Crakelt.

Construction. Upon ED describe a semicircle, and inscribe therein EI $=$ ER, and divide ED in S in the given ratio of m to n; then take BA (in FB produced) $=$ a fourth proportional to ES, ST ($\perp$ to ED and meeting the circumference in T) and DI, and draw ACaE for the line required.

Demonstration. By Euc. 6, 19, the $\triangle$ EDC : $\triangle$ ERa :: ED^2 : ER^2; $\therefore$ by division, permutation, &c. the trapezium CDRa : ED^2 — ER^2 :: $\triangle$ EDC :: ED^2 : $\triangle$ ABC : BA^2; whence by permutation, &c. the trapezium CDRa : $\triangle$ ABC :: ED^2 — ER^2 or ED^2 — EI^2 $=$ (Euc. 1, 47,) DI^2 : BA^2 :: (by construc.) ES^2 : ST^2 $=$ (by prop. of the circle) ES $\times$ SD : : (Euc. 6, 1,) ES : SD :: m : n, by construction.

VII. QUESTION 648, *by Mr.* Isaac Dalby.

In the right-angled plane triangle, there is given one of the legs,

also a line drawn parallel thereto and terminated by the other leg and hypothenuse; to determine geometrically, the triangle so, that the rectangle under the hypothenuse and a line drawn from the acute angle next the given leg, to the point where the said parallel line meets that other leg, may be of a given magnitude.

Answered by Mr. Isaac Dalby.

Construction. At right angles to the given base *cv* take CB so, that their rectangle may be = the given magnitude. Upon BC let a semicircle be described, and produce BC so, that BD may be a fourth proportional to the given parallel line, the base *cv* and BC; also, produce *cv* to make *v*W a fourth proportional to BD, CD, and *cv*: Bisect CW in K, and draw the line DK; with which as rad. and cent. K, describe another semicircle meeting CW produced both ways in G and *g*; from D to the former semicirc. on BC, apply DP = CG, producing it to meet the circumference again in R, and on DR produced let fall the perpendic. BQ, cutting the said semicircle in *m*; join B, R, and BQR will be the triangle required.

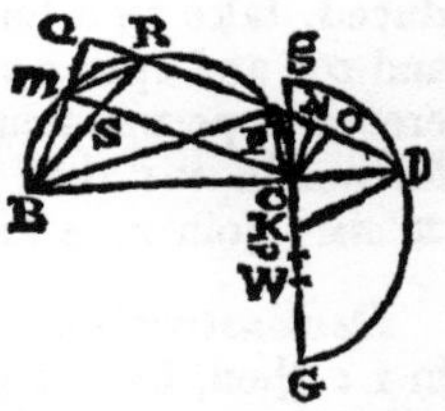

Demonstration. Join R*m*, *m*C, CP, BP and draw CN, CO ∥ BQ, BR respectively. By Euc. 3, 35, *g*C : CD :: CD : CG (=DP, by construc.) and by sim. △s, DO : DC :: DC : DP; ∴ (5, 9,) DO = *g*C; but, because K*g*, KG are equal, and CW is bisected in K (by construc.), WG is = C*g*; ∴ OP = CW. Moreover, the ∠s B*m*C, BQP being right ones, C*m* is ∥ PQ; consequently the arcs *m*R, CP, and their chords *m*R, CP are equal to each other respectively, and thence QR = PN: but, by sim. triangles, BD : CD :: BQ : CN :: QR (PN) : NO, and by construc. BD : CD :: *cv* : *v*W; whence by equality, PN : NO :: *cv* : *v*W, and, compounding, OP : CW :: NO : *v*W; but we have already proved that OP = CW; ∴ (5, 9,) NO = *v*W, and consequently PN (=QR) = *cv*, the base. Again, by similar triangles, B*m* : BQ (:: *m*S : QR) :: BC : BD; ∴ *m*S is = the given parallel line, by construction. Lastly, by sim. triangles, BC : CP (*m*R) :: BR : QR; whence *m*R × BR = BC × QR (*cv*), the given magnitude, by construction.

This construction, though less concise than some others, may perhaps be found acceptable to some of our readers, as it is derived from first principles only, without either transforming it to another prob. or calling in the assistance of any other prob. already known in order to its solution, &c. whereby it appears to be an original, independent problem.

VIII. QUESTION 649, *by Mr.* S. Ogle.

Two lines AB, AC being given in position, and a point P in one of them (AB); it is required to draw from thence a right line PH meeting the other given line AC in H so, that if another right line HI be drawn to meet AB and make a given angle therewith, the perimeter of the triangle PHI so formed may be the least possible.

Answered by Mr. Burrow.

Construction. Draw PK making with the given line AB, the given angle APK, and describing on AK a semicirc. take AW = the diff. between AP and PK, and apply it therein from A, and join K, W: then draw PS ⊥ to KW, cutting AK in H, the point required; for drawing HI ∥ to PK, PHI will be the triangle whose perimeter is a minimum.

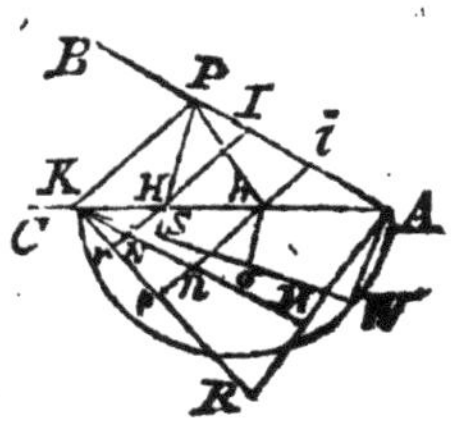

Demonstration. Draw AR ∥ to PK and = AP, and join K, R, and draw KM ∥ AP, meeting AR in M and continue IH to meet KM in N and KR in *r*: then I say the perimeter of the △PHI is always = PH + HS + IN; for, KH : KA :: PI : PA :: H*r* : AR; but PA = AR (by construc.); ∴ PI = H*r*, and HS is always = N*r*: for since HS and AW are each ⊥ to KW, AW : HS :: AK : HK :: AP : IP :: MK : NK :: MR : N*r*; but AW = MR, each being by construc. = AP — PK: whence HS = N*r*; therefore HS + HN = PI, and PH + HS + IN = the perimeter of the △PHI, as was affirmed. Now, the same holds true wherever the point H is taken betwixt A and K; but of the three lines which constitute the perim. IN is always the same, and therefore the perim. will be a min. when the sum of the two variable lines PH and HS is so, and that will evidently be when they make one straight line PHS ⊥ to KW.

IX. QUESTION 650, *by Mr.* J. Chipchase.

On a certain day last summer, at a quarter past midnight, when the sun was just rising, a person set forward on a journey along a parallel of latitude, which he ended at sun-set, having by his watch performed it in 24 hours; at noon the greatest shadow of his walking-stick (which is $3\frac{1}{2}$ feet long) exceeded its shadow when placed perpendicular to the horizon by 1·522267 feet; from whence is required the latitude he was in, the day of performing the journey, and distance travelled?

Answered by the Rev. Mr. Crakelt.

It will readily appear that the staff will cast the greatest shad. at the given time, when the sun's rays fall upon it in a perpendicular direction; therefore, if TSB and SAZ be supposed two right-angled triangles, having ST (the given staff = $3\frac{1}{2}$ feet) perpendicular to the horiz. SO, and = SA so inclined to it as to make its shad. SZ a max. as likewise TB ∥ to AZ, they will be sim. and equal, and of course SZ — SB = BZ = 1·522267 feet = TB — SB, and thence the ∠TBS or AZS = 47° 0′ 42″ = the sun's merid. altitude. Now, with ZA rad. describe the circumference of a circle intersecting SZ in H, and when produced in O, and having let fall AI ⊥ on HZ, drawn ZZ ∥ thereto, and made OF = the versed sine of 3° 45′ the given time of the ⊙'s rising after midnight, perpendicularly to HO apply to the circumf. of the said circ. LD a fourth proportional to HF, FO, and AI; then draw AD, cutting HO in G, and perpendic. thereto ZP, and PO will be = the latitude required.

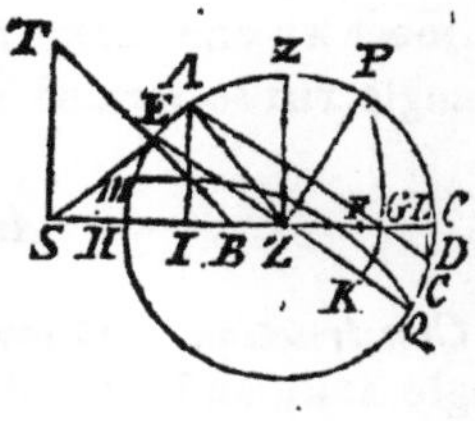

For, if PG be the representation of a merid. arc passing through G and meeting EQ drawn through the centre Z ∥ to AD, in the point K, by the nature of the orthograph. project. sim. triangles, and construction, it will be EK : KQ :: AG : GD :: AI : LD :: HF : FO, or, by composition, EQ : KQ :: HO : FO, and consequently, KQ (corresponding to the semi-nocturnal arc GD) = FO, answering to 3° 45′ as it ought.

The calculation will from hence be very easy; for, since ZK and KQ are respectively = the cosine and versed sine of 3° 45′, and AI = sine of 47° 0′ 42″ the ⊙'s merid. alt. by supposing ZA or ZE = 1, will be had DL (= KQ × AI ÷ EK) = ·0007839, the nat. sine of 0° 2′ 42″ the ⊙'s depress. at midnight; therefore by the nat. of the sphere, the comp. of the required lat. HE or ZP (= $\frac{1}{2}$ × (HA + OD)) = 23° 31′ 42″, and the ⊙'s declin. = EA (= $\frac{1}{2}$ × (HA — OD)) = 23° 29′, giving June 21st for the day on which the traveller performed his journey.—— Lastly, if a degree on the equator be supposed to measure $69\frac{1}{2}$ miles, we shall, by sim. sectors, have 1 : sine of 23° 31′ 42″ :: $69\frac{1}{2} \times 7\frac{1}{2}$: 208·0843, &c. miles, the length of the required journey.

X. QUESTION 651, *by Mr.* W. Chartreux.

The latitude of the place, and the position of two hour-circles, with respect to the meridian, being given; it is required to determine what the declination of a star must be, so that in passing over the interval contained between those hour circles, the change in altitude may be the greatest possible.——This question has been proposed before, but never answered.

Answered by the Rev. Mr. Crakelt.

When the change in alt. or ZS — Z⊙, in the given time is a max. the ∠S Z⊙ P, ZSP will, by art. 293 of my translation of Mauduit's Trigonometry, be equal; therefore, if h and h' be put for the hour angles ZP⊙ and ZPS, L for the lat. or comp. of ZP, and D for the required declin. or comp of P⊙ = PS, we shall have, by art. 214, sin. h ÷ (cos. D × tang. L — sin. D × cos. h) = tang. Z⊙P = tang. ZSP = sin. h' ÷ (cos. D × tang. L ∓ sin. D × cos. h'), according as the ∠ZPS is acute or obtuse; or, by multiplying, dividing by cos. D, substituting tang. for sin. ÷ cos. and transposition, sin. h' × cos. h × tang. D ∓ cos. h' × sin. h × tang. D = (by art. 85,) sin. (h' ∓ h) × tang. D = (sin. h' — sin. h) × tang. L; and therefore, by art. 87 and 92, cos. $\frac{1}{2}$ × (h' — h) : cos. $\frac{1}{2}$ × (h' + h) :: tang. L : tang. D, and sin. $\frac{1}{2}$ × (h' + h) : sin. $\frac{1}{2}$ × (h' — h) :: tang. L : tang. D; which are the two theorems given on p. 189 of the above-quoted book.

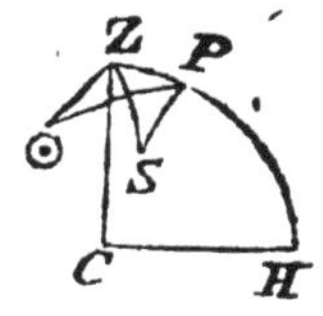

XI. QUESTION 652, *by Mr.* Cha. Hutton.

If water runs through a pipe $1\frac{1}{4}$ inch diameter with a constant velocity of 6 feet per second into an empty conical vessel, having a hole of 1 inch diameter in its bottom: Required the time when the surface of the water in the vessel will be just 1 foot above the bottom of it; also, what will be the greatest height to which the water will rise, and the time in which it will rise to the said greatest height, the diameter of the bottom of the vessel being 3, the diameter of the top 5, and the altitude 6 feet?

Answered by Mr. Cha. Hutton, *the Proposer.*

Let x denote any variable height of the fluid in the vessel, and z the corresponding time, $m = 32\frac{1}{6}$, p = the area of the section of the pipe, h = that of the hole in the vessel, and a = the velocity of the water in the pipe. Then $h\sqrt{mx}$ = the quantity running out of the vessel per second (see p. 4 of my Mathem. Miscel.), and ap = the quantity running in per second: Wherefore $ap - h\sqrt{mx}$ = the rate of the vessel's filling per second, and consequently $ap\dot{z} - h\dot{z}\sqrt{mx}$ will be the increase in $\dot{z}$ time, and is therefore = the flux. of the quantity in the vessel. Now, the vessel being given, by the rules of mensuration, the content of the part whose height is x, will be found $= \frac{1}{27} \times (xx + 27x + 243) \times \cdot 7854x$, whose fluxion, or $\frac{1}{9} \times (xx + 18x + 81) \times \cdot 7854\dot{x}$, must therefore be $= ap\dot{z} - h\dot{z}\sqrt{mx}$: Whence $\dot{z} = \frac{xx + 18x + 81}{2s - 2r\sqrt{x}} \times \dot{x}$, by putting the given numbers instead of h, m, a and p, and making $r = \frac{\sqrt{32\frac{1}{6}}}{32}$, and $s = \frac{75}{256}$; or $\dot{z} = \frac{v^5 + 18v^3 + 81v}{s - rv} \times \dot{v}$, by putting vv for x; the fluent of which is $z = -\frac{v^5}{5r} - \frac{sv^4}{4rr} - \frac{ss + 18rr}{3r^3} \cdot v^3 - \frac{ss + 18rr}{2r^4} \cdot svv - \frac{(ss + 9rr)^2}{r^5} \cdot v +$

$\frac{(ss+9rr)^2}{r^6} \cdot s \times$ hyp. log. of $\frac{s}{s-rv}$; which is a general expression of the time for any alt. vv (x).——Now, when $x = 1$ foot, v is $= 1$ also, and then the above expression brings out $4' \; 47 \cdot 555''$ for the time when the water will be one foot high in the vessel.

Also, since the greatest height is evidently when the water runs as fast out of the vessel as it runs in by the pipe, $s - rv$ must, in that case, be $= 0$; therefore $v = s \div r = 1 \cdot 652983$, and $x \; (= vv) = 2 \cdot 732352$ feet, the greatest height of the fluid. But when $s - rv$ vanishes, the expression $s \div (s - rv)$ becomes infinite, and the value of x infinite also of consequence; so that the water will be an infinite time in rising to the said greatest height $2 \cdot 732352$, that height being the limit to which it continually approaches from nothing, but to which it can never attain in any given finite time.

XII. QUESTION 653, *by* Astronomicus.

To investigate or demonstrate the nature of the curve which a fixed star by means of the aberration would appear to describe, if the earth, instead of revolving in an ellipsis, was to move in a parabolic or an hyperbolic orbit.

***Note,* This curious question has been proposed before, but as it was thought to have been neither investigated nor fully demonstrated, we have been requested to re-propose it.**

Answered by Mr. W. Wales; at whose Request it was re-proposed.

Let VER represent the conic section in which the earth is supposed to revolve, F the focus, V the place of the perihelion, E the place of the earth at any assigned time, and S that of a star: Draw the tangent ET, which will give the direction of the earth's motion at E, take Eα to ES, as the velocity of the earth when at E, is to the velocity of light, and V$\beta \perp$VF, as the velocity when at V. From S, on a plane passing thro' the star parallel to that of the earth's orbit, draw SB and SA, $=$ and $\parallel$ to Vβ and Eα, and A and B will be the apparent places of the star in that plane, when the earth is at E and V, on account of aberration. Let fall on ET the perpendicular FP, which, it is well known, will be reciprocally as Eα, or its equal SA; moreover, FV will be reciprocally as SB. Now, if the trajectory VER be a parabola, the tang. which is drawn parallel to the axe will be at an infinite distance therefrom, and the perpendicular thereon from F will be so likewise; its reciprocal therefore is $= 0$, and so the path in question passes through the points S, A, and B. Draw AB. By a prop. of the parab. FV : FP :: FP : FT, and because FV $\times$ SB $=$ FP $\times$ SA, FV : FP :: SA : SB, and by equality, FP : FT :: SA : SB; whence, the $\angle$s F and S being equal by construction, the $\triangle$s FPT and SAB are similar; and

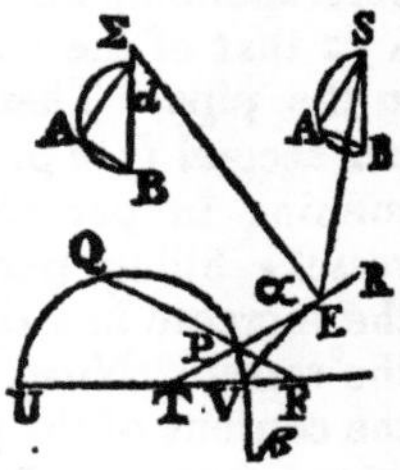

consequently, F being a right angle, A will be a right angle also, and that wherever the point A may fall; the curve SAB (or path in this case, required) is therefore a circle.

But if the earth's orbit VER be an hyperbola, and Σ the true place of the star, describe on the transverse axe UV the circle UQPV, &c. then it is well known that all perpendiculars from the focus F will meet the tangents to this orb on which they are demitted, in the circumference of this circle, and if FP be produced to meet it again in Q, by the prop. of the circle FP and FV will be reciprocally as FQ and FU; but they are also reciprocally as AΣ and BΣ; therefore FU : FQ :: BΣ : AΣ; and as the ∠s QFU and AΣB are equal, and that wheresoever A and Q fall, it is evident that the path of the point A will be similar to that of the point Q; but that of the point Q is known to be a circle: consequently that of the point A (or the curve apparently described by the star on account of the aberration, in this case) is also a circle.

Scholium 1. Because the apparent path of the star will not be on the plane above specified, unless its true place be the pole of the ecliptic, but on one which is perpendicular to the line joining the eye and star, it follows from the nat. of the orthographic projection, that the apparent path will be in every other case an ellipsis, excepting that when the star is in the ecliptic, and then it will be a right line.

2. Because the asymptote becomes a tangent to the hyperbola, when both are produced out to an infinite distance, it follows that the perpendicular from the focus on the asymp. will be greatest; therefore its reciprocal will be least, and the star can apparently approach no nearer to its true place than by that dist. Moreover, because the asymptote passes through the centre of the circle UQPV, the said perpendicular will be a tang. to it; therefore if from Σ, ∥ to the asymp. a line be drawn = to the said reciprocal, it will be a tang. to the path at its nearest approach to the star's true place; and therefore if the circle was compleated, the true place of the star would be without it.

Answer to the same by the Rev. Mr. Wildbore.

Let AMD represent the earth's orbit, AB its transverse axis when an hyperbola, F, *f* its foci, &c. MT a tang. to it at M, FO a perpendic. let fall from F upon MT, *a*F another upon F*f*, and AGBH the circumf. of a circ. described about the diam. AB; and suppose when the earth is at A, that F and *a* are the real and apparent places of the star reduced to the plane of the ecliptic. From F ∥ to the tang. MT, take F*d* : F*a* :: FA : FO, and from *a* upon F*a* take *ca* = *dc*, with which as rad. describe the semicircle *adb*, and the arc *ad* will be the apparent locus of the star reduced to the ecliptic, whilst the earth moves from A to M in an hyperbola, but when the orbit is a parabola, *ca* the rad. of the circle must be taken = F*c* = ½F*a*; which circle will likewise be the

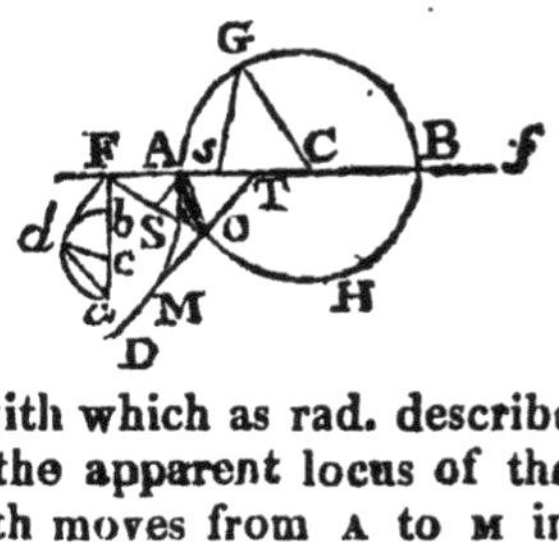

locus in this case.——For, by the nature of aberration and central motion (vide prop. 2, p. 3, of Simp. Essays), *d* is the apparent place of the star, reduced to the ecliptic, when the earth is at M (as by construction), because the velocity of the earth at M is to the velocity when at A as FA, is to FO; per cor. 2d to prop. 1st on p. 24 of Simp. Essays aforesaid, and (drawing AS to meet FO so that AS and SO may be equal, and joining the points A, O and *a*, *d*) the ∠s *dFa*, AFO being equal, and their including sides proportional (by construction) the △s AFO, *dFa*, and *dca*, ASO will be similar, and consequently F*a* : FC :: FO : FS; but the ratio of FO to FS is always the same (being in the constant ratio of FB to FC, as is easily demonstrated, the circumference AGBH being the locus of the point O, per prop. 21, part 3, of Steel's Conic Sections), and therefore the ratio of F*a* and FC will be always the same likewise, and F*a* being constant, FC and *ca* will be always the same, and the point *d* be always found in the circumf. of the same circle *adb*, as per construction.——But in the case of a parabolic orbit, AC being infinite, the locus of the point O becomes a right line ⊥ to FA, and as the triangle FAO may be then circumscribed by a semi-circle, AS = SO will be = FS, and consequently FC = *ca*.

Corollary. The proportion FB : FC :: FO : FS :: F*a* : FC also holds when the orbit is an ellipsis; only AC being then greater than FC, SO must be greater than FS, and *ca* than FC; wherefore the circ. arc *ad* will in this case fall partly beyond the axis, and be compleated in one revolution of the earth. But in the case of the hyperbola above, it is evident that the point *d*, in moving along the arc *ad*, can never arrive at the position wherein F*d* would be ∥ to the asymptote, nor at the point F when the orbit is a parabola.

Another Solution by Mr. J. Landen, taken from the Appendix to Dr. Hutton's *Edition of the Diaries.*

Lemma. 1. If $\frac{2d}{1 \backsim ee}$ and $\frac{2d}{\sqrt{(1 \backsim ee)}}$ be the axes of a conic section, the curve will be a circle if e be $= 0$; and according as e is less, equal to, or greater than 1, the curve will be an ellipsis, a parabola, or a hyperbola.

The distance of the focus f from the vertex a, of any conic section, will be $= \frac{d}{1+e}$: and, y being put to denote the ray fp from the said focus to any point p of the curve, the perpendicular fq, from the same focus upon the tangent pq, will be $= d\sqrt{\frac{y}{2d+(ee-1)y}}$.

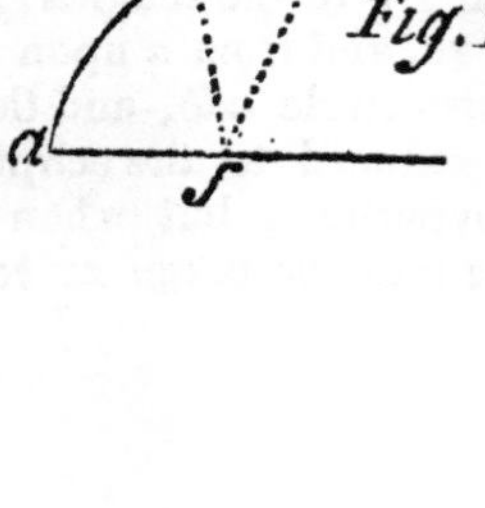

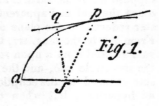

Fig. 1.

Moreover $\sqrt{\frac{2dy+(ee-1)yy-dd}{2dy+(ee-1)yy}}$

will be the sine of the angle pfq, radius being 1; and, that sine being always to the sine of the angle afq in the constant ratio of e to 1, the sine of the last-mentioned angle will be $= \frac{1}{e}\sqrt{\frac{2dy + (ee-1)yy - dd}{2dy + (ee-1)yy}}$.

Lemma 2. The earth being supposed to describe a conic section whose axes are as expressed in the preceding lemma, let its velocity at the vertex a be denoted by $b \times (1 + e)$: then will its velocity at any other point p of its trajectory be $= b\sqrt{\frac{2d + (ee-1)y}{y}}$, being reciprocally as the perpendicular from the focus upon the tangent.

Proposition.

The velocity of the earth in its trajectory being to the velocity of light as $b\sqrt{\frac{2d + (ee-1)y}{y}}$ to c; it is proposed to determine the linear aberration of a star from its true place, measuring such aberration on a plane passing through the star, parallel to the plane of the said trajectory?

Let s be the star, and p the earth in its trajectory moving from the vertex a; and, qkp being a tangent to the trajectory, let pk be to ph as the velocity of the earth at p to the velocity of light, *i. e.* as $b\sqrt{\frac{2d + (ee-1)y}{y}}$ to c; then phs being the line joining the earth and the star, if the angle spt be equal to the angle phk, st, parallel to pk, will be the linear aberration sought, at the time the earth is at p, the star appearing to be at t instead of its true place s.

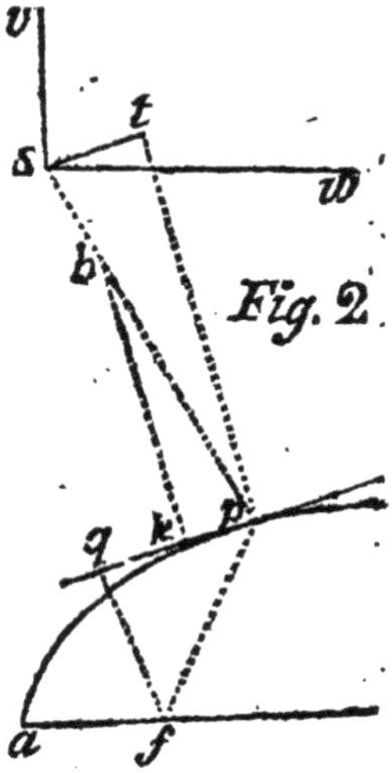

Fig. 2.

Hence it follows that ps (the distance of the star from the earth) being denoted by D, and st by z, z will be $= \frac{b\text{D}}{c}\sqrt{\frac{2d + (ee-1)y}{y}} = r\sqrt{\frac{2d + (ee-1)y}{y}}$, r being put for $\frac{b\text{D}}{c}$.

Moreover, if sw be parallel to the axis af, and sv perpendicular to sw; the angle tsv will, it is obvious, be equal to the angle afq made by af and the perpendicular fq on the tangent pq.——Let the sine of the angle tsv (or afq) be denoted by v: Then, by lemma 1, v will be

pear to an observer on the earth; which, from what is done above, may be readily computed, when the situation of the earth with respect to the star is known.

It is observable, that, in a parabola, as the earth recedes from the vertex, the angle of aberration will after some time (if not immediately) continually decrease, and at length may become less than any assignable angle: and it will decrease in like manner when the trajectory is a hyperbola, if c be greater than $b\sqrt{(ee - 1)}$. But if c be $= b\sqrt{(ee - 1)}$, the angle of aberration may, as the earth so recedes, increase nearly to 90°; and if c be less than $b\sqrt{(ee - 1)}$, that angle may increase still more.

XIII. QUESTION 654, *by the Rev. Mr.* Wildbore.

Abstracting from refraction, at what time on the 21st of June, at Salton, in Norway, lat. 67° N. will the veloc. of the shad. of the summit of an erect object be the greatest possible on an horizontal plane?

Answered by the Rev. Mr. Wildbore, *the Proposer.*

Suppose PO to represent the object, BE the path of the shadow of P from midnight to the time required: draw PS ∥ to the earth's axis, till it cuts the transverse diameter BD. With PO (1) rad. describe the arc OS cutting the axis in S, and having joined PD, PB, PE, and OE, let fall oa ⊥ to PE, ob to PS, and sr to PE. Then 'tis well known that rs is the sine of the co-declination BPS, ob that of the co-lat. OPS, ao the sine, EO the tang. PE the secant of the co-alt. EPO, and BOE the azim. from the north at the time required. Moreover, the distance described at E, during the infinitely small instant of time that the velocity can be considered as uniform, is, it is well known, $= \sqrt{(\dot{EO}^2 + EO^2 \times \angle \dot{BOE}^2)}$. But, by Simp. Flux. p. 282, as $\dot{BOE} : \dot{EPO} = \dot{EO} \div EP^2$ :: (in the oblique spherical triangle terminated by P the pole, Z the zenith, and ⊙ the sun, *see fig.* quest. 651) the co-tang. $\angle ⊙ : s.\, Z⊙$ (the co-alt. = EPO) $= ao = EO \div EP$; consequently $\dot{BOE} = \dot{EO} .\, \text{co-t.}\, ⊙ \div EO . PE$: Moreover, as $s.\, P⊙ \times \cos.\, ⊙ : s.\, Z⊙ :: \angle \dot{Z} : \dot{P}$ the fluxion of the time $=$

$$\frac{\dot{Z} \times s.\, Z⊙}{s.\, P⊙ \times \cos.\, ⊙} = \frac{\dot{BOE} . ao}{rs . \cos.\, ⊙} = \frac{\dot{EO} .\, \text{co-t.}\, ⊙}{rs . PE^2 . \cos.\, ⊙}.$$

But, by the laws of motion, the distance $\sqrt{(\dot{EO}^2 + EO^2 \times \dot{BOE}^2)} = \dot{EO} \times \sqrt{\left(1 + \frac{\text{co-t.}\, ⊙^2}{PE^2}\right)}$ ÷ by the time $\frac{\dot{EO}}{rs . PE^2 . s.\, ⊙}$ is = to the velocity of the uniform motion during that time $= rs . PE \sqrt{(PE^2 . s.\, ⊙^2 + \cos.\, ⊙^2)}$, which by the quest. is to be a maximum, or, because $PE^2 = 1 + EO^2$, $s.\, ⊙^2 + \cos.\, ⊙^2 = 1$, $PE\sqrt{(1 + EO^2 . s.\, ⊙^2)} = PE\sqrt{(1 + EO^2 -}$

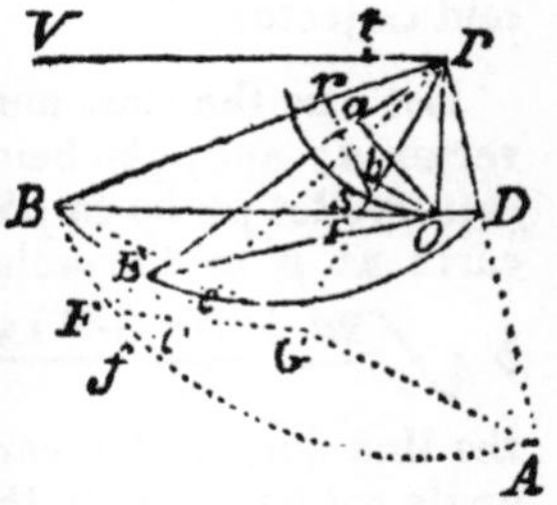

$EO^2 . \cos. ☉^2$), $PE^2 \times (PE^2 - EO^2 . \cos. ☉^2)$ is a maximum. But it is proved by writers on spherics, that the cos. $\angle ☉ \times$ s. $Z☉ \times$ s. $P☉ =$ cos. $ZP -$ cos. $Z☉ \times$ cos. $P☉ =$ cos. $☉ \times ao \times rs = Pb - rP \times aP$; but $ao = EO \div EP$, and $aP = 1 \div EP$ (or $PO^2 \div EP$), therefore, cos. $☉ \times EO = (Pb . EP - rP) \times rs^{-1}$: Consequently $PE^2 \times (PE^2 - (Pb . PE - rP)^2 . rs^{-2}) = PE \times (PE + (PE . Pb - rP) . rs^{-1}) \times PE \times (PE - (PE . Pb - rP) . rs^{-1})$ is a maximum, or $PE \times (PE . (rs + Pb) - rp) \times PE \times (PE . (rs - Pb) + rP)$ a maximum, take any where, $Pt = rP \div (Pb + rs)$ and $PV = rP \div (Pb - rs)$, and $PE \times (PE - Pt) \times PE \times (PV - PE)$ is a maximum, wherefore, per Euc. II. 5, and v. 4, $(2PE - Pt) \times PE \times (PV - PE) = (2PE - PV) \times PE \times (PE - Pt)$; whence PE may be readily found by a geometrical construction, or otherwise, $= 89{\cdot}79712 =$ the co-secant of 38′ 43″, the ☉'s altitude at the time required.

Additional Solution.

As several of our correspondents have complained of the obscurity of the solution above given to this question, we shall here, at their request, subjoin another solution a little different from it, which may, perhaps, be better understood by some readers than the above original one.

In order to this, let the following additions, in dotted lines, be made to the figure above, viz. complete the cone PBA, and produce its axe PS to cut BD in T and the diameter AB in the centre G of its base; also produce PE to the circumference at F, and draw another line Pef indefinitely near it, on which demit the ⊥ Ec; and, lastly, join G, F.

Then, conceiving the right-angled △ PGB to revolve with an uniform motion about the axe PG till it comes to the position PGF, it is evident that BGF is the angle of time corresponding to the ∠ BOE of azimuth and PEO of altitude; also Ff is as the constant fluxion of the time or arc BF, Ee the fluxion of the arc BE described by the shadow of P on the horizontal plane BED, ec the fluxion of PE, and Ec the fluxion of an arc whose rad. is PE. And when the shadow of P moves quickest, then Ee will be greatest in respect of Ff, and therefore we are to find $Ee \div Ff$ a maximum.

Now put a and b for the sine and cosine of the lat. PTO or BTG, c and d for the sine and cosine of the declin. PBG, and z the sine of the alt. PEO to the rad. 1. Then, in the spherical △ $PZ☉$ or PZS in the fig. to the solution of question 651, all the sides are expressed, to find the ∠ P of the time, whose cosine therefore, or the cosine of BGF in our figure, will, by spherics, be found $= (ac - z) \div bd$, and therefore the flux. of BGF, or of the arc whose cosine is this quantity and radius 1, is $\dot{z} \div \sqrt{(bbdd - (ac - z)^2)}$, which drawn into the rad. BG, or m, gives $Ff = m\dot{z} \div \sqrt{(bbdd - (ac - z)^2}$. Again, $d : 1$ (rad.) $: : m$ (BG) $: m \div d =$ PB $=$ PF, and $z : 1 : : n$ (PO) $: n \div z =$ PE; hence (PF : PE : : Ff :) $Ec = nd\dot{z} \div z \sqrt{(bbdd - (ac - z)^2)}$; also $ce = P'E =$

$-n\dot{z}z^{-2}$; $\therefore$ $Ee\ (=\sqrt{(Ec^2+ce^2)})=\frac{n\dot{z}}{zz}\sqrt{\frac{bbdd-aacc+2acz-cczz}{bbdd-(ac-z)}}$; consequently $Ee \div Ff = (n \div mzz)\sqrt{(bbdd - aacc + 2acz - cczz)}$ a minimum; whose flux. being equated to 0, we obtain $z = (3a - \sqrt{(8dd + aa)}) \div 2c =$ the sine of 38′ 17″, supposing the declination to be 23° 29′. But if the true present declination 23° 28′ be used, z will come out $=$ the sine of 36′ 58″ for the altitude when the shadow of the summit moves fastest.—And hence the time from midnight is easily found, its cosine being $= (ac - z) \div bd$. H.

THE PRIZE QUESTION, *by* Peter Puzzlem.

Suppose a body to be fastened to one end of a string, the other end whereof is fastened to a fixed point c, and suppose it to be impelled from a certain given point b with a given velocity in a direction at right-angles to bc: To find the law and direction of the force which must continually act on such body so, that its velocity shall vary according to any proposed law, and itself be always found in a circle (whose centre is c) revolving with any proposed velocity about the given fixed diameter bcd.

Answered by Peter Puzzlem, *the Proposer.*

Let bp, bq be circular arcs described from the cent. c in different planes cbp, cbq; and, the $\angle pbq$ being supposed indefinitely small, let pq be an indefinitely small particle of the curve described by the body. Put r, y, and s to denote the sines of the arcs bp, bq, and the angle bpq respectively; also, put u for the angular velocity of the circ. in which the body is always to be found, about bcd, measured at a distance from $c =$ to the rad. cb, and v for the velocity of the body along the said circle. Call the said rad. a, and considering the required force as compounded of two forces, one acting at right angles to the plane of the circ. in which the body is always found, and the other in the direct. of the tang. to the said circ. at the point where the body at any time is; let f and g denote those two forces respectively.

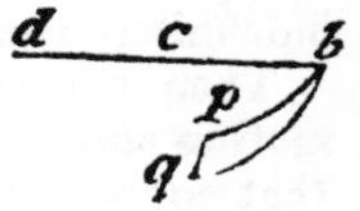

Then $a : u :: y : w\ (= uy \div a)$, the velocity at q at right angles to bq being denoted by w. Also, $y : s :: r : rs \div y =$ sine of bqp, and $rs \div y : w :: a : awy \div rs$ the absolute velocity of the body at q along the curve pq; which would be invariable if no force were to act on the body but the tension of the string. Therefore in that case, a, r, and s being considered as invariable whilst w and y vary, $\dot{w}y + w\dot{y}$ would be $= 0$, and $\dot{w} = -w\dot{y} \div y = -u\dot{y} \div a$. But the force f continuing to act on the body, w will be $= \dot{u}y \div a + u\dot{y} \div a$, consequently $\dot{u}y \div a + 2u\dot{y} \div a$, the excess of $\dot{u}y \div a + u\dot{y} \div a$ above $-u\dot{y} \div a$, will be the flux. of the velo. of the body at right-angles to bq, occasioned by the action of the force f; and therefore, the fluxion of the time being manifestly $= \frac{a\dot{y}}{v\sqrt{(aa-yy)}}$, $f \times \frac{a\dot{y}}{v\sqrt{(aa-yy)}}$ will be $= \frac{\dot{u}y}{a} + \frac{2u\dot{y}}{a}$ and $f = \frac{v\sqrt{(aa-yy)}}{aa} \times \frac{\dot{u}y + 2u\dot{y}}{\dot{y}}$. Moreover $\sqrt{(vv + ww)}$, the velocity in the curve pq, would be invariable, if the forces f and g were to cease acting; therefore $v\dot{v} + w\dot{w}$ would in such case be $= 0$, and $\dot{v} = -\frac{w\dot{w}}{v} = \frac{uuy\dot{y}}{aav}$: consequently we have $g \times \frac{a\dot{y}}{v\sqrt{(aa-yy)}} = \dot{v} - \frac{uuy\dot{y}}{aav} = \frac{aav\dot{v} - uuy\dot{y}}{aav}$, and $g = \frac{\sqrt{(aa-yy)}}{a^3} \times \frac{aav\dot{v} - uuy\dot{y}}{\dot{y}}$.

Thus is the question solved generally, the required force being compounded of the forces f and g found above; from whence its quantity and direction may be known, and the solution may be easily adapted to particular cases. I shall here only take notice of one particular case, and that is when u is invariable, which is the same as lem. 2, p. 3, of the late Mr. Simpson's Miscell. Tracts. In which case $\dot{u}$ being $= 0$, f appears to be $= 2vu\sqrt{(aa-yy)} \div aa$, and its greatest value (when $y = 0$) $= 2vu \div a$. The gent. I have just now mentioned has considered v as invariable, without taking any notice that it will not be so, unless the body be acted on by a force $g = \mp uuy\sqrt{(aa-yy)} \div a^3$. Indeed he has considered the velocity u as very small, and then g will be very small, but not absolutely $= 0$, nor yet indefinitely small, for u (though small) being finite, g will be also finite. The solution here given is true whether the velocity u be small or great. If, u being invariable, g be $= 0$, $aav\dot{v}$ will be $= uuy\dot{y}$, and $v = \sqrt{(aaee + uuyy)} \div a$, e being put for the velocity of the body at the commencement of the motion. Therefore it appears that f, will in that case be $= 2u\sqrt{(aa-yy)} \times \sqrt{(aaee + uuyy)} \div a^3$; which, when u is very small, is nearly $= 2eu\sqrt{(aa-yy)} \div a^2$, the value of f as computed by Mr. *Simpson*.

If u be invariable, and only one force act on the body, that force must be $= \sqrt{(4aavv + uuyy)} \times \sqrt{(aauu - uuyy)} \div a^3$; and the direction in which it must act, inclined to the plane of the circle in which the body is, in an angle whose sine is to rad. as $2v$ to $\sqrt{(4aavv + uuyy)} \div a$. Mr. Simpson seems not to have been aware that the force g must necessarily act to keep the velocity v from varying!

Mr. *De la Lande*, in his Astronomy (art. 3457) propos ng to explain Mr. Simpson's solution, has observed, that, only the force f acting, v will not be invariable when u is so. But, without computing the necessary force g, or the exact value of the force f when g is $= 0$, he neglects a part of the force f, and intirely neglects the force g, as being what he calls *infiniment petits du troisème ordre*; whereas they are not generally *infiniments petits* of any order whatever, being assign-

$= \frac{1}{e}\sqrt{\frac{2dy + (ee - 1)\,yy - dd}{2dy + (ee - 1)\,yy}}$; whence $y = \frac{d}{1 - ee} \mp \frac{de\sqrt{(1 - vv)}}{(1 - ee)\sqrt{(1 - eevv)}}$.

But from the equation $z = r\sqrt{\frac{2d + (ee - 1)\,y}{y}}$ (found above) we get $y = \frac{2drr}{zz + (1 - ee)\,rr}$. From which two values of y it appears that z will be $= r\sqrt{(1 - eevv)} \pm er\sqrt{(1 - vv)}$.

Let AEB be a semicircle whose centre is C; and in the diameter ACB (or the continuation thereof) take any point D; moreover let CE, DE be drawn and denoted by r and z respectively: Then, the sine of the angle CDE being denoted by v, and the distance CD by m, $z\sqrt{(1-vv)} \mp \sqrt{(rr - vvzz)}$ will be $= m$. Hence $z = r\sqrt{\left(1 - \frac{mmvv}{rr}\right)} \pm m\sqrt{(1 - vv)}$. Comparing this equation with that which we just now found above, we may observe that they will be exactly alike if m be equal to er.——It follows therefore, that, s being the true place of the star, sw parallel to the axis (af) of the earth's trajectory, sv perpendicular to sw, and st so drawn that the sine of the $\angle tsv$ be equal to v: if sc be taken (upon sv) equal to er $\left(= \frac{bev}{c}\right)$; and from the centre c, with a radius equal to r $\left(= \frac{bv}{c}\right)$, an arc be described intersecting st in t; the point t so determined will be the apparent place of the star in the plane passing through the star parallel to the earth's trajectory, when the earth is at such a point p thereof that the sine of the angle afq (fig. 2,) be $= v$.

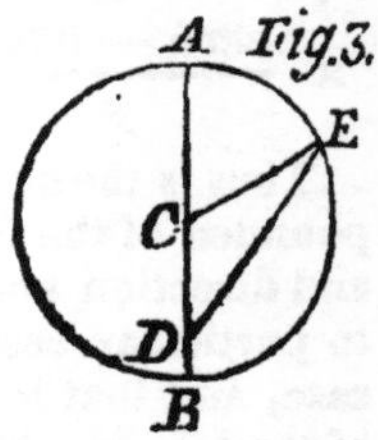

Fig. 3.

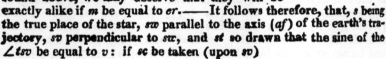

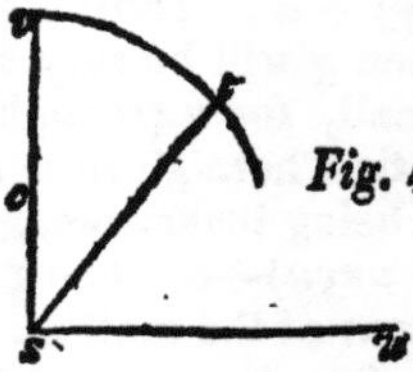

Fig. 4.

If D be invariable (which can only be when the earth's trajectory is a circle, and the star is situated in a line erected perpendicular to the plane of the orbit from the centre thereof) the curve of aberration, on our said plane parallel to the trajectory, will be a circle whose centre is c, e being then $= 0$.

If the trajectory be an ellipsis not very eccentric, and the distance of the star from the earth be at any time *very great* in comparison with the distance of the earth from the focus; D may be considered as continuing of *nearly* the same value whilst the earth describes its whole orbit: our curve of aberration will therefore, in such case, be *nearly* (but not exactly) a circle,

Whether D be considered as invariable or variable, z (or the star's apparent place) will be truly determined by our construction; and by means thereof the star's true place may be readily found from its observed apparent place.

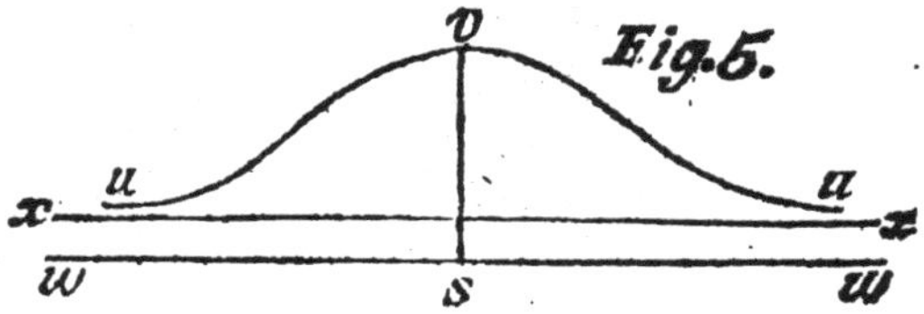

Fig. 5.

Supposing the trajectory to be a parabola, and g the distance of the star from the vertex thereof; let s be the star's true place, and $sv = \frac{2bg}{c}$: then will sw, at right angles to sv, be parallel to an asymp. (xx) to the curve of aberration uvu, which will always have two branches extending ad infinitum from the point v.——The distance of the asymptote from s depends on the situation of the star with respect to the earth's trajectory.

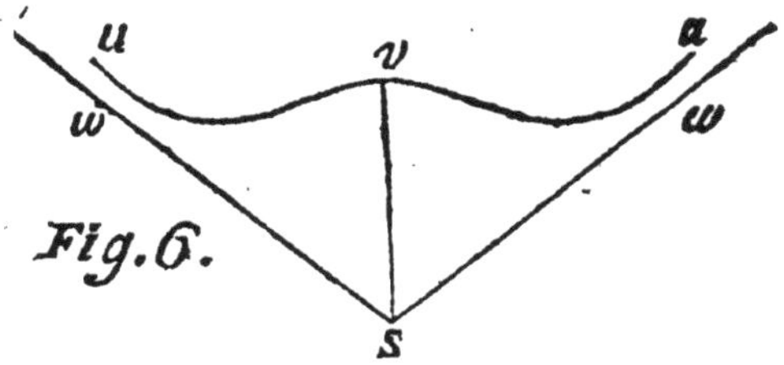

Fig. 6.

If the trajectory be a hyperbola; and, s being the star's true place, sv be $= \frac{bg}{c} \times (1 + c)$; the right lines sw, sw, so drawn that the sine of the angle vsw be $= \frac{1}{e}$, will be asymptotes to the curve of aberration uvu.

The two branches of the curve of aberration excurring ad infinitum from the point v (fig. 5 and 6) will be similar when the star is situated any where in a plane at right angles to the plane of the earth's trajectory, the common section of the two planes being the principal axis of the trajectory. If the star be not so situated, the two branches will take different forms; but their asymptotes will make equal angles with the line sv on different sides thereof, and form one continued right line (as in fig. 5) or intersect each other at s (as in fig. 6).

These conclusions respecting the curve of aberration are so obvious, upon considering the values of st and its position in fig. 4, that it is unnecessary to be more explicit on that head. But it may be worth while to add a remark or two concerning *the angle of aberration*, i. e. the angle under which the line of aberration above found would ap-

parallel to the horizon and the liquor just covering its ends; the bung diameter being 7 and the length 11 feet?

Answered by Mr. Alex. Rowe.

The resistance of any body moving in a fluid being as the surface against which the particles strike drawn into the square of the sine of the ∠ of incidence, and the quantity of the fluid acting against the slant side of the cone being as the diff. between the areas of the greater and less ends; if a be put for ($5\frac{1}{2}$) half the length of the cask, b for ($3\frac{1}{2}$) half its bung diameter, and x for half the dif. between the bung and head diameters, we shall have $(b-x)^2 + x^2(b^2-(b-x)^2) \div (a^2+x^2)$ as the resistance in direction of the length, a min. the fluxion of which being made $= 0$, we have $x = (\sqrt{(a^2+4b^2)} - a)\ (a \div 2b) = 2 \cdot 67319$; and hence the head diameter $(2(b-x)) = 1 \cdot 6536$ feet $= 19 \cdot 8432$ inches. Then, per Hutton's Mens. p. 526, $(103 \cdot 8432^2 \times 3 + 64 \cdot 1568^2) \times 132 \times \cdot 00023209 = 1117 \cdot 18$ ale gallons = the whole content of the cask. Moreover $\frac{1}{2}(84 \pm 19 \cdot 8432) = 51 \cdot 9216$ and $32 \cdot 0784 =$ the wet and dry inches: hence (per. p. 545 and 546 of the same book) $32 \cdot 0784 \div 84 = \cdot 3819$ nearly, corresponding to which, in the table of circular segments at the end of the same book, is $\cdot 2757066$; then $1117 \cdot 18 \times \cdot 2757066 \times 1 \cdot 2732395 = 392 \cdot 1755$ gallons = the cont. of the vacuity; which taken from the whole, leaves 725 gallons contained in the cask.

IV. QUESTION 658, *by Mr.* Moss.

To determine all the different ways it is possible to pay 50l. with pistoles at 17s. each, guineas, moidores, and six-and-thirties.

Answered by Mr. Moss, *the Proposer.*

Let the pistoles, guineas, moidores, and thirty-six shilling pieces be denoted by v, x, y, and z respectively. Then (by the prob.) $17v + 21x + 27y + 36z = 1000$. Now if a table be first formed of all the possible values of x, that will bring out the expressions $\frac{4x-0}{17}$, $\frac{4x-1}{17}$, $\frac{4x-2}{17}$, $\frac{4x-3}{17}$, &c. to $\frac{4x-16}{17}$, whole numbers it will greatly facilitate the process. For, if z and y be each taken unity, we shall have $17v + 21x = 937$, whence $v = x + \frac{4x}{17} - 55\frac{2}{17}$, and therefore, in the table formed as above, against $\frac{2}{17}$ (or $\frac{4x-2}{17}$) we shall have $\left\{\begin{matrix} x = 9, 26, 43 \\ v = 44, 23, 2 \end{matrix}\right\}$. Again, if $z = 1$, and $y = 2$, we shall have $17v + 21x = 937 - 27 (= 937 - 1\frac{10}{17})$, and therefore it is plain the above expression $\frac{4x}{17} - 55\frac{9}{17}$ (or $\frac{4x-9}{17}$) must always be diminished by $1\frac{10}{17}$ (or $\frac{10}{17}$) when y is increased by unity. Thus, when $z = 1$, and $y = 2$, $\frac{4x}{17} - 53\frac{9}{17}$, or $\frac{4x-9}{17}$, must be a whole number, and ∴, in the table formed as above, $\begin{cases} x = 15, 32 \\ v = 35, 14 \end{cases}$; when $z = 1$, and $y = 3$, $\frac{4x-16}{17}$ will be a whole number, whence, by the table $\begin{cases} x = 4, \ 21, \ 38 \\ v = 47, 26, \ 5 \end{cases}$; and when $z = 1$, and $y = 4$, then $\frac{4x-6}{17}$ will be a whole number, whence $\begin{cases} x = 10, \ 27 \\ v = 38, \ 17 \end{cases}$: by the same method of proceeding, viz. by continually subtracting $\frac{10}{17}$ from the expressions $\frac{4x-2}{17}$, $\frac{4x-9}{17}$, $\frac{4x-16}{17}$, $\frac{4x-6}{17}$, $\frac{4x-13}{17}$, &c. we shall obtain all the values of x and v when $z = 1$, and $y = 2, 3, 4, 5$, &c. to its utmost limit 31. Also, when $z = 2$, and 3, and y (in each case) $= 1, 2, 3, 4$, &c. to their utmost limits 30 and 31, we shall, by the foregoing method, readily obtain all the possible values of x and v. Whence the following table:

z	y	x	v	z	y	x	v	z	y	x	v
	1	9, 26, 43	44, 23, 2		1	17, 34	32, 11		1	8, 25	41, 20
	2	15, 32	35, 14		2	6, 23, 40	44, 23, 2		2	14, 31	32, 11
	3	4, 21, 38	47, 26, 5		3	12, 29	35, 14		3	3, 20, 37	44, 23, 2
	4	10, 27	38, 17		4	1, 18, 35	47, 26, 5		4	9, 26	35, 14
	5	16, 33	29, 8		5	7, 24	38, 17		5	15, 32	26, 5
	6	5, 22	41, 20		6	13, 30	29, 8		6	4, 21	38, 17
	7	11, 28	32, 11		7	2, 19	41, 20		7	10, 27	29, 8
	8	17, 34	23, 2		8	8, 25	32, 11		8	16	20
	9	6, 23	35, 14		9	14, 31	23, 2		9	5, 22	32, 11
	10	12, 29	26, 5		10	3, 20	35, 14		10	11, 28	23, 2
	11	1, 18	38, 17		11	9, 26	26, 5		11	17	14
	12	7, 24	29, 8		12	15	17		12	6, 23	26, 5
	13	13	20		13	4, 21	29, 8		13	12	17
	14	2, 19	32, 11		14	10	20		14	1, 18	29, 8
1	15	8, 25	23, 2	2	15	16	11	3	15	7	20
	16	14	14		16	5, 22	23, 2		16	13	11
	17	3, 20	26, 5		17	11	14		17	2, 19	23, 2
	18	9	17		18	17	5		18	8	14
	19	15	8		19	6	17		19	14	5
	20	4	20		20	12	8		20	3	17
	21	10	11		21	1	20		21	9	8
	22	16	2		22	7	11		23	4	11
	23	5	14		23	13	2		24	10	2
	24	11	5		24	2	14		26	5	5
	26	6	8		25	8	5		31	1	2
	28	1	11		27	3	8				
	29	7	2		30	4	2				
	31	2	5								

The rest of the answers may be readily put down, by inspection only, from the preceding table. For, since four moidores are = 3 thirty-six shilling pieces, we have $17v + 21x - 5y - z$ ($= 17v + 21x - y - 4z$) $= 937$, and therefore the values of v and x will be the very same, whether $z=4$ and $y=1, 2, 3, 4, 5,$ &c. or $z = 1$ and $y=5, 6, 7, 8, 9,$ &c. Whence the number of different answers will be found to be 412.

V. QUESTION 659, *by Mr.* Wm. Wilkin.

In the plane triangle ABC, there is given the angle at C, and the parts or segments of the base AD, AE, to construct the triangle so, that, if BD be drawn, the ∠ABD may be a maximum, and BC to EC as m to n.

Answered by Messrs. Burrow, Cole, Coughron, Cottom, Crakelt, Dalby, Edwards, Sewell, *and* Turner.

Construc. Make AD, AE = the given parts of the base; produce the same to any distance, and draw cb forming the ∠ at c = the given ∠C, and make cb : CE :: m : n; draw Eb; then (by prob. 43, Simp. Algeb.) describe a circle to pass through the points A, D, and touch Eb in the point B; draw AB and BD, and BC ∥ bc, and it is done.

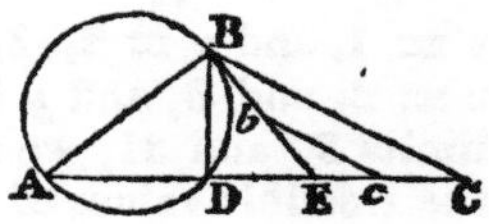

Demonstration. For, by sim. △s, CB : CE :: cb : CE :: m : n; and the ∠ABD is a maximum, by problem 44, Simpson's Algebra.

VI. QUESTION 660, *by Mr.* Stephen Hodges, *at the Right Honorable the Earl* Spencer'*s at Althorp.*

Of all the plane triangles having the same given base and perpendicular, to determine geometrically, that whose vertical angle shall be the greatest?

Answered by Mr. R. Burrow.

Construction. Upon C, the middle of the given base AB, raise the given ⊥ CD; draw DA, DB; and ADB shall be the greatest △ required.

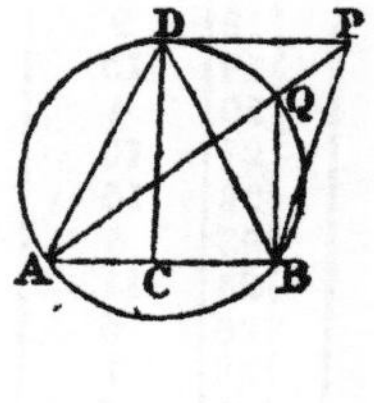

Demonstra. Draw DP ∥ AB, and of any length; also draw AQP and QB; and about ADB describe the circle ADQB. Then ∠ADB = ∠AQB (Euc. 3, 21) greater than the ∠APB (Euc. 1, 21).

Cor. 1. The △ADB is isosceles; for AD = DB (by Euc. 1, 4). Also, of all △s having the same base and altitude, that whose vertex is nearer to the vertex of the isosceles △ has a greater vertical ∠ than that whose vertex is more remote.

Cor. 2. Hence also it may be observed that the △ is likewise isosceles when the base and vertical ∠ are given, and the ⊥ a maximum, or when the ⊥ and vertical ∠ are given, and the base a minimum.

Mr. *Crakelt*, at the end of his solution, observes that, we are hence naturally led to a correction of the solution to question 615, pa. 321, Diary for 1771; for though the construction there exhibited answers, I apprehend, the proposer's intentions, yet it by no means fulfils the literal meaning of the question, it being manifest from what is above done, that in order to have the urn appear the greatest possible, it must be placed half above and half below the level of the observer's eye: in consequence of which the column's height will be only 3 feet, instead of 35·09751.

VII. QUESTION 661, *by Mr.* Moss.

The difference of the sides including a known angle of a plane triangle being given, and also the sum of one of those sides and that opposite the given angle, to construct the triangle.

Answered by Mr. John Turner.

Analysis. Suppose the thing done, and that ABC is the △, of which are given the ∠A, the sum AC + CB, and the difference AB — AC. Then also will be given AB + BC, it being evidently = AC + CB + the given difference AB — AC; and therefore, if AD and AE be made respectively = the two given sums AB + BC and AC + CB, the △ ADE will also be given; and then between the sides of this given △ we have only to apply BC so as to cut off BD and CE each = to it; whence appears this easy

Construction. From the given ∠DAE, and make DA and AE = the two given sums of each side and the base; draw any line FG ∥ DE, and make EH and HI each = DF; then draw EIB, and lastly BC ∥ HI cutting off the △ ABC required.

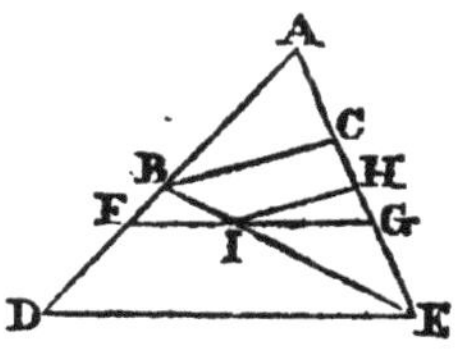

Demonstration. Because of the ∥s DE, FIG, and HI, BC, we have (EI : EB : :) DF : DB : : EH : EC : : HI : CB; but, by the construction, the antecedents DF, EH, HI, are all equal; therefore, the consequents DB, EC, and CB, are also equal; and consequently AB + BC = the given sum AD, and AC + CB = AE.

An Algebraical Solution by Draconarius.

Put AB + BC $= a$, AC + BC $= b$, and $x = \frac{1}{2}$AB — $\frac{1}{2}$AC; then AB

— AC' $= a - b = 2d$, AB $= x + d$, AC $= x - d$, and BC $= \frac{1}{2}a + \frac{1}{2}b - x = s - x$. Hence $((s - x)^2 - 4d^2) \div (2x^2 - 2d^2) = m$ the vers. s. ∠A; then the value of x is easily found from this quadratic equation.

VIII. QUESTION 662, *by Mr.* Isaac Dalby.

Given the vertical angle, the line bisecting it, and the difference of the segments of the base made thereby, to construct the triangle.

Answered by the Rev. Mr. Crakelt.

Construction. Constitute the rhombus BEDF, which may have the ∠EBF = the given one, and its diagonal BD = the given bisecting line; and through F draw a line in such a manner, that the part GI, intercepted by DE and BE produced, may be = the given difference of the segments of the base; made by the said bisecting line: then thro' D draw a parallel to GF to meet BE and BF produced in A and C; and ABC will be the △ required. For, since (Euc. 1, 34) AD = GF, and DC = IF, therefore will AD — DC = GF — IF = GI.

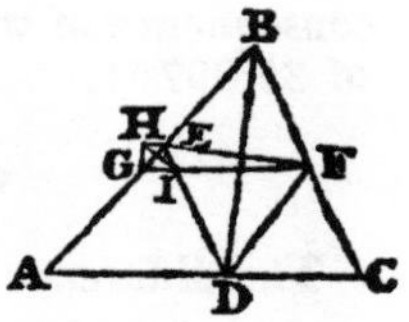

The same answered by Mr. Geo. Coughron.

Upon any line AB describe the segment of a circle capable of containing the given vertical ∠; bisect AB with the ⊥ DE, and draw EC so that 2DF may be to FC as the given difference of the segments to the given bisecting line; and ABC will evidently be similar to the required triangle.

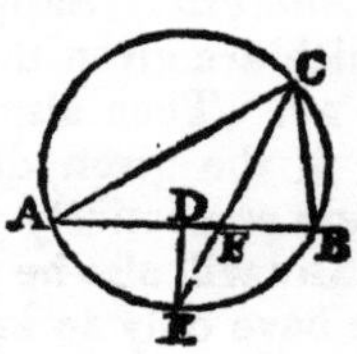

Additional Solution by Mr. John Haycock, *taken from the Miscellanea Mathematica.*

Constr. Like as in Mr. *Crakelt*'s construction, constitute the rhombus BEDF, which may have the ∠EBF = the given one, and its diag. BD = the given bisecting line. Draw the diagonal FE which produce till the rectangle FHE be = the square described on a fourth proportional to FE, EB, and the given difference of the segments, from H apply HI = the said fourth proportional, and through I draw FIG; then ADC being drawn ∥ FG, and meeting BE, BF, produced, in A, C, will form the △ ABC required.

Demonstration. Join G, H. Then, since FH : HI :: HI : HE, the triangles, FHI, EHI; and also FEG are similar, because ∠BEF = ∠FEI; then the triangles GHI, EBF also appear to be similar, because a circle may pass through E, H, G, I, and consequently ∠HGI = ∠BEF; then EF : EB :: GI : GH (HI) :: given diff. seg. base : HI by construction; therefore GI = the given difference of the segments of the base. I suppose I scarce need add AD = GF, DC = IF and AD — DC = GF — IF = GI.

IX. QUESTION 663, *by Mr.* Steph. Ogle.

While a given circular wheel is trundling on an horizontal plane, it is required to determine its point of contact therewith such, that the sum of the altitudes of any four given points in its circumference above the said plane, may be equal to a given quantity M; and to ascertain the limits within which the solution is possible.

Answered by Messrs. Cole, Crakelt, and Dalby; *the Principles of their Constructions and Demonstrations being exactly alike.*

Construction. Join the four given points A, B, C, D; and from the middle, G, of EF bisecting the opposite sides AB, CD, with $\frac{1}{4}$ of the given line, M, as radius descr. a circ. arc; to touch which and the given wheel draw an indefinite line by prob. 40, Simp. Algeb. and the required position of the wheel will be obtained.

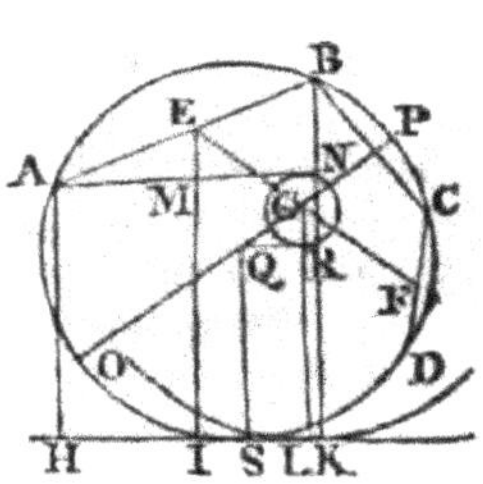

Demonstration. For, on the said line demit the ⊥s AH, EI, BK, and GL; also from A draw a ∥ to HL to cut EI and BK in M and N; then, by sim. △s, we shall have AB : AE :: BN : EM; but AB, is by construction = 2AE, therefore, BN = 2EM; and consequently as (by Euc. 1, 34) MI = AH = NK, 2EI will be = AH + BK. For the same reason, twice the ⊥ from F will be = the sum of the ⊥s from D and C; and twice the ⊥ from G, or 2GL = the sum of the ⊥s from E and F; and, of course, the sum of the ⊥s from the four points A, B, C, D, = twice the sum of the ⊥s from E and F = four times the ⊥ GL = M by the construction.

Scholium (by Mr. *Crakelt*). Thro' G and the centre of the wheel draw the diameter OP; and it will appear (by Euc. 3, 7) that M must not be given greater than 4GO, nor less than 4GP.——The centre of a second circle for four points being thus found, that for five may be determined by dividing the line joining G and the 5th point in the ratio of 1 to 4 (the less part commencing at G); and if from thence a circ. arc be described with the fifth part of M, an indef. tang. to it and the given wheel will give the position of the said wheel in that case. Divide the line joining this centre and a 6th point in the ratio of 1 to 5; and from the point of division with a 6th part of M describe a circ. arc; and an indef. tang. to it and the given wheel, will give the position of the said wheel for six points. And, by proceeding in a similar manner, the construction may be performed for any number of ⊥s from any number of points, either within, on, or without the circumference of the given wheel; the last line being always divided in the ratio of 1 to $n - 1$, and the last radius = the nth part of M.

Messrs. Burrow, Sewell, *and* Ogle, *the Proposer, Construct it thus:*

Having determined G as above; with it as a centre, and radius = the difference of the radius of the wheel and $\frac{1}{4}$M, describe a circle; and from the centre, Q, of the given circle, draw QR to touch it in R; then QS drawn ∥ to GR determines the point S required. Which is evident from what is done above without further demonstration.

X. QUESTION 664, *by the Rev. Mr.* Crakelt.

Through the point of intersection of two given circles it is required to draw a right line in such a manner, that the sum of the respective rectangles under the parts thereof intercepted between the said point and their peripheries, and given lines M and N, may be equal to a given square R^2.

Answered by Mr. W. Sewell.

Construction. From B the point of intersection let the diameters BH, BG, be drawn, and produced if necessary; on BH take F so that BH : BF :: M : N; draw GF, about which as a diameter describe the circle GEF, in which apply GE a third proportional to M and R; ∥ to GE draw ABC, and it will be the line required.

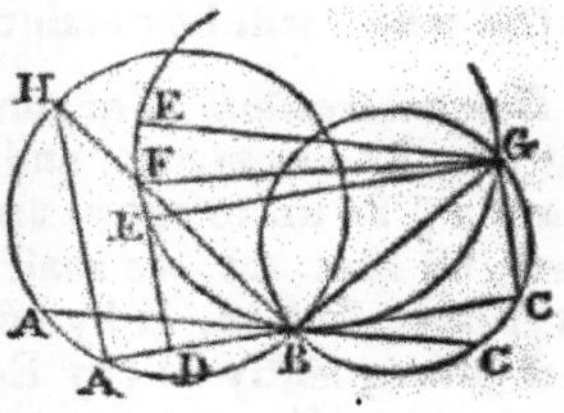

Demonstration. Draw HA, GC, and FED. The ∠HAB = ∠FEG (being each a right angle, by Euc. 3, 31) = ∠FDB by Euc. 1, 29; therefore HA is ∥ FD, by Euc. 1, 28; therefore, by sim. △s, BA : BD :: BH : BF :: M : N, by construction, therefore M × BD = N × BA: Again, by construction, M : R :: R : GE = DB + BC; therefore, R^2 = (M × BD + M × BC =, by substit.) N × AB + M × BC.

Corollary. The sum of the rectangles will be a maximum, when AC is drawn ∥ FG; for GF is the greatest line in the circle.

The same answered by the Rev. Mr. Crakelt, *the Proposer.*

Construction. Through the point of intersection, A, and the centres of the given circles draw the diameters AF, AG; and having determined a third proportional Q to N and R, and also a fourth proportional S, to N, M, and AG, on AG produced take AI = S, and describing on it a semicircle, through A draw a line DAK to terminate in it and the periph. of the circ. descr. on AF which may be = Q, and it will be that required.

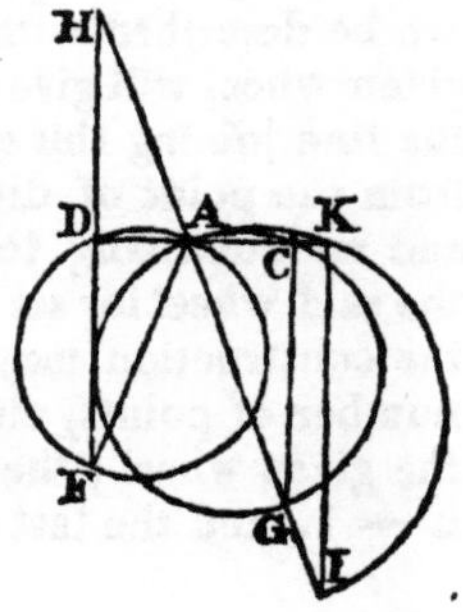

Demonstration. Let the said line intersect the periph. of the circle on AG in the point C, and draw the lines GC, IK, and FD meeting GA, when both are produced in H : then, by construction and sim. triangles, S (AI) : AH :: Q — AD (AK) : AD; or (Euc. 6, 1) $S \times N = AG \times M$ (by construction) : $AH \times N$ :: Q — AD : AD; or, by compos. $AG \times M + AH \times N : AH \times N$ :: Q : AD :: (Euc. 6, 1) $Q \times N = R^2$ (by construction) : $AD \times N$; but by sim. $\triangle$s, AG : AH :: AC : AD; or (Euc. 6, 1) $AG \times M : AH \times N :: AC \times M : AD \times N$; or, by compos. $AG \times M + AH \times N : AH \times N :: AC \times M + AD \times N : AD \times N$; and consequently, by equality, $AC \times M + AD \times N = R^2$.

Scholium. Had it been required to make $AC \times M - AD \times N = R^2$, then AK — AD must have been determined = Q, and division used in the demonstration instead of comp.

XI. QUESTION 665, *by the Rev. Mr.* Lawson.

Having given the base and vertical angle of a plane triangle, it is required to find the locus of the extremity of the line that continually bisects the said vertical angle, and is an arithmetical mean between the two sides comprehending the same.

Answered by Mr. Isaac Dalby.

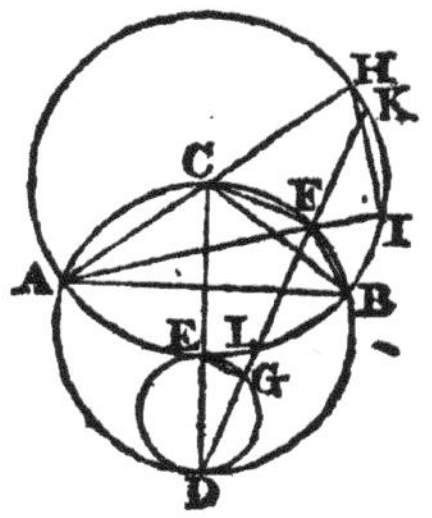

Construction. On the given base AB describe a circle whose segment ACB may contain the given angle; bisect AB with the diameter CD, and with the centre C and rad. CA, describe the circle AEBIH cutting it in E; then the circle DEG described about the diameter DE will be the locus required.

Demonstration. For, let AFB be any position of the $\triangle$, and draw the bisecting line FGD; also, produce AC and AF to H and I, and draw HI, CF, EG, CB. Then the $\angle$CAF = $\angle$CDF (Euc. 3, 21) and $\angle$AIH = $\angle$DFC = $\angle$DGE (being each a right $\angle$ by Euc. 3, 31) therefore the $\triangle$s AIH, DCF, DEG are similar, and AC + CB (= AH) : AF + FB (= AI, by q. 7, Hutton's Math. Misc. :: CD : FD ::) CE : FG; but 2CE = AC + CB by construction; therefore 2FG = AF + FB.

Very nearly in the same manner is the solution given by Mr. *Wildbore,* who subjoins this

Corollary. If DF, produced, cut the circle AEB in K and L; then FL being = FK (Euc. 3, 3) and $FL \times FK$ or $FL^2 = AF \times FI = AF \times FB$ (Euc. 3, 35,) consequently FL is a geometrical mean, between AF, FB like as FG is the arithmetical mean; and the part LG intercepted between the two circles is the difference between the arith. and geom. means.

The *Analysis* is very evident. For the circle ACB being drawn about the △ AFB, and the bisecting line FD drawn, as also the diameter CD, and the line CB, and any curve EG be supposed the locus; then, if AC and AF be produced till CH = CB or CA, and FI = FB, C will be the centre of the circle passing thro' A, B, I, H, and therefore the ∠ AIH right as well as DFC, but the ∠ CDF = ∠ CAF, as being on the same segment, therefore the two △ s AHI, DCF are similar and hence A = HAC + CB : AI = AF + FB or 2CE : 2FL, or CE : FL :: CD : FD; therefore the right line EG must be ∥ to CF, and the curve EG sim. to the curve CF, or EGD a semi-circle as well as CBD.

XII. QUESTION 666, *by Mr.* Todd.

To determine the nature of the curve which will cut at right angles, any number of parabolas having the same vertex and axis?

Answered by Mr. Wm. Sewell.

Geomet. Analysis. Let ABC be any one of the parabolas, AD their common axe, and DBE the curve required. Draw the tang. BF, which will be normal to the curve DBE. Now by the doctrine of tang. AG : FG :: $m : n$ (m and n being the exponents of the terms of the general equ. to parabolas $AG^m = GB^n$) but when the abs. GA is as the subnormal GF of a curve DB, that curve is an ellipse, whose two semi-axes, AD, AE are in the const. duplicate ratio of GA to GF; that is DBE is the quad. of an ellipse, and $AD^2 : AE^2 :: m : n$. In the com. parab. where $m = 1$, and $n = 2$; then AE = AD $\sqrt{2}$; or AD : AE :: side of a square to its diagonal.

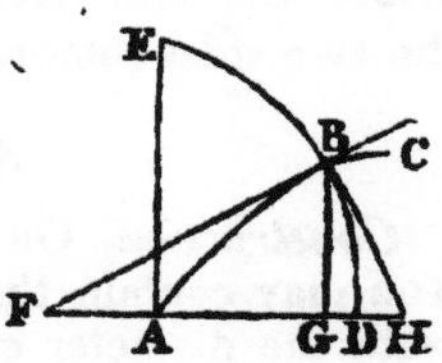

The same answered by Mr. R. Burrow.

Construc. In the common axe AD assume any point D, and take AE to AD :: $\sqrt{n} : \sqrt{m}$ (that is, as $\sqrt{2} : 1$, or as the diag. of a sq. to its side in the com. parab.) and with the semiaxes AD, AE descr. the ellipse DBE, which will be the curve required.

Demonstration. Let ABC be any parab. whose axe is AGD: draw the absc. BG, and tang. to it BF, which ought to be a normal to the ellipse. Now, by the parab. GA : GF :: $m : n$; but, by the constr. $AD^2 : AE^2 :: m : n$, ∴ $AD^2 : AE^2$ :: GA : GF; consequently BF is ⊥ DBE at B.

A Fluxionary Investigation of the same by Mr. Wm. Cole.

Put AD = a, DG = x, and GB = y; then will AG = $a - x$, and (in the com. parab.) GF = $2a - 2x$, also in the right-angled △ FBH (BH being drawn ⊥ BF) we have $2a - 2x : y :: y : yy \div (2a - 2x)$ = GH, which, being, the sub-tang. of the curve, is $= y\dot{x} \div \dot{y}$, whence $y\dot{y} = 2a\dot{x} - 2x\dot{x}$; the fluents of which being taken, we get $yy = 4ax - 2xx$, an equa. to the ellipse, whose axes are as 1 to $\sqrt{2}$.

XIII. QUESTION 667, *by the Rev. Mr.* Wildbore.

To redintegrate (or find the whole fluent of) the expression $\frac{\dot{x}^n}{1-xx}$; where n is any whole positive number whatever.

Answered by the Rev. Mr. Wildbore, *the Proposer.*

First. The fluent of $\frac{2\dot{x}}{1-x^2}$ is the hyp. log. of $\frac{1+x}{1-x}$ = hyp. log. of $1 + x$ — hyp. l. of $1 - x$ = hyp. l. of v — hyp. l. of u, putting $1 + x = v$, and $1 - x = u$; ∴ the fluent of $\frac{\dot{x}^{n-1} 2\dot{x}}{1-x^2}$ is $\dot{x}^{n-1} \times$ (h. l. of v — h. l. of u) $= \dot{x}^{n-2} \times \dot{x}$ l. $v - \dot{x}$ l. $u = \dot{x}^{n-2} \times \dot{v}$ l. $v + \dot{u}$ l. u (because $\dot{x} = \dot{v} = -\dot{u}$, and l. v and l. u stand for the hyp. log. of v and u respectively.)

2dly. The fluent of this last expression, by Simp. Flux. Art. 341, is $\dot{x}^{n-2} \times : v$ l. $v - u$ l. $u - u - a$, where a is the necessary correction, which, because $v + u = 2$, putting $A = 2 + a$, becomes $\dot{x}^{n-2} \times : v$ l. $v + u$ l. $u - A = \dot{x}^{n-3} \times : v\dot{v}$ l. $v - u\dot{u}$ l. $u - A\dot{x}$.

3dly. The fluent of this last is $\dot{x}^{n-3} \times : \frac{vv}{2}$ l. $v - \frac{vv}{2.2} - \frac{uu}{2} \times u + \frac{uu}{2.2} - Ax - b$; which being ordered in the same manner as those above, we obtain

4thly. Its fluent $= \dot{x}^{n-4} \times : \frac{v^3}{2.3}$ l. $v - \frac{v^3}{2.3.3} - \frac{v^3}{2.2.3} + \frac{u^3}{2.3}$ l. $u - \frac{u^3}{2.3.3} - \frac{u^3}{2.2.3} - \frac{Ax^2}{2} - bx - c$. In the same manner the 5th or fluent of this last will be found; and thence the 6th fluent; and in like manner the 7th will be found $= \dot{x}^{n-7} \times$ into all the following quantities, viz.

$$-\frac{v^6\,\text{l.}\,v}{2.3.4.5.6} - \frac{v^6}{2.3.4.5.6} \times : \tfrac{1}{2} + \tfrac{1}{3} + \tfrac{1}{4} + \tfrac{1}{5} + \tfrac{1}{6},$$

$$-\frac{u^6\,\text{l.}\,u}{2.3.4.5.6} + \frac{u^6}{2.3.4.5.6} \times : \tfrac{1}{2} + \tfrac{1}{3} + \tfrac{1}{4} + \tfrac{1}{5} + \tfrac{1}{6},$$

$$-\frac{Ax^5}{2.3.4.5} - \frac{bx^4}{2.3.4} - \frac{cx^3}{2.3} - \frac{dx^2}{2} - ex - f.$$

And now it is evident that the redintegration or whole fluent of $\frac{2\dot{x}^n}{1-x^2}$ is $\frac{v^{n-1}\,\mathrm{l}.\,v}{2.3.4,5\ldots n-1} - \frac{v^{n-1}}{2.3.4\ldots n-1} \times : \frac{1}{2} + \frac{1}{3} + \frac{1}{4} +$ &c. to $\frac{1}{n-1}, \mp \frac{u^{n-1}\,\mathrm{l}.\,u}{2.3.4\ldots n-1} \pm \frac{u^{n-1}}{2.3.4\ldots n-1} \times : \frac{1}{2} + \frac{1}{3} + \frac{1}{4} +$ &c. to $\frac{1}{n-1}, - \frac{Ax^{n-2}}{2.3\ldots n-2} - \frac{bx^{n-3}}{2.3\ldots n-3} - \frac{cx^{n-4}}{2.3\ldots n-4}$ &c. where the upper or lower sign takes place according as n is odd or even, and A, b, c, &c. may be any constant quantities at pleasure, and the half of the expression is evidently the fluent required.

Corol. 1. As, generally speaking, all these fluents must be corrected to be $=0$ when $x=0$, unless otherwise limited by the prob. to which the equation belongs; therefore, taking $x=0$, or $v=u=1$ in the fluents above, it will appear that in this case $A=0$, $b=0$, $c = -\frac{2}{2.3} \times : \frac{1}{2} + \frac{1}{3}$, $d=0$, $e = \frac{2}{2.3.4.5} \times : \frac{1}{2} + \frac{1}{3} + \frac{1}{4} + \frac{1}{5}$, $f=0$, $g = -\frac{2}{2.3.4.5.6.7} \times : \frac{1}{2} + \frac{1}{3} + \frac{1}{4} + \frac{1}{5} + \frac{1}{6} + \frac{1}{7}$, $h=0$, &c. whence the law of continuation is manifest.

Corol. 2. The fluents found above include several others, as those of $\dot{x}^m \times$ hyp. l. of $\frac{1+x}{1-x}, \frac{x^m \dot{x}^n}{1-x^2}$, &c. Moreover, the fluent of $\dot{x}^n z^m$ where $z =$ the hyp. l. of $\frac{1+x}{1-x}$, may be found in the same manner, tho' not perhaps within the limits allotted to a Diary solution.

THE PRIZE QUESTION, *by* Peter Puzzlem.

Let ABC be a given triangle, and let the line BD meet AC in D, so that BC being greater than AB, AD may be greater than CD: It is required to find with what velocity the said triangle, considered as a very thin plate of heavy metal, must revolve about BD as an axis, that, whilst it is so revolving, the said line BD shall (if possible) always keep, in an upright position, with the angular point B resting on an horizontal plane?

Answered by Peter Puzzlem, *the Proposer.*

AQ being ∥ to the horizon, call the same m; also call BQ, n, BD, a;

AD, e; CD, f; and Dt (any variable part of DA) x: and let the velocity of any point of the Δ whose distance from BD is r be denoted by v.

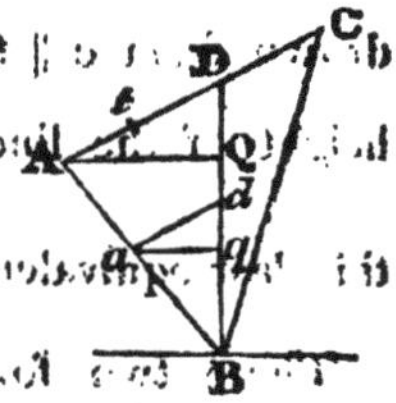

Then $e : m :: x : mx \div e$, the distance of t from BD; $r : v :: mx \div e : mvx \div er$, the velocity of t; and $\left(\frac{mvx}{er}\right)^2 \div \frac{mx}{e} \left(= \frac{mv^2x}{er^2}\right)$ will be the centrifugal force of the point t. Moreover, the height of t above the horizontal plane supporting the Δ being $= a - \frac{a-n}{e}x$, $h : a - \frac{a-n}{e}x :: \frac{mv^2x}{er^2} : \frac{amv^2x}{ehr^2} - \frac{a-n}{e^2hr^2}mv^2x^2$, the force which acting $\parallel$ to the horizon at a point in BD whose height above B is $= h$, would be equivalent to the centrifugal force of t. Therefore, $\frac{amv^2e}{2hr^2} - \frac{a-n}{3hr^2}mv^2e \left(= \frac{amv^2e}{6hr^2} + \frac{mnv^2e}{3hr^2}\right)$ the whole fluent of $\frac{amv^2x\dot{x}}{ehr^2} - \frac{a-n}{e^2hr^2}mv^2x^2\dot{x}$, is the force which, acting at the height h, would be equivalent to the centrifugal force of the whole line AD. Now $gme \div 2h$ being, by the well known principles of mechanics, the force which, acting parallel to the horizon at the height h, would be equivalent to the gravity of the said line AD, g denoting (32·16 feet) the accelerative force of gravity; it is evident that the whole equivalent force, with respect to the revolving line AD is $= \frac{amv^2e}{6hr^2} + \frac{mnv^2e}{3hr^2} + \frac{gme}{2h}$.

Let the $\perp$ from B upon AC be denoted by p; and ad be $\parallel$ AD, let the distance of that line ad from B (measured on the said $\perp$) be expressed by y. Then, will ad be $= ey \div p$; B$d = ay \div p$; and aq $\parallel$ AQ, $= my \div p$; B$q = ny \div p$. Which values of ad, Bd, aq, Bq, being respectively substituted for e, a, m, n, in the expression for the last mentioned equivalent force, the same becomes $\frac{aemv^2y^3}{6hp^3r^2} + \frac{emnv^2y^3}{3hp^3r^2} + \frac{gemy^2}{2hp^2}$. It is obvious then, that $\frac{1}{6} \times : \frac{aempv^2}{4hr^2} + \frac{emnpv^2}{2hr^2} + \frac{gemp}{h}$, the whole fluent of $\frac{aemv^2y^3\dot{y}}{6hp^3r^2} + \frac{emnv^2y^3\dot{y}}{3hp^3r^2} + \frac{gemy^2\dot{y}}{2hp^2}$, will express the force which, acting $\parallel$ to the horizon, at the height h above B, would be equivalent to the whole force of the revolving Δ ABD urging it to turn about the fixed point B.

In the same manner the equivalent force respecting the Δ BCD is found = to $\frac{1}{6}$ × : $\frac{afmpv^2}{4hr^2} + \frac{fm'npv^2}{2hr^2} + \frac{gfm'p}{h}$, m being = the line drawn from C ∥ to AQ to meet the continuation of BD, and n = the height of the line so drawn above B. But, m' being = $fm \div e$, this last equivalent force is = $\frac{1}{6}$ × : $\frac{af^2mpv^2}{4ehr^2} + \frac{f^2mn'pv^2}{2ehr^2} + \frac{gf^2mp}{eh}$.

These two forces being equated, we obtain $v^2 = \frac{e^2 - f^2}{(a + 2n')f^2 - (a + 2n)e^2} \times 4gr^2$; and it appears that BD cannot possibly keep in an upright position unless $(a + 2n')f^2$ be greater than $(a + 2n)e^2$.

If AB, BD, and AD be each = 5, and CD = 4; n will be = 2·5, n' = 7, and $v = r\sqrt{\frac{2}{3}g}$.

Questions proposed in 1774, *and answered in* 1775.

I. QUESTION 668, *by* Agriculturus.

The difference $\frac{a}{a+c} \backsim \frac{b}{b+c}$ is $= \frac{c}{a+c} \backsim \frac{c}{b+c} = c \times \left(\frac{1}{a+c} \backsim \frac{1}{b+c}\right)$. Required the investigation or demonstration?

Answered by Mr. Hugh Byron, *of Manchester.*

The quantity $\frac{a}{a+c}$ is evidently $= 1 - \frac{c}{a+c}$, and $\frac{b}{b+c} = 1 - \frac{c}{b+c}$; the difference of which two is $\frac{c}{a+c} \backsim \frac{c}{b+c}$.

The same answered by Mr. James Nicholson, *of Horton Grange.*

The given difference, $\frac{a}{a+c} \backsim \frac{b}{b+c}$, reduced to a common denominator, is $\frac{ac \backsim bc}{(a+c)\,.\,(b+c)}$ = (by adding c^2 to both parts) $\frac{c\,.\,(a+c)}{(a+c)\,.\,(b+c)} \backsim \frac{c\,.\,(b+c)}{(a+c)\,.\,(b+c)} = \frac{c}{b+c} \backsim \frac{c}{a+c}$.

Mr. John Fatherly, *of Washington, thus answers the same.*

The difference, $\frac{a}{a+c} \backsim \frac{b}{b+c}$, reduced to a common denominator is $\frac{ac \backsim bc}{(a+c).(b+c)}$; and the difference $\frac{c}{a+c} \backsim \frac{c}{b+c}$, also reduced to a common denominator, is $\frac{ac \backsim bc}{(a+c).(b+c)}$, which is the same as the other.

II. QUESTION 669, *by Mr.* Isaac Dalby.

Jack, who's a true good natured blade,
Practis'd so long th' tippling trade,
'Till, toper like, both free and willing,
He'th spent his every crown and shilling.
Yet is not of all help bereft,
For now the lucky rogue's had left
A thousand pounds by an old grannum,
To buy one hundred pounds per annum.
Jack now does for the same engage,
Until he's sixty years of age;
Being at this time as we are told,
No more than twice a dozen old.
How long is't ere the time will come
For Jack to have that yearly sum;
If cash at compound int'rest lent,
Will bring exactly five per cent.

Answered by Mr. John Aspland, *Master of the Free School, at Soham, Cambridgeshire.*

Put $a = 100$, $p = 1000$, $r = 1{\cdot}05$, $T = 36$ $(= 60 - 24)$, and t = the required time till the annuity commences. Then, by the doctrine of annuities, we shall have $(a - a \div r^T) \div (r-1) =$ the present worth of a annuity for T time, and $(a - a \div r^t) \div (r-1) =$ the present worth of the same for t time; the difference of which two must therefore be = to p; that is, $(a \div r^t - a \div r^T) \div (r-1) = p$. Hence $r^t = a \div (p(r-1) + a \div r^T)$, and then $t =$ (log. of a — log. of $(p(r-1) + a \div r^T)) \div$ log. of $r = 8{\cdot}127$ years, the time required.

III. QUESTION 670, *by* Coquetarius.

Given the continual products of every $n - 1$ of n quantities; to find those quantities.

Answered by Mr. David Cunningham.

Let z, y, x, w, v, &c. represent the required quantities, and a, b, c, d, e, &c. the given continual products of every $n-1$ of them; the number of quantities being n. Then, by the question, we have

$$\left.\begin{array}{l} yxwv, \text{ \&c. } (n-1) = a, \\ zxwv, \text{ \&c. } (n-1) = b, \\ zywv, \text{ \&c. } (n-1) = c, \\ zyxv, \text{ \&c. } (n-1) = d, \\ zyxw, \text{ \&c. } (n-1) = e, \\ \text{\&c.} \qquad \text{\&c.} \end{array}\right\}$$

Hence, by taking the continual product of all these equations, we have $(zyxwv(n))^{n-1} = abcde(n)$: or, by extraction, $zyxwv(n) = (abcde(n))^{1 \div (n-1)}$, which call p. Then, by dividing this last equation by each of the first given ones in order, we have $z = p \div a, y = p \div b, x = p \div c, w = p \div d, v = p \div e$, &c.

Scholium. So that, in general, it may be remarked, that each required quantity is always reciprocally as that particular one of the given products into which it did not enter as a factor; and any such required quantity is equal to the quotient resulting from the division whose divisor is the product above-mentioned, and its dividend the $n-1$ root of the continual product of all the given products.

IV. QUESTION 671, *by Mr.* Wm. Spicer.

There are two spire steeples, which stand on the banks of a river opposite to each other: Now if lines be drawn from the vertex of each steeple to the base of the other, their lengths will be 1170·6342 and 1151·13705 feet respectively; also two heavy bodies being let fall, at the same time, from the vertex of each steeple, it was observed that the sound of the body which fell down the lower steeple, arrived at the base of the higher, just at the same moment of time with the other body. Required the breadth of the river, and the altitude of each steeple?

Answered by Mr. Wm. Smart, *of Wicken.*

Put $a = 1151{\cdot}13705$, $b = 1170{\cdot}6342$, $d = 1142$, $s = 16\frac{1}{12}$, and $x =$ the altitude of the lower steeple. Then, by 47, Euc. 1, $\sqrt{(a^2 - x^2)} =$ breadth of the river, and $\sqrt{(b^2 - a^2 + x^2)} =$ the altitude of the higher steeple; also $\sqrt{(\sqrt{(b^2 - a^2 + x^2)} \div s)}$ and $\sqrt{(x \div s)}$ are the times of falling down the two steeples, and $\sqrt{(a^2 - x^2)} \div d$ = the time of sound's passing over the river: Hence $\sqrt[4]{(b^2 - a^2 + x^2)} = \sqrt{x} + \sqrt{(a^2 s - s x^2)} \div d$. This put into numbers, &c. x is found $= 144\frac{1}{4}$ feet, the lower altitude, and then $257\frac{1}{3} =$ the greater, and $1142 =$ the breadth of the river.

V. QUESTION 672, *by Mr.* Stephen Hodges.

In Althorp park there's an elliptical piece of water, with a circular island in the middle of it, the diameter of which and the two axes of the ellipse are in geometrical progression; and the transverse axe of the the ellipse exceeds the diameter of the island by 24 poles; also the difference of the squares of the transverse axe of the ellipse and diameter of the island, is to the sum of the squares of the two axes of the ellipse, as 5 to 7. Required the dimensions and area of each?

Answered by Mr. Mic. Taylor.

Let x, xz, xz^2 denote the diameter of the circle and the axes of the ellipse; then, by the question, $x^2z^4 - x^2 : x^2z^4 + x^2z^2 :: 5 : 7 ::$ (abbreviating by x^2) $z^4 - 1 : (z^2 + 1)\, z^2 ::$ (abbreviating by $z^2 + 1$) $z^2 - 1 : z^2$; hence $7z^2 - 7 = 5z^2$, and $z = \sqrt{(7 \div 2)}$ the ratio of the terms or progression. Again, by the question, $xz^2 - x = 24 = a$; hence $x = a \div (z^2 - 1) = \frac{2}{5}a$, the diameter of the circle; and consequently $xz = \frac{1}{5}a\sqrt{14}$, and $xz^2 = \frac{7}{5}a$, the axes of the ellipse. And therefore the two areas are $\frac{4}{25}a^2n$, and $\frac{7}{25}a^2n\sqrt{14}$, or, in numbers, 72·3824 and 473·9531.

The same, otherwise by Mr. Rob. Marshall, *of Hartley.*

Let $a = 12 =$ half the difference, and $x =$ half the sum of the transverse and the diameter of the circle; then $x + a =$ the former, and $x - a =$ the latter; also, by the question, $\sqrt{(x^2 - a^2)} =$ the conjugate of the ellipse; then, again, by the question, as $5 : 7 :: (4ax : 2x^2 + 2ax ::)\ 2a : x + a$; hence $5x + 5a = 14a$, or $x = \frac{9}{5}a$, and the diameters and areas as above.

VI. QUESTION 673, *by Mr.* Alex. Rowe.

A lady having bought a piece of land (for which she is to pay 3000 guineas) to be cut out, in the form of a right-angled triangle, from a large common: now, the only limitation besides is, that the sum of the hyp. and ⊥ is to be 100 chains; she requests the diarian correspondents will inform her what dimensions to take that she may have the most land for her money, with the price per acre it will cost her.

Answered by Gemini.

Put $a = 100$, the sum of the hypothenuse and perpendicular, and $x =$ the perpendicular of the triangle; then is $a - x =$ the hypothenuse, and $\sqrt{(a^2 - 2ax)} =$ the base; hence $x\sqrt{(a^2 - 2ax)} =$ (double the area) a maximum. Which put in fluxions, &c. there results $x = \frac{1}{3}a$ the perpendicular, $a - x = \frac{2}{3}a$ the hypothenuse, $\sqrt{(a^2 - 2ax)} = a\sqrt{\frac{1}{3}}$ the base of the $\triangle$, and $\frac{1}{6}a^2\sqrt{\frac{1}{3}} = 962{\cdot}25$ square chains $= 96{\cdot}225$ acres; which produce 31·1769 guineas an acre, or 32l. 14s. 8½d. nearly.

Corollary. The hypothenuse being double the perpendicular as appears in the solution, the one acute angle will be double the other, viz. the angle at the perpendicular double the angle at the base, the former being 60° and the latter 30°.

VII. QUESTION 674, *by Mr.* Wm. Cole.

On a horizontal plane is a garden which was surrounded by a very high wall; but the owner, thinking the wall too high, had its top cut off, by a plane oblique to the horizon, so as to leave the highest point 16, and the lowest only 8 feet high: now the horizontal distance of these two parts is 20 yards; and if the foot of a ladder, of a certain length, be placed in a certain point in the garden, its top will just reach the top of the wall all around. Required the length of the ladder, and area of the garden?

Answered by Mr. Tho. Bosworth.

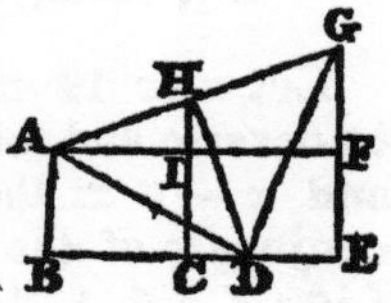

Draw the horizontal line BE = 60 feet, and ⊥ to it draw BA = 8, and EG = 16; join A, G, and draw AF ∥ BE; bisect AG in H, and draw HC ⊥ BE, and HD ⊥ AG; and, lastly, join D, A, and D, G. Then will D be the point where the foot of the ladder is to be placed, and DA = DG the length of it; moreover, the form of the garden is an ellipse whose conjugate axe is = AF, and its transverse axe = AG: all which is too evident to need any demonstration.

Calculation. Because H bisects AG, and HC is ∥ EG, also CI = BA = EF, therefore HC = $\frac{1}{2}$BA + $\frac{1}{2}$EG = 12; and by sim. △s, AF : FG :: HC : CD = 12 × 8 ÷ 60 = 1·6; hence ED = $\frac{1}{2}$BE — CD = 28·4, and then (by 47, Euc. 1) DG = $\sqrt{(DE^2 + EG^2)}$ = 32·597 feet, the length of the ladder, also AG = $\sqrt{(AF^2 + FG^2)}$ = 60·531 feet; whence ·7854 AF × AG = 2852·462 square feet = 316·94 square yards, the area of the garden.

VIII. QUESTION 675, *by* Geometricus.

Given one angle, a side adjacent to the said angle, and the diff. of the other two sides, to determine the △.

Answered by Mr. Alex. Rowe.

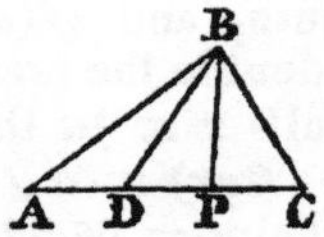

Draw AB = the given line, and the indefinite line AC making therewith the given ∠, on which take AD = the given difference, and draw BD; then make the ∠DBC = ∠BDC; so shall ABC be the △ sought. For, since ∠DBC = ∠BDC, the side DC = BC (6, Euc. 1), and therefore AC — CB = AD the given diff. by the construction.

Calculation. In the $\triangle$ ABD are given AB, AD, and the $\angle$ A; to find BD and the $\angle$ ADB. Then, in the isosceles $\triangle$ DBC, are given DB and all the angles; to find CB = CD, &c.

An Algebraic Solution by Mr. Tho. Robinson.

Put a = AB; d = the difference of the sides AC, BC; c = cos. $\angle$ A; and x = AC: Then BC = $x \pm d$, and, by a known theorem, $AB^2 + AC^2 - 2c \cdot AB \cdot AC = BC^2$, that is, $a^2 + x^2 - 2acx = x^2 \pm 2dx + d^2$; therefore $x = (a^2 - d^2) \div (2ac \pm 2d)$ = AC.

IX. QUESTION 676, *by Mr.* Joseph Edwards.

Given the axe and greatest ordinate of a semi-parabola = 16 and 8 respectively; supposing a tangent drawn from the extremity of the curve to cut the axe produced, it is required to find the area of the greatest rectilineal triangle that can be cut off, by another tangent from the space contained by the curve, tangent, and continuation of the axe.

Answered by Mr. Wm. Hedley.

Let Acc be the parabola, BC its ordinate, TC the tangent meeting the axe AB produced, tcE the other tangent required, and ED and cb ∥ BC.

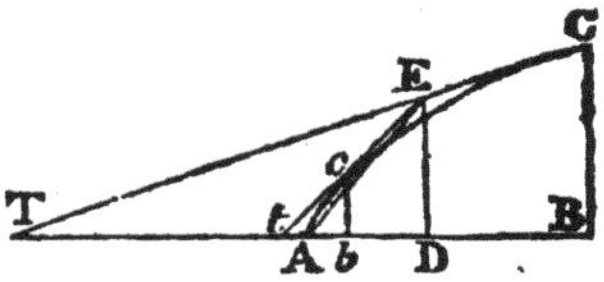

Put $a = 16$ = AB = AT (by the prop. of the parab.), $b = 8$ = BC, x = At = Ab. Now AD = $\sqrt{ax}$, by the prop. of the parab. also, by sim. $\triangle$s, TB : BC :: TD : DE $= (a + \sqrt{ax})\, b \div 2a$; then the area $\frac{1}{2}$Tt × DE $= \frac{1}{2}(a - x)(a + \sqrt{ax})\, b \div 2a$ = a maximum, or $(a - x)(\sqrt{a} + \sqrt{x})$ a maximum. This being put into fluxions and reduced, we have $x = \frac{1}{9}a$; and therefore the greatest area TEt = $\frac{8}{27}ab$ = $37\frac{25}{27}$.

Corollary. Hence it appears, that when the $\triangle$ TEt is a maximum Tt is to TA as 8 to 9. Also, since the $\triangle$ TBC is $\frac{1}{2}$TB × BC = ab, and the parabola ABC = $\frac{2}{3}ab$, and of course their difference or the trilineal TAC = $\frac{1}{3}ab$; which being to the $\triangle$ TEt ($\frac{8}{27}ab$) as 9 to 8, the same proportion as the above bases, it is Tt : TA :: $\triangle$ TEt : trilin. TCA; but, if the right-line AE be drawn, the $\triangle$ TEt : $\triangle$ TEA :: Tt : TA; therefore the $\triangle$ TEA = the trilin. TCA. Also the four spaces, TEt, TCA, ABC, TBC, are in proportion as the numbers 8, 9, 18, 27. Moreover, Ab, AD, AB are in proportion as the numbers 1, 3, 9; and bc, DE, BC are as 1, 2, 3; so that tE is bisected in the point of contact c.

X. QUESTION 677, *by Mr.* Wm. Wilkin.

Suppose a body P move from the end A, along the diameter AB of a semicircle, and at the same instant another body Q begin to move from the other end B along the circumference BCA, and both to arrive at the same instant at the opposite ends of the diameter. To find the diameter of the semicircle, the whole time the bodies are in motion, together with the time of their nearest approach, and their position and distance at that time; the former body P moving with an accelerated velocity thus, one foot the first second of time, two feet the second, three feet the third, &c. and the latter body Q with an uniform velocity of 5 feet in a second.

Answered by Mr. W. Sewell, *Teacher of the Mathematics, at Walsham.*

Put a for the semi-circumf. AQB, $p = 3{\cdot}1416$, $r =$ the radius, and $c = 5$ the measure of the uniform velocity at Q. Now, it is manifest, from arithmetical progression that P will describe $\frac{1}{2}s\,(s + 1)$ feet, and Q (by uniform motion) cs feet, in s of time; therefore, since a and $2r$ are to be described in the same time, it will be, $cs : \frac{1}{2}s\,(s + 1) :: pr : 2r$, or $2c : s + 1 :: p : 2$; and hence $s = 4c \div p - 1 = 5{\cdot}3662''$ the whole time of motion; consequently $2r = 17{\cdot}0811 =$ AB the diameter, and $a = 26{\cdot}834$, &c. $=$ AQB.

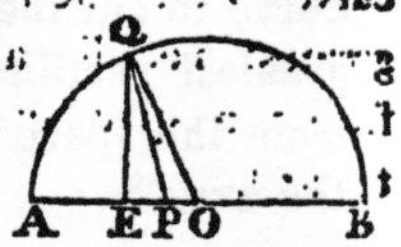

Now let Q and P be the co-temporary places of the bodies when their distance PQ is a minimum, which, it is evident, cannot happen till P has passed the ordinate EQ; put AE $= x$, EQ $= y$, and EP $= z$; then $z^2 + y^2$ being $=$ PQ^2, the same is a minimum; therefore $2z\dot{z} - 2y\dot{y} = 0$, or $z\dot{z} = y\dot{y}$: but $y\dot{y}$ is $= r\dot{x} - x\dot{x}$, for y^2 is $= 2rx - x^2$; also $z\dot{z} = \frac{1}{2}(s^2 + s - 2x) \times \frac{1}{2}(2s\dot{s} + \dot{s} + 2\dot{x})$, for z ($=$ AP $-$ AE) is $= \frac{1}{2}s\,(s + 1) - x$; therefore, by substitution, $r\dot{x} - x\dot{x} = \frac{1}{2}(s^2 + s - 2x) \times \frac{1}{2}(2s\dot{s} + \dot{s} + 2\dot{x})$: but $\dot{s}$ is $= r\dot{x} \div cy$, for A$\dot{}$Q $= -r\dot{x} \div y$, and therefore $c\dot{s}$ ($=$ B$\dot{}$Q $= -$ A$\dot{}$Q) is $= r\dot{x} \div y$; whence, again, by substitution, &c. $4(r - x) \div (s^2 + s - 2x) = 2 + (2s + 1)\,r \div cy$. Now to avoid the trouble of an infinite series in solving this equation, s may be taken $= a \div c - (2r \div c)\sqrt{(3x \div (6r - x))}$, (for $a - 2r\sqrt{(3x \div (6r - x))}$) is a near approximate value of the arc BQ (cs) by page 92 of Mr. Hutton's excellent Treatise on Mensuration); and thence by trial and error, $x = 5{\cdot}5178$; therefore $s = 3{\cdot}3207''$ the time of the nearest approach, AP $= 7{\cdot}1738$, and PQ $= 8{\cdot}1576$ the nearest distance.

XI. QUESTION 678, *by Mr.* Thomas Moss.

Two lines BP, BQ, forming a given $\angle$PBQ, and the length of one of them, BP being given; so to draw a right line from a given point F without those lines, cutting the given line BP produced in G, and the indef. line BQ in E, that the part PG produced, and the segment BE of the indefinite line, may be in the ratio of two given lines m, n.

Answered by the Rev. Mr. C. Wildbore.

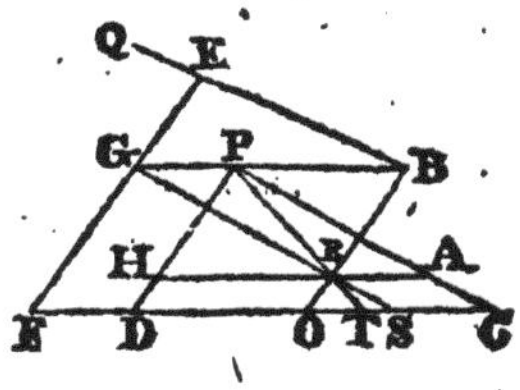

Construction. PB being the given line, PBQ the given ∠, and F the given point; draw PC and FC ‖ to QB and PB, and take FC : CT :: n : m, divide FT (by Simpson's Geometry 5, 17,) in D, so that TC : FD :: TD : PB; join P, D, and draw FE ‖ PD, and it shall be the line cutting BQ in E and BP (produced) in G, as required.

Demonstration. Join P, T; draw BO ‖ PD cutting PT in R; and draw RA ‖ TC and cutting PC in A and PD in H. Then (Simp. Geom. 4, 21,) as TC : RA :: DT : HR = PB; and (Constr.) TC : FD :: DT : PB; ∴ RA = FD = GP; ∴ RA being = and ‖ GP, GR must be so to PA, and GS to PC: in like manner it is evident that GR = EB = PA, and that (by sim. △s) TC : CP :: RA = GP : AP = EB :: m : n.

XII. QUESTION 679, *by Mr.* John Turner.

The diameter of a circle, and the position of two points, with regard to the centre of it, being given; to determine a point in the periphery of the said circle, from which lines drawn to the two given points may obtain a given ratio.

Answered by Mr. Wm. Sewell.

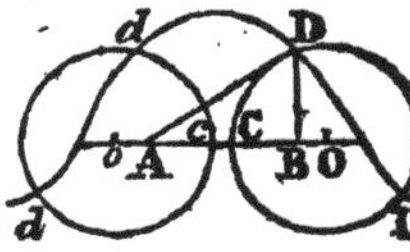

Construc. Draw an indefinite right-line thro' the two given points A, B; and let AB be divided in C in the given ratio of m to n; then take CO a 4th propor. to AC − CB, CB, and AC; and with the centre O and radius OC describe a circle; so shall either of the points D, D of intersection with the given curve, be the point required, let that curve be what it may.

Demonstration. For, from any one of those points, as suppose D, let DA, DB be drawn: then, by Lemma, pa. 336 Simpson's Algeb. it appears that AD : BD (:: AC : BC) :: m : n by the constr.

Mr. *Wildbore* remarks, that if there be set off Ac = BC, and with the radius oc = OC another equal circle be described, its intersections d, d, with the given curve, will be other points answering the conditions of the question.

XIII. QUESTION 680, *by the Rev. Mr.* Lawson.

In a plane △, having given the vertical ∠, the difference of the base and one side, and the sum of the ⊥, from the ∠ at the base contiguous to that side upon the opposite side, and the segment thereby cut off from that opposite side contiguous to the other ∠ at the base; to construct the △.

Answered by the Rev. Mr. Wildbore.

Construction. Upon any right-line SH take SD = ½ the given sum

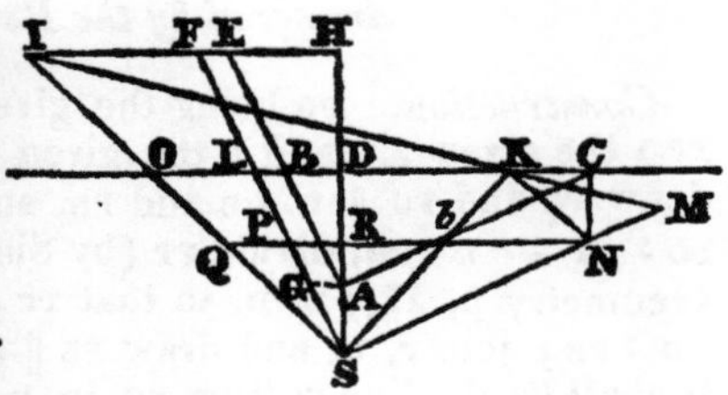

of the ⊥ and segm. ⊥ to it draw the indefinite line LC; then, having made the ∠DSL = the comp. of the given one, produce SL till LF = the given diff. of the base and side; draw HEFI ⊥ and = HS, and join I, S; take DK = DS, join S, K, and draw IKM, also KN ∥ IS; to IM set off SM = SF cutting KN in N, and draw NC ⊥ DK; then draw CA ∥ SM, AG ∥ LD, and lastly AE ∥ SF and cutting LD in B; so shall ABC be the △ required.

Demonstration. The ∠ABC = ∠SLC = the comp. of LSD is evidently the given one. Moreover KN being ∥ IS, and NC to SH, the △s IHS, KCN are sim. and IH being = HS, KC must be = CN; and AC being ∥ SN, and AS to CN, ACNS must be a parallelogram, ∴ AC = SN, and AS = CN; and because KN is ∥ IS, the △s IMS, KMN are sim. conseq. SI : NK :: SM = SF : NM :: HS : CN = AS :: SF = SM : SG; and the antecedents SM, SF being equal, the consequents NM, SG must be so too, ∴ FG = SN = AC; but, by reason of the ∥ lines, it is evident that FG = EA, and EB = FL, conseq. AE = AC, and AC − AB = FL the given diff. by constr. Moreover, KC being = CN = AS, DS + DK = DA + AS (KC) + DC − KC = AD + DC = 2DS the given sum by constr.

Corol. 1. If the lines AD, DC, whose sum is given, make any other given ∠ than a right one; then, since the ∠s B and D of the △ ADB are given, the ∠A is given likewise: therefore making the △ LDS equiangular to it, the method of constr. will be still the same as before, as is evident from the dem. which still holds *verbatim et literatim*.

Corol. 2. The solution of this problem thus given, includes that of another of at least equal difficulty, viz. When there are given the vert. ∠, the sum of its including sides; and the base being supposed divided *ad libitum*, the length of one of its segments, and the ratio of the other to its adjacent side, are likewise given; to construct the △. For, the ∠ at D, and the ratio of AD to A*b* = AB, being given, the ∠s A, B, D are given; and consequently the constr. is the same as in the former Corol.

The same answered by Mr. Henry Clarke, *of Salford.*

Construc. Draw any line AD = the given sum; make DK ⊥ and = AD; join A, K, and draw KE making the ∠DEK = the given vert. ∠; with rad. KE and centre K describe the arc BEC, to which draw the tangents AB, AC; with centre D and rad. = the given diff. describe the arc LI, to touch which and the two lines AB, AC (by Simpson's Geom. Prop. 44, or by Lawson's Appolonius on Tangencies, Prob. 9,) describe a circle FI cutting AD in F; lastly, join D, H, and H, F; so shall DHF be the △ required.

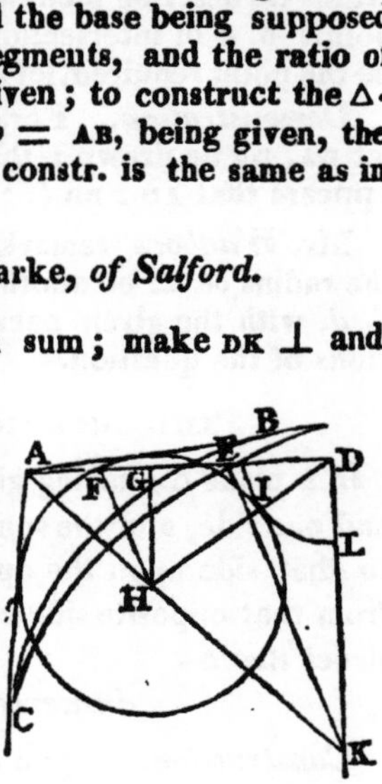

Demonstration: Draw HG ⊥ AD. Then, since DK is ⊥ and = AD, and the △s AHG, AKD similar, HG will be = AG, and ∴ HG + GD = AD the sum of the ⊥ and seg. by constr. Again, since AK bisects the ∠CAB, HF is ∥ KE, ∴ the ∠HFD = ∠KED = the given vert. ∠ by constr. Also, since HF = HI, ∴ ID = HD — HF = the given diff. by construction.

Additional Solutions taken from the Miscellanea Mathematica.

1st, *By Mr.* Lawson, *the Proposer.*

Construction. Draw AD = the given sum, and DF = the given difference forming the ∠ADF = the supplement of the given vert. angle: Draw EFG ∥ AD, and EDC making half a right angle with the same; draw AE to which apply DH = DF, and respectively parallel to these draw AC, CB; and ABC will be the triangle required.

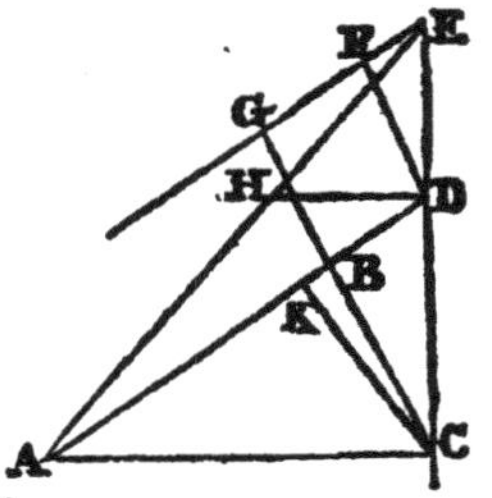

Demonstration. Produce CB to G, and draw the ⊥ CK. Then, because of the parallels we shall have DF : CG :: ED : EC :: DH : CA; but, by constr. DH = DF, ∴ CA = CG, and BG = DF will be the difference between the base AC, and the leg CB. Again, because the ∠DCK = half a right one, CK = KD, and AD = the sum of the ⊥ CK and seg. KA. Lastly, because BDFG is a parallelogram, the ∠ABC or its vertical GBD is the sup. of BDF, and therefore the given vertical one.

Corol. If CK makes given oblique angles with AD; then making CDK = half the sup. of CKD, the construction will evidently be as before.

2nd, *By Mr.* Charles Wildbore.

Construction. Make DK ⊥ DS (see the figure to the first solution), and each = half the given sum, producing both these lines through D; make ∠DSL = the complement of the given one, and on SL produced take LF = the given diff. draw IFH ∥ DK and = HS; draw IS, and KN parallel to it; draw also IKM, and apply to it SNM = SF; lastly, draw QPN ∥ ODK, and PNS will be the triangle required.

Demonstration. Because IH = HS, OP = DS = DK, and OK = 2DS; and OKNQ being a parallelogram by constr. OQ = KN, and OK = QN = RN + RS (RQ) = 2DS the given sum by constr. also because of the parallels we have IS : OQ = KN :: SF : LP :: SM : MN, but SF = SM, ∴ LP = MN, and SN = SP + FL, consequently the two sides SN, SP have the given diff. FL. Moreover the ∠SPR is that given, being the complement of DSL, and the △ that required.

XIV. QUESTION 681, *by Mr.* Geo. Coughron.

Suppose a ball connected to a cord hanging over a pulley, be made to vibrate at any given distance below the pulley, at the same instant that it is drawn up or down with a velocity varying according to any given law (the *n*th power of the distance of the ball from the pulley) required the number of vibrations it will make in a given time: the cord being considered as without weight, passing freely over the pulley, the vibrations made in small similar arcs, and the ball a point.

Answered by Mr. Wm. Sewell.

In general, suppose a pendulum a long, to make b vibrations in c time: put d for the first length of the string, x for the length of it at the end of any variable time t, and v the number of vibrations performed during that time. By mechanics, $\sqrt{a} : \sqrt{x} :: c \div b : c\sqrt{x} \div b\sqrt{a}$ = the time of 1 vibration of a pendul. x long; and, supposing the law of the velocity to vary in the constant ratio of $r : x^n$, we have $\dot{t} = \pm\dot{x} \div rx^n$, $\therefore c\sqrt{x} \div b\sqrt{a} : 1$ (vibration) $:: \pm\dot{x} \div rx^n : \dot{v} = \pm m\dot{x} \div x^{n+\frac{1}{2}}$, ($m$ being put for $b\sqrt{a} \div cr$); the fluents of these two equations being taken and corrected, we shall obtain $t = \pm(x^{1-n} - d^{1-n}) \div r(1-n)$ and $v = \pm(x^{\frac{1}{2}-n} - d^{\frac{1}{2}-n}) \times 2m \div (1-2n)$; from the 1st of these two x is found $= (d^{1-n} \pm (1-n)\,rt)^{\frac{1}{1-n}}$, which substituted in the latter, it becomes $v =$

$$\pm\frac{2m}{1-2n}\times\left((d^{1-n}\pm(1-n)\cdot rt)^{\frac{1-2n}{2-2n}} - d^{\frac{1-2n}{2}}\right)$$

the number of vibrations required: the upper or under sign taking place according as the string is supposed to lengthen or shorten.

Scholium. When $n = 1$, or $= \frac{1}{2}$, the above expression (tho' otherwise general) is found to fail: the reason of which is very evident; for let either of these values be written instead of n in the equations $\dot{t} = \pm\dot{x} \div rx^n$, $\dot{v} = \pm m\dot{x} \div x^{n+\frac{1}{2}}$ above given, and it will appear that the particular fluent of the one or the other is a log. expression: and thus, when $n = \frac{1}{2}$, then $v = \pm m \times$ hyp. log. of $(\sqrt{(d \pm \frac{1}{2}rt)^2} \div d)$; and when $n = 1$, then $v = \pm(2m \div \sqrt{d})(1 - N^{\mp\frac{1}{2}rt})$, where N is the number whose hyp. log. is 1.

Corol. In the case when the string is lengthened or shortened with an uniform velocity, n must be $= 0$, in order that the general veloc. rx^n may become a const. quant. r; substit. then 0 for n in the general expression for the veloc. we obtain $v = \pm 2m(\sqrt{(d \pm rt)} - \sqrt{d})$ for the number of vibr. when the velocity is uniform; but $rt = x - d$, therefore $v = \pm 2m(\sqrt{x} - \sqrt{d})$, where $x =$ the length of the string at the end of the time. And, again, in this case, when the string is shortened, and $x = 0$, or the ball arrives at the pulley, this expression becomes barely $2m\sqrt{d} =$ (by restoring the value of m) $2b\sqrt{ad} \div cr$ $=$ (when a is the second pendulum) $2\sqrt{ad} \div r$, which is also known from other principles.

The Rev. Mr. Wildbore *thus answers the same from First Principles.*

I have not yet seen a solution to any quest. of this nature, that has been sufficiently elementary, which makes me hope that the following, if it be not too long, will not be unacceptable.

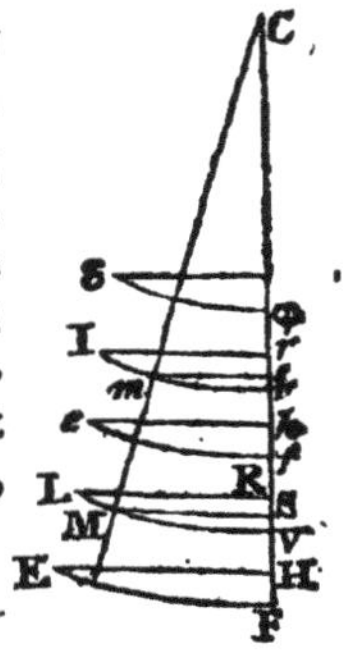

Suppose CF to be the first vertical position of the bal and string at rest; and that it receives an impulse at F sufficient to make it rise to E, or to the height HF, if the length of the string were invariable. But, the string being continually shortened by drawing the body up towards the pulley C, suppose that when the length of the string is CV, or the ball is drawn up the dist. FV, that M is the place of the body, and CM $=$ CV of the string vibrating. Take RV $=$ HF, and erect the $\perp$ RL; and let us suppose, in the first place, that the body is drawn up, or the string shortened, with an uniform velocity, or thro' the dist. d in the time m; and let $32\frac{1}{6}$ feet $=$ the power of gravity $= 2s$, CF $= a$, CV $=$ CM $= u$, RV $=$ HF $= b$, RS $= x$, MV $= z$, and $t =$ the time the body has been in motion. Then, the veloc. of M along the circle being $= 2\sqrt{sx}$, $\dot{t} = \dot{z} \div 2\sqrt{sx}$; but $\dot{z} = \text{CV} \times \text{S}\dot{\text{V}} \div \sqrt{(2\text{CV} \cdot \text{SV} - \text{SV}^2)}$, or, when the arc of semi-vibration is indefinitely small, $\dot{z} = \dfrac{\text{CV} \cdot \text{S}\dot{\text{V}}}{\sqrt{(2\text{CV} \cdot \text{SV})}} = \dfrac{\text{S}\dot{\text{V}}\sqrt{\text{CV}}}{\sqrt{2\text{SV}}} = \dfrac{-\dot{x}\sqrt{u}}{\sqrt{(2b - 2x)}}$; $\therefore \dot{t} = \frac{1}{2}\sqrt{\dfrac{u}{2s}} \times \dfrac{-\dot{x}}{\sqrt{(bx - x^2)}} = \dfrac{-m\dot{u}}{d}$ when the velocity is uniform; therefore $\dfrac{d\dot{x}}{2m\sqrt{2s}\sqrt{(bx - x^2)}} = \dfrac{\dot{u}}{\sqrt{u}}$. Let A $=$ the arc whose versed sine is $2x \div b$ and rad. 1, $p = 3{\cdot}14159$; then, taking the correct fluents of each side of the equation, $d(p - \text{A}) \div 2m\sqrt{2s} = 2\sqrt{a} - 2\sqrt{u}$; and when the body arrives at e, so that $hf =$ HF, having completed one semi-vibration, x being $= 0$, and A $= 0$, $dp \div 4m\sqrt{2s} = \sqrt{a} - \sqrt{u} = q$, and $\sqrt{a} - q = \sqrt{u}$; whence $u = a - 2q\sqrt{a} + q^2$, and $t = m(a - u) \div d = m(2q\sqrt{a} - q^2) \div d = p(2\sqrt{a} - q) \div 4\sqrt{2s} =$ the time of the first semi-vibration.

Now u, the length of the string at e when one semi-vibration is completed, is $= a - 2q\sqrt{a} + q^2$, the string being then shorter than at first by $2q\sqrt{a} - q^2$; and, the body being still drawn up with the same uniform velocity, suppose m to be the place of it when the length of the string is $=$ CV $= w$; then, rv being $=$ FH $= b$, let $rs = x$, $mv = z$, $cf = u = e$; then $\frac{-m\dot{w}}{d} = \frac{-\dot{z}}{2\sqrt{sx}} = \frac{1}{2}\sqrt{\frac{w}{2s}} \times \frac{\dot{x}}{\sqrt{(bx - x^2)}}$, and $\frac{-\dot{w}}{\sqrt{w}} = \frac{\dot{A}}{2m\sqrt{2s}}$; therefore, taking the fluents, $2\sqrt{e} - 2\sqrt{w} = d\text{A} \div 2m\sqrt{2s}$; and, when the body has completed the next semi-vibration, or the string arrives again in a vertical position Cϕ, Cϕ being $= w$, $\sqrt{e} - \sqrt{w} = q$, $w = e - 2q\sqrt{e} + q^2 =$ Cϕ $= (\sqrt{e} - q)^2 = (\sqrt{a} - 2q)^2$. And now it is evident that, if the operation be continued, the same quantity q will always be the diff. of the square roots of the lengths of the string at the beginning and end of each semi-vibration, and consequently the length of the pendulum at the end of any number of semi-vibrations, will be $= (\sqrt{a} - cq)^2$, where c is $=$ the said number, therefore when the length of the string at the end of any time is given, the number of vibrations is given likewise. And when the body arrives at C, $\frac{1}{2}c = \sqrt{a} \div 2q = 2m\sqrt{2sa} \div dp$ the number of whole vibrations $=$ twice the number that would be performed in the same time $(ma \div d)$ by a pendulum whose invariable length is a. So that this method agrees with the common one when the arcs of vibration are indefinitely small, (similar they cannot be). For, if $u =$ any variable length of a pendulum, the time of its vibration is $= p\sqrt{u} \div \sqrt{2s}$, and as this time : 1 vibration :: $\dot{t} = -m\dot{u} \div d$: $-m\dot{u}\sqrt{2s} \div dp\sqrt{u} =$ the fluxion of the number of vibrations in the time $m(a - u) \div d$; the corrected fluent gives the vibra. performed in that time $= 2m\sqrt{2s}(\sqrt{a} - \sqrt{u}) \div dp = (\sqrt{a} - \sqrt{u}) \div 2q$; and if this $= \frac{1}{2}$ or a semi-vibration, then $\sqrt{a} - \sqrt{u} = q$, &c. as before.

Now, if the velocity be not uniform, but variable, and proportionable to u^n, as per question, it is evident, from what is done above, that $\dot{t} = \frac{1}{2}\sqrt{\frac{u}{2s}} \times \frac{-\dot{x}}{\sqrt{(bx - x^2)}} = \frac{-m\dot{u}}{du^n}$, or $\frac{\dot{A}}{2\sqrt{2s}} = \frac{m\dot{u}}{du^{n+\frac{1}{2}}}$ whence $\frac{d}{2m\sqrt{2s}}(p - \text{A}) = \frac{2}{1-2n}a^{\frac{1}{2}(1-2n)} \backsim \frac{2}{1-2n}u^{\frac{1}{2}(1-2n)}$, which at the end of the semi-vibration is $= \frac{dp}{2m\sqrt{2s}} = 2q$, at which time therefore, the shortening of the pendulum $= a - u = a - (a^{\frac{1}{2}(1-2n)} \pm (1-2n)q)^{2 \div (1-2n)}$. In like manner $w = (e^{\frac{1}{2}(1-2n)} \pm (1-2n)q)^{2 \div (1-2n)} = (a^{\frac{1}{2}(1-2n)} \pm (1-2n)2q)^{2 \div (1-2n)}$, because $e =$ the value of u when $\text{A} = 0$; therefore at the end of c semi-vibrations the length of the pendulum will be $(a^{\frac{1}{2}(1-2n)} \pm (1

$-2n)cq)^{(2\div 1-2n)}$. And when this $= 0$, then $c = \pm \frac{a^{\frac{1}{2}(1-2n)}}{(1-2n)q}$.

By the common method, $\frac{\mp \dot{u}}{4qu^{n+\frac{1}{2}}}$ = the fluxion of the number of vibrations up or down, whence the fluent $= \frac{1}{2q\,.\,(1-2n)} \times :$ $u^{\frac{1}{2}(1-2n)} \backsim a^{\frac{1}{2}(1-2n)}$ = the number of vibrations. And when $u = 0$, the number of vibrations is either infinite or $\pm a^{\frac{1}{2}(1-2n)} \div 2q\,(1-2n)$ ($= \frac{1}{2}c$ as before) according as the index $\frac{1}{2}(1-2n)$ is negative or positive: The number of vibrations performed in the descent being always = those performed in the ascent through the same line, and with the same velocity at every point. I shall only add, that if the arcs of vibration be not indefinitely small, the common method fails, and the solution must be wrought through with the first value of $\dot{x}$ here given.

THE PRIZE QUESTION, *by the Rev. Mr.* Wildbore.

To determine the position of the asymptote of the curve which is the locus of all the angular points formed by drawing three lines, from three given poles, so that the angle formed by the same corresponding two may always be bisected by the third.

Answered by C. Bumpkin.

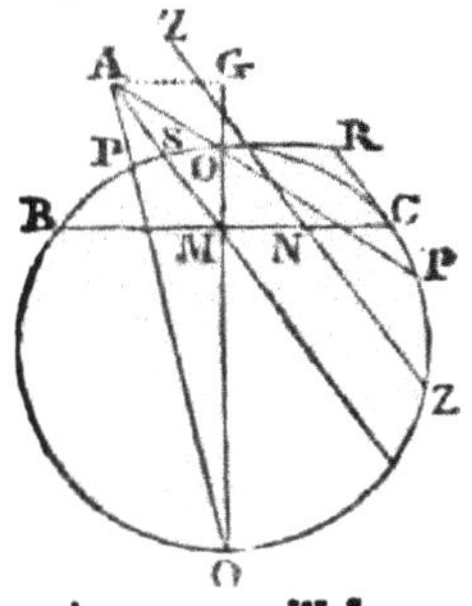

Let the poles be A, B, C; and let the ∠ formed by the lines from B and C be bisected by a line from A. Through B and C describe the circle BOCoB, and take BO = CO. If then the right line AoP be drawn intersecting the said circle in o and P, the point P will be in the curve to which the asymp. is required to be drawn. It is obvious therefore from the generation of the curve, that if BC be bisected in M, and AM be drawn, this line AM will be ∥ to the asymptote sought. Draw PR ∥ AM; and let SoR, ∥ BC, meet PR in R and AM in S; draw also MoG, meeting AG, ∥ BC, in G. Call BM (= CM) b; MG, d; AG, e; and Mo, y. The diameter OMo will be $= b^2y^{-1} + y$; Ao $= \sqrt{((d-y)^2 + e^2)}$: and, by sim. △s, $d : e :: y : (ey \div d =)$ oS; Ao : oG :: oO : oP $= (d-y)\,.\,(b^2 + y^2) \div y\sqrt{((d-y)^2 + e^2)}$; and Ao : oS :: oP : oR $= e(d-y)(b^2 + y^2) \div d((d-y)^2 + e^2)$. Which expression, when $y = 0$, becomes $= b^2e \div (d^2 + e^2)$.——To this last quantity make MN equal, (N being in MC); then will ZNZ, ∥ AM, be the required asymptote.

The same answered by Plus Minus.

Let A, B, and C be the three given points (A being that from which the bisecting line is to be drawn); draw CB, producing it indefinitely, and draw HAF ∥ to it; bisect CB in D, and draw DE and AI ⊥ to it. Let M be a point of the curve, from which draw MPQ ⊥ BC; and draw MC, MA, MB. Let DI (= EA) = a, CD (= DB) = d, ED = f, AP = x, and PM = y. Then, by sim. △s, we have, MQ $(y+f)$: QB $(d-x-a)$ :: PM (y) : $y(d-x-a)\div(y+f)$ = PN; therefore AN = $(dy+fx-ay)\div(y+f)$. Again, MQ $(y+f)$: QC $(d+x+a)$ :: PM (y) : $y(d+x+a)\div(y+f)$ = PO; therefore AO = $(dy-fx+ay)\div(y+f)$. But, by the nature of the question, AN : AO $(dy+fx-ay : dy-fx+ay)$ :: BM : CM $(\surd((y+f)^2+(d-x-a)^2) : \surd((y+f)^2+(d+x+a)^2))$; this proportion furnishes the equation, which being divided by $y+f$ and brought to order, is $ay^3+(af-fx)y^2+(ax^2-f^2x+a^2x-d^2x)y=fx^3+afx^2$. Equate all the highest terms, viz. $ay^3-fxy^2+ax^2y-fx^3$, to 0, and the roots of that equation will be $y=fx\div a$, $y=+\surd-x^2$, and $y=-\surd-x^2$, the two last, being impossible, shew that the curve has only one asymptote; and the other, that if you set off AF = AE the other way from A, and erect the ⊥ FG = ED the contrary way to ED, and join A, G; or, barely joining D, A; this line DG will be ∥ to the only possible asymptote.—Transform the original equation by putting $fv\div r$ for y, and $z+av\div r$ for x, (where $r=\surd(f^2+a^2)$) and you will get the following equation not having the cube of the ordinate, viz. $zv^2+(ad^2\div rr)v^2+(2a\div r)z^2v+((f^2+d^2+a^2)\div r)zv=-z^3-az^2$. From this equation take the highest powers of v (viz. $zv^2+(ad^2\div rr)v^2$), equate them to 0, and you will have $z=-ad^2\div rr=-ad^2\div(ff+aa)$; therefore, take AH, towards E, $=add\div(ff+aa)$, through H draw an indefinite right line ∥ DG, and it will be the asymptote required.

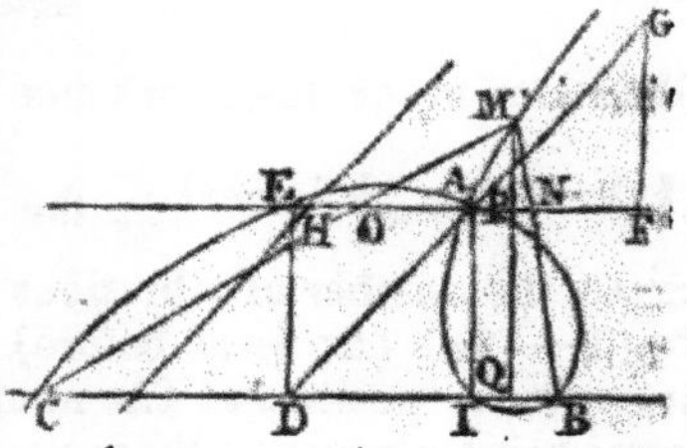

Remark 1. This curve, so long as it continues to be of the second order, will always be a *defective nodated hyperbola*; having its double point in A, and passing through the points B, I, C, E.

2. If $d=(aa+ff)\div\surd(aa-ff)$, it will have a diameter (i. e a diameter of the second order) otherwise not.

3. If either f or a be = 0, the locus becomes a circle with a right-line.

4. If the point M be any where in the portion of the curve BAEC, the line AM will not bisect the ∠CMB, but its comp. to two right ones; in all other points of the curve, it bisects CMB.

Questions proposed in 1775, *and answered in* 1776.

I. QUESTION 682, *by Mr.* Stephen Williams, *of Truro.*

Ingenious Artists, pray declare,
From what is here subjoin'd, *
The age and fortune of a fair,
Who's virtuous, loving, kind.

$$\left.\begin{array}{l} {}^{*}x^3 + x^2y + xy^2 + y^3 = a = 94921632, \\ 2xy + 2y^2 + x - y = b = 17280. \end{array}\right\}$$ Where x denotes her fortune in pounds, and y her age in years.

Answered by Mr. Rob. Abbatt, *Jun.*

From the second equation $x = (b + y - 2yy) \div (2y + 1)$; which being written for it in the first equation the value of y is found $= 18$, her age, from the resulting equation, and then $x = 450$, her fortune. Or, from the consideration of the above expression for x being a whole number, y and x are easily found by a few trials.

II. QUESTION 683, *by Mr.* Wm. Spicer.

Two men, Charles and George, and their wives Frances and Susannah, went to the market to buy sheep: Cha. bought 8 more than Geo. and Geo. bought 16 more than Frances; also the number of sheep the two women bought was 72; moreover, one of the women bought 32 sheep more than her husband; now, the sum of the squares of the number of sheep that each person bought, being a minimum, it is required to find what number each person bought, together with the name of each man's wife?

Answered by Mr. Geo. Perrott.

Put x = number of sheep bought by Frances: then, by the question, $x + 16$ = George's, $x + 24$ = Charles's, and $2 - x$ = Susannah's; also the sum of their squares $4xx - 64x + 6016$, or $xx - 16x$ = a minimum. This put in fluxions, &c. we have $x = 8$. Hence the sheep they severally bought, are for Charles 32, George 24, Frances 8, and Susannah 64. Consequently, Charles and Susannah are man and wife, also George and Frances.

III. QUESTION 684, *by Mr.* John Shadgett.

A gentleman has a garden 60 yards square, which is divided into two equal parts by a walk parallel to the side of it, the breadth of the walk being 8 feet: now in the one part is a circular fountain, and in the other an elliptic one, the transverse and conjugate of the ellipse and the diameter of the circle being in geometrical progression, of which the ratio is that of 3 to 2; also the remaining area of the part in which is the circular fountain is to the remaining area of the other part, as 2 is to 1. Required the diameter of the circle and the axes of the ellipse?

Answered by Mr. John Shadgett.

Put $a = 1720$ the whole area on each side of the walk, $n = \cdot 7854$, and $4x =$ diameter of the circle; then $6x$ and $9x$ are the axes of the ellipse. Hence $16nxx =$ area of the circle, and $54nxx =$ area of the ellipse; and, by the question, $a - 16nxx = 2a - 108nxx$; hence $x = \sqrt{(a \div 92n)} = 4\cdot 878933$. Consequently the diameter and axes are 19·51573, 29·2736, and 43·9104. Hence it may be observed that the numbers were not quite consistently proposed; because the conjugate axe is a little longer than the breadth of the piece.

IV. QUESTION 685, *by Mr.* Wm. King.

Required the dimensions and solidity of the least cone that will circumscribe the segment of a sphere, whose chord is 39 inches, and the axe of the whole sphere and the versed sine of the half segment, are in the ratio of 1·5686 to 1?

Answered by Mr. Wm. Cole.

By the nature of the circle $2\sqrt{(1 \times \cdot 5686)} = 2\sqrt{\cdot 5686} = 1\cdot 50812$ = the chord of a sim. segment of a circle whose diameter is 1·5686; hence as 1·50812 : 39 :: 1·5686 : 40·567 (= the axis 2OE) :: ·2157 ($= 1 - \frac{1}{2} \times 1\cdot 5686$) : 5·5785 = OF. Now, to find the least cone, put $a =$ OF, $r =$ OE $=$ OD, and $x =$ OB; then compute the expression for the solidity of the cone, and its fluxion made = 0, will determine x, &c. Or, by Simpson's Geom. p. 209, because BH = 2HF, we have HO = BO — BH $= \frac{1}{3}x - \frac{2}{3}a$; but DO^2 = BO × OH, or $rr = \frac{1}{3}x\,(x - 2a)$; hence $x = \sqrt{(3rr + aa)} + a$, and consequently the altitude of the cone BF $= \sqrt{(3rr + aa)} + 2a = 46\cdot 7285$. Hence the other dimensions are easily found.

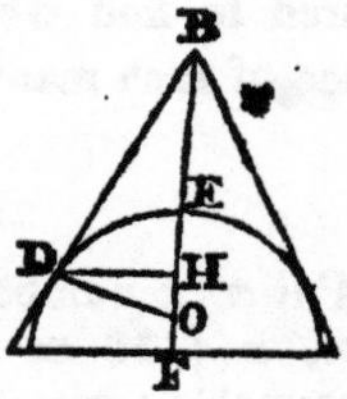

Cor. 1. When a is negative, or the segment is less than a hemisphere, then will $x = \sqrt{(3rr + aa)} - a$, and the altitude of the cone $= \sqrt{(3rr + aa)} - 2a$.

2. When $a = 0$, or when the segment becomes a hemisphere, then will $x = r\sqrt{3} =$ the altitude of the cone.

3. When $a = r$, or the segment becomes a whole sphere, then

will $x = 3r$, and the altitude of the cone $= 4r =$ double the axis of the sphere.

V. QUESTION 686, *by Mr.* Edw. Reed, *of West Alvington.*

Two circum-navigators depart from the same place at the same time, the one east, the other west, on the same parallel, till they meet: It is required to find their latitude, and the longitude each has made, the one of them sailing 1 mile (or minute) the first day, 2 the second, 3 the third, &c. and the other 1 mile the first day, 4 the second, 16 the third, &c. also the difference of their distances or longitudes was 1000 geographical miles or minutes?

Answered by the Proposer Mr. Edw. Reed.

Put x for the number of days sailed, and $a = 1000$ miles. Then, by arithmetical progression, $1 + 2 + 3$, &c. to x terms $= \frac{1}{2}x(x+1)$; and, by geometrical progression, $1 + 4 + 16$, &c. to x terms $= \frac{1}{3}(4^x - 1)$; hence, by the question, $\frac{1}{3}(4^x - 1) - \frac{1}{2}x(x + 1) = a$, or $4^x - \frac{3}{2}x(x + 1) = 3a + 1$; from which is found $x = 5{\cdot}789649$. Then the two distances become 19·65483, and 1019·65483, and their sum or the whole circumf. =1039·30966; which being divided by 21600 (the equator), the quotient ·0481161 is the cosine of 87° 14′ 31″ 31‴ the required latitude. Moreover 19·65483 miles in that latitude are = 6° 48′ 29″ the longitude where they will meet; found by this proportion, as 1039·30966 : 19·65483 :: 360° : 6° 48′ 29″.

VI. QUESTION 687, *by Mr.* Wm. Cole, *of Colchester.*

Being in a garden on the 10th of April at 9 o'Clock in the morning, I saw a tower at a small distance, whose shadow projected into the garden, and fixed my staff upright at the extremity of the shadow; and 2 hours afterwards, being in the garden again, I measured from the extremity of the shadow to my staff 40 yards, at which time I observed that the shadow of my staff was equal to its length. Whence I would know the height of the tower?

Answered by Mr. James Young.

The declination at the two times were 8° 2′ 14″ and 8° 4′ 4″, found by proportioning for the given hours from noon. Now, because the length of the shadow was equal to the length of the object at the second observation, therefore the sun's altitude then was 45°. Consequently there are given the declination, altitude and hour; to find the latitude = 51° 20′, and the sun's azimuth from the south at the second observation = 21° 14′ 52″. Also, there are now

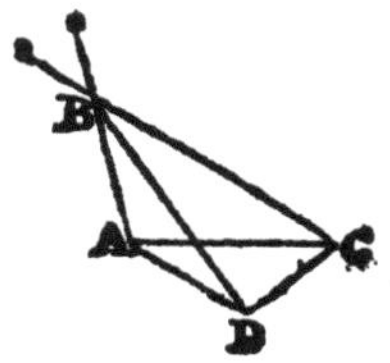

given the latitude, declination, and hour at the first observation; to find the altitude 33° 8′ 7″ and azimuth 56° 44′ 1″ at that time. Hence the difference of azimuths is = 35° 29′ 9″ = the ∠DAC made by the shadows AC, AD on the horizon. But the lengths of those shadows are to each other as the co-tangents of the respective altitudes, that is, as the cot. of 33° 8′ 7″ to the cot. of 45°, or as 1·5319344 to 1; then as 2·5319344 (the sum of the terms of the ratio) : ·5319344 (their difference) : : cot. 17° 44′ 34″ ($\frac{1}{2}$∠DAC) : tang. 33° 17′ 20″ = $\frac{1}{2}$ the difference of the angles D and C, which taken from their half sum leaves the ∠ACD = 38° 58′ 5″; and hence as s. ∠DAC : s. ∠DCA : : DC = 40 : DA = AB = 43·334 yards, or 130 feet nearly = the height of the tower required.

VII. QUESTION 688, *by Mr.* Geo. Beck, *of Coventry.*

Given the base, the sum or difference of the sides, and the ratio of the difference of the segments of the base to the perpendicular; to construct the triangle.

Answered by the Proposer Mr. Geo. Beck.

Construction. To the given base AB make AC ⊥, and in the given ratio of the difference of the segments to the perpendicular of the triangle, and draw CS bisecting the base in S: from A and B draw (by prob. 15, page 223, Simpson's Geom.) AD, BD meeting in CS, so that BD ± AD = the given sum or difference, and ADB will be the required △.

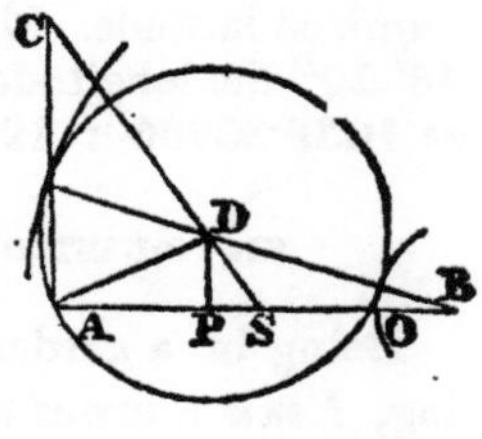

Demonstration. Demit the ⊥ DP, and take PO = PA. Then SP = SA — AP = $\frac{1}{2}$BA — $\frac{1}{2}$AO = $\frac{1}{2}$BO, or 2SP = BO the difference of the segments BP, PA; and, by sim. triangles, 2SP (BO) : PD : : 2SA (BA) : AC, in the given ratio by construction. Also, by construction, AB is the given base, and the sum or difference of AD, DB is = the given sum or difference.

Scholium. The point D is evidently the centre of a circle passing through A, and touching another circle whose centre is B and radius = the given sum or difference.

VIII. QUESTION 689, *by Mr.* Henry Clarke, *of Salford.*

In the mound of an elliptical garden, whose transverse is to the conjugate as 3 to 2, a pedestal, 50 yards high, is so placed, as that the apparent magnitude of an Herculean figure, 10 feet high, on the top of it, is the greatest possible to an eye situated on the transverse and 20 yards from the centre of the ellipse. Required the area f the garden?

Answered by Mr. Alex. Rowe.

Let ADB represent half the ellipse, E the place of the eye, FG the pedestal, and GH the statue. Draw EF, EG and EH; and put $a =$ FG $= 150$ feet, $b =$ FH $= 160$, and $x =$ EF: then $x \div a =$ tang. ∠G, and $x \div b =$ tang. ∠H; ∴ $(a - b)\ x \div (ab + xx) =$ the tang. of their diff. or the ∠GEH, which is a maximum by the question; the flux. of which being made $= 0$, we obtain $x = \sqrt{ab} =$ EF the dist. of the eye from the foot of the pedestal.

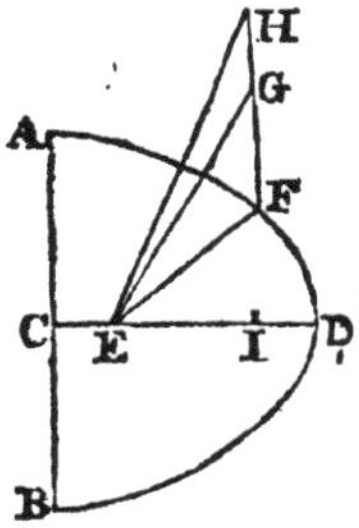

Let now I denote the focus; and put $c =$ CE $= 60$ feet, and $z = 2$CD the trans. consequently $\frac{2}{3}z =$ AB; also CI$^2 =$ CD$^2 -$ CA$^2 = \frac{5}{36}zz$, and 2CD : AB :: AB : $\frac{4}{9}z =$ the latus rectum; now EF must be ⊥ to the curve, because it must be at the shortest dist. from FH; hence, by Cor. 1, prop. 26, Emerson's Conics, we have $z - \frac{4}{9}z : \frac{4}{9}z$ :: CI$^2 -$ CE^2 : EF^2, or $5 : 4 :: \frac{5}{36}zz - cc : ab$, from which proportion z is found $= 3\sqrt{(ab + \frac{4}{5}cc)} =$ trans. Consequently the area is $= 6 \times \cdot 78539$, &c. $\times (ab + \frac{4}{5}cc) = 126669$ sq. feet $= 14074\frac{1}{3}$ sq. yards $= 2\cdot 40792$ acres.

IX. QUESTION 690, *by Mr.* Wm. Wilkin.

Required the nature, area, and rectification of the curve, of which the rectangle of the abscissa and ordinate is every where equal to the rectangle under the constant quantity a and the sum of the tangent and subtangent?

Answered by Mr. John Gould.

If x, y, and z denote the abscissa, ordinate, and curve respectively, then is $y\dot{x} \div \dot{y}$ the subtang. and $y\dot{z} \div \dot{y}$ the tang. Hence, by the question $xy = ay\,(\dot{x} + \dot{z}) \div \dot{y}$. But $\dot{z} = \sqrt{(\dot{x}\dot{x} + \dot{y}\dot{y})}$; ∴ $xy = ay\,(\dot{x} + \sqrt{(\dot{x}\dot{x} + \dot{y}\dot{y})}) \div \dot{y}$; from which equation we obtain $\dot{y} = 2ax\dot{x} \div (xx - aa)$; and the correct equation of the fluents is $y = a \times$ hyp. log. of $((xx - aa) \div (bb - aa))$, where b is the value of x when $y = 0$. Moreover the flux. of the curve is readily found $= \dot{x} + 2aa\dot{x} \div (xx - aa)$; the fluent of which is $x + a \times$ hyp. log. $((x - a) \div (x + a))$; which, when corrected, is $(x - b) + a \times$ hyp. log. $((x - a)\,(b + a) \div (x + a)\,(b - a))$.

X. QUESTION 691, *by Mr.* G. Cetii.

At the end of Ch. 2, part 3, Ward's Introduc. to the Mathematics, he has given two problems, viz. the 19th and 20th, for the geom. description of a regular pentagon, which have long been received and used without any examination or proof of their truth, that I know of:

It is required therefore to demonstrate or investigate the truth or falsity of the same?

Answered by Mr. H. Clarke.

In the 1st case. The constr. being made according to Ward's 19th prob. it is evident, from the like situation of the circles, &c. that wherever the point E is taken in GK, the lines AB, FC will be always parallel to each other, and the ∠FCB = CFA, as also FAB = CBA. The sides CB, BA, and AF are likewise equal, being each a radius of the same or equal circles. What remains then to be proved, is whether the ∠FCB be = the greater angle, in a regular pentagon, formed by the diag. and a side. The most direct (if not the only) method of investigating, which appears to be by a calculation of that angle.

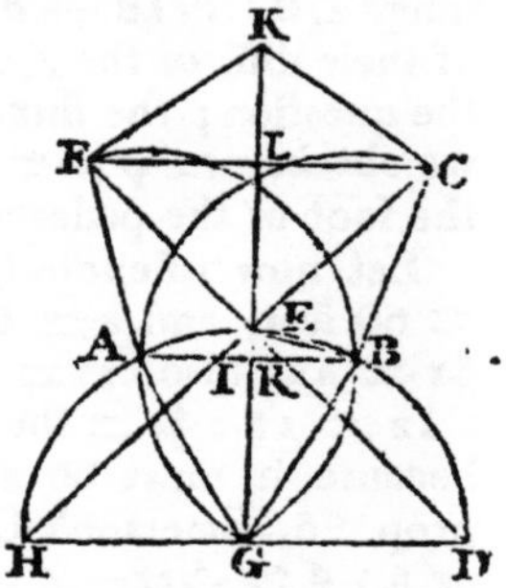

Having drawn such lines as appear by the fig. AB (= BG) will be the side of a regular hexagon, and BE the side of a regular duodecagon inscribed in the same circle, and they are therefore in the ratio of 1 to $\sqrt{(2 - \sqrt{3})}$. And since BI is ⊥ GK and LC ∥ BI, the ∠ILC is a right ∠; and the ∠LEC = HEG is evidently ½ a right ∠. Also, as BE is the side of a duodecagon, the ∠BGE = ⅙ of a right ∠, and consequently the ∠BEG = ⅚ of a right ∠; but the ∠CEB is the supplement of LEC + BEG, therefore the ∠BEC = ⅔ of a right ∠ = 60°. Then, as the sines of the ∠s BEC, BCE are in the ratio of their opposite sides BC, BE, we have as $1 : \sqrt{(2 - \sqrt{3})}$ :: s. 60° : s. 26° 38′ 2″ nearly; hence the ∠FCB = FCE + ECB = 71° 38′. 2″. But the ∠FCB in a regular pentagon is 72°. Therefore this construction is false.

Again it is evident that the construc. of the 20th prob. is on the same principle with the globular projection of the sphere, DA being the plane of projection, and CE the distance of the eye. But as it is obvious, from the principles of this projection, that if the semicircle DBA be divided into any number of equal parts, there will be required a different distance of the eye to project each point at equal distances on the line DA, it can only be accidental that the line CE, found by this construction, is nearly the true dist. to project the point B in F.

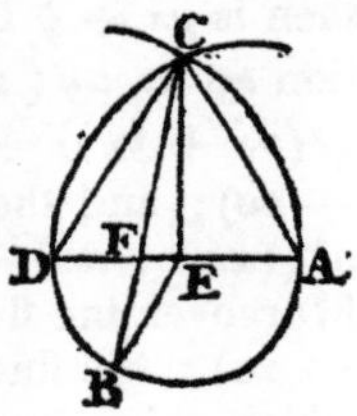

This will farther appear from the following calculation.——Let the diam. DA be supposed to be divided into 180 equal parts, answering to the degrees in the semicircle, of which let the arc DB be ⅖ or 72°, the arc subtended by the side of a regular pentagon; also take DF = 72 = ⅖ of DA. Put r = rad. (= 100 suppose), and s and c = the sine and cosine of the arc DB: Then, by a well-known process, we have $EC = EF \times s \times r \div (DE \times c - EF \times r) = 18 \times 9510{\cdot}565 \div (90 \times 30{\cdot}9017 - 1800) = 174{\cdot}478$, the true dist. to project B and F. But as DEC is a right-angled Δ, $\sqrt{(DC^2 - DE^2)} = \sqrt{(200^2 - 100^2)}$

$= 100\sqrt{3} = 173{\cdot}2 =$ BC the dist. by this construction. Which is therefore false also, but may, as well as the former, serve for common uses, where the rad. is not very large. But, however, it was wrong in Mr. Ward to call them true constructions, as it has been in all his editors since to continue them as true ones.

Remark. The falsity of this latter construction, with other curious properties related to it, are demonstrated in Art. 48 of Professor Hutton's Math. Miscel. I believe this latter constr. was first given by Mr. Cha. Renaldine, as the former was by Mr. Albert Durer.

Mr. Wm. Wilkin *also, from the late Mr.* Coughron's *Papers,*

Brings out exactly the same conclusions, in nearly, the same manner; and also adds the following method to discover what figures the latter method of construction will truly determine. Put n for the number of sides of the polygon, $a =$ sine of $60°$ (rad. 1), $d =$ DA, s and c for the sine and cosine, of $360° \div n$; then $2a + s : a :: cd : acd \div (2a + s) =$ EF. But, by the above method, the same EF is $= \frac{1}{2}d - 2d \div n = d \times (n - 4) \div 2n$; therefore $ac \div (2a + s)$ must be $= (n - 4) \div 2n$ when this construction is true; and if they are not equal, the construction is false. By this method we readily find that the construction is true for the trigon, tetragon, and hexagon, but false for the octagon.

XI. QUESTION 692, *by Mr.* D. Cunningham, *of Alnwick.*

Let any right-line AB of a given length (10) be moved round, from a horizontal position on its end A, with an uniform angular motion, while the same end A is carried uniformly forward in the direction AB: It is required to determine the nature and area of the curve described by the end B, supposing the end A to have moved over twice AB or QQ while 180 degrees is described by the angular motion; when AB will again have come into a horizontal position, and the point B will have arrived at the place from whence it first set out.

Answered by Mr. John Gould.

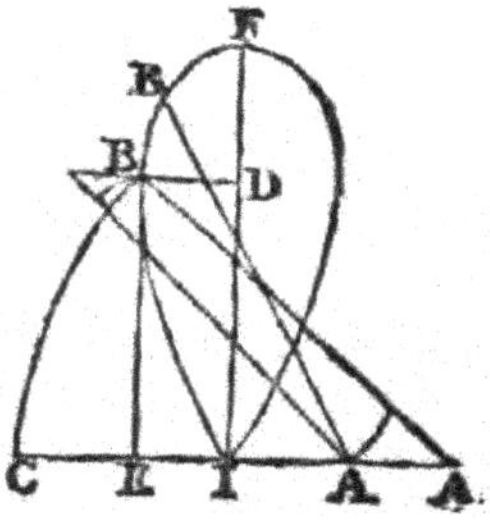

Put AB $= a$; the velocity on the base to the angular velocity as 1 to c; v the space passed over by A; x the sine $\angle$BAC, rad. a; and z the arc corresponding. Then we have $1 : c :: \dot{v} : c\dot{v} = \dot{z}$, and the flux. of the decrease of the area, considering the line AB as keeping its inclination, is $x\dot{v} = x\dot{z} \div c = ax\dot{x} \div c\sqrt{(aa - xx)}$; also the fluxion of the increase of the area by the angular motion is $\frac{1}{2}a\dot{z} = \frac{1}{2}aa\dot{x} \div \sqrt{(aa - xx)}$; consequently

$\frac{1}{2}a a\dot{x} \div \sqrt{(aa - xx)} - a x\dot{x} \div c\sqrt{(aa - xx)}$ is the flux. of the area of the curve ; and the fluent is $= (a\sqrt{(aa - xx)} - aa) \div c +$ the sector whose radius is a and sine x. The curve is part of *a curtate cycloid.*

XII. QUESTION 693, *by Mr.* W. Sewell.

To determine the area of a curve whose equation is $y =$ l. 2 + l. $4a$ + l. $8b$ + l. $16c$, &c. to x terms : where l. denotes the hyp. log. and a, b, c, &c. are the natural values of the preceding terms, viz. $a = 2$, $b = 4a$, $c = 8b$, &c.

Answered by Mr. Stephen Williams, *of Truro.*

There is given $y =$ l. 2 + l. 2^3 + l. 2^6 &c....(x) = l. 2 + 3 l. 2 + 6 l. 2 + 10 l. 2 &c....(x) = l. 2 × : 1 + 3 + 6 + 10 &c.... $(x) = (x^3 + 3x^2 + 2x) \times \frac{1}{6}$ l. 2 ; hence $y\dot{x} = (x^3 + 3x^2 + 2x) \times \frac{1}{6}\dot{x}$ × l. 2, whose fluent $(x^4 + 4x^3 + 4x^2) \times \frac{1}{24}$ l. 2 is the area of the curve, as required.

XIII. QUESTION 694, *by Mr.* Cullen O'Conner.

Walking, one day, and carelessly moving my cane up-and-down in my hands behind my back, the shadow of the curve described by the end of it, on a wall by the side of which I was walking, drew my attention ; and having amused myself with observing the variations of curvature according to several different degrees of velocity with which I walked forward, I had the curiosity once to measure as accurately as I could, the space I moved while the cane vibrated freely in my hand, by its own weight, from a horizontal position, in which it was held directly out from behind my back, till it struck my heel in a vertical position, and found I had moved just 3 feet horizontally forward. Now, the length of the cane being 4 feet, at what rate per hour did I move, and what is the nature and quadrature of the curve described by the end of the cane while it so vibrated from the horizontal to the vertical position ?

Answered by Mr. O'Conner, *the Proposer.*

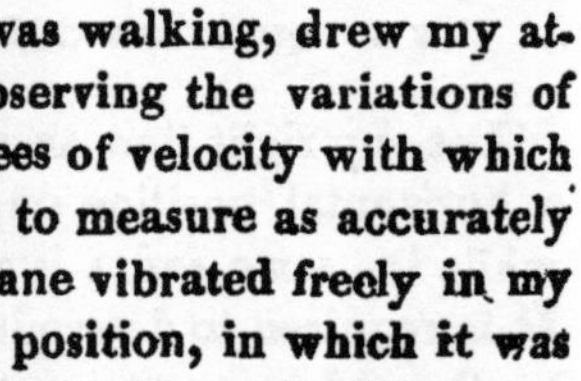

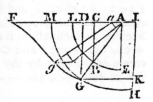

I. *To find the time of the vibration of any pendulum in a whole quadrant.* Put $p =$ AB = AE the length of any pendulum from its axis to its centre of oscil. $c = 16\frac{1}{12}$, $v =$ BC any ⊥ dist. descended when the pendulum is in the position AB, and $z =$ the corresponding arc LB ; also $t =$ the time of describing LB. Now $2\sqrt{cv} =$ velocity acquired in falling thro' CB = the velocity of B in the arc at

B; but $\dot{t} = \frac{\dot{z}}{2\sqrt{cv}}$, and $\dot{z} = \frac{p\dot{v}}{\sqrt{(pp - vv)}}$, $\therefore \dot{t} = \frac{p\dot{v}}{2\sqrt{(cv)}.\sqrt{(pp\ vv)}}$ the whole fluent of which, when B arrives at E, is (by pa. 142 of Mr. Landen's Lucub.) $\frac{1}{2}\sqrt{(p \div c)} \times (e + \sqrt{(ee - 2f)})$, where $e = \frac{1}{4}$ of the periphery of an ellipse whose semi-axes are 1 and $\sqrt{2}$, and $f = \frac{1}{4}$ of the periph. of the circle whose radius is 1.

Now $f = \frac{1}{2}$ of $3·1416 = 1·5708$, and (by Rule 6, p. 233, of Mr. Hutton's Mensur.) e is found $= 1·9100$; $\therefore e + \sqrt{(ee - 2f)} = 2·6217$; also, as the centre of oscil. of the stick is $\frac{2}{3}$ of its length from the axis, therefore in this case $p = \frac{2}{3}$ of $4 = \frac{8}{3}$: hence then $\frac{1}{2}\sqrt{(p \div c)} \times (e + \sqrt{(ee - 2f)})$ is $= ·2035947 \times 2·6217 = ·533764'' = 32''' \ 1^{\text{iv}} \ 33^{\text{v}}$ the whole time of the pendulum's vibration.

But in this time I walked 3 feet, $\therefore$ $·533764'' : 1'' :: 3 : 5·62046$ feet, the constant horiz. velocity in a second, which is at the rate of 3·83213 miles an hour.

II. *To find the nature, &c. of the curve* FGH *described by the compound motion.* Put $a = 4 =$ AG $=$ IH, $b = 5·62046$ the const. veloc. with which A is moved along FI, $c = 16\frac{1}{12}$, $x =$ abs. FD, and $y =$ ord. DG. Then, since the velocity of B in the arc at B is $2\sqrt{(c \times \text{BC})}$, the velocity of G (AG being $= \frac{3}{2}$ AB) is $\frac{3}{2} \times 2\sqrt{(c \times \text{BC})} = 3\sqrt{(c \times \frac{2}{3}\text{DG})} = \sqrt{6cy}$; and hence $a : y :: \sqrt{6cy}$ (veloc. ⊥ AG) : $(y \div a)\sqrt{6cy}$ = the velocity of G in the horizontal direction GK arising from the vibration; $\therefore b + (y \div)a\sqrt{6cy} =$ the velocity of D along FI: then $1''$ time : $b + (y \div a)\sqrt{6cy}$ velocity :: $\dot{t}$ time $\left(= \frac{\text{B}\dot{\text{L}}}{\text{velocity of B}} = \frac{\text{B}\dot{\text{L}}}{2\sqrt{(c \times \text{BC})}} = \frac{a\dot{y}}{\sqrt{(cy)}.\sqrt{(aa - yy)}}\right) : \frac{y\dot{y}}{\sqrt{(aa - yy)}} + \frac{ab}{\sqrt{6c}} \times \frac{y^{-\frac{1}{2}}\dot{y}}{\sqrt{(aa - yy)}} = \dot{x}$; and the fluents being taken, the resulting equation will express the relation between the abscissa and ⊥ ord. But the flu. of the first term is evidently $-\sqrt{(aa - yy)}$, and that of the latter will be expressed by elliptic and hyperbolic arcs, as found by Mr. Maclaurin in art. 802 of his fluxions.

Lastly, the flux. of the area FDG, or $y\dot{x}$, is $\frac{y^2\dot{y}}{\sqrt{(aa - yy)}} + \frac{ab}{\sqrt{6c}} \times \frac{y^{\frac{1}{2}}\dot{y}}{\sqrt{(aa - yy)}}$. Now the fluent of the former of these terms is MGD the cir. seg. whose sine is y and rad. a, and the fluent of the latter term is universally expressed by hyp. arcs, as in art. 799 of Mr. Macluarin's Fluxions. But when the stick arrives at the ⊥ position IH, the whole fluents (the latter term being then found by pa. 142 of Mr. Landen's Lucubr.) become $\frac{1}{2}aaf + ab\sqrt{(a \div 6c)} \times (e - \sqrt{(ee - 2f)})$, where the values of e and f are as above specified; which expression in numbers gives 18·0512 for the whole area of the curve.

Corol. 1. The point of contrary flexure in this curve will be found by putting equal to nothing either the value of $\ddot{x}$, viz. the fluxion of $\frac{y\dot{y}}{\sqrt{(aa-yy)}} + \frac{ab}{\sqrt{6c}} \times \frac{y^{-\frac{1}{2}}\dot{y}}{\sqrt{(aa-yy)}}$, when $\dot{y}$ is supposed constant, or else the fluxion of $\frac{ab + y\sqrt{6cy}}{\sqrt{(6cy)}\sqrt{(aa-yy)}}$, where the numerator and denominator express the ratio of the horizontal and vertical velocities of the moving point G; from either of which results this equation $2ay\sqrt{(6cy)} = aab + 3byy$; from which y is found $= \cdot 963$ nearly at that point.

Cor. 2. By art. 460 of Mr. Simpson's Flux. the whole fluent of $\dot{t}$, or of $\frac{p\dot{v}}{2\sqrt{(cv)}\sqrt{(pp-vv)}}$ is $= f\sqrt{\frac{p}{2c}} \times : 1 + \frac{1^2}{2^2 . 2} + \frac{1^2 . 3^2}{2^2 . 4^2 . 2^2} + \frac{1^2 . 3^2 . 5^2}{2^2 . 4^2 . 6^2 . 2^3} + \frac{1^2 . 3^2 . 5^2 . 7^2}{2^2 . 4^2 . 6^2 . 8^2 . 2^4}$, &c. which flu. was also found above $= \frac{1}{2}\sqrt{(p \div c)} \times (e + \sqrt{(ee - 2f)})$; and therefore the sum of the infinite series $1 + \frac{1^2}{2^2 . 2} + \frac{1^2 . 3^2}{2^2 . 4^2 . 2^2}$ &c. is $= \frac{e + \sqrt{(ee - 2f)}}{f\sqrt{2}} = 1\cdot 1802$, &c.

The same answered by Mr. John Gould.

1st. The time of descent through a quad. of a circle whose rad. is r, being $= \sqrt{(r \div s)} \times (\frac{1}{2}e + \frac{1}{2}\sqrt{(ee - 2c)})$ by Mr. Landen's theor. p. 308, Philos. Trans. for the year 1771, where $c = \frac{1}{4}$ of the periphery of a circle whose rad. is 1, and $e = \frac{1}{4}$ of the periphery of an ellipsis whose semi-axes are $\sqrt{2}$ and 1. Therefore taking $r = \frac{2}{3}$ of 4 feet, we have 32‴ 1⁗ for the time the proposer was in moving 3 feet. Consequently the rate per second is 5·62 feet nearly; and the rate per hour 3·835 miles.

2ndly. Putting $a = 4$ feet, the length of the stick; $b = 5\cdot 62$, the rate per second found above; x (= DG) the sine of the angle made with the horizontal line by the stick after falling any space; z the arc (MG) corresponding to it, rad. being $a =$ AG; and v any space passed over by the end A carried towards I; then the velocity on the horizontal line AI to the angular velocity of the lower end G of the stick, will be as b to $\sqrt{(6sx)}$, s denoting $16\frac{1}{12}$ feet. Hence as $b : \sqrt{6sx} :: \dot{v} : \frac{\dot{v}\sqrt{(6sx)}}{b} = \dot{z} = \frac{a\dot{x}}{\sqrt{(aa-xx)}}$; whence $\dot{v} = \frac{ab\dot{x}}{\sqrt{(6sx)} . \sqrt{(aa-xx)}}$, and the fluent, taken by Mr. Maclaurin's theorem, pa. 655 of his Fluxions, will give the value of v when x is given less than a. It appears then that $x\dot{v}$, the fluxion of the increase of the area, considering the stick as keeping its inclination, is $= \frac{abx^{\frac{1}{2}}\dot{x}}{\sqrt{(6s)}\sqrt{(aa-xx)}}$; moreover, the fluxion of the increase of the area by the angular motion is $\frac{\frac{1}{2}aa\dot{x}}{\sqrt{(aa-xx)}}$; consequently $\frac{\frac{1}{2}aa\dot{x}}{\sqrt{(aa-xx)}} + \frac{abx^{\frac{1}{2}}\dot{x}}{\sqrt{(6s)}\sqrt{(aa-xx)}}$ = the fluxion of the area or sector AGF of the curve. The fluent of the first term is the sector of a circle whose radius is a and sine x; the fluent of the second term may be found when x is less than a, by Mr. Maclaurin's theorem, in his Fluxions, page 653; and, when x is $= a$, the fluent is $(ba^{\frac{3}{2}} \div \sqrt{6s}) \times (e - \sqrt{(ee-2c)})$ by Mr. Landen's Lucub. page 142. By means of which fluents, the whole area, when the stick becomes ⊥ to the horizon, is found = 18 feet, nearly.

XIV. QUESTION 695, *by the Rev. Mr.* Wildbore.

Given the difference of the segments of the base; and the difference of the angles at the base, to construct the triangle when the area is a maximum.

Answered by the Rev. Mr. Wildbore, *the Proposer.*

Construction. Upon IC, the given difference of the segments, describe a segment of a circle IABC to contain an angle = the given diff. bisect IC with the indefinite ⊥ DG passing through the centre of the circle; bisect GO in P, and make, by Mr. Simpson's Geom. v. 18, as OM : $\frac{1}{2}$OB :: OB : MP; draw AMB ∥ IC, and draw AC, BC; so is ACB the required △.

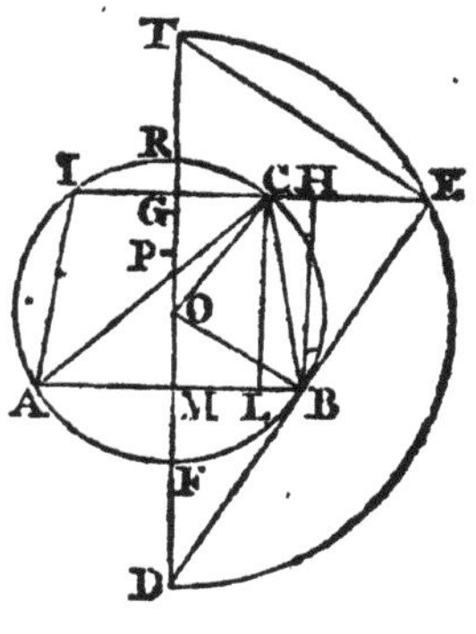

Demonstration. Let fall DH, CL, and join A, I. Then the ∠IAC = ABC (BAI) — BAC; and GC = ML = MB — LB (CH); therefore IC = AL — LB. Draw the tangent DBE: then since, by construction, OM : $\frac{1}{2}$OB :: OB : MP = PO + OM, OM : OB :: OB : 2MP :: (by sim. triangles) OB : OD, therefore, OD = 2MP = 2OP + 2OM = MG + OM, by construction; hence, DO — OM = DM = MG; and, DG being bisected in M, the tang. DE must be bisected in B, and consequently the rectangle MH = △ ABC is a maximum by theor. 8 of Mr. Simpson on the Maxima and Minima of geometrical quantities.

The same answered by Mr. George Sanderson.

Construction. Having bisected the given difference IC of the segments with the indef. ⊥ DOGT, make the ∠GCO = the comp. of the given difference of the ∠s at the base; find (by Mr. Simpson's Geom. v. 18) the rectangle GTO = $2OC^2$, and with the centre O describe the semicircle TLD meeting IC produced in E; draw TE, DE, as also OB ∥ TE meeting DE in B; with centre O describe a circle through B, and draw the chord AB ∥ IC; lastly, draw AC, BC, and ABC is the triangle required.

Demonstration. Since TD = 2TO, and TO × TG = $2OC^2$, by construction, therefore TD × TG = (by Euc. 6, 8 cor.) $TE^2 = 4OC^2$; hence TE = 2OC: But by sim. △s, TE = 2OB; therefore the circle with OB rad. will pass through C. But the ∠GOC is = the given one, by construction = ∠ABC — ∠BAC by the nature of all △s; and IC is evidently = the difference of the segments of the base. Now as DE is a tangent to the circle, and is bisected in the point of contact B, it appears by Mr. Simpson on the Maxima and Minima, theor. 8, that the rectangle MH is the greatest that can be applied in the ∠G, whose opposite ∠ shall be in the circumference CBF; but MH = △ABC, therefore ABC is a maximum.

THE PRIZE QUESTION, *by* Plus Minus.

A gentleman's house stands close to a road too narrow for carriages to turn in; so that, to widen it, he intends to make a sweep; but there is a pretty fir-tree in the present fence which he would save, because it is just fronting his door, and is besides in a line with the door and two other firs in the field over the way. He also intends to lay out a small spot in that field for a shrubbery. Now he is so whimsical, that he will have the fences both of sweep and shrubbery, such that three lines being drawn from any point of either to those three trees, that to the middlemost shall be a geometrical mean between the other two. The distance between the two trees next the house is 76, and between the other two 37 feet. But he will not put his project in execution till he can see the fences described, and whether they be to his liking. Pray, ladies, help us out at a dead lift?

Answered by the Proposer Plus Minus.

Let B, A, G be the three trees, of which B is next to the house; draw BAG, producing it as far as is necessary. Supposing M to be a point sought, draw MB, MA, MG, and let fall MP ⊥ BG. Let AG = a = 37, AB = b = 76, AP = x, and PM = y. Then, by the question, GM $(\sqrt{(yy + xx - 2ax + aa)})$ × BM $(\sqrt{(yy + xx}$

$+ 2bx + bb)) = $ AM$^2 (xx + yy)$. Hence $xy^2 + ((bb + aa) \div (2b - 2a))y^2 = -x^3 + x^2 (4ab - bb - aa) \div (2b - 2a) + abx - aabb \div (2b - 2a)$.——To describe the curve which is the locus of this equation set off from the vertex of the abscissa A, and towards B, AD $= (bb + aa) \div (2b - 2a)$, and through D draw an indef. ⊥ to AD for the asymptote. Set off also AK, towards B, and $= \frac{1}{4}b - \frac{1}{4}a$, then draw the indefinite line KC ‖ to the asymptote. Set off from K, both ways, KV, Kv $= \frac{1}{4}\sqrt{(aa + 6ab + bb)}$; make also KE = KD. Now draw any line at pleasure from the point E cutting KC in C; set one foot of your compasses in C, and extend the other to V or v, and about the centre C describe an arch cutting EC in $\mu\mu$, which will be two points of the curve required.

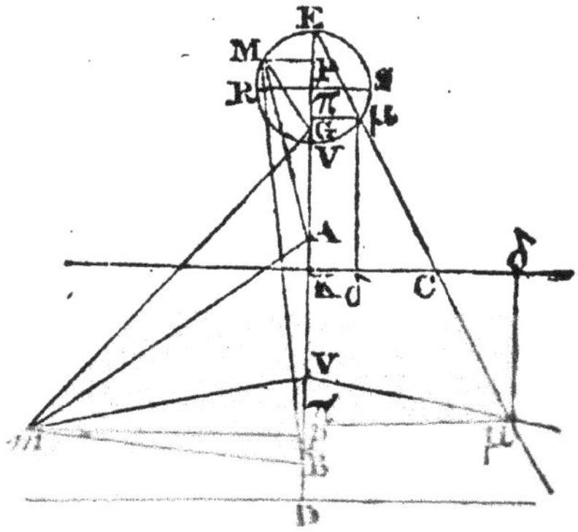

For, drawing the ordinate $\mu\pi$ also $\mu\delta$ ‖ AP, meeting KC in δ; we have Eπ $\left(\frac{ab}{b-a} - x, \text{ or } \frac{ab - bx + ax}{b - a}\right)$: $\pi\mu$ (y) :: EK$\left(\frac{ab}{b-a} + \frac{b-a}{4}, \text{ or } \frac{(b+a)^2}{4b - 4a}\right)$: $\frac{bby + 2aby + aay}{4ab - 4bx + 4ax}$ = KC. But Cδ^2 (= (KC $- \pi\mu)^2 = \left(\left(\frac{bby - 2aby + aay + 4bxy - 4axy}{4ab - 4bx + 4ax}\right)^2\right) + \delta\mu^2$ $\left(\left(\frac{4x + b - a}{4}\right)^2\right)$ = Cμ^2 = CV^2 = KC^2 + KV^2 $\left(\left(\frac{bby + 2aby + aay}{4ab + 4ax - 4bx}\right)^2 + \frac{aa + 6ab + bb}{16}\right)$; and this is the original equation, when brought to order; therefore μ is a point of the curve required.

The locus of the above equation will always be a *conchoidal defective hyperbola with an oval at its convexity, and having a diameter*, so long as it continues to be of the third order. But if $a = b$, it becomes a *conic equilateral hyperbola*, whose foci are B and G. If either a or $b = 0$, it becomes a *right line with a conjugate point.*

Scholium. By a method similar to the above you may describe the curve of the prize question solved in the last year's Diary; and indeed any *defective nodated hyperbola* may be described, the asymptote and three points (one of which is the double point) being given. Thus, through the double point A [See the figure to the second solution to the prize in last year's Diary] draw AG ‖ to the asymptote producing it indefinitely; let the other two given points be B and C; find a point Z in the line AG, which is equidistant from A and E, and draw EZ, producing it ad libitum; then find another point X of the line AG equidistant from C and A, and draw CX, producing it till it meet XZ; this intersection, for distinction sake, I call the *pole*. Now from the pole draw any line at pleasure cutting AG in some point R; then with the centre R and radius RA sweep an arch cutting the line from the pole in two points of the curve.

But note, that the two single points must not, with the *pole* hereafter mentioned, lie all in one right line. If these points be both in the cissoidal legs, there is no danger of this happening; for the pole is always a point of the knot.

The same universally for any three Points by the Rev. Mr. Wildbore.

Let T, S, F be *any* three given points, and A a point of intersection such, that $AS^2 = AT \times AF$. Join the three given points, let fall the $\perp$s Ss, AP, and find O the centre of the circle passing thro' T, S, F; draw $O\Pi \parallel TF$ meeting AP, produced if necessary, in Π, also the $\perp$ or bisecting TF; through S and r draw SL, to which draw the $\perp$ OL, and produce it till it meets a $\perp$ from A in g; and produce PA to C till the $\triangle$ $CO\Pi$ is similar to FSs.

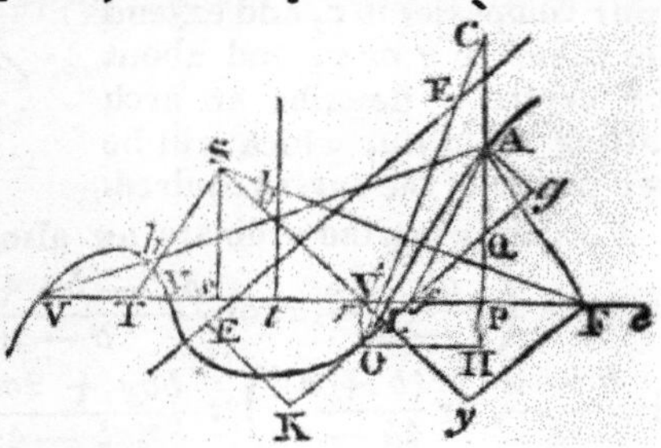

Then, since $AS^4 = AF^2 . AT^2 = AF^2 \times ((TF - FP)^2 + PA^2) = AF^2 \times (AF^2 + TF^2 - 2TF . FP)$, and $AS^2 = (sF - FP)^2 + (AP - Ss)^2 = AF^2 + Fs^2 + Ss^2 - 2Ss . AP - 2PF . sF$, the square of this must be $= AS^4$, and consequently $AF^2 \times (TF^2 - 2TF . FP) = AF^2 . FT . 2rP$ must be $= AS^4 - AF^4 = 2AF^2 \times (sF^2 + Ss^2 - 2Ss . AP - 2sF . FP) + (sF^2 + Ss^2 - 2Ss . AP - 2sF . FP)^2 = 4AF^2 . rF . rP$: Now, $4rF . rP - 2sF^2 - 2Ss^2 + 4Ss . AP + 4sF . (Fr - rP) = 4Ss . AP - 4sr . rP + 2Ts : sF - 2Ss^2 = 4Ss \times (AP - Q\Pi + ro) = 4Ss . AQ$, because $Ss : sr :: rP = O\Pi : \Pi Q$, by sim. $\triangle$s, and $Ss \times (2ro + Ss) = Ts . sF$ by the nature of the circle; hence we obtain $4Ss . AQ . AF^2 = (sF^2 + Ss^2 - 2Ss . AP - 2sF . FP)^2 = (2sF . rP - Ts . sF + Ss^2 - 2Ss . AP)^2 = 4Ss^2 \times (C\Pi - ro - AP)^2$, or $AF^2 . AQ = Ss . (C\Pi - \Pi A)^2$. Draw Af, AF then $Af^2 = AP^2 + Pf^2$, and AF^2 (Euc. 2, 13) $= Af^2 + fF^2 - 2fF . fP = Ag^2 + fg^2 + fF^2 - 2fF . fP$, and $Ss : Ag :: Sr : AQ$, consequently $(Sr \div Ss) \times (Ag^3 + fg^2 . Ag + fF^2 . Ag - 2fF . Ag . fQ = AF^2 . AQ$ (fQ being $= fg - Qg = fg - sr . Ag \div Ss$), and therefore $fP = \frac{fg . Ss}{Sr} - \frac{sr . Ag}{Sr}$, and, by sim. $\triangle$s, $Ss : rF :: O\Pi = rf + fP : CQ$,

and $CQ - AQ = C\Pi - A\Pi = \frac{rF . rf}{Ss} + \frac{fg . rF}{Sr} - \frac{rF . sr . Ag}{Ss . Sr} -$

$\frac{sr \cdot Ag}{ss}$; but, producing sr till it meets the $\perp$ Fy, $sy = \frac{rF \cdot sr}{sr} + sr$, consequently $CQ - AQ = \frac{rf \cdot rF}{ss} + \frac{rF \cdot fg}{sr} - \frac{sy \cdot Ag}{ss}$, and the general equation $AF^2 \cdot AQ = ss \cdot CA^2$ becomes $(sr \div ss) \times (Ag^3 + fg^2 \cdot Ag + fF^2 \cdot Ag) - 2fF \cdot Ag \times (fg - \frac{sr \cdot Ag}{ss}) = ss \times \left(\frac{rf \cdot rF}{ss} + \frac{rF \cdot fg}{sr} - \frac{sy \cdot Ag}{ss}\right)^2 = \frac{rf^2 \cdot rF^2}{ss} + \frac{2rF^2 \cdot rf \cdot fg}{sr} + \frac{rF^2 \cdot fg^2 \cdot ss}{sr^2} - 2sy \cdot Ag \times \left(\frac{rf \cdot rF}{ss} + \frac{rF \cdot fg}{ss}\right) + \frac{sy^2 \cdot Ag^2}{ss}$. This equat. by making $Ag = \frac{ss^2 \cdot rF^4}{rs^3} - x = EK - x$, and $fg = y - \frac{sy \cdot rF \cdot ss}{rs^2} + \frac{Ff \cdot ss}{sr} = y - \frac{rF^2 \cdot sr \cdot ss}{sr^3}(LK) - \frac{rF \cdot ss}{sr} + \frac{Ff \cdot ss}{sr} = y - LK - Lf = y - Kf$, will be reduced to the Newtonian form and determine the curve to be the pure anguineous defective hyperbola without oval, decussation, cusp, or conjugate point, and of the 37th species, having its only asymptote EE passing through E and $\parallel$ to Kg; and it will continue of this same species so long as the three given points are at the angles of a triangle, and TS and SF unequal. But, if they be equal, then $sr = 0$, $sr = ss$, $fF = rF$, $rf = 0$, and the general equation becomes $fF^2 \times Ag + Ag^3 + Ag \cdot fg^2 = rF^2 \cdot gf^2 \div sr + sr \cdot Ag^2$, which answers to the curves of the 45th species. If the given points be not at the angles of a triangle, but in a right line as in the question, then, S coinciding with s, $ss = 0$, $sr = sr$, and $2AF^2 \times (sF^2 - 2sF \cdot FP) + (SF^2 - 2SF \times FP)^2 = 4AF^2 \cdot rF \cdot rP$, or $sF^2 \times (sF - 2FP)^2 = AF^2 \times ((4rF - 4sF) \cdot rP + 4sF \cdot rF - 2sF^2) = AF^2 \times (2Ts \cdot sF - 4rs \cdot (rF - FP))$: take $rc = Ts \times SF \div 2rs$, and $FD = sF^2 \div 2rs$, then will $\frac{1}{2}FD \times (SF - 2FP)^2 = (AP^2 + PF^2) \times (rc - rF + FP) = (AP^2 + PF^2) \times PC$; or (bisecting FS in d) $2FD \times (dF - FP)^2 = 2FD \times (dc - CP)^2 = PC \times (AP^2 + (PC - CF)^2)$, which is an equation of the Newtonian form, and answers to the defective conchoidal hyperbola with an oval at the convexity, which is the 39th species, c being the centre and cv the diameter of the curve, and the asymptote a $\perp$ to cv at c, the points V′, V, v, where the conchoid and oval cross the diameter at right angles, being easily found by geometrical construction, so that $V'S^2 = TV' \cdot V'F$, $VS^2 = TV \times VF$, and $vS^2 = Tv \cdot vF$; and if VV′ be bisected in t, a $\perp$ to VV′ erected there, and any line applied from v to cross it in any point b, and thereon be taken both ways bA and bh each $= bV$, the point A will be in the fence, and h in the boundary of the shrubbery as is evident from Mr. Maclaurin's Geometria Organica, page 38, corol. 9; and thus may points be determined *ad libitum*. There is still one case remaining, and which is simpler than any of the foregoing, viz. when the the points are in the same right line, and TS = SF; and here the loca will be two equal and opposite equilateral hyperbolas to the common diameter VV′.

Scholium. I have been the more particular concerning these curves, because there seems some probability of their being the long-lost *Loca ad Medietates* of Eratosthenes.

END OF THE SECOND VOLUME.

Printed by W. Glendinning, 25, Hatton Garden, London.

Lightning Source UK Ltd.
Milton Keynes UK
UKHW021320231118
332790UK00009B/1070/P

9 780266 591146